Innovations in Mobile Multimedia Communications and Applications:

New Technologies

Ismail Khalil
Johannes Kepler University Linz, Austria

Edgar Weippl
SBA Research, Austria

Information Science
REFERENCE

Senior Editorial Director: Kristin Klinger
Director of Book Publications: Julia Mosemann
Editorial Director: Lindsay Johnston
Acquisitions Editor: Erika Carter
Development Editor: Myla Harty
Production Coordinator: Jamie Snavely
Typesetters: Jennifer Romanchak, Milan Vracarich Jr. & Michael Brehm
Cover Design: Nick Newcomer

Published in the United States of America by
Information Science Reference (an imprint of IGI Global)
701 E. Chocolate Avenue
Hershey PA 17033
Tel: 717-533-8845
Fax: 717-533-8661
E-mail: cust@igi-global.com
Web site: http://www.igi-global.com/reference

Library of Congress Cataloging-in-Publication Data

Innovations in mobile multimedia communications and applications : new technologies / Ismail Khalil and Edgar Weippl, editors.
p. cm.
Includes bibliographical references and index.
Summary: "This book provides an in-depth coverage of next-generation mobile computing paradigm, including mobile wireless technologies, mobile services and applications, and research and development challenges surrounding backend systems, network infrastructure, and mobile terminals including smart phones and other mobile devices"-- Provided by publisher.
ISBN 978-1-60960-563-6 (hardcover) -- ISBN 978-1-60960-564-3 (ebook) 1. Multimedia communications. 2. Mobile computing. I. Khalil, Ismail,
1960- II. Weippl, Edgar R.
TK5105.15.153 2011
621.3845'6--dc22
2011008139

British Cataloguing in Publication Data
A Cataloguing in Publication record for this book is available from the British Library.

Table of Contents

Detailed Table of Contents

This chapter proposes a scenario-based and diameter-bounded algorithm for cluster formation and management in mobile ad hoc networks (MANETs). A (k, r) −Dominating Set is used for the selection of clusterheads and gateway nodes depending on the topology of the network. Here k is the minimum number of clusterheads per node in the network and r is the maximum number of hops between the node and the clusterhead. The non-clusterhead node selects the most qualified dominating node as its clusterhead from among the k dominating nodes. The quality of the clusterhead is a function of various metrics, which include connectivity, stability and residual battery power. The long-term service as a clusterhead depletes its energy, causing it to drop out of the network. Similarly, the clusterhead with relatively high mobility than its neighbors leads to frequent clusterhead election process. This perturbs the stability of the network and can adversely affect the network performance. Load balancing among the clusterheads and correct positioning of the clusterhead in a cluster are vital to increase the lifespan of a network. The proposed centralized algorithm periodically calculates the quality of all dominating nodes in the network and if it goes below the threshold level it resigns the job as the clusterhead and sends this message to all other members in the cluster. Since these nodes have k dominating nodes within the r -hop distance, it can choose the current best-qualified node as its clusterhead. Simulation experiments are conducted to evaluate the performance of the algorithm in terms of the number of elements in the (k,r)-DS, the load balancing factor, the number of re-affiliations per unit time and the number of dominating set updates per unit time. The results establish the potential of this algorithm for use in MANETs.

Conservation of resources such as bandwidth, energy and memory are of a concern in Resource Constrained Networks (RCNs). Wireless mobile devices, especially low cost devices are stifled by the limited resources such as battery power, screen size, input, memory and processors. The low cost wireless mobile devices penetrating the developing world market demand for a cost effective messaging format that fits within the constrained wireless environment. Reduction of verbosity is considered to be the most effective step in controlling the resource consumption in RCNs. This chapter presents a method for optimizing resource consumption by the use of a new messaging format with less verbosity. The proposed format is based on YAML Ain´t Markup Language (YAML), which is further enhanced with message level security specifications.

The nodes are expected to transmit at different power levels in heterogeneous mobile adhoc networks, thus leading to communication links of different length. Conventional MAC protocols that unconditionally presume that links are bi-directional and with unvarying energy distribution may not succeed or execute badly under such circumstances. Interference and signal loss resulting out of distance and fading diminish the entire throughput attained in heterogeneous networks to a greater extent. This chapter presents a MAC protocol, which adaptively transmits data frames using either the energy efficient nodes or a list of high data rate assistant nodes. In addition, a cross-layer based energy level on-demand routing protocol that adaptively regulates the transmission rate on basis of congestion is projected as well. Simulation results illustrate that the proposed protocols considerably diminish energy consumption and delay, and attain high throughput in contrast with the Hybrid MAC and traditional IEEE 802.11 protocols.

This chapter proposes to use virtual backbone structure to handle control messages in ad hoc networks. This structure is effective in reducing the overhead of disseminating control information. In the first part, the approach to build the virtual backbone on the setup phase is presented. The construction of backbone is based on the Minimum Connected Dominating Set (MCDS). The novelty is in the way on finding the MCDS. A Linear Programming approach is used to build a Minimum Dominating Set (MDS). Then, a spanning tree algorithm is applied to provide the MCDS. A theoretical analysis based on probabilistic approach is developed to evaluate the size of MCDS. Different techniques of diffusion in ad hoc networks are presented and compared. The flooding technique is simple and efficient, but it is expensive in term of bandwidth consumption and causes broadcast storm problem. Simulation results show that technique using virtual backbone performs flooding and it is compared to MPR (Multipoint Relay). The second part of this chapter presents a distributed procedure to maintain the backbone when the mobility of terminals is introduced. A maintenance procedure will be executed by the node which changes its position. This procedure is distributed and guarantees the node connectivity to the backbone. The authors believe that the maintenance of the backbone with small size will be more effective. Simulation results show the performance of this procedure when mobility and scalability are considered.

Mobile Ad Hoc Network (MANET) is a centerless packet radio network without fixed infrastructure. In recent years tremendous attentions have been received because of capabilities of self-configuration and self-maintenance. However, attenuation and interference caused by node mobility and wireless channels sharing weaken the stability of communication links especially in ubiquitous MANET. A mathematical exploring model for next-hop node has been established. The negative impact of wireless routes discontinuity on pervasive communication is alleviated by a novel route reconstructed scheme proposed in this chapter based on restricting the route requirement zone into a pie slice region on intermediate nodes according the solution of the exploring equation. The scheme is an effective approach to increase survivability and reduce average end-to-end delay during route maintenance as well as allowing continuous packet forwarding for fault resilience so as to support mobile multimedia communication. The ns-2 based simulation results show remarkable packets successful delivery rate and end-to-end delay improvements of source-initiated routing protocol with route reconstructing scheme, and especially in the case of high dynamic environments with heavy traffic loads, more robust and scalable performance will be obtained.

Cellular IP networks deal with the concepts of micro-mobility. Buffer management in Cellular IP networks is very crucial as its proper usage not only increases the throughput of the network but also results in the reduction of the call drops. This chapter proposes a model for buffer management in Cellular IP network using Particle Swarm Optimization (PSO), an evolutionary computational method often used to solve hard problems. The model considers two kinds of buffers; Gateway buffer and Base Station buffer. In the proposed two-tier model, the first tier applies a prioritization algorithm for prioritizing real-time packets in the buffer. In the second tier PSO algorithm is used on a swarm of cells in the network. PSO is applied for a given time slot, called window. In each window period the swarm can store number of packets depending on the window size and the total number of packets. The effect of various parameters e.g. number of packets, size of packets, window size, and a threshold value on buffer utilization has been studied by conducting the simulation experiments.

Cooperative communications among group of teleoperated unmanned ground vehicles (UGVs) allows to exploit spatial diversity in wireless fading channels by relaying signals between each other. Due to the high speed of the UGVs, the nature of the channel environments and the possible co-channel interference, the effect of multipath propagation and the Doppler spread are more pronounced. In this chapter, we proposed a low complexity dynamic channel assignment (DCA) technique with adaptive modulation and coding (AMC) strategy to allocat the available bandwidth over a number of communications links in a cooperative UGV network. In many processing algorithms and transmission protocols reported in the literature, performance improvement in terms of system throughput and reliability has been demonstrated. The proposed DCA with AMC in a cooperative UGV network has two objectives. First, to maximize the overall throughput of the cooperative UGV network and second, to significantly reduce the probability of outage in the system. In this chapter, the outage is defined as the percentage of time the links are incapable of supporting a minimum required transmission rate which is determined by the application. The DCA approach is formulated in terms of a binary optimization problem that is solved using the branch-and-bound method. The authors assum the links in the network to be Rayleigh faded and we used a finite state Markov chain (FSMC) for their modeling. Using Monte Carlo simulation, we showed that the proposed DCA approach in a cooperative UGVs provides significant gain in the overall throughput and reduction in the outage probability compared to the static channel assignment (SCA).

The efficient dimensioning of cellular wireless access networks depends highly on the accuracy of the underlying mathematical models of user distribution and traffic estimations. The optimal placement/deployment of e.g. UMTS, IEEE 802.16 WiMAX base stations or IEEE 802.11 WLAN access points is based on user distribution and traffic characteristics in the service area. In this chapter we focus on the tradeoff between the accuracy and the complexity of the mathematical models used to describe user movements in the network. We propose a novel Markov chain based model capable of utilizing user's movement history thus providing more accurate results than other models in the literature. The new model is applicable in real-life scenarios, because it relies on information effectively available in cellular networks (e.g. handover history). The complexity of the proposed model is analyzed, and the accuracy is justified by means of simulation.

In the present chapter an attempt is made to compare multi-carrier and single carrier modulation schemes for wireless communication systems with the utilization of fast Fourier transform (FFT) and its inverse in both cases. With the assumption that in OFDM (orthogonal frequency division multiplexing), the inverse FFT transforms the complex amplitudes of the individual sub-carriers at the transmitter into time domain, the inverse operation is carried out at the receiver. In case of single carrier modulation, the FFT and its inverse are used at the input and output of the frequency domain equalizer in the receiver. Different single carrier and multi-carrier transmission systems are simulated with time-variant transfer functions measured with a wideband channel sounder. In case of OFDM, the individual sub-carriers are modulated with fixed and adaptive signal alphabets. Furthermore, a frequency-independent as well as the optimum power distribution are used. Single carrier modulation uses a single carrier, instead of the hundreds or thousands typically used in OFDM, so the peak-to-average transmitted power ratio for single carrier modulated signals is smaller. This in turn means that a SC system requires a smaller linear range to support a given average power. This enables the use of cheaper power amplifier as compared to OFDM system.

Fixed Mobile Convergence is an important challenge for telecommunication operators given the heterogeneity of access networks technologies and the variety of terminals. Fixed Mobile Convergence, which introduces the concept 'Being always best connected', is considered to be the next step in the evolution of telecommunication networks and should increase the operators' revenues. In order to enforce the concept 'Being always best connected', this chapter presents and analyzes an architecture for enterprises, named 'Business Zone'. After defining the Business Zone, we present its architecture and we analyze its main components while limiting our study to the transport of VoIP flows. Then we present two methods that we have patented: the first method authorizes a VoIP flow to be transmitted according to the available resources in the Business Zone; the second method enhances the decision process during a handover.

There are many mobile applications currently available on the market, which have been developed specifically for smart phones. The operating system of these smart phones is flexible enough to facilitate the high level application development. Similar to other pervasive devices, mobile phones suffer from limited amount of resources. These resources vary from the power (battery) consumption to the network bandwidth consumption. In this research the mobile resources are identified and classified. Furthermore, a monitoring approach to measure resource utilization is proposed. This monitoring tool generates traces about the resource usage which is followed by a benchmarking model which studies monitoring traces and enables users to extract qualitative information about the application from quantitative trace of resource usage.

Vector quantization (VQ) is a well-known compression method. In the encoding phase, given a block represented as a vector, searching the closest codeword in the codebook is a time-consuming task. In this chapter, two fast encoding algorithms for VQ are proposed. To reduce the search area and accelerate the search process, the first algorithm utilizes three significant features of a vector that are, the norm, and two projection angles to two projection axes. The second algorithm uses the first two features as in the first algorithm with the projection value of the vector to the second projection axe. The algorithms allow significant acceleration in the encoding process. Experimental results are presented on image block data. These results confirm the effectiveness of the proposed algorithms.

All-IP networks are under development with multimedia services in mind. Video multicast is an efficient way to deliver one video simultaneously to many users over such heterogeneous wired-to-wireless networks, such as in wireless IP applications where a mobile terminal communicates with an IP server through a wired IP network in tandem with a wireless network. Unicast video streaming is also an attractive way to deliver time-shifted TV to mobile devices. This Chapter presents a simple cross-layer model that leads to the optimal throughput to multiple users for multicasting video over a heterogeneous network. An adaptive forward-error-correction scheme is applied at the byte-level as well as at the packet-level to reduce channel errors. The results show that a server can significantly adapt to the bandwidth and FEC codes to maximize the video quality of service. For unicast streaming, the Chapter presents a single negative acknowledgment scheme in which a video stream is transmitted over a heterogeneous network from a streaming server to a mobile device in a WiMAX network. The broadband streaming system is compared to several candidate solutions based on originally wired network congestion controllers. Multi-connection streaming is also investigated.

This chapter deals with the problem of speaker characterization, for which the principal interest is the improvement of the techniques of talker identification. For this purpose, we investigate the effect of spectral resolution in the speaker identification performance. This investigation employs an approach based on the second order statistical measures using the Mel Frequency Spectral Coefficients (MFSC) and looks for the best spectral resolution (optimal number of MFSC). In fact, researchers do prefer using low spectral resolutions for many justifiable reasons, but we do not know what is the best resolution to adopt, especially in talker identification and we do not know what are the performances got with high spectral resolutions either. To find that optimal resolution, in microphonic and telephonic bandwidth, we have experimented several dimensions for the MFSC coefficients and several types of additive noises, at several SNR ratios. Results show the importance of the high spectral resolution in noisy environment and telephonic bandwidth, while the current research works have always favoured the low resolution of 24 coefficients in such tasks. For example, we notice an improvement of about 11% on the identification score, since we increase the resolution from 24 to 48 MFSC, in the telephonic bandwidth.

This chapter presents a very simple and efficient algorithm for codeword search in the vector quantization encoding. This algorithm uses 2-pixel merging norm pyramid structure to speed up the closest codeword search process. The authors first derive a condition to eliminate unnecessary matching operations from the search procedure. Then, based on this elimination condition, a fast search algorithm is suggested. Simulation results show that, the proposed search algorithm reduces the encoding complexity while maintaining the same encoding quality as that of the full search algorithm. It is also found that the proposed algorithm outperforms the existing search algorithms.

In future communication networks, video is expected to represent a large portion of the total traffic, given that especially variable bit rate (VBR) coded video streams, are becoming increasingly popular. Consequently, traffic modeling and characterization of such video services is essential for the efficient traffic control and resource management. Besides, providing an insight of video coding mechanisms, traffic models can be used as a tool for the allocation of network resources, the design of efficient networks for streaming services and the reassurance of specific QoS characteristics to the end users. The new H.264/AVC standard, proposed by the ITU-T Video Coding Expert Group (VCEG) and ISO/IEC Moving Pictures Expert Group (MPEG), is expected to dominate in upcoming multimedia services, due to the fact that it outperforms in many fields the previous encoded standards. This article presents both a frame and a layer (i.e. I, P and B frames) level analysis of H.264 encoded sources. Analysis of the data suggests that the video traffic can be considered as a stationary stochastic process with an autocorrelation function of exponentially fast decay and a marginal frame size distribution of approximately Gamma form. Finally, based on the statistical analysis, an efficient model of H.264 video traffic is proposed.

This chapter describes a new Speaker Discrimination System (SDS), which is a part of an overall project called Audio Documents Indexing based on a Speaker Discrimination System (ADISDS). Speaker discrimination consists in checking whether two speech segments come from the same speaker or not. This research domain presents an important field in biometry, since the voice remains an important feature used at distance (via telephone). However, although some discriminative classifiers do exist nowadays, their performances are not enough sufficient for short speech segments. This issue led us to propose an efficient fusion between such classifiers in order to enhance the discriminative performance. This fusion is obtained, by using three different techniques: a serial fusion, parallel fusion and serial-parallel fusion. Also, two classifiers have been chosen for the evaluation: a mono-gaussian statistical classifier and a Multi Layer Perceptron (MLP). Several experiments of speaker discrimination are conducted on different databases: Hub4 Broadcast-News and telephonic calls. Results show that the fusion has efficiently improved the scores obtained by each approach alone. So, for instance, we got an Equal Error Rate (EER) of about 7% on a subset of Hub4 Broadcast-News database, with short segments of 4 seconds, and an EER of about 4% on telephonic speech, with medium segments of 10 seconds.

Section 4
Innovations in Mobile Multimedia Applications and Services

Preface

We are living in a world of mobile multimedia. The field of mobile computing and multimedia is expanding in an unprecedented pace. Indicators are the rapidly increasing penetration of the smart phones and other mobile devices market around the world, which is growing nearly twice as fast as the desktop market. In addition, technological advancements and the exponential growth and globalization of communication infrastructures have significantly enhanced the usability of mobile communication and computer devices. From the first CT1 cordless telephones to today's smart phones and laptops/netbooks with wireless Internet connection, mobile tools and utilities have made the life of many people at work and at home much easier and more comfortable. As a result, mobility and wireless connectivity are expected to play a dominant role in the future in all aspects of economy. The addition of mobility to data communications systems has not only the potential to put the vision of "being always on" into practice, but has also enabled new generation of services, from reading books, playing games, taking photos, listening to music, checking the weather to sophisticated business, education, entertainment, finance, productivity, social networking travel, navigation applications. For these reasons, we believe that this is the right time to introduce this book in the area of mobile computing and multimedia communications.

Mobile multimedia is the set of protocols and standards for multimedia information exchange over wireless networks. It enables information systems to process and transmit multimedia data to provide the end user with services from various areas, such as the mobile working place, mobile entertainment, mobile information retrieval, user-generated content and context based services.

Multimedia information as combined information presented by more than one media type (text [+pictures] [+graphics] [+sounds] [+animations] [+videos]) enriches the quality of the information and is a way to represent reality as adequate as possible. Multimedia allows the user to enhance his/her understanding of the provided information and increases the potential of person to person and person to system communication.

The special requirements coming along with the mobility of users, devices, and services and specifically the requirements of multimedia as traffic type don't only bring the need of new paradigms in software-engineering and system-development but also in non-technical issues such as the emergence of new business models and concerns about privacy, security or digital inclusion to name a few.

The primary goal of the book is to provide researchers and academic communities around the world with the highest quality articles while reporting the state-of-the-art research results and scientific findings allowing students, developers, engineers, innovators, research strategists and IT-managers in this field to gain greater insight into mobile multimedia as they relate to applications, management, and opportunities within any given construct.

The book provides an in-depth coverage of next-generation mobile computing paradigm, including mobile wireless technologies, mobile services and applications, and research and development challenges surrounding backend systems, network infrastructure, and mobile terminals including smart phones and other mobile devices. It substantially emphasizes the following components organized into four sections.

SECTION 1: INNOVATIONS IN WIRELESS AND MOBILE NETWORKS MANAGEMENT

The first section is about "Wireless and Mobile Networks Management" and consists of 7 chapters. The first chapter entitled "*Multi-Purpose DS-Based Cluster Formation and Management in Mobile Ad Hoc Networks*" by V. S. Anitha and M. P. Sebastian from India proposes a scenario-based algorithm for cluster formation and management in mobile ad hoc networks. In this algorithm, the clustering set up phase is accomplished by a distributed (k, r) − Dominating Set finding algorithm for choosing some nodes that act as coordinators of the clustering process.

The second chapter titled "*Optimizing Resource Consumption for Secure Messaging in Resource Constrained Networks*" by P. P. Abdul Haleem and M. P. Sebastian from India presents a method for reducing the verbosity of messages in the constrained wireless mobile networks. Wireless mobile devices, especially low cost devices are stifled by the limited resources such as battery power, screen size, input, memory and processors. The relevance of low cost wireless mobile devices in penetrating to the third world market demands for a cost effective messaging format that fits in the constrained wireless environment. The proposed scheme is based on YAML Ain′ Markup Language (YAML), a user friendly and lightweight messaging format. Measures to reduce the message size and energy consumption together with secure processing are proposed.

The third chapter "*Improving Energy Efficiency and Throughput in Heterogeneous Mobile Ad Hoc Networks*" by Manu. J. Pillai and M. P. Sebastian from India presents a MAC protocol, which adaptively transmits data frames using either the energy efficient nodes or a list of high data rate assistant nodes. In addition, a cross-layer based energy level on-demand routing protocol that adaptively regulates the transmission rate on basis of congestion is projected as well. Simulation results illustrate that the proposed protocols considerably diminish energy consumption and delay, and attain high throughput in contrast with the Hybrid MAC and traditional IEEE 802.11 protocols.

The fourth chapter in this section titled "*Performance Enhancement of Routing Protocols in Mobile Ad hoc Networks*" by Kais Mnif from Tunisia and Michel Kadoch from Canada, proposes the use of virtual backbone structure to handle control messages in ad hoc networks. The structure is effective in reducing the overhead of disseminating control information. The construction of the backbone is based on the Minimum Connected Dominating Set (MCDS). The novelty is in the way to find the MCDS. A Linear Programming approach is used to build a Minimum Dominating Set (MDS). A spanning tree algorithm is applied to provide the MCDS. A theoretical analysis based on probabilistic approach is developed to evaluate the size of MCDS. Different techniques of diffusion in ad hoc networks are presented and compared.

The fifth chapter in this section "*Realization of Route Reconstructing Scheme for Mobile Ad Hoc Network*" by Qin Danyang, Ma Lin, Sha Xuejun and Xu Yubin from China presents a mathematical exploring model for next-hop node in mobile ad hoc networks. The negative impact of wireless routes discontinuity on pervasive communication is alleviated by a novel route reconstructed scheme proposed in this chapter based on restricting the route requirement zone into a pie slice region on intermediate nodes according the solution of the exploring equation. The scheme is an effective approach to increase survivability and reduce average end-to-end delay during route maintenance as well as allowing continuous packet forwarding for fault resilience so as to support mobile multimedia communication. The ns-2 based simulation results show remarkable packets successful delivery rate and end-to-end delay improvements of source-initiated routing protocol with route reconstructing scheme, and especially in the case of high dynamic environments with heavy traffic loads, more robust and scalable performance will be obtained.

The sixth chapter "*Buffer Management in Cellular IP Network Using PSO*" by Mohammad Anbar and Deo Prakash Vidyarthi from India proposes a model for buffer management in Cellular IP network using Particle Swarm Optimization (PSO), an evolutionary computational method often used to solve hard problems. The model considers two kinds of buffers: Gateway buffer and Base Station buffer. In the proposed two-tier model, the first tier applies a prioritization algorithm for prioritizing real-time packets in the buffer. In the second tier PSO algorithm is used on a swarm of cells in the network. PSO is applied for a given time slot, called window. In each window period the swarm can store number of packets depending on the window size and the total number of packets. The effect of various parameters e.g. number of packets, size of packets, window size, and a threshold value on buffer utilization has been studied by conducting the simulation experiments.

The last chapter in this section "*Throughput Optimization of Cooperative Teleoperated UGV Network*" by Ibrahim Y. Abualhaol and Mustafa M. Matalgah from USA proposes a low complexity dynamic channel assignment (DCA) technique with adaptive modulation and coding (AMC) strategy to allocate the available bandwidth over a number of communications links in a cooperative Unmanned Ground Vehicles (UGVs) network. The proposed DCA with AMC in a cooperative UGV network has two objectives. First, to maximize the overall throughput of the cooperative UGV network and second, to significantly reduce the probability of outage in the system.

SECTION 2: INNOVATIONS IN MOBILITY ENGINEERING, PERFORMANCE AND OPTIMIZATION

The second section is about "Mobility Engineering, Performance and Optimization" and consists of 4 chapters. The first chapter "*The Accuracy of Location Prediction Algorithms Based on Markovian Mobility Models*" by Péter Fülöp, Sándor Imre, Sándor Szabó, Tamás Szálka from Hungary proposes a novel Markov chain based model capable of utilizing user's movement history thus providing more accurate results than other models in the literature. The new model is applicable in real-life scenarios, because it relies on information effectively available in cellular networks (e.g. handover history). The complexity of the proposed model is analyzed, and the accuracy is justified by means of simulation.

The second chapter titled "*Frequency Domain Equalization and Adaptive OFDM vs. Single Carrier Modulation*" by Inderjeet Kaur from India, compares multi-carrier and single carrier modulation schemes for wireless communication systems with the utilization of fast Fourier transform (FFT) and its inverse in both cases. With the assumption that in OFDM (orthogonal frequency division multiplexing), the inverse FFT transforms the complex amplitudes of the individual sub-carriers at the transmitter into time domain, the inverse operation is carried out at the receiver. In case of single carrier modulation, the FFT and its inverse are used at the input and output of the frequency domain equalizer in the receiver.

The third chapter titled "*Definition and Analysis of a Fixed Mobile Convergent Architecture for Enterprise VoIP Services*" by Joel Penhoat, Olivier Le Grand, Mikael Salaun, and Tayeb Lemlouma from France presents and analyzes an architecture for enterprises, named 'Business Zone' in order to enforce the concept of "Being always best connected". After defining the Business Zone, the chapter shows its architecture and analyzes its main components while limiting the study to the transport of VoIP flows. Then two patented methods are presented: the first method authorizes a VoIP flow to be transmitted according to the available resources in the Business Zone; the second method enhances the decision process during a handover.

The last chapter in this section "*A Qualitative Resource Utilization Benchmarking for Mobile Applications*" by Reza Rawassizadeh, Amin Anjomshoaa, and A Min Tjoa from Austria identifies and classifies mobile resources and proposes a monitoring approach to measure resource utilization. It provides a monitoring tool, which generates traces about the resource usage and proposes a benchmarking model which studies traces and enables users to extract qualitative information about the application from quantitative resource usage traces. Results of the study could assist quality operators to compare similar applications from their resource usage point of view, or profile a single application resource consumption.

SECTION 3: INNOVATIONS IN MULTIMEDIA ANALYSIS, MODELING, PROCESSING AND TRANSFORMATION

The third section is about "Graphics, Video, Audio, Analysis, Modeling, Processing and Transformation" and consists of 6 chapters. In the first chapter entitled "*Fast Vector Quantization Encoding Algorithms for Image Compression*" by Ahmed Swilem from Egypt, two fast encoding algorithms for VQ are proposed. To reduce the search area and accelerate the search process, the first algorithm utilizes three significant features of a vector that are, the norm, and two projection angles to two projection axes. The second algorithm uses the first two features with the projection value of the vector to the second projection axe. The algorithms allow significant acceleration in the encoding process. Experimental results are presented on image block data. These results confirm the effectiveness of the proposed algorithms.

While the second chapter "*Mobile Video Streaming Over Heterogeneous Networks*" by Ghaida A. Al-Suhail, Martin Fleury, and Salah M. Saleh Al-Majeed present a simple cross-layer model that leads to the optimal throughput of multiple users for multicasting MPEG-4 video over a heterogeneous network. For heterogeneous wired-to-wireless network, at the last wireless hop, there are bit errors associated with the link-layer packets that are arising in the wireless channel, in addition of overflow packet dropping over wired links. The authors employ a heuristic TCP function to optimize the cross-layer model of data link and physical (radio-link) layer.

The third chapter "*Automatic Talker Identification Using Optimal Spectral Resolution: Application in Noisy Environment and Telephony*" by Siham Ouamour, Halim Sayoud and Mhania Guerti from Algeria deals with the problem of speaker characterization, for which the principal interest is the improvement of the techniques of speaker authentication. For this purpose, authors investigate the effect of spectral resolution in the speaker authentication performance. This investigation employs an approach based on the second order statistical measures using the Mel Frequency Spectral Coefficients (MFSC) and looks for the best spectral resolution (optimal number of MFSC).

The fourth chapter titled "*A Fast Image Encoding Algorithm Based on the Pyramid Structure of Codewords*" by Ahmed A. Radwan, Ahmed Swilem, Mamdouh M. Gomaa from Egypt presents an algorithm for codeword search in the vector quantization encoding. This algorithm uses 2-pixel merging norm pyramid structure to speed up the closest codeword search process. The algorithm first derives a condition to eliminate unnecessary matching operations from the search procedure. Then, based on this elimination condition, a fast search algorithm is suggested. Simulation results show that, the proposed search algorithm reduces the encoding complexity while maintaining the same encoding quality as that of the full search algorithm. It is also found that the proposed algorithm outperforms the existing search algorithms.

The fifth chapter titled "*Analysis and Modeling of H.264 Unconstrained VBR Video Traffic*" by Harilaos Koumaras, Charalampos Skianis, and Anastasios Kourtis from Greece presents and analyzes both a frame and a layer level of H.264 encoded sources. Analysis of the data suggests that the video traffic can be considered as a stationary stochastic process with an autocorrelation function of exponentially fast decay and a marginal frame size distribution of approximately Gamma form. Finally, based on the statistical analysis, an efficient model of H.264 video traffic is proposed.

The last chapter in this section "*Speaker Discrimination on Broadcast News and Telephonic Calls Based on New Fusion Techniques*" by Siham Ouamour and Halim Sayoud from Algeria describes a new Speaker Discrimination System (SDS), which is part of an overall project called Audio Documents Indexing based on a Speaker Discrimination System (ADISDS). Speaker discrimination consists of checking whether two speech segments come from the same speaker or not. This research domain presents an important field in biometry, since the voice remains an important feature used at distance (via telephone).

SECTION 4: INNOVATIONS IN MOBILE MULTIMEDIA APPLICATIONS AND SERVICES

The fourth section is about "Mobile Computing and Multimedia Applications and Services" and consists of 5 chapters.

The first chapter "*FCVW: Experiments in Groupware*" by Ivan Tomek, and Elhadi Shakshuki from Canada surveys several groupware products and describes FCVW (Federated Collaborative Virtual Workspace), an experimental project designed to explore certain groupware aspects that are not sufficiently addressed by existing products.

The second chapter "*A Model for Mobile Learning Service Quality in University Environment*" by Nabeel Farouq Al-Mushasha, and Shahizan Hassan from Jordan and Malaysia proposes a service quality model for m-learning in a university environment. A questionnaire survey was conducted which measured ten dependent variables and three independent variables. The dependent variables were meant to measure service quality, information quality, and system quality. The dependent variables were meant to measure the causal relationship between overall learners' perceived service quality, learner satisfaction, and learner behavioral intention to use the service in future. The findings revealed that the factors that lead to service quality of m- learning in a university environment were interface design, reliability, trust, content usefulness, content adequacy, ease of use, accessibility, and interactivity. The findings also indicates that there are causal relationships between learner satisfaction with overall service quality, and learner satisfaction with learner behavioral intention.

The third chapter is "*Evaluating E-Communities of Wireless Networks Worldwide*" by Theodoros I. Kavaliotis and Anastasios A. Economides from Greece. This chapter presents an evaluation framework and analyzes the current status of such Electronic Communities of Wireless Networks (ECWNs) in four continents: Africa, America, Europe and Oceania. The evaluation framework contains fifty criteria categorized into four categories: Usability, Technical Characteristics, Community's Commitment, and Members' Commitment. The results show that there are large differences among ECWNs with respect to the forum structure, archives accessibility, interactivity, services, members' commitment, participation and relationships.

The fourth chapter in this section "*Typology and Challenges in Developing Mobile Middleware Based Community Network Infrastructure*" by Vijayan Sugumaran, and Shriram Raghunathan from India presents the factors contributing to the proliferation of mobile communities and places the mobile community networks in a current and future perspective. An architecture for the mobile community network is proposed and the challenges in implementing such a network are also discussed.

The fifth chapter titled "*A Secure and Trustworthy Framework for Mobile Agent-Based E-Marketplace with Digital Forensics and Security Protocols*" by Qi Wei and Ahmed Patel from Malaysia, proposes a framework which includes safe, secure, trusted and auditable services, as well as forensic mechanisms to provide audit trails for digital evidence of transactions and protection against illegal activities. The proposed framework and protocols provide a secure communication for mobile agents when they move to different security environments to deal with e-marketplace activities such as search information, negotiation and payments. The paper concludes by highlighting and discussing further research work to build viable systems.

In closing, we would like to thank all of the authors for their insights and excellent contributions to this book, in addition to all those who assisted in the review process.

Ismail Khalil
Johannes Kepler University Linz, Austria

Edgar Weippl
Secure Business Austria Research, Austria

Section 1
Innovations in Wireless and Mobile Networks Management

Chapter 1
Multi-Purpose DS-Based Cluster Formation and Management in Mobile Ad Hoc Networks

V. S. Anitha
Govt. Engineering College, India

M. P. Sebastian
Indian Institute of Management Kozhikode, India

ABSTRACT

This chapter proposes a scenario-based and diameter-bounded algorithm for cluster formation and management in mobile ad hoc networks (MANETs). A ***(k, r)*** *−Dominating Set is used for the selection of clusterheads and gateway nodes depending on the topology of the network. Here* ***k*** *is the minimum number of clusterheads per node in the network and* ***r*** *is the maximum number of hops between the node and the clusterhead. The non-clusterhead node selects the most qualified dominating node as its clusterhead from among the* ***k*** *dominating nodes. The quality of the clusterhead is a function of various metrics, which include connectivity, stability and residual battery power. The long-term service as a clusterhead depletes its energy, causing it to drop out of the network. Similarly, the clusterhead with relatively high mobility than its neighbors leads to frequent clusterhead election process. This perturbs the stability of the network and can adversely affect the network performance. Load balancing among the clusterheads and correct positioning of the clusterhead in a cluster are vital to increase the lifespan of a network. The proposed centralized algorithm periodically calculates the quality of all dominating nodes in the network and if it goes below the threshold level it resigns the job as the clusterhead and sends this message to all other members in the cluster. Since these nodes have* ***k*** *dominating nodes within the* ***r*** *-hop distance, it can choose the current best-qualified node as its clusterhead. Simulation experiments are conducted to evaluate the performance of the algorithm in terms of the number of elements in the* ***(k, r)-DS,*** *the load balancing factor, the number of re-affiliations per unit time and the number of dominating set updates per unit time. The results establish the potential of this algorithm for use in MANETs.*

DOI: 10.4018/978-1-60960-563-6.ch001

INTRODUCTION

A MANET is a self configuring wireless, multi-hop, and dynamic network consists of mobile nodes that operate without the need for any established infrastructure. This type of network is highly demanding due to the lack of infrastructure and easiness and cost effectiveness in installation. Each node in the network will be able to communicate directly with any other node that resides within its transmission range, whereas intermediate nodes are required for communicating with the nodes that reside beyond the transmission range. All nodes in the ad hoc network take part in the communication process. They are free to move randomly, and arbitrarily organized themselves to create unpredictable network topologies. A MANET can operate in stand-alone fashion or can be connected to a fixed infrastructure with the help of gateway devices to facilitate Internet connectivity even in remote places.

The MANET is useful in situations where the development of infrastructure is expensive or inconvenient to use or set up, or if it is absent. It has many emerging applications, which include commercial, industrial and war front applications, search and rescue operations, sensor networks and vehicular communications. Nodes in the MANETs are generally battery operated with limited processing power and memory capacity. Wireless networks generally have limited bandwidth and nodes with high mobility. Such networks often experience high partitioning rate due to the increase in the link breakage rates. If the two hosts that want to communicate are outside their wireless transmission ranges, they could communicate only if the other hosts between them in the ad hoc wireless network are willing to forward packets for them. Instead of using specialized routers for path discovery and traffic routing, each node in the ad hoc network acts as a router and take part in the communication process. Each node operates in distributed peer-to-peer mode, acts as an independent router and generates independent data. Designing efficient routing protocols for MANETs is a very challenging task and it is an active area of research. The major issues in cluster based MANETs are mobility management, topology assignment, clustering overhead, frequent clusterhead re-election, overhead of clusterhead, depletion of battery power, security and Quality of Service (QoS).

Destination Sequenced Distance-Vector (DSDV) (Perkins, 1994), Dynamic Source Routing (DSR) (Johnson, 1996) and Ad hoc On demand Distance Vector (AODV) (Perkins, 1999) are some of the popular protocols proposed for multi-hop routing. These flat routing schemes encounter scalability problems with increased network size. The signaling overhead of the routing algorithms based on reactive and proactive routing schemes increases with the size and mobility of the network. The expensive message flooding schemes for route discovery and maintenance can be reduced in hierarchical routing. Hierarchical architectures help in increasing the lifetime of the network and also increase the network scalability. With clustering, the mobile nodes are divided into a number of virtual groups called clusters. Nodes in a cluster can be of type clusterhead, gateway or ordinary node. The clusterhead is the coordinator for the operations within the cluster. Cluster based virtual network architecture requires many information exchanges to perform routing as well as to form and maintain clusters. A stable clustering algorithm should not change the cluster configuration frequently. The advantages of clustering include (i) efficient handling of mobility management (ii) provision for optimization in routing mechanism (iii) shared use of application within the group (iv) spatial reuse of resources (v) better bandwidth utilization (vi) aggregation of topology information (vii) virtual circuit support (viii) making a dynamic topology appear less dynamic (Mc Donald, 1999) and (ix) minimizing the amount of storage for communication (Basagni, 2006).

This chapter presents a Scenario-based and bounded-distance clustering algorithm for MA-

NETs. It works in two phases. The first phase is the *Clustering Setup* phase and the second phase is *Cluster Formation and Management*. The first phase is accomplished by a centralized ***(k, r)***-Dominating set computation algorithm for choosing some nodes that act as coordinators of the clustering and routing process. While selecting dominating nodes, redundancy is achieved by choosing the value of parameter ***k*** greater than one and parameter ***r*** allows increasing local availability. These two parameters can be conveniently set depending on the requirement. The dominating nodes are potential nodes to become clusterheads and during the *Cluster Formation and Management* phase, the nodes which are not in the dominating set select the best node in the dominating set within ***r***-hop distance as its clusterhead. The selection is based on quality, which is a function of parameters such as stability of the dominating node with respect to its neighbors, remaining energy with the node, and the connectivity. The selection of a clusterhead based on these parameters helps in maintaining the structure of the created cluster, as stable as possible. This minimizes the topology changes and the associated overheads during clusterhead changes.

The rest of the chapter is organized as follows. The previous work done in the area of cluster based routing is reviewed in the next section. Section three discusses the basic concepts and design parameters of scenario-based clustering algorithms. Section four presents the details of ***(k, r) –DS*** algorithm and clusterhead association with an example. Section five discusses cluster formation and management. The performance of the algorithm is evaluated in Section six. Section seven concludes the chapter.

RELATED WORKS

Many clustering algorithms for MANETs have been proposed in the literature to select clusterheads. In Link Cluster Algorithm (LCA) (Baker, 1981), the clusterhead selection is based on the highest identity number among a group of nodes (each node is identified by a unique number). In LCA, the cluster radius is one hop as all cluster members are connected directly to the clusterhead. The lowest-ID algorithm (Ephremides, 1987) and Maximum-connectivity algorithm (Parekh, 1994) are two earlier popular algorithms in which the clusterhead selection is on the basis of the lowest virtual identification number and maximum number of neighbor nodes, respectively. These two metrics alone are not sufficient for the selection of a clusterhead in a dynamic environment due to the high overhead associated with clusterhead change over. Many modifications to these algorithms were proposed to make the clusterhead selection and cluster management more stable and power efficient. The Least Cluster Change (LCC) (Chiang, 1997), 3hBAC (3-hop Between Adjacent Clusterhead) (Yu, 2003) and Lin and Gerla's algorithm (Lin, 1997) are examples in this category.

One of the prominent characteristics of MANET is its mobility (Choi, (2006); Mc Donald, (1999)). Basu, Prithwish (2001) propose the MOBIC routing algorithm in which each node calculates the relative mobility values with respect to the neighboring nodes and these values are considered for the clusterhead selection. Another major challenge in the MANET performance is the energy limitation. Energy depletion leads to partitioning of networks and interruption in communication. Ryu (2001) proposes an energy conservation-clustering scheme. To provide pseudo optimum power saving clustering, he proposes two heuristic schemes, namely, single phase and double phase clustering schemes. All the above algorithms create clusters in such a way that the maximum distance between any two nodes in the cluster is at most two hops. Such types of algorithms are more suitable for MANETs, where the nodes are densely organized and they form a large number of clusters.

A weight based criteria is introduced in DCA (Distributed Clustering Algorithm) (Basagni, 1999) for the selection of clusterhead. In DCA, each node is assumed to have a unique weight but the technique of assigning weights has not been discussed. Stefano Basagni (1999a) introduces a distributed algorithm that partitions the nodes of a multi-hop wireless network into clusters. The algorithm is proved to be adaptive to changes in the network topology. Distributed Mobility Adaptive Clustering (DMAC), Generalized-DMA (Basagni, 1999b), and Backbone Protocol (Basagni, 2001) are evolutions of the above category. Basagni (2006) provides a thorough simulation based comparison of some of the most representative ad hoc clustering protocols. Basagni (2005) explores the impact of node mobility on DMAC. He considers many mobility models such as Random Way Point, Random Walk and Manhattan models. The weighted Clustering Algorithm (WCA) (Chatterjee, 2002) is a combined weight metric-based clustering approach to form single hop clusters. All nodes in the network have to compute their weights and have to know the weights of all other nodes before starting the clustering process. This process can take a lot of time. Also, limiting the cluster dimensions to single hop reduces the scalability for large scale networking environment.

The selection of nodes to form a virtual backbone that supports routing is an important issue in wireless ad hoc and sensor networks. Selection of clusterhead and gateway nodes is crucial since they are the service providers of the network. Single clusterhead concept enforces over work for the clusterhead in the communication process. Using more than one clusterhead for a single cluster can reduce this overloading. In dominating set based clustering, the dominating nodes work as the clusterheads to relay the routing information and data packets. Wu (1999) proposes a distributed algorithm to find a Connected Dominating Set to design efficient routing schemes for a MANET. Chen et al. (2002) present independent dominating sets for computing clusters such that the minimum distance between the clusterheads is $k+1$ hops from each other.

Some algorithms create variable diameter clusters based on the mobility pattern of nodes to ensure maximum stability. Mobility-based D-Hop clustering algorithm (MobDHop) (Inn, 2006) forms variable diameter clusters that may change its diameter adaptively with respect to the mobile node´s moving patterns.

From the discussions above, we can see that most of the clustering algorithms discussed in the literature except the combined weight metric algorithms, take only one of the factors such as degree of the node, nodeID, mobility, transmission range, transmission rate or battery power for the selection of the clusterhead. In most of the algorithms, non-clusterhead nodes select a single clusterhead and failure of that clusterhead leads to clusterhead re-election or clusterhead re-affiliation. Due to the dynamic nature of nodes in the network, the nodes as well as the clusterheads may move in different directions. The system then has to adapt to the new resulting topology from time to time. This results in the formation, removal or merging of clusters. The (k, r) – DS-based algorithm improves reliability, provides variable degree of clusterhead redundancy and avoids frequent clusterhead re-elections. This algorithm allows flexibility in the determination of clusterhead density. Use of suitable cluster maintenance mechanisms help to preserve the network topology as stable as possible. The main motivation of this work is the design of a better general purpose parameterized clustering protocol supporting scalability, adaptability, load balancing, and stability.

DESIGN OF SCENARIO BASED CLUSTERING

The basic concepts and design of the Scenario-based Clustering Algorithm for MANETs (SCAM) are explained in this section. In a dominating set based algorithm, the dominating set computation is crucial because the overall performance of the algorithm depends heavily on the (*k*, *r*) − DS problem solution (Amis, 2000).

Basics of Scenario Based Clustering

A MANET is represented using a graph ***G = (V, E),*** where ***V*** represents a set of wireless mobile nodes and ***E*** represents a set of edges. An edge between two nodes indicates that both are within their transmission range. The first step in this algorithm is finding out the ***(k, r) −DS***. The ***(k, r) -Domination problem*** seeks to determine a minimum number of nodes called dominating set nodes ***DS,*** such that any node ***u*** not in ***DS*** is within the ***r*** -hop distance with at least ***k*** nodes in the ***DS***. The two parameters used here are ***k*** and ***r***, where ***k*** represents the minimum number of clusterheads per node and ***r*** is the maximum distance between the nodes and their clusterheads. There are many approximation algorithms for finding the dominating set (Spohn et al., 2007) and connected dominating set (Guha, 1998; Dai, 2005) of nodes in a MANET. The problem of finding the ***(k, r) −DS*** of minimum cardinality for an arbitrary graph is *NP-complete* (Garey, 1978). A centralized solution for finding the ***(k, r) −DS*** can be used if all nodes in the network know the topology of the network. Distributed algorithms are applicable for both synchronous and asynchronous networks and are suitable for dynamic networks with large number of nodes. Nodes in the ***(k, r) −DS*** are the potential candidates to become clusterheads.

Determining the quality of potential candidates as dominating nodes use various metrics such as degree of the dominating node, battery power and stability of the dominating node with respect to the neighboring nodes. Bi-directional connectivity between nodes and capability of nodes in measuring its signal strength are assumed for all nodes.

The degree of the node ***v*** is the total number nodes within the transmission range of ***v***. The degree of ***v***, ***DG***$_v$ is computed using the formula

$$\sum_{u,v \in V,\ u \neq v} \{ D_{u,v} \prec T_x, \text{Transmission range} \} \qquad (1)$$

Mobile nodes in a MANET normally depend on battery power. In order to increase the network life span, each node should reduce the energy conception. A dominating node with high residual battery power, ***B***$_v$, can perform well as clusterhead for a longer duration. Hence residual battery power is a better measure than the consumed battery power (Choi, 2006) or the cumulative time (Chatterjee, 2002) during which the node acts as clusterhead. But long-term service as a clusterhead can cause reduction in the battery power. If the battery power of the current clusterhead goes below a certain threshold level, it sends this message to all the neighboring nodes. In this case, all member nodes select the next qualified dominating node as the new clusterhead. This is to avoid the total collapse of the current network topology and reduce the number of clusterhead elections and cluster formations.

The third metric is the mobility of a node. To compute this, each node in the dominating set needs to find out the distance from its neighboring nodes. For distance computation, the Friis (Friis, 1946) free space propagation model is used where the received power, ***P***$_r$ ***is computed as***

$$P_r = P_t * G_t * G_r * \frac{\lambda^2}{(4 * \pi * D)^2} \qquad (2)$$

P$_r$ is the power received by the receiving antenna, ***P***$_t$ is the power input to the transmitting antenna, G_t and ***G***$_r$ are the gains of the transmitting and the receiving antennas, respectively, λ is the wavelength, and ***D*** is the distance. ***P***$_r$ is inversely proportional to the square of the distance.

Finding the exact physical location needs much of computation. However, the approximate distance between nodes ***v*** and ***u*** at time ***t*** is calculated from the equation (2) as

$$D_t^{v,u} = \frac{k}{\sqrt{P_r}}, \qquad (3)$$

where $\boldsymbol{v} \in$ dominating set ***DS***, ***u*** is an element of the set of neighboring nodes of ***v*** and ***k*** is a constant. The relative mobility between ***v*** and ***u*** indicates whether they are coming closer to or moving away from each other. The relative mobility of node ***u*** with respect to ***v*** at time ***t*** is given as

$$RM_t^{v,u} = D_t^{v,u} - D_{t-1}^{v,u} \qquad (4)$$

$RM_t^{v,u}$ is positive if the node ***u*** is moving away from ***v*** and negative if ***u*** is coming closer to ***v***. The distance from ***v*** to ***u*** is measured at certain time interval for ***T*** times and $RM_1^{v,u}, RM_2^{v,u}, \ldots\ldots RM_T^{v,u}$ are calculated. The standard deviation of the relative mobility gives the variation of distances over a time period as

$$SDRM = \sqrt{\frac{1}{T}\sum_{i=1}^{T}(RM_i - \overline{RM})^2} \qquad (5)$$

where

$$\overline{RM} = \frac{1}{T}(RM_1^{v,u} + RM_2^{v,u} + \ldots. + RM_T^{v,u})$$

The local stability of a node ***v*** dominating set ***DS, LSTAB,*** with respect to all its neighbors is the mean of standard deviation of relative mobilities of all its neighboring nodes. A low value of this is an indication of a stable node. This stability is either due to less mobility or due to group mobility (when node ***v*** and all its neighboring nodes move in the same direction with more or less same velocity).. The quality of a node in the dominating set to work as a clusterhead is computed as

$$Q_v = W_1 * DG_v + W_2 * B_v + W_3 * LSTAB \qquad (6)$$

where W_1, W_2 and W_3 are the weights associated with various factors affecting the quality. Suitable values can be assigned to W_1, W_2 and W_3 based on the required application such that $W_1 + W_2 + W_3 = 1$. DG_v B_v and ***LSTAB*** are in the normalized form.

Design Principles

The major factors to be considered while designing and implementing the proposed clustering algorithm are

1. Selection of an optimal number of clusterheads to yield high throughput, but with as low latency as possible
2. Scalability
3. Efficiency and stability (if clustering structure becomes too complex, the number of messages needed to maintain the routing structure would cause congestion in the network)
4. Mechanism to prevent the clusters from growing too large (if the clusters grow large, the load on the clusterhead becomes too large)
5. Availability of a maintenance mechanism for the existing clusters (most of the existing clustering algorithms create new clustering structures from scratch after a specified time interval)
6. Quick adaptation with the change of topology.

A stable cluster formation avoids the frequent clusterhead election process, thereby reducing the total overhead. For creating a stable cluster, the clusterhead election scheme must be designed in such a way that the elected clusterhead can do its job for a longer period. To select the most powerful node as clusterhead, various parameters are to be considered.

The first parameter is the degree of a node. The degree of a node is high if it has more number of nodes within its transmission range. The position of the clusterhead is relevant in the cluster-based approach. A node with high degree is an indication that it is centrally located. A large cluster size may put too heavy load on the clusterhead and reduces the system throughput. However, a small cluster size results in large number of clusters in the network. Hence, setting upper and lower limits on the number of nodes connected to a clusterhead are important for load balancing. The second parameter is the residual battery power of the clusterhead. Since the clusterheads play the leading role in the communication process, their energy consumption is more compared to the ordinary nodes. The "early death" of a clusterhead increases the number of clusterhead elections, which in turn increases the traffic of control packets in the network. Therefore a node with sufficient battery power is to be selected as clusterhead to reduce the amount of overhead incurred due to the clusterhead re-election and to avoid nodes dropping out of the network prematurely. Stability of the clusterhead is another important parameter in selecting the clusterhead. Selection of a less mobile node as clusterhead considerably reduces the number of re-affiliations. The weighted sum of these metrics determines the quality of a dominating node. Depending on the requirement, different weight values can be assigned to stability, node degree and residual battery power. Thus, a dominating node can compute its own quality.

The parameter ***r*** determines the diameter of the cluster and the parameter ***k*** determines the number of clusterhead per node. Among the ***k*** dominating nodes, each node selects the most powerful dominating node as its clusterhead. In the event of failure of a clusterhead due to mobility, low battery or link failure, each member node re-affiliates itself to the next powerful dominating node. This local re-affiliation increases the lifespan of the network before forcing a clusterhead re-election.

SCENARIO BASED CLUSTERHEAD ELECTION IN MANETS

The Cluster Creation Algorithm

1. **Compute***(k, r) –Dominating Set***using***(k, r) –Dominating Set***algorithm.**
2. Nodes in the ***(k, r) –Dominating Set*** compute its quality by assigning suitable weights to the parameters such as degree of the node, battery power and mobility.
3. Dominating nodes send the message containing the quality to other nodes within the r -hop distance.
4. Each node in the network other than the dominating nodes, selects the most qualified dominating node within the ***r***-hop distance as its clusterhead by sending the NODE_JOIN_REQ (NJ) message.
5. On receiving the NJ message, the dominating node accepts the request by sending NJ_ACK packet if the degree (number of accepted cluster members) of that dominating node does not exceed the threshold. If a dominating node does not receive any NJ message for a specified time interval it can select the most qualified dominating node within the ***r*** -hop distance as its clusterhead and can join with that cluster (merging of clusters). But the status of this node is quazi-dominating and can change to dominating node when required. This is possible only if that dominating node has ***k*** other dominating nodes within the ***r*** -hop distance. This will reduce the total number of clusters created.
6. If the ordinary node does not receive any NJ_ACK messages within the stipulated time it can send the NJ message to the next

qualified dominating node within the ***r*** -hop distance.

7. If the node neither receive any NJ_ACK messages even after ***k*** attempts nor any new CLUSTER HEAD ADVERTISEMENT (CHA) message during the above period then initiate the ***(k, r) − DS*** algorithm for clusterhead re-election.
8. In case of a clusterhead failure or if the battery power of the clusterhead goes below the minimum desired level, the clusterhead sends this using the CHA message. All the nodes attached to that clusterhead select the next qualified dominating node as their new clusterhead.

The *Clustering Setup* Phase

The *clustering setup* phase is accomplished by a centralized ***(k, r)*** - Dominating set computation algorithm for choosing the dominating nodes. Dominating nodes are potential nodes to become clusterheads. Algorithm 1 follows a centralized approach for the ***(k, r) - DS*** problem. This is a greedy approximation algorithm which repeatedly selects the most weighted node in the priority heap as the dominating node. Any node, ***u*** is said to be ***(k, r)*** - Dominated if node ***u*** has at least ***k*** neighbors within the ***r*** - hop distance from the elements in the DS. We assume that the nodes have unique identifiers and that they are capable of knowing their neighbors.

To compute dominating a set with parameter ***k*** and ***r***, the *Required_Ch* value of each node ***u*** in the network is calculated. The value of the variable *u.Required_Ch* indicates the number of required clusterheads within the ***r***-hop distance of the node ***u***. Initially the *Required_Ch* value for each node is assigned a value ***k***. If node ***u*** is not covered (a node is said to be covered or dominated if it has ***k*** nodes in the dominating set within its ***r*** - hop distance), the value of *u.Required_Ch* is ***k*** minus the number of nodes in DS within the distance ***r*** from ***u***. Once node ***u*** has been covered, its *u.Required_Ch* is set to zero. Along with the *Required_Ch*, another variable called *Weight_Value* is also used to measure the capability of a node to act as a member in the dominating set. The *Weight_Value* of node ***u*** depends on the residual battery power and the *Required_Ch* value of ***r***-hop neighborhood. *The Required_Ch* value of ***r***-hop neighborhood is calculated as $\sum_{v \in Vrd} u.Required\_Ch$, V_{nl} is the set of the ***r*** - hop neighborhood of node ***u***. Thus, the *Weight_value* helps in selecting an efficient node with most number of uncovered nodes as the clusterhead. The selection of node ***u*** as clusterhead affects the *Weight_Value* of any node within 2***r*** - hop distance from ***u***. This is because selection of node ***u*** reduces the *Required_Ch* value of any node ***v*** in node ***u***'s ***r***-hop distance by one, if ***v*** is not covered.

A priority heap is used as the data structure and all nodes are inserted in the heap such that the node with the highest Weight_*Value* is at the root. The root node is selected as the dominating node and is removed from the heap and inserted in the dominating set, ***DS***. After selecting the root node as a member of the dominating set, the *Required_Ch* value of all nodes within the ***r*** - hop distance from the root node is reduced by one. When it becomes zero, the node is covered and is removed from the heap. These nodes are dominated nodes. After these updates, the heap is sorted so that the node with the highest *Weight_Value* is at the root. Ties are broken using the highest residual battery power as criterion. The process repeats until the heap is empty.

Illustrative Example

Consider the example shown in Figure 1, which is for an arbitrary network, consisting of 15 nodes to which we apply the clustering setup algorithm. Using the ***(k, r) -DS*** algorithm with parameter (2, 2), the dominating nodes obtained are C, F, L, H and O. These nodes can act as clusterhead nodes and each non-clusterhead node has at least 2 clus-

*Algorithm 1. Algorithm to find the (**k, r**) - Dominating Set*

```
Data: G = (V, E) and parameters r and k.
Result: DS, Dominating Set.
{
        Initialization;
        DS = φ;
        for each u ∈ V do
        {            n = Find_Weighted_ Numberof_Rhop_Neighbor (u);
                /* 1 + Number of single hop nodes * r + Number of 2-hop nodes * (r-1)
                +.... +                                  Number of r-hop nodes * 1
*/
                Compute (n, u.Weight_Value) ;/* Compute the Weight_value of node u */
                u.Required_Ch = k;
                Insert_Priority_Heap(H, u);
        }
        while (H ≠ φ) do
        {            d = del_root(H);
                /* Root node has the largest weight and is the best choice for dominating
                node */
                d.Required_Ch = 0;
                DS = DS U d;
                /* append dominating set */
                Find_Rhop_Neighbours(d, G, V_rd);/* Set of  r-hop neighbourhood of
        node d */
                for (u ∈ V_rd) and (u ∈ H) do
                {            u.Required_Ch = u.Required_Ch - 1;
                        if u.Required_Ch = = 0 then
                        {            Delete (H, u)
                        /* A node is said to be (k, r) dominated if that node has at least
k
                        neighbors within r-hop distance */
                        }
                }
                f =Find_2rhop_Neighbor (d, G, V_2rd);
                /* Required Ch of all nodes within r -hop distance from dominating node d
                is reduced by one and that affects the weight value of nodes within 2r
                distance from d */
                for each (u ∈ H) and (u ∈ V_2rd) do
                {            Compute (f, u.Weight_Value) ;
                        Find_ Rhop_ Neighbous(u, G, V_urd);
                        for each (w ∈ V_urd) do
                        {         u.Weight_Value = u.Weight_Value + w.Required_Ch ;
                         }
                }
                makeheap(H);
        }
}
```

terhead nodes within the 2-hop distance. Node A has three dominating nodes (H, C and F) within the 2-hop distance. Instead of considering H, C and F as clusterhead nodes, only the node with the best quality is selected as the clusterhead for node A. Thus, node A selects F as its clusterhead. The advantage of this approach is that it prevents reaching more than one copy of the same message routed by various clusterhead nodes, thus saving the energy of dominating nodes. The nodes H and C can be in sleep-mode or can act as an ordinary node until it receives a NODE_JOIN_REQ from any other node. Similarly, other non-clusterhead nodes are associated with a single clusterhead. The quality of the dominating nodes is obtained as shown in Table 1. A system of 15 nodes on a

Algorithm 2. Cluster formation Algorithm

```
Algorithm - Scenario_ cluster_ formation()
{
        // Find the dominating nodes using distributed k, r) algorithm
        call Find_Dominating_Set(G, k, r)
        if (STATUS == Dominating) then  /* Dominating nodes are clusterhead nodes
*/
        {         Broadcast FIND_NEIGHBOUR packet with TTL=r;
                Start Find_Neighbour_Timer();
        }
        if ((Packet_Received == FIND_NEIGHBOUR) and (TTL(Packet_Received) >=
        0) and (STATUS = = Dominated)) then/* Dominated nodes have at least k
        dominating nodes within r-hop distance*/
        {         Send FN_ACK  packet through the route in CACHE_PATH;
                Start Fn_Ack_Timer();
        }
        while (Find_Neighbour_Timer() <= FN_TH) do
        {       if ((STATUS= = Dominating) and (Packet_Received = =
        FIND_NEIGHBOUR_ACK)) then
                {       Update_ Neighbour_ List();
                }
                if (Find_Neighbour_ Timer() >FN_TH) and ((STATUS == Dominating))
then
                {          Compute_Quality();
                        Create_CHA_Packet();
                        Multicast_ CHA_Packet(Neighbor_List);
                }
        }
        while (Fn_Ack_Timer() <= FN_TH) do
        {       if (((Packet_ Received == CHA) and (STATUS == Dominated)) then
                {         update_ Cluster_Head_List();
                }
                if ((Fn_ Ack_ Timer() >FN_ ACK TH) and ((STATUS == Dominated))
then
                {         Cluster_ H = Find_Most_Qualified_Node(Cluster_Head_List);
                Send_ NJ_ Packet(Cluster_ H);
                }
        }
        if ((Packet Received = = NJ) and (STATUS = = Dominated)) then
        {        if (DEGREE <Thresh_ degree) the
                {         Send_  NJ_ ACK(); Update(Cluster_ member);
                        Increment DEGREE; Start_ NJ_ Ack_ Timer ();
                }
        }
        while (NJ_ Ack_ Timer() <= NJ_ ACK_ TH) do
        {        if ((Packet_ Received == NJ_ ACK) and (STATUS == Dominated)) then
                {         Update(Current_ CH); Start_ CH_ Timer();
                 }
        }
        if (NJ_ Ack_ Timer() >NJ_ ACK_ TH) then
        {        Cluster_ H=Find_ Next_ Qualified(Cluster Head List);
                Send_ NJ_ Packet(Cluster_ H);
        }
}//end of cluster formation
```

100X100 grid is considered. The nodes could move in different directions with displacement varying uniformly between 0 to a maximum of 10 per unit time. W_1, W_2 and W_3 are 0.5, 0.3 and 0.2, respectively. The maximum number of nodes a clusterhead can handle is limited to five in this case.

Table 2 shows the dominating nodes of each node and the selected clusterhead at time T_1. During the time slot T_1, node A selects node F as its clusterhead. The clusters formed after selecting the suitable clusterhead are as shown in Figure 2. At time T_2 when the clusterhead node L resigns its job due to low battery power, all the nodes attached to L are re affiliated to the next qualified dominating node of each member node of L.

Analysis

Consider the graph $G=(V, E)$ of n nodes. The first *for-loop* of **-(k, r) - DS***algorithm* takes *nlogn* steps for initialization and heap creation. The *while–loop* executes n times and the innermost for loop executes $O(\rho^2)$ time, where $\rho < n$. The value of ρ increases with ***r*** and can have a maximum value of n. Thus the algorithm runs in $O(n^3)$ complexity. Initially the heap is empty. After the initialization all nodes are inserted into the heap and hence the size of the heap is n after the initialization. The heap is created in such a way that the root node has the highest weight. This node is a dominating node and is getting deleted. Nodes other than the root nodes are also getting deleted if it has enough number of clusterheads within the ***r***-hop. The algorithm terminates when the heap is empty and the deleted nodes are either dominating nodes or a node with ***k*** dominating nodes within the ***r*** -hop distance.

Table 1. Quality of Clusterheads

Clusterheads	Quality
C	3.4052
F	5.9310
H	4.6562
L	4.7717
O	3.0021

Figure 1. (2, 2) –DS of the network with 15 nodes

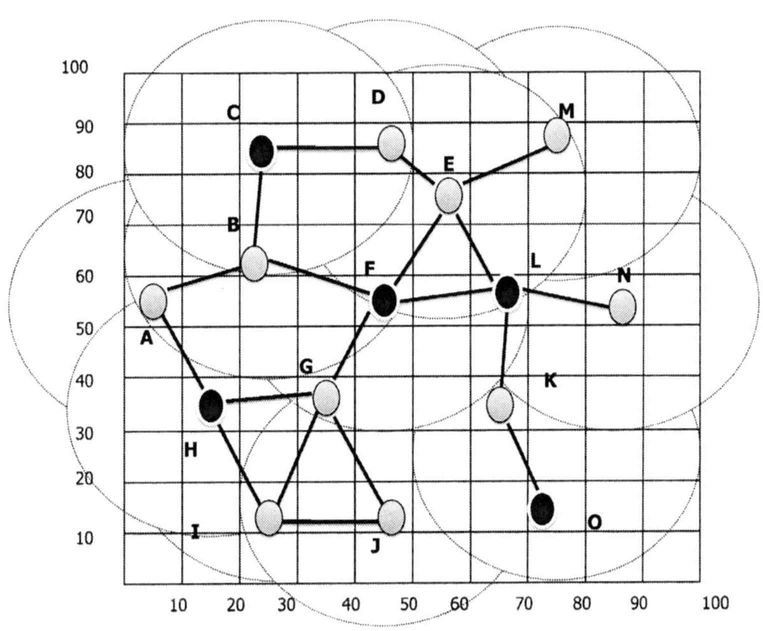

Table 2. Clusterhead association at time T_1

Nodes	Clusterheads (within *r*-hops)	T_1
A	F, H, C	F
B	F, L, H, C	F
C	F	F
D	F, L, C	F
E	F, L, C	F
F	L, H, C	CH
G	F, L, H	L
H	F	CH
I	F, H	H
J	F, H	H
K	F, L, O	L
L	F, O	CH
M	F, L	L
N	F, L, O	L
O	L	L

CLUSTER FORMATION AND MANAGEMENT

The issues of clusterhead computation, clusterhead association and cluster maintenance are discussed in this section.

Figure 2. Scenario based Cluster Formation

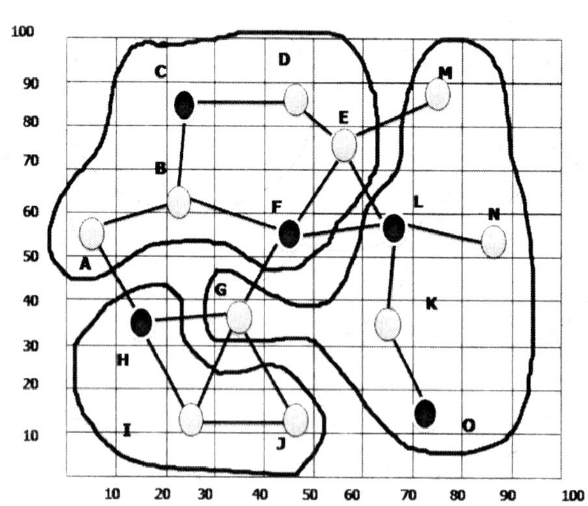

Quality Computation and Advertisement of Clusterheads

The dominating nodes send FIND_NEIGHBOUR packets to all neighboring nodes within the ***r*** -hop distance. On receiving the FIND_NEIGHBOUR packet from a dominating node, each node within the ***r*** -hop distance sends *T* number of *FN_ACK* packets one in each round. The entries of these packets are useful for the calculation of the quality of the dominating node. The formats of FIND_NEIGHBOUR and FN_ACK packets are as shown in Table 3 and Table 4. The dominating nodes wait for $2r + (T-1)$ rounds (where *T is* the number of time slots which is used for calculating local stability of dominating nodes) and then compute their quality. A dominating node uses a list to store the details of all neighboring nodes within the ***r*** -hop distance and creates the Neighbour_List data structure. After computing the quality of the clusterhead, it advertises its quality by sending the CHA (CLUSTER_HEAD_ADVERTISEMENT) packets. The format of this packet is as shown Table 5.

Clusterhead Assignment

Clusterheads periodically send CHA messages to prove its presence. On receiving the CHA messages, the ordinary nodes create/modify its Clusterhead_List. Nodes those are able to hear the CHA select the clusterhead with the highest quality and send NODE_JOIN (NJ) request to the selected clusterhead. If the total number of nodes attached with that clusterhead is below the maximum allowable number then the clusterhead accepts this request and sends NJ_ACK packet. The NJ packet with STATUS true indicates successful clusterhead assignment. Otherwise, the node should send the NJ message to the next qualified node.

Cluster Maintenance

Whenever a new node is switched on, it waits for CHA packets. There are two possible cases depending on the number of CHA packets received.

- Case 1: If the number of CHA packets received with different N_ID is greater than or equal to $\boldsymbol{k}$, then the new node selects the most qualified node as its clusterhead and sends NJ packets to the selected clusterhead. If the new node receives at least one pair of CHA packets with different Cluster_ID then this node acts as a gateway node.
- Case 2: If the number of CHA packets received is less than $\boldsymbol{k}$ then the new node acts as a clusterhead and enters into the quality finding state.

All dominating nodes have a list of neighbors within the $\boldsymbol{r}$-hop (stored in its Neighbor_List data structure) and they can multicast their quality. Non-dominating nodes select the most qualified dominating node among $\boldsymbol{k}$ of such a set as its clusterhead. The selected dominating node continues to act as clusterhead only if it has the best quality. The member nodes of any cluster re-affiliate to another clusterhead if its quality is greater than the current clusterhead. This improves the load balancing among the clusterheads.

PERFORMANCE EVALUATION

The performance of the algorithm is evaluated through simulations. The simulation parameters are listed in Table 6. The IEEE 802.11 has been used as MAC layer. Nodes in the simulation move according to the Random Waypoint Model. The maximum number of nodes that a clusterhead can handle is suitably adjusted depending upon the capability of the clusterhead and the system requirement. The values assumed for $\boldsymbol{k}$ and $\boldsymbol{r}$ are $\boldsymbol{k} = 1$ or 2 and $\boldsymbol{r} = 1, 2, 3$ or 4. Appropriate values are assigned for W_1, W_2 and W_3. The highest weight is assigned for mobility and the lowest weight for degree. These values are adjustable depending upon the scenario. The nodes are randomly placed over a terrain. To ensure that the network is connected, a connectivity test is performed.

Table 3. FIND_NEIGHBOUR packet

TYPE	STATUS	S_ID	TTL	C_PATH

Table 4. FN_ACK packet

TYPE	STATUS	S_ID	H_COUNT	C_PATH	N_ID	C_DIST

Table 5. CHA packet

TYPE	STATUS	N_ID	QUALITY	CLUSTER_ID

Table 6. Parameters used

Parameter	Value
Mobility Model	Random way point
Number of Nodes	10-500
Network Size	1000 X 1000 m^2, 500 X 500 m^2
Transmission range	20m - 150m
Max speed of node	20 m/s
Traffic model	128 byte packets, 4 packets/sec
Simulation time	600s

Experiments are repeated for 30 trials with various network topologies and varying number of nodes. The results show the average over 30 trials.

The first experiment provides a report of the average number of clusterheads or clusters created for different values of the number of nodes in the network. The number of dominating nodes strongly influences the latency, communication overhead and, inter and intra communication strategies. The number of clusterheads defines the dominating set. It is observed that the number of dominating nodes increases with the increase in total number of nodes in the proposed Scenario-based Clustering Algorithm for Mobile ad hoc networks (SCAM) is as shown in Figure 3. The number of dominating nodes created decreases with increase in the cluster diameter. This is because with larger values of ***r***, more number of nodes is covered. When the value of ***r*** is increased from 1 to 2, the number of dominating nodes is reduced by 51%, which results in lower number of clusters created. The rate of reduction decreases with increasing diameter.

Figure 4 shows that the average number of clusters is relatively high when the transmission range is small. When the transmission range increases, more and more nodes are connected to the same clusterhead, thereby reducing the number of clusters created. A smaller backbone is desirable for minimizing the routing overhead. Hence transmission power of a node can also be a parameter finding computing the quality of dominating nodes.

Three popular algorithms including H-DEGREE, Ruan's Algorithm and WCA are used for comparing the performance of SCAM. The H-DEGREE algorithm is one of the earlier popular algorithms which elects the clusterhead based on the highest connectivity. It creates clusters with single-hop radius. Ruan's algorithm is a central-

*Figure 3. Number of dominating nodes for **k** =1 and **r** =1, 2, 3 and 4*

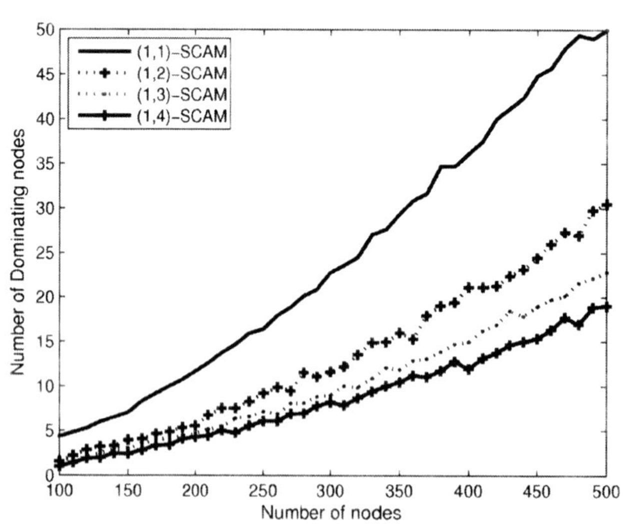

ized dominating set based algorithm whereas WCA is a combined weight metric algorithm. As SCAM is a combination of DS-based and combined weight metric-based algorithms, those two algorithms are worth for comparison purposes.

Figure 5 shows the number of clusters formed for SCAM, Ruan, WCA and H-DEGREE as a function of transmission range. The results show that the average number of clusterheads created with SCAM (with $k=2$ and $r=2$) is less compared with WCA, Ruan and H-degree. This is because SCAM creates less number of clusters when the cluster radius increases and makes use of the cluster merging process. But increased r leads to increased cluster size, which adversely affects the performance. So the selection of r is critical.

Figure 4. Transmission range vs. number of of clusters

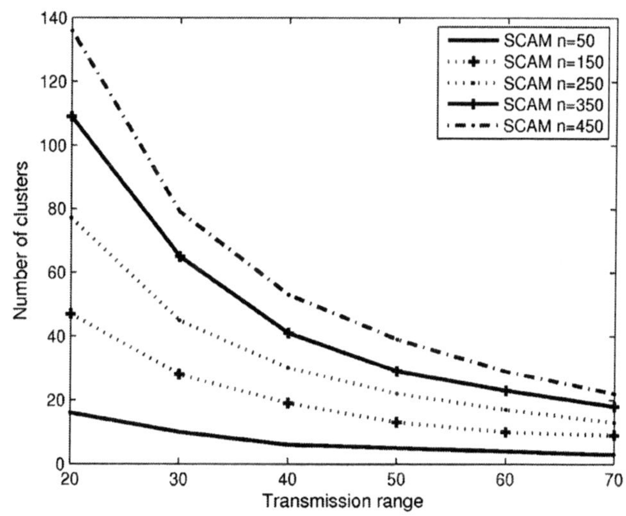

Figure 5. Number of clusters in Scenario-based algorithm, Ruan, H-DEGREE and WCA

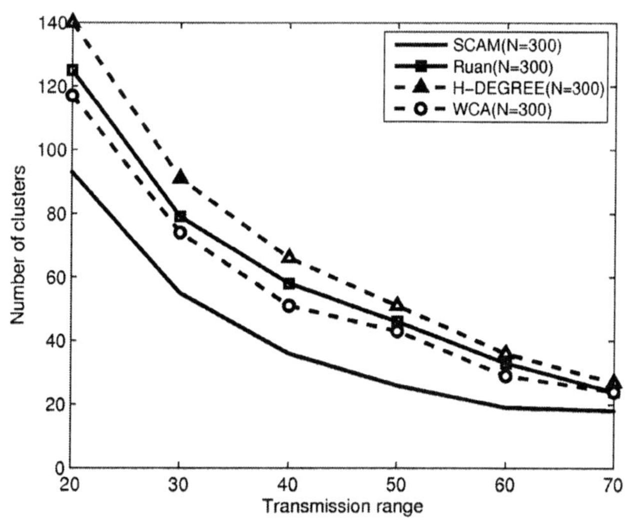

The algorithm achieves load balancing by assigning various time slots for the different clusterheads based on its quality and by specifying an agreeable threshold on the number of nodes that a clusterhead can handle. Figure 6 shows the Load Balancing Factor (LBF) for Scenario-based algorithm with $k = 1$ and $r = 2$, Scenario-based algorithm with $k = 1$ and $r = 1$ and H-degree (Maximum connectivity). The LBF is the inverse of the variance of the cardinality of the clusters (Chatterjee, 2002). A higher value of LBF signifies a better load distribution. Figure 7 shows the effect of putting a limit on total number of nodes that a clusterhead can handle. This mechanism is used to prevent the clusters from growing too large. SCAM outperforms other algorithms be-

Figure 6. Load Balancing Factor without LBF-heuristics

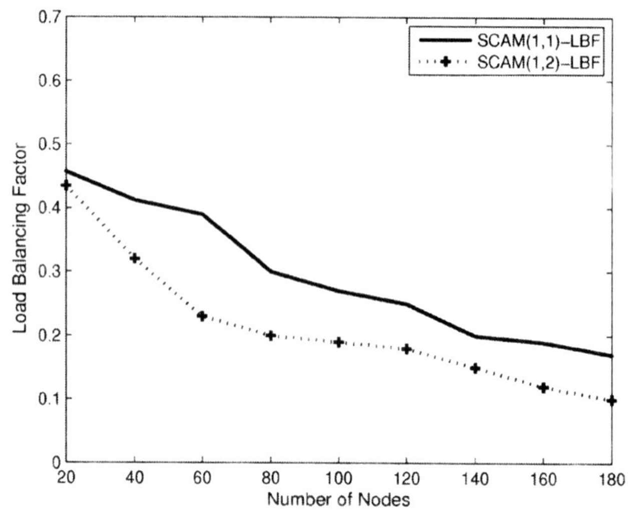

Figure 7. Load balance factor in Scenario-based algorithm with LBF-heuristics

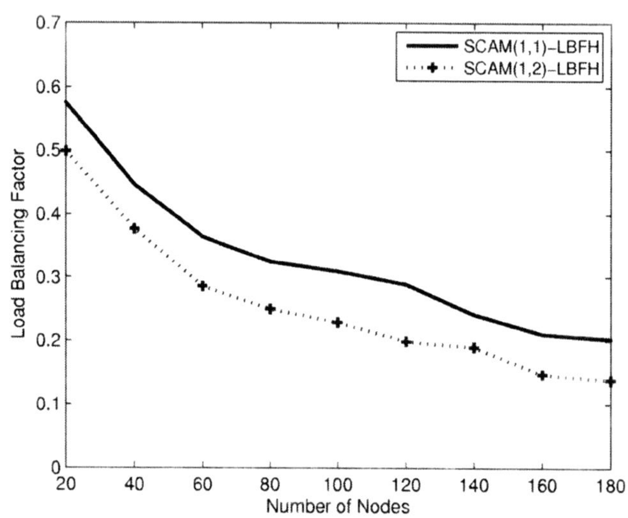

cause it is multi-clusterhead and bounded distance clustering algorithm. This is established in Figure 8.

The created clusters preserve its structure for a longer period than the H-DEGREE algorithm. It is observed that the number of re-affiliations increases with the increase in number of nodes and higher values of displacement as shown in Figure 9. Hence the number of cluster election process invoked is less in SCAM compared with WCA and H-DEGREE. The effect of number of dominating node updates on mobility is as shown in Figure 10.

Figure 8. Comparison of Load balancing Factor

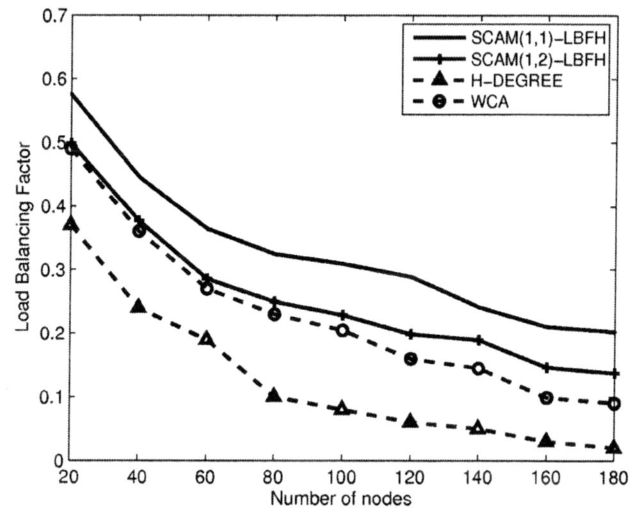

Figure 9. Number of re-affiliations per unit time

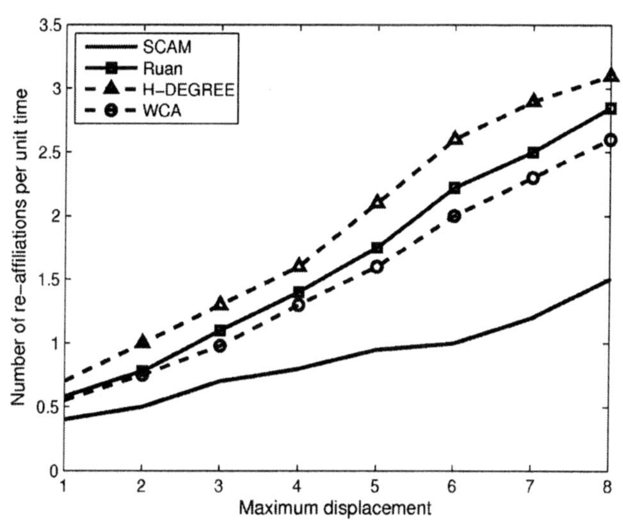

Figure 10. Number of dominating set updates

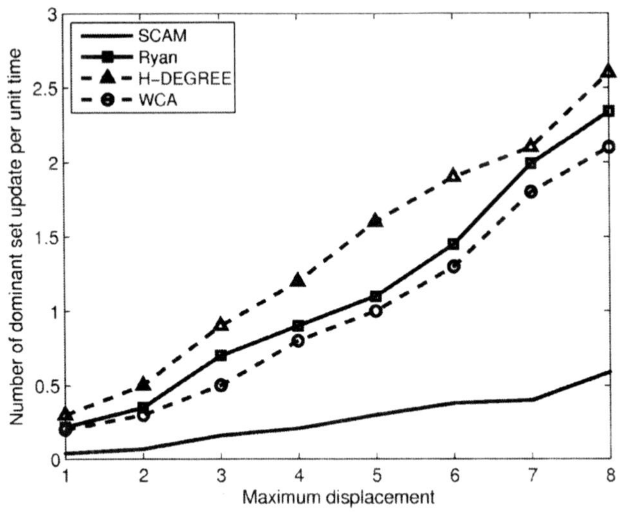

CONCLUSION

The proposed SCAM algorithm is a dominating set based, weighted, adaptive and centralized clustering algorithm. It creates clusters in such a way that the created clusters remain stable over a long period of time. The clusterhead election in this algorithm is based on the (k, r)−Dominating Set computation. To achieve scalability, SCAM allows variable diameter clusters (with the parameter ***r***) and clusterhead redundancy (with the parameter ***k***). Suitable values are assigned to parameters ***k*** and ***r*** depending on node density of the network and application. The selection of clusterhead is based on its quality, which is a function of its degree, residual battery power and stability. Different weights are assigned to each parameter depending upon the application. This approach is adaptive in nature as it adapts easily to the dynamic topology of the network as well as the node density of the network. Proper load balancing is achieved by putting a limit on the maximum number of nodes a clusterhead can handle and re-affiliation with another clusterhead based on the quality. The algorithm is scenario based as it creates stable clusters by accepting various parameters and weights.

The proposed algorithm uses techniques to maintain the cluster structure as stable as possible with less control messages. In this algorithm, intra-cluster routing requires a table driven strategy and inter-cluster routing is on demand basis. SCAM outperforms H-DEGREE, Ruan's algorithm and WCA in terms of number of re-affiliations, number of dominating set updates and number of clusters created. The reduction in number of re-affiliations improves the stability of the network topology under different scenarios.

Incorporating security, providing guaranteed quality of service and enhancement for use in heterogeneous mesh networks are topics suggested for further research.

REFERENCES

Amis, A. D., Prakash, R., Vuong, T., & Huynh, D. (2000). Max-min-d-cluster formation in wireless ad hoc networks. In *IEEE INFOCOM* (pp. 1293–1302). Washington, DC: IEEE.

Baker, D. J., & Ephremides, A. (1981). The architectural organization of a mobile radio network via a distributed algorithm. *IEEE Transactions on Communications*, *29*(11), 1694–1701. doi:10.1109/TCOM.1981.1094909

Basagni, S. (1999a). *Distributed clustering for mobile ad hoc network*. In Proc. Of Int'l Symp. Parallel Architectures, Algorithms and Networks (I-SPAN '99), (pp. 310–315).

Basagni, S., Mastrogiovanni, M., Panconesi, A., & Petrioli, C. (2006). Localized Protocols for Ad Hoc Clustering and Backbone Formation: A Performance Comparison. *IEEE Transactions on Parallel and Distributed Systems*, *17*(4), 292–306. doi:10.1109/TPDS.2006.52

Basagni, S., Turgut, D., & Das, S. K. (2001). Mobility adaptive protocols for managing large ad hoc networks. *In IEEE International Conference on Communications, 5*, 1539–1543.

Basagni (1999b). *Distributed and mobility adaptive clustering for multimedia support in multi-hop wireless networks*. In Proceedings of vehicular Technology Conference(oo.1 9-22).

Basu, P., Khan, N., & Little, T. D. C. (2001). *A mobility based metric for clustering in mobile ad hoc networks*. In Proceedings IEEE ICDCW 01, (pp. 413–18).

Chatterjee, M., Das, S. K., & Turgut, D. (2002). WCA: a weighted clustering algorithm for mobile ad hoc networks. *Cluster Computing*, *5*(1), 193–204. doi:10.1023/A:1013941929408

Chen, G., Nocetti, F., Gonzales, J., & Stojmenovic, I. (2002). *Connectivity based k-hop clustering in wireless networks*. In HICSS 02 Proceedings of the 35th Annual Hawaii International conference on System Sciences. IEEE Computer Society, (pp. 2450-2459).

Chiang, Ching-Chuan & Gerla, M. (1997). *Routing in clustered multi-hop, mobile wireless networks with fading channel*. In Proc. IEEE SICON 97 (pp. 197-211).

Choi, W., & Woo, M. (2006). *A distributed weighted clustering algorithm for mobile ad hoc networks*. In Proc. IEEE ICIW. (pp.73-78).

Dai, F., & Wu, J. (2005a). On constructing k-connected k-dominating set in wireless networks. *Journal of Parallel and Distributed Computing*, *66*, 947–958. doi:10.1016/j.jpdc.2005.12.010

Ephremides, A., Wieseltheir, J. E., & Baker, D. J. (1987). A design concept for reliable mobile radio networks with frequency hopping signaling. *Proceedings of the IEEE*, *75*(1), 56–73. doi:10.1109/PROC.1987.13705

Er, I. I., & Seah, W. K. G. (2006). Performance analysis of mobility-based d-hop (mobdhop) clustering algorithm for mobile ad hoc networks. *Computer Networks*, *50*(17). doi:10.1016/j.comnet.2005.12.013

Friis, H. T. (1946). A note on a simple transmission formula. *Proceedings of IRE*, *34*(5), 254–256. doi:10.1109/JRPROC.1946.234568

Ghosh, R., & Basagni, S. (2005). *Limiting the impact of Mobility on Ad Hoc Clustering*. Proceedings of Second ACM International Workshop on Performance Evaluation of Wireless Ad Hoc, Sensor and Ubiquitous Network, 197-204.

Guha, S., & Khuller, S. (1998). Approximation algorithms for connected dominating sets. *Algorithmica*, 20.

Johnson, D., & Maltz, D. (1996). Dynamic source routing ad hoc wireless networks. *Mobile Computing, 1*(1):153–181. Boston, MA: Kluwer Academic Publishers

Lin, C., & Gerla, M. (1997). Adaptive clustering for mobile wireless networks. In *IEEE JSAC, 15*, 1265–75.

Mc Donald, A. A., & Znati, T. F. (1999). Mobility Based framework for Adaptive Clustering in Wireless Ad Hoc Networks. *IEEE Journal on Selected Areas in Communications*, *17*(8), 1466–1486. doi:10.1109/49.780353

Parekh, A. (1994). *Selecting routers in ad hoc wireless networks*. In Proceedings of IEEE international Telecommunications symposium (Pp. 420-424).

Perkins, C., & Bhagwat, P. (1994). Highly dynamic destination-sequenced distance vector routing (DSDV) for mobile computers. *Computer Communication Review*, *1*(1), 234–244. doi:10.1145/190809.190336

Perkins, C., & Royer, E. M. (1999). *Ad hoc on demand distance vector routing*. In Proceedings 2nd IEEE Workshop on Mobile Computing Systems and Applications(pp. 90-100).

Ryu, J., Song, S., & Cho, D.-H. (2001). *New clustering schemes for energy conservation in two tiered mobile ad-hoc networks*. In Proc. IEEE ICC 01, pages 862–66, Vol.3. IEEE.

Spohn, M. A., & Garcia-Luna-Aceves, J. J. (2007). Bounded-distance multi-clusterhead formation in wireless ad hoc networks. *Ad Hoc Networks*, *5*, 504–530. doi:10.1016/j.adhoc.2006.01.005

Wu, J., & Li, H. (n.d.). *On calculating connected dominating set for efficient routing in ad hoc wireless networks*. In Proc. of Proceedings of the 3rd ACM International Workshop on Discrete Algorithms and Methods for Mobile Computing and Communications, pages 7–14.

Yu, J. Y., & Chong, P. (2003). *3hbac(3-hops between adjacent clusterheads): a novel non-overlapping clustering algorithm for mobile ad hoc networks*. In Proc. IEEE Pacrim 03, (Vol. 1, pp. 318–21).

Chapter 2
Optimizing Resource Consumption for Secure Messaging in Resource Constrained Networks

P. P. Abdul Haleem
National Institute of Technology, India

M. P. Sebastian
Indian Institute of Management, India

ABSTRACT

Conservation of resources such as bandwidth, energy and memory are of a concern in Resource Constrained Networks (RCNs). Wireless mobile devices, especially low cost devices are stifled by the limited resources such as battery power, screen size, input, memory and processors. The low cost wireless mobile devices penetrating the developing world market demand for a cost effective messaging format that fits within the constrained wireless environment. Reduction of verbosity is considered to be one of the most effective steps in controlling the resource consumption in RCNs. This chapter presents a method for optimizing resource consumption by the use of a new messaging format with less verbosity. The proposed format is based on YAML Ain't Markup Language (YAML), which is further enhanced with message level security specifications.

INTRODUCTION

The abundant growth of Internet services and the increased number of wireless mobile users resulted in the penetration of wireless mobile devices into the realm of wired networks. This penetration raised many issues related to wireless mobile networks and devices. Wireless mobile devices (with limited resources like battery power, memory, processing power, input and screen) together with wireless networks (with unreliable channels, low bandwidth, increased latency,

DOI: 10.4018/978-1-60960-563-6.ch002

increased rate of retransmission of lost packets, and weak security features) impose a number of limitations. However, users expect to access much wider range of applications in wireless mobile devices than in the conventional devices.

A major constraint in the wireless mobile devices is power. Even with the latest processors and memory chips, the rechargeable battery attached to the mobile devices still tends to fall behind the expectations. The RF part of cellular engine, which is responsible for transmission and reception of messages, is the biggest consumer of energy in a mobile device (Michael, 2005). Security protocols also consume power – the number of packets transmitted or received and the size of the keys are two important factors to be taken care to conserve energy. Wireless networks are generally less reliable, less secure and with more message latency.

Verbosity reduction of messages is a major concern in the constrained wireless mobile environment (Kangasharju, Lindholm, & Tarkoma, 2008). Reducing the verbosity of messages may have an impact cutting across several layers of the wireless networking protocols. XML is the default protocol for messaging ("SOAP-Tutorial", 2007). However XML faces many problems when used in the wireless mobile environment (Kangasharju, 2005). The number of bytes required for data representation is huge in XML. Due to this verbosity, XML buffers need to be flushed more often at the time of input and output, leading to lesser throughput. Also, larger messages are vulnerable to retransmissions. The highly textual nature of XML makes the string parsing compulsory for further processing. XML documents are structured and this adherence of the document to the accompanying structure is to be verified by the parsers. XML has been one of the easy targets for hackers, due to its long term use and universality.

Distributed computing and systems have changed significantly with the introduction of wireless communication and mobile nodes. These differences are not only quantitative (in that the links and nodes have inferior performance), but also qualitative (in that the characteristics also differ from fixed links and stationary computers). In tune with the widespread use and popularity of wireless mobile devices, complex operations in distributed and collaboration technologies allow people to move across organizational boundaries and to collaborate among/in organizations and communities with mobile devices. As a result, user communities demand for increased flexibility, inter-connectivity, and autonomy of involved systems as well as new coordination and interaction styles for collaboration among people. Three emerging trends in this context are (i) support for virtual web communities, (ii) support for web recommender systems that are adaptive to the ubiquitous devices, and (iii) adaptive content generation and delivery. Present day popular applications like Web services and agent communications languages (ACLs) also make use of a data format.

In many of the above approaches, an XML based dynamic content creation mechanism is employed to create, maintain and provide multiple variants of the content depending upon the type of devices from which the request is originated. The device (the capacity of the device) and information (the principle of poly-representation based on the document surrogates and the data source characteristics) "contexts" play decisive roles in the modeling of effective contextual information retrieval systems (Tamine-Lechani, Boughanem, & Daoud, 2009). Clearly, an alternative scheme with the merits of XML and with less verbosity can be a catalyst in the performance gain of the data management and sharing services, which is one of the most important and challenging part of a middle-ware layer in a distributed storage system that allows group members to share the data in Collaborative Working Environments (CWEs).

Reduction of message size results in the reduction of number of transmissions, which in turn reduces the data transmission cost and storage space (Lai, Chen, & Chen, 2008). It is interesting to note that, in the wireless world, reducing the message size is of primary importance than the processing efficiency gains through an alternative format (Kangasharju et al., 2008). Hence it is clear that the verbosity of the data format is the major factor which directly influences all the constraints, including the energy consumption, of wireless mobile devices and networks.

Coupled with this, the application of a lightweight cryptographic system suitable to the wireless mobile environment can reduce the power consumption considerably. ECC is already in use in the wireless mobile security arena due to its dramatic decrease in the key size and running time without compromising the security. A 160-bit ECC key offers the same level of security as a 1024-bit RSA key (Chou, 2003). The energy analysis on various cryptographic algorithms and key exchange protocols reveals considerable savings with ECC without compromising in security (Potlapally, Ravi, Raghunathan, & Jha, 2003).

There is an argument that the designers of wireless mobile devices are competing to pack more facilities into the devices and hence there is no scope for energy conserving measures, especially for messaging. But in reality, these features are used only by the top 10% of the customers. At the same time, low-cost devices are steadily increasing their market share, especially in developing countries. The projected shipments of ultra-low cost devices is estimated to be 24.3 million in the Asia Pacific region, as opposed to 0.6 million in the North America region (Gokran, 2007).

Thus, a messaging standard that performs well within the constraints of wireless mobile environments is a need. This chapter presents a messaging standard that avoids some of the limitations of XML (pertaining to the wireless mobile environment), retaining the good aspects. The suggested format has the following features: (i) simple and flexible, both for the user and the application programmer, (ii) editable and easily readable, (iii) less verbose without using compression/ binary encoding (i.e., without affecting the readability), (iv) facility for a schema definition, (v) consumes less bandwidth and needs short transmission time, (v) support for secured transmission of messages in transit, and (vi) support for ECC compatibility.

RELATED WORKS

Data Formatting Schemes

Design of alternative messaging formats to XML has received attention among the researchers and many proposals are available in the literature. The standard compression techniques, alternative formats for XML, caching (Devaram & Andresen, 2003) and binary encoding are the notable works in this area. It is worth mentioning that these methods propose good alternatives for the wired world. For instance, standard compression techniques such as Millau (Girardot & Sundaresan, 2000), Gzip (Cai, Ghandeharizadeh, Girardot, Schmidt, & Song, 2002), XMill (Cai et al., 2002) yield better performance with larger messages with a high redundancy rate. With SOAP, this is a working solution, as HTTP supports generic compression tools. But, it may not perform well with large sets of short messages, typical in the wireless mobile scenario. Compression routines will have to employ different compression techniques on different parts of the XML document (for elements and attributes). Also, a compression/decompression layer will be an additional overhead to the resource scarce wireless mobile devices (Schmelzer, 2004).

The binary encoding technique was employed to reduce the size of the messages. Wireless Application Protocol Binary XML (WBXML), also called Binary XML ("WAP Binary XML Content Format", 1999), is the oldest known format of this kind. The lack of support to the XML namespaces is considered to be a problem in utilizing WBXML as an alternative to XML (Sandoz & Pericas-Geertsen, 2005). The Xebu ("Xebu Serialization Format", 2007) format builds up its tokenization during a stream of messages, instead of considering each XML document in isolation, as in most of the other formats. But this format remains neutral to human language as WBXML. Self descriptive nature and easy readability properties of XML are hampered due to the introduction of the above enhancements. Also the user always has to depend on a conversion process to convert to binary format and revert back to the original format – ease of preparation and readability are affected.

Several alternative serialization formats were proposed to reduce the verbosity of XML which include YAML, JavaScript Object Notation (JSON) (Crockford, 2006), Open Node Syntax (ONX) (Open Node Syntax, 2005) and Simple Outline XML (Simple Outline XML: SOX, 2002). These formats suggest methods for conserving memory. They also tend to maintain the agility of XML. Among these standards, we have chosen YAML as the base format due to the following reasons (Clark, 2002; YAML, 2007) (i) YAML's versatile block indent syntax allows formatting of structured data in a visually uncluttered way, making it exceptionally human readable, (ii) YAML packs the information with less number of bytes among majority of the existing standards, (iii) YAML is a superset of JSON syntax, (iv) YAML offers extensible data types beyond primitives (i.e., beyond strings, floats, ints and bools) which can include class-type declarations or Unicode types, (v) YAML is resistant to delimiter collision, (vi) it is easier to embed a YAML document or other type of documents within a YAML document due to YAML's insensitivity to quotes and braces in scalar values, (vii) YAML handles indents as small as a single space, and this may offer better compression than markup languages, (viii) YAML offers a simple relational scheme that allows repeats of identical data to be referenced from two or more points in the tree rather than entered redundantly at those points, and (ix) YAML's lack of an associated command language (seen as a relative security benefit as parsers at least should be safe to apply to tainted data without fear of a latent command-injection security hole).

Message Level Security

In addition to reducing the verbosity of the messages, secure transmission of the messages is also to be guaranteed. Security need to be ensured for the messages in transit either at transport level or at message level. The transport level security is based on the Secure Sockets Layer (SSL) or the Transport Layer Security (TLS) that runs beneath the HTTP. Message level security (Bertino & Martino, 2006) includes all the security benefits of SSL, with additional features. It is an application layer service and facilitates the protection of data between applications. Its main features include (i) does not use the heavy SSL, (ii) data chunks are protected, and (iii) intended to work with intermediaries. It can be more effective and has the added flexibility that the message can be sent over any transport protocol.

The advantages of message level security over transport layer security are: (i) provision for end-to-end security as opposed to the point-to-point security(End-to-end security is ideal when multiple intermediary nodes exist between the two endpoints), (ii) support for message security while in transit and at rest (as opposed to only while in transit on the wire in transport layer security), and (iii) support for element-wise signing and encryption. Hence, it is evident that message level security specifications need to be included along with the format specification to make it secure.

As stated earlier, resource conservation is of primary concern in RCNs. Security algorithms and protocols generally consume the major chunk of the limited available resources in RCNs. The main sources of power drain during a secure wireless session are the number of packets transmitted/received and the size of messages required for establishing a session.

Implementation of ECC based security techniques are discussed in several papers (J.-H. Han, Kim, Jun, Chung, & Seo, 2002) (Edoh, 2004). Application of ECC cuts down the computational and communication costs without compromising the complexity of the process.

ECC is used in the design of the elliptical curve variations of Integrated Encryption Scheme (IES) and Signcryption. Elliptic Curve Integrated Encryption Scheme (ECIES) is a public-key encryption scheme based on ECC which provides semantic security against chosen-plaintext and chosen-ciphertext attacks (Certicom-Research, 2000). Signcryption is a method to reduce the cost of the signature-then-encryption method, by combining the functions of digital signature and public key encryption in a logically single step (Zheng, 1997a). Signcryption costs on the average 50% less in computation time and 91% less in message expansion than signature-then-encryption with RSA (Zheng, 1997b). Also, Signcryption technique provides faster and better security than conventional methods. Many papers are available making use of the effectiveness of Signcryption, in the wireless world (Peng & Li, 2005; Zhang et al., 2005; Park et al., 2006). Implementation of ECC based techniques are discussed in several papers (J.-H. Han et al., 2002; Edoh, 2004 for instance). Application of ECC cuts down the computational and communication costs without compromising the complexity of the process. Thus, it can be seen that IES and Signcryption with ECC appear to be better mechanisms in the constrained wireless mobile environment.

Generic specifications for encryption, signature and Signcryption are formulated in (Haleem & Sebastian, 2008). These specifications are derived from the W3C specifications for XML encryption ("XML Encryption Syntax and Processing", 2002) and signature ("XML-Signature Syntax and Processing", 2008). The generic specifications are to be made compatible with ECC based algorithms.

A digital signature is used for representing the document's identity. Any change in the document alters the message digest. Hence it is necessary to have a uniform standard for the messages and their digests that are being signed as there can be similar YAML messages, varying only in white spaces or line breaks. Also, there are possibilities of serializing the elements within a node in different orders. Even though reordering preserves the semantics, the signature derived will be distinct for these semantically equivalent messages. These effects are to be nullified in order to ensure that the message digest being created is not affected by the textual variations. This processing is popularly known as canonicalization, and is effectively implemented for XML files. W3C published the specifications for two types of canonicalization algorithms: Inclusive XML Canonicalization (Canonical XML, 2001) and Exclusive XML Canonicalization (Exclusive XML Canonicalization, 2002). The inclusive canonicalization process facilitates a case to case conversion of elements and seems to be the simple compared to the exclusive canonicalization process.

From the literature survey, it can be concluded that most of the existing message formatting schemes including XML are verbose in nature and are not suitable for efficient use in RCNs. Hence, there is a need for developing a lightweight and compact messaging format for RCNs. It is observed that YAML is the most promising existing data format which can serve as a base format for developing the lightweight compact message format.

It is learnt that message level security is better than transport level security for messaging. To implement message level security for the new format, specifications must be developed for encryption, signature and Signcryption of the messages. A canonicalization process also must be developed to minimize the effect of textual variations on the message digest. However it is observed that provision of security to messages is a heavyweight process. So it is necessary to minimize the security overheads without compromising the security. ECC allows phenomenal reduction in key sizes for the same level of security and hence there is a need for developing specifications for ECC compatiblity with the new format.

THE PROPOSED METHOD

In order to reduce the consumption of resources such as memory, bandwidth and battery power, we mainly concentrate on reducing the verbosity of messages. Care is taken to see that reducing the verbosity is achieved without compromising the proved features of XML such as self descriptive nature, readability, ease of preparation, schema awareness and security.

Our work uses YAML as the base format, a data serialization language devised by Brian Ingerson, Clark Evans, and Oren Ben-Kiki ("YAML", 2005). It has a human-readable format that takes features from languages such as XML, C, Python, Perl, as well as the format for electronic mail as specified by RFC 2822 ("YAML", 2005). YAML is optimized for data serialization, configuration settings, log files, Internet messaging and filtering ("YAML", 2005).

The advantages of YAML are summarized as follows: (i) easily readable and editable by any standard word processor, (ii) interacts well with scripting languages, (iii) uses host languages' native data structures, (iv) has a consistent information model, (v) enables stream-based processing, (vi) as expressive and extensible as XML, and (vii) its serialization reduces the verbosity of the message considerably, that too without affecting its readability ("YAML", 2005).

Our proposal consists of two parts: (i) lightweight format, and (ii) security specifications to achieve message level security. The proposed format based on YAML is designated as Thinned YAML (TYAML).

The TYAML Format

The steps in developing TYAML are (i) addition of schema awareness, (ii) verbosity reduction Phase I, and (iii) verbosity reduction Phase II.

Addition of Schema Awareness to TYAML: - Removing clutter from contents is an important step in verbosity reduction. This helps in conserving memory, when there are a group of messages with a common structure to be transmitted. This process results in the creation of a schema definition for TYAML – we designate it as YASchema.

There are three kinds of nodes in YAML - scalar, sequence, and mapping. Sequence means an ordered series of entries (Ben-Kiki, Evans, & Dot, 2008); mapping means an unordered association of unique keys to values (Ben-Kiki et al., 2008); and scalar means any datum with opaque structure presentable as a series of Unicode characters (Ben-Kiki et al., 2008). Combining these primitives generate a directed graph structure. Apart from this, there can be mappings of sequences and sequences of mappings.

The entire document is scanned one node at a time. The message structure is to be evaluated to decide about the type of schema to be constructed. The type can be of any of the three kinds - scalar, sequence and mapping. Mappings of sequences and sequences of mappings are considered as special cases. Primitive details like name of the element, its data type and an ID value are added to the schema for every member in a node. Once this process is completed, we get the YASchema description for the message to be transferred.

Advantages of YASchema: - YASchema offers the following improvements (i) squeezing techniques are applied to define schema definitions in flow styles, (ii) elements of schema and their properties are designed in such a way that they require only minimum bytes (this is done without hampering the readability), (iii) facility to define ID codes for the elements of YAML are given (this arrangement significantly reduces the size of the original message), and (iv) the process of reusing the same schema across multiple sessions are envisioned (a special directive is added in the YAML file to specify the schema definition to be referred).

Verbosity Reduction Phase I: - In this process, the built-in features of plain YAML is used in cutting the verbosity of the message. It can be seen that YAML's natural approach of representing information provides high degree of readability but also boosts the memory requirements needed for representing the information. A method of representation known as "Flow style" method in YAML (Ben-Kiki, Evans, & Dot, 2008) helps one to squeeze the size considerably. There are no significant gains in converting scalars using this method, as the effort made in this regard may not produce good results. In certain cases, reorganization of the contents to this style yields better performance. For e.g., a complex message consisting of a customer invoice containing many product details can be subjected to this stage for better conservation of message size. Reorganization of elements into flow style is done in this stage.

Verbosity Reduction Phase II: - This stage uses YASchema for reducing the message size. All primitive details regarding each and every element in the message are defined in the schema, in the schema creation stage itself. In addition to this, special ID codes are provided for each element.

The message is scanned and the elements are encoded with the ID codes. The ID codes could be referred from the accompanying schema, or if the sender likes to reuse a schema that had been sent earlier, the schema to be referred could be specified in the message itself. This makes the decoding easy, making the message schema aware and is a useful value addition to YAML. The process is as shown in Figure 1.

Figure 1. Verbosity Reduction Phase II

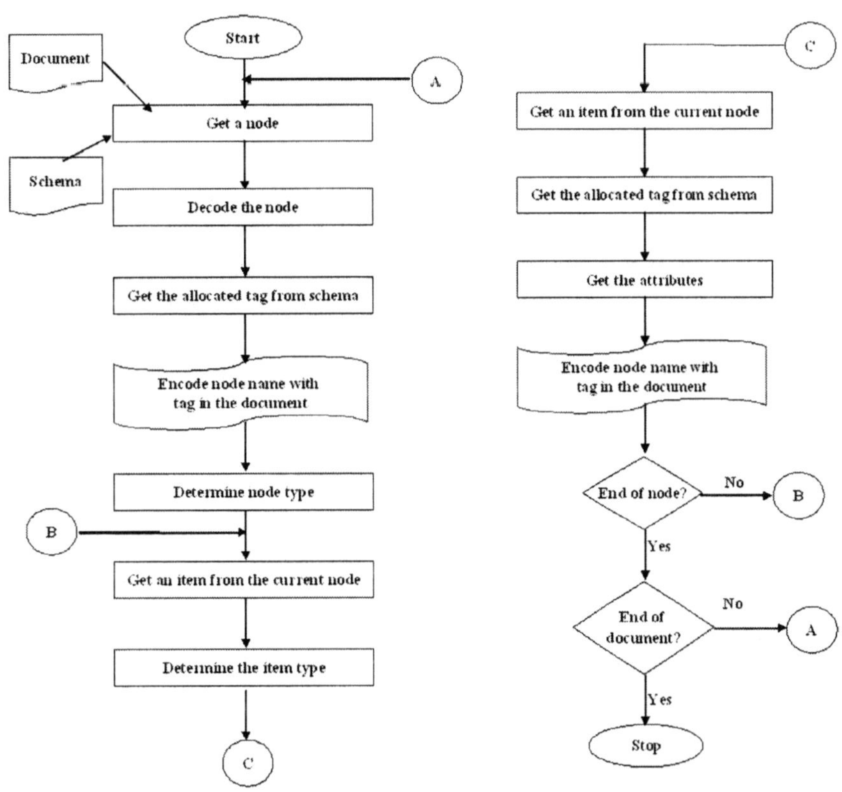

Incorporating Message Level Security

In addition to optimizing the message size, secure transfer of messages is a must. Exchanging the information without disclosing its contents is vital in the present-day networked world. Apart from ensuring the confidentiality of messages, authentication, data integrity and support for non-repudiation also must be ensured. Incorporating these mechanisms makes TYAML safe and secure.

There is a need for a uniform specification to describe and exchange the parameters and associated information. The security specification defines the formulation of document structure and the associated schema for a particular purpose. This is essential to maintain uniformity among the different platforms and application program interfaces.

Security protocols and algorithms play a limiting or negative role in process efficiency and energy efficiency. So, restricting the number of packets transmitted or received and using smaller keys are important to overcome the above limitations. Analysis on the energy consumption for the processing of various cryptographic algorithms underlines the need to have messages with reduced size and small key sizes (R. Potlapally Ravi & Niraj, 2006).

Accordingly in this chapter, the generic specifications in (Haleem & Sebastian, 2008) are refined to include ECC based algorithms (ECIES, ECDSA, and EC Signcryption), to make TYAML compatible with ECC based algorithms.

Canonicalization Process: - All the causes of variations that occur for XML messages will not be applicable to the sliced form of YAML, as there are subtle differences in the methodology adopted for the representation of information. YAML specification (Ben-Kiki et al., 2008) gives an insight into the need for canonicalization. Also, the methods to compare the equivalences of two YAML structures are discussed in the specification. YAML supports the need for scalar equality by insisting that every scalar tag must specify a mechanism to produce the canonical form of any formatted content.

The inclusive canonicalization process is formulated for TYAML. The canonicalization process ensures that conversions are done to make a unified format for messages to be signed. We have taken a subset of the canonicalization process (relevant to YAML) suggested by W3C (Canonical XML, 2001) for our purpose, as follows: (i) the document is represented in UTF-8 encoded format, (ii) line breaks are normalized to "#xA", (iii) default attributes (specified in the YASchema) are added to each element if no values are provided, (iv) white spaces are normalized, (v) name space declarations except superfluous in nature are preserved, (vi) attributes are included in ascending lexicographic order if the order of occurrence of the attributes are not specified in the accompanying YASchema, and (vi) facilities to include or exclude comments after canonicalization.

Specification for ECIES: - The skeleton structure of the specifications for ECIES is as shown in Figure 2. *EncrData* is the root element. The encrypted part of the document is initiated with the *EncrData* mapping. It is the core element where all encrypted data are enclosed. Elements specific to ECIES are *KDF*, *MAC* and *SharedInfo*. These elements house the key derivation function, message authentication code function and the shared info called *S1* and *S2*, respectively. Also, the *CipherValue* element encompasses the results *v* (the public key), *c* (the cipher key) and *t* (the message tag).

Specification for ECDSA: - The generic specification for digital signature is modified to include the elements needed for ECDSA operations. The minimal structure is as shown in Figure 3. The specification includes provisions to maintain information such as signature algorithm, digest method, signature value and key information.

Specification for EC Signcryption: - The skeleton structure of the specification and YASchema for EC Signcryption is as shown in Figure 4. The specification has provision to include all the parameters needed to perform EC Signcryption - Signature Algorithm, Encryption Method, One-way Hash, Keyed Hash, Key Information and Signcryption Values.

PERFORMANCE EVALUATION

Messages differing in size and complexity are normally exchanged between the users. We classify the messages into five types: *Short* type represents the simple messaging format with only text, *Small* category consists of a single record with String, Float and DateTime types of data, *Medium* category contains the details of 25 customer records, and *Large* and *Composite* categories are constituted as invoice records with varying degrees of complexity. These categories are designed in such a way to provide consistent test data scheme.

YAML, XML and TYAML versions are serialized for each category of messages for performance evaluation.

VERBOSITY

Verbosity: TYAML vs. Other Non Binary Formats: - Verbosity of TYAML data sets are compared against ONX, JSON and SOX formats (Figures 5, 6 and 7, respectively). These three formats are chosen as representatives of the alternative non-binary serializations.

The observations from Figures 5, 6 and 7 are: (i) for *Small* category, all formats exhibit more or less similar performance level; the SOX format has a slight advantage over TYAML (ii) when the complexity of the data set increases, TYAML

Figure 2. (a) Specifications for ECIES and (b) Corresponding YASchema

```
EncrData:
  Type:
  EncrMethod:
    Algorithm:
  KDF:
    Algorithm:
  MAC:
    Algorithm:
  SharedInfo:
    S1:
    S2:
  KeyInfo:
    n:
    g:
      x:
      y:
    a:
    b:
    p:
    h:
  CipherData:
    CipherValue:
      vValue:
      cValue:
      tValue:
```

(a)

```
tp:B2
  EncrData:(tg:Ed, tp:B2, rd:T, nsl:, ns:)
   Type:(tg:_Tp_, tp:1, rd:T)
   EncrMethod:(tg:_Em_, tp:B2, rd:T)
     Algorithm:(tg:_Al_, tp:1, rd:T)
     KDF:(tg:, tp:B2, rd:T)
       Algorithm: (tg:_Al_, tp:1, rd:T)
     MAC:(tg:, tp:B2, rd:T)
       Algorithm: (tg:, tp:1, rd:T)
     SharedInfo:(tg:_ShI_, tp:B2, rd:T)
       S1: (tg:, tp:1, rd:T)
       S2: (tg:, tp:1, rd:T)
     KeyInfo:(tg:_KI_, tp:B2, rd:T)
       n: (tg:, tp:1, rd:T)
       g: (tg:, tp:B2, rd:T)
         x: (tg:, tp:1, rd:T)
         y: (tg:, tp:1, rd:T)
       a: (tg:, tp:1, rd:T)
       b: (tg:, tp:1, rd:T)
       p: (tg:, tp:1, rd:T)
       h: (tg:, tp:1, rd:T)
     CipherData:(tg:_Cd_, tp:B2, rd:T)
       CipherValue:(tg:_CV_, tp:B2, rd:T)
         vValue: (tg:_vv_, tp:1, rd:T)
         cValue: (tg:_cv_, tp:1, rd:T)
         rValue: (tg:_rv_, tp:1, rd:T)
```

(b)

Figure 3. (a) Specifications for ECDSA and (b) Corresponding YASchema

```
Signature:
  SignedInfo:
    SignatureMethod:
      Algorithm :
    Reference:
      URI:
      DigestMethod:
        Algorithm:
      DigestValue:
  SignatureValue:
    rValue:
    sValue:
  KeyInfo:
    ECDSAKeyValue:
      n:
      g:
        x:
        y:
      a:
      b:
      p:
      h:
```

(a)

```
tp:B2
Signature:(tg:Sg, tp:B2, rd:T, nsl:, ns:)
  SignedInfo:(tg:_SI_, tp:B2, rd:T)
    SignatureMethod:(tg:_Sm_, tp:B2, rd:T)
      Algorithm : (tg:_Al_, tp:1, rd:T)
    Reference:(tg:_Rf_, tp:B2, rd:T)
      URI: (tg:_Ur_, tp:1, rd:T)
      Transforms:(tg:_Tr_, tp:B2, rd:F)
        TransAlg: (tg:_Ta_, tp:1, rd:F)
      DigestMethod:(tg:_Dm_, tp:B2, rd:F)
        DigAlg: (tg:_Da_, tp:1, rd:T)
      DigVal: (tg:_Dv_, tp:1, rd:T)
  SignatureValue:(tg:_SI_, tp:B2, rd:T)
    rValue: (tg:_rV_, tp:1, rd:T)
    sValue: (tg:_sV_, tp:1, rd:T)
  KeyInfo:(tg:_Ki_, tp:B2, rd:T)
    ECDSAKeyValue: (tg:_dsaKv_, tp:B1, rd:T)
      n: (tg:, tp:1, rd:T)
      g: (tg:, tp:B2, rd:T)
        x: (tg:, tp:1, rd:T)
        y: (tg:, tp:1, rd:T)
      a: (tg:, tp:1, rd:T)
      b: (tg:, tp:1, rd:T)
      p: (tg:, tp:1, rd:T)
      h: (tg:, tp:1, rd:T)
```

(b)

Figure 4. (a) Specifications for EC Signcryption (b) Corresponding YASchema

```
ECSigncryption:
  Type:
  SignedInfo:
    SignatureMethod:
      SigAlgorithm:
    EncryptionMethod:
      EncAlgorithm:
    Reference:
      URI: ""
      OnewayHash: One way hash function
      KeyedHash: Keyed Hash Function
    KeyInfo:
      n:
      g:
        x:
        y:
      a:
      b:
      p:
      h:
  SigncryptionValue:
    CValue:
    RValue:
    SValue:
```

(a)

```
tp:B2
ECSigncryption:(tg:_EcYs_, tp:B2, rd:T)
  Type:(tg:_tp_, tp:1, rd:T)
  SignedInfo:(tg:_Si_, tp:B2, rd:T)
    SignatureMethod:(tg:_Sm_, tp:1, rd:T)
      SigAlgorithm:(tg:_SAg_, tp:1, rd:T)
    EncryptionMethod:(tg:_Em_, tp:1, rd:T)
      EncAlgorithm:(tg:_EAg_, tp:1, rd:T)
    Reference:(tg:_Rf_, tp:B2, rd:T)
      URI:(tg:_Ui_, tp:1, rd:T)
      OnewayHash:(tg:_Oh_, tp:1, rd:T)
      KeyedHash:(tg:_Kh_, tp:1, rd:T)
      KeyInfo:(tg:_Kf_, tp:B2, rd:T)
        n:(tg:, tp:1, rd:T)
        g:(tg:, tp:1, rd:T)
          x:(tg:, tp:1, rd:T)
          y:(tg:, tp:1, rd:T)
        a:(tg:, tp:1, rd:T)
        b:(tg:, tp:1, rd:T)
        p:(tg:, tp:1, rd:T)
        h:(tg:, tp:1, rd:T)
  SigncryptionValue:(tg:_Sv_, tp:B2, rd:T)
    CValue:(tg:_C_, tp:1, rd:T)
    RValue:(tg:_R_, tp:1, rd:T)
    SValue:(tg:_S_, tp:1, rd:T)
```

(b)

needs only less number of bytes than the other serializations. Hence TYAML has maximum advantage in the *composite* category.

Verbosity: TYAML vs. XML, YAML (XML in WFF):- The verbosity of TYAML is compared against plain YAML and XML. For this compari-

Figure 5. Verbosity: TYAML vs. Other Non Binary Formats (short category)

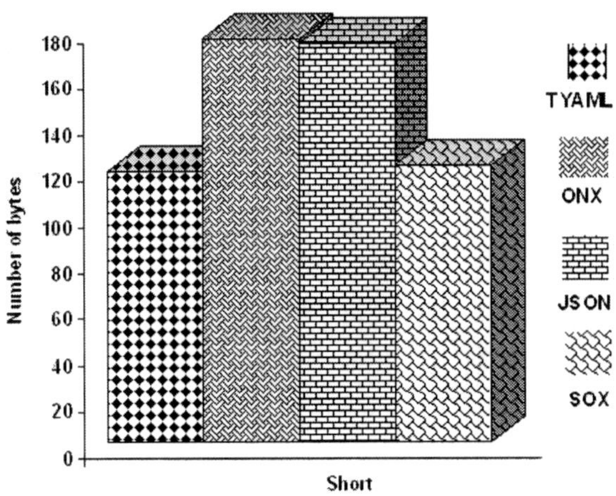

Figure 6. Verbosity: TYAML vs. Other Non Binary Formats (medium category)

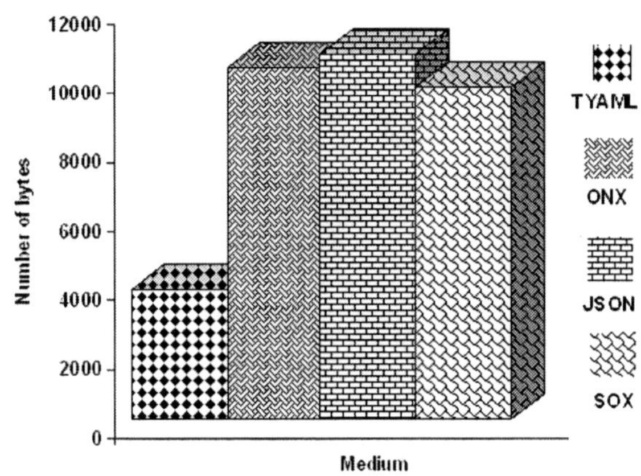

Figure 7. Verbosity: TYAML vs. Other Non Binary Formats (composite category)

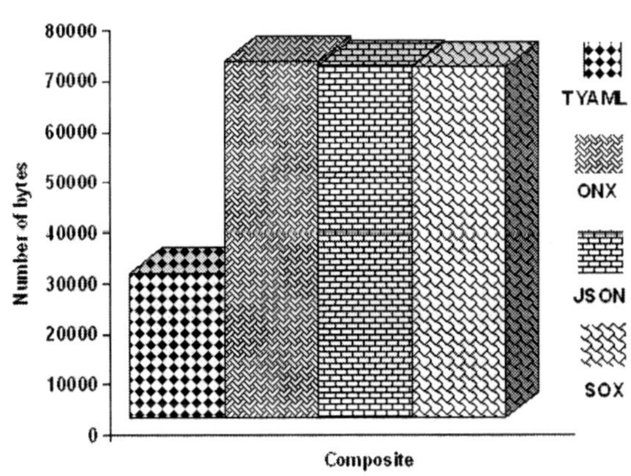

son XML is serialized in the well formed form (WFF), as WFF is accepted universally by all the browsers. It can be observed from Figure 8 that TYAML is performing better than YAML and XML in all categories. The advantage of XML is more visible in the *medium*, *large* and *composite* categories.

Verbosity: TYAML vs. YAML (both without schema):- YAML and TYAML are compared for verbosity in different categories. It can be observed from Figure 9 that TYAML has a marginal advantage in *short* and *small* categories than YAML. In the other categories, TYAML has an advantage over YAML. This demonstrates the effectiveness of the thinning measures for the verbosity reduction.

Packets Needed: TYAML vs. XML:- Maximum Segment Size (MSS) in TCP/IP on an Ethernet network is taken as the reference for the calculation of the number of packets (1460 bytes). The number of packets needed for the transmission of data is crucial in the constrained wireless mobile environment because of the limitations such as packet loss, retransmission, energy consumption, etc. It can be observed from Figure 10 that TYAML data sets require minimum number of packets, especially in the larger categories (*large* and *composite*).

SECURITY SPECIFICATIONS

The performance of TYAML specifications are compared with that of XML. A key size of 283 bits is used for the EC algorithms. The parameters and other details of the process and processed values are represented as per the specifications in TYAML and XML, for comparison.

Figure 8. Verbosity: TYAML vs. XML (in WFF) and YAML

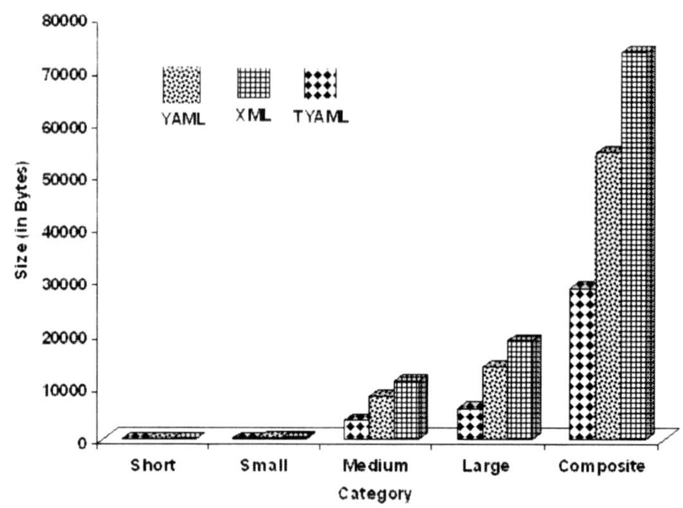

Figure 9. Verbosity: TYAML vs. YAML (both without schema)

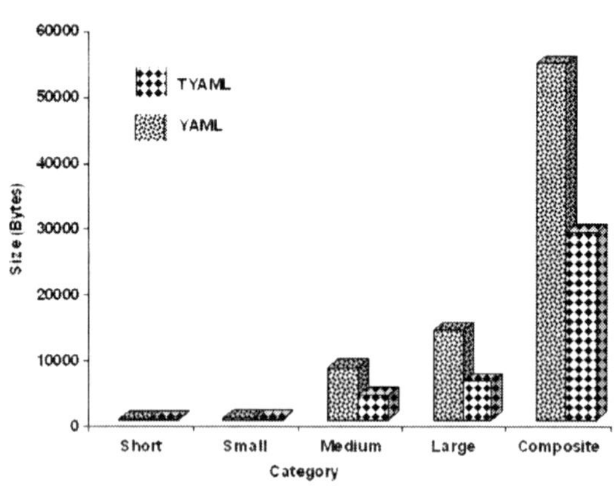

Figure 10. Packets Needed: TYAML vs. XML

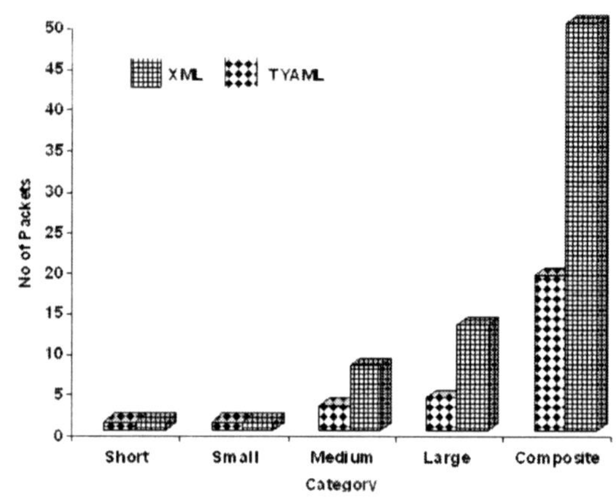

Verbosity for ECIES Specification: TYAML vs. XML: - The verbosity of the XML specification and XML Schema for ECIES are compared with the corresponding TYAML specification and YASchema in Figure 11. It can be seen that TYAML/YASchema combination is performing better than the XML/XML Schema combination.

Verbosity for ECDSA Specification: TYAML vs. XML: - The verbosity of XML specification and XML Schema for ECDSA are compared with the corresponding TYAML specification and YASchema as shown in Figure 12. In this case also, TYAML/YASchema combination is performing better than the XML/XML Schema combination.

Verbosity for EC Signcryption Specification: TYAML vs. XML: - The specifications (XML and TYAML) and corresponding schemas (XML Schema and YASchema), after applying the EC version of Signcryption, are compared for verbosity in Figure 13. It can be seen that the TYAML/YASchema combination is performing better than the XML/XML Schema combination.

Figure 11. Verbosity for ECIES Specification: TYAML vs. XML

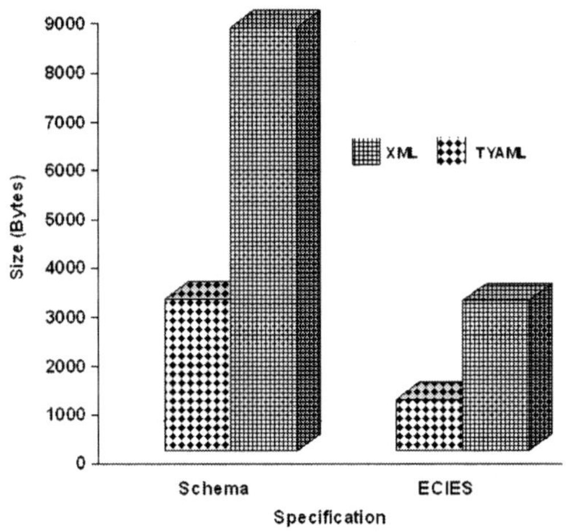

Figure 12. Verbosity for ECDSA Specification: TYAML vs. XML

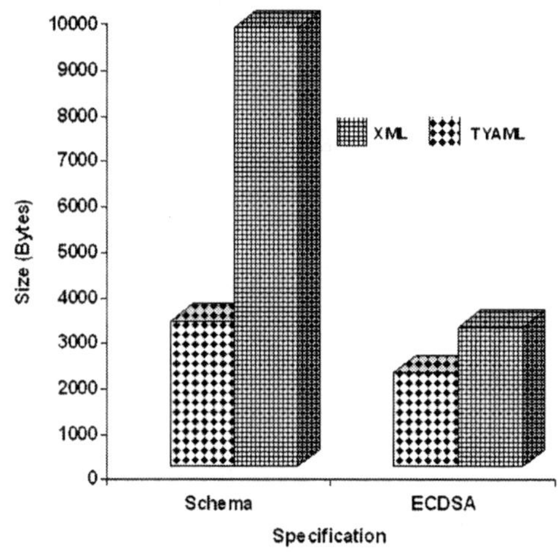

The results of the verbosity comparisons for EC based specifications between XML and TYAML are as shown in Figure 14. The TYAML format reports gain in all categories.

CONCLUSION

The challenges posed by the networks and devices due to the increased penetration of wireless mobile devices calls for the need to have a messaging standard suitable for constrained wireless mobile environments. This chapter proposed a solution to overcome the limitations imposed by latency, bandwidth requirement and low battery backup in wireless networks.

XML, in its new incarnation as a messaging protocol, is the default format for the purpose. But the inherent limitations of the constrained wireless mobile environment make XML unsuitable in this environment. The suggested method is based on YAML, a user friendly lightweight

Figure 13. Verbosity for EC Signcryption Specification: TYAML vs. XML

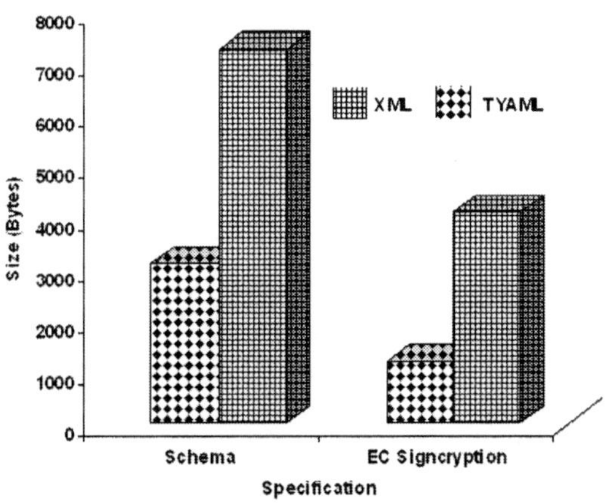

Figure 14. Verbosity of EC based Specifications: TYAML vs. XML

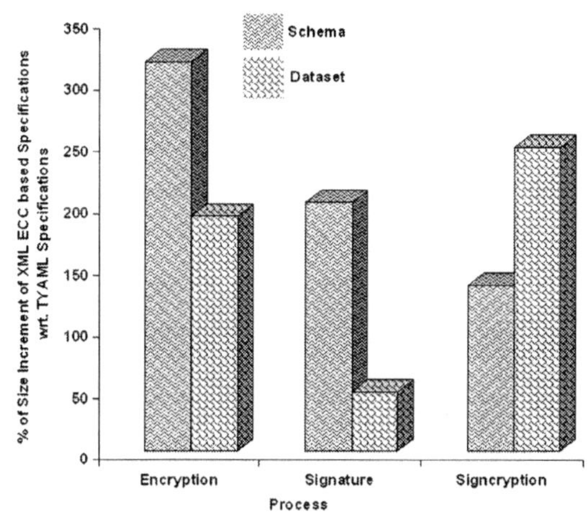

messaging format. It is observed that message size is reduced using YAML in its natural form itself. This is further enhanced by verbosity reduction measures. To optimize the results, we have proposed a schema definition also. Apart from the verbosity reduction, proposals for fast and energy conserving security processing are also included. Improvements are observed in the case of message size, packets needed, and in terms of security specifications.

Several enhancements are needed to furnish this as a messaging standard. TYAML/YASchema combination can be fully geared up to contain the attribute based access control mechanism as envisaged in policy languages like eXtensible Access Control Markup Language (XACML). These are topics suggested for further research. The improvement over the message size can save battery life, transmission time and energy consumption.

REFERENCES

Ben-Kiki, O., Evans, C., & Dot, I. (2008). *YAML 1.2 Specification*, (Available at http://yaml.org/spec/1.2/).

Cai, Ghandeharizadeh, Girardot, Schmidt, & Song. (2002, September). *A Comparison of Alternative Encoding Mechanisms for Web Services*. In 13th International Conference on Database and Expert Systems Applications, Aix en Provence, France.

Canonical, X. M. L. (2001, March). Retrieved from http://www.w3.org/TR/xml-c14n

Certicom-Research. (2000). *Standards for Efficient Cryptography, sec 1: Elliptic Curve Cryptography (Tech. Rep.)*. Foster City, CA: Certicom Research.

Chou, W. (2003). *Elliptic Curve Cryptography and Its Applications to Mobile Devices (Tech. Rep.)*. College Park, MD: University of Maryland.

Clark, K. G. (2002, July). *Look Ma, No Tags*. Retrieved from http://www.xml.com/pub/a/2002/07/24/yaml.html.

Crockford, D. (2006, July). *The application/json Media Type for JavaScript Object Notation (JSON)*. Retrieved from http://tools.ietf.org/html/rfc4627

Devaram, & Andresen. (2003). *SOAP Optimization via Parameterized Client-Side Caching*. Retrieved from http://java.sun.com/ developer/technicalArticles/ xml/fastinfoset/

Edoh, K. D. (2004). *Elliptic Curve Cryptography: Java Implementation*. In INFOSECCD '04: Proceedings of the 1st Annual Conference On Information Security Curriculum Development (pp. 88–93). New York, NY: ACM.

Exclusive, X. M. L. Canonicalization. (2002, July). Retrieved from http://www.w3.org/ TR/xml-exc-c14n/).

Girardot, & Sundaresan. (2000). *Millau: An Encoding Format For Efficient Representation And Exchange Of XML Over The Web*. In Proceedings of the 9th International World Wide Web Conference On Computer Networks (p. 747-765). Amsterdam: North-Holland Publishing Co. Retrieved from citeseer.ist.psu.edu/article / widener01open.html

Gokran, P. (2007, November). *Making your Low-Cost Handsets Plans Succeed*. In Telecom Era (pp. 23–32).

Haleem, P. P. A., & Sebastian, M. P. (2008). An Alternative Approach For Slicing Down The Message Size And Enhancing The Security In Wireless Mobile Network. *The Mediterranean Journal of Computers and Networks*, *5*, 148–149.

Han, J.-H., Kim, Y.-J., Jun, S.-I., Chung, K.-I., & Seo, C.-H. (2002). *Implementation Of ECC/ECDSA Cryptography Algorithms Based On Java Card.* In ICDCSW '02: Proceedings of the 22nd International Conference On Distributed Computing Systems (pp. 272–278). Washington, DC: IEEE Computer Society.

Han, Y., Yang, X., Wei, P., Wang, Y., & Hu, Y. (2006). ECGSC. *Elliptic Curve Based Generalized Signcryption, LNCS4159*, 956–965.

Johnson, & Menezes. (1999, August). *The Elliptic Curve Digital Signature Algorithm (ECDSA) (Tech. Rep.)*. Technical report CORR 99-34, Department of C&O, University of Waterloo, 1999.

Kangasharju, J. (2005). *Mobile XML Messaging (Tech. Rep.). Department of Computer Science.* Finland: University of Helsinki.

Kangasharju, J., Lindholm, T., & Tarkoma, S. (2008). XML Security with Binary XML for Mobile Web Services. *International Journal of Web Services Research*, *5*, 1–19. doi:10.4018/jwsr.2008070101

Lai, Y.-X., Chen, Y.-L., & Chen, H. (2008, November). PEJA: Progressive Energy-Efficient Join Processing For Sensor Networks. *Journal of Computer Science and Technology*, *23*(6), 957–972. doi:10.1007/s11390-008-9191-2

Michael, M. P. (2005). *Research Seminar On Energy Awareness, University Of Helsinki.* Energy Awareness for Mobile Devices.

Open Node Syntax. (2005, June). Retrieved from http://www.seairth.com/ web/onx/onx.html

Park, N., Kim, H., Chung, K., Sohn, S., & Won, D. (2006). *Xml-Signcryption Based LBS Security Protocol Acceleration Methods In Mobile Distributed Computing.* In ICCSA 2006: International Conference On Computational Science And Its Applications (LNCS 3984, pp. 251–259). Glasgow, UK: Springer-Verlag

Peng, C., & Li, X. (2005, September). *Threshold Signcryption Scheme Based On Elliptic Curve Cryptosystem And Verifiable Secret Sharing.* In 2005 International Conference On Wireless Communications, Networking And Mobile Computing (Vol. 2, pp. 1182–1185).

Potlapally, N., Ravi, S., Raghunathan, A., & Jha, N. (2003, August). *Analyzing The Energy Consumption Of Security Protocols.* In Proceedings Of The 2003 International Symposium Of Low Power Electronics And Design, Seoul, Korea.

Sandoz, P., & Pericas-Geertsen, S. (2005). *Fast Infoset@Java.net.* In Xtech 2005: XML, The Web And BeyondRetrieved from http://idealliance.org/proceedings/xtech05/ papers/04-01-01/

Schmelzer, R. (2004, November). *Will Binary XML Solve XML Performance Woes?* Retrieved from http://searchwebservices.techtarget.com / tip1,289483, sid,26 gci1027726,00.html

Simple Outline, X. M. L. SOX. (2002). Retrieved from http://www.langdale.com.au/ SOX/

Soap-tutorial. (2007). Retrieved from http://www.w3schools.com/ soap/default.asp

Tamine-Lechani, L., Boughanem, M., & Daoud, M. (2009, July). *Evaluation Of Contextual Information Retrieval Effectiveness: Overview Of Issues And Research.* Knowledge and Information Systems.

WAP Binary XML Content Format. (1999, June). Retrieved from http://www.w3.org/ TR/wbxml/

Xebu Serialization Format. (2007, September). Retrieved from https://hoslab.cs.helsinki.fi/homepages/xebu/

XML Encryption Syntax and Processing. (2002, December). Retrieved from http://www.w3.org/TR/xmlenc-core/

XML-Signature Syntax and Processing. (2008, June). Retrieved from http://www.w3.org/ TR/ xmldsig-core/

YAML. (2005)Retrieved from http://yaml.org/

Yaml. (2007, November). Retrieved from http:// en.wikipedia.org/ wiki/YAML)

Zhang, F., Mu, Y., & Susilo, W. (2005). *Reducing Security Overhead For Mobile Networks*. In AINA '05: Proceedings Of The 19th International Conference On Advanced Information Networking And Applications (pp. 398–403). Washington, DC: IEEE Computer Society.

Zheng, Y. (1997a). *Digital Signcryption or How to Achieve Cost (Signature & Encryption) << Cost(Signature) + Cost(Encryption)*. LNCS 1294, pp.165–179). Retrieved from citeseer.ist.psu.edu/ zheng97digital.html

Zheng, Y. (1997b). *Signcryption And Its Applications In Efficient Public Key Solutions*. In Proc. 1st International Information Security Workshop (pp. 291–312). Retrieved from citeseer.ist.psu.edu/ zheng97signcryption.html

Chapter 3
Improving Energy Efficiency and Throughput in Heterogeneous Mobile Ad Hoc Networks

Manu J. Pillai
National Institute of Technology Calicut, India

M. P. Sebastian
National Institute of Technology Calicut, India

ABSTRACT

The nodes are expected to transmit at different power levels in heterogeneous mobile adhoc networks, thus leading to communication links of different length. Conventional MAC protocols that unconditionally presume that links are bi-directional and with unvarying energy distribution may not succeed or execute badly under such circumstances. Interference and signal loss resulting out of distance and fading diminish the entire throughput attained in heterogeneous networks to a greater extent. This article presents a MAC protocol, which adaptively transmits data frames using either the energy efficient nodes or a list of high data rate assistant nodes. In addition, a cross-layer based energy level on-demand routing protocol that adaptively regulates the transmission rate on basis of congestion is projected as well. Simulation results illustrate that the proposed protocols considerably diminish energy consumption and delay, and attain high throughput in contrast with the Hybrid MAC and traditional IEEE 802.11 protocols

INTRODUCTION

An autonomous group of mobile users that normally communicates over bandwidth constrained wireless links is referred to as MANET (Nasipuri, Zhuang, & Das, 1999). Due to the mobility of nodes, the network topology might change hastily in an impulsive manner over time. MANET is a temporary infrastructure-less multi-hop wireless network where in the nodes are permitted to move arbitrarily. These networks broaden the restricted wireless transmission range of each node by multi-hop packet forwarding, therefore, are appropriate for scenarios where in pre-deployed infrastructure

DOI: 10.4018/978-1-60960-563-6.ch003

facility is not accessible. The set of applications for MANETs is varied, ranging from small, static networks that are inhibited by power sources, to large-scale, mobile, highly dynamic networks. In case of ad hoc networks, every node forwards packets for its peer nodes, and every flow surpasses several hops of wireless links for a source destination pair.

Given that the network performance rapidly degrades with the escalating number of nodes, a homogeneous ad hoc network is considered poor in terms of scalability. Nodes are commonly heterogeneous in case of realistic ad hoc networks. For instance, in a battlefield network, soldiers carry portable wireless devices, and further powerful and dependable communication devices are carried by vehicles, tanks, aircraft, and satellites; these devices/nodes have diverse communication characteristics in terms of transmission power, data rate, processing capability, reliability, etc. Thus, modeling these network elements as different types of nodes would appear more realistic (Xiaojiang, Dapeng, Wei, & Yuguang, 2006).

The nodes are expected to transmit at different power levels in these heterogeneous mobile adhoc networks, thus leading to communication links of different lengths. Owing to link asymmetry, the transmission of a low power node might not arrive at a higher power node whereas the communication in the reverse direction could be probable (Xiaojiang, & Dapeng, 2007). As a consequence, conventional MAC and routing protocols that unconditionally assume bi-directional links are bound to either be unsuccessful or perform unsuccessfully. Additionally, interference and signal loss resulting out of distance and fading diminish the entire throughput attained in heterogeneous networks to a greater extent.

In this article we project a novel protocol on basis of the IEEE 802.11 Distributed Coordination Function (DCF), which adaptively transmits the data frames either with the aid of energy efficient nodes a list of high data rate assistant nodes. The rest of the article is organized as follows: Section 2 briefs the related works. Section 3 presents the proposed protocol. Section 4 provides a prelude on the enhanced on-demand routing protocol based on the energy levels. Section 5 assesses the performance of the proposed protocols in contrast with the Hybrid MAC and the standard 802.11 MAC protocols through simulation. Section 6 summarizes the conclusion.

RELATED WORKS

An alternate methodology to collision resolution in a CSMA protocol with the introduction of spatial back off was projected by Charles J. Colbourn et al. (2007). They illustrate that collision resolution with the aid of power back off can be extremely successful, outperforming IEEE 802.11 in both static and mobile ad hoc network situations. Nevertheless, these results are obtained on basis of an optimistic centralized power-aware routing strategy and not the distributed power-aware strategy. Aran Bergman et al. (2006) projected a novel utility function which reflects the tradeoff amid the energy consumption brought about by à MAC protocol and its throughput, thereby representing the energy efficiency of the algorithm. They alter the "0.487" algorithm to enhance its energy efficiency.

A new routing protocol called multiclass (MC) routing, which is exclusively designed for heterogeneous MANETs, was presented by Xiaojiang et al. (2006). A new medium access control (MAC) protocol for heterogeneous MANETs, which is more proficient in comparison with the IEEE 802.11b, was introduced as well. Nevertheless, the necessity of information from various layers make the selection of a node which has larger transmission range (power) and data rate, better processing capability, and more reliable and robust than the other type a complicated one.

Vasudev Shah et al. (2007) created a cross-layer framework to deal with the link asymmetry problem at both MAC and routing layers, and they carried out extensive simulations in assessing the performance of their proposed framework in diverse settings. They illustrate that the general throughput in power heterogeneous networks is improved by as much as 25% over conventional layered approaches. Selection of B nodes is a pain stacking process here as well. Ian D. Chakeres et al. (2003) put forth a simple way to amend the route discovery phase of an on-demand routing protocol to select the «best» route with consideration for heterogeneity. They provide HBRD, a simple method to bias on-demand route discovery in the direction of resource-rich routes by means of delay, during the route discovery propagation.

A medium access control (MAC) protocol for ad hoc wireless networks that utilizes multiple channels dynamically to improve the performance was presented by Jungmin So et al. (2005). A few schemes that determine the initial contention window size for a transmission based on the distance traveled by the flow were presented by Bright Chu (2003) and a momentous enhancement in performance was illustrated through his the simulation results. A comprehensive solution for power control in MANETs was presented by Alaa Muqattash et al. (2004), which accentuates the interplay between the MAC and network layers, whereby the MAC layer manipulates the selection of the next-hop by proper adjustment of the power of the route request packets, in an indirect manner.

Backoff strategies for multiple access protocols have characteristically been studied with the aid of statistical assumptions on the distribution of problem inputs. Even though these analyses have offered important insights into the efficacy of various backoff strategies, they do not seem to answer question as to which backoff algorithm executes best in the worst case or on inputs, like the bursty inputs, which the statistical models failed to deal with. Bender et al. (2004) study the randomized backoff strategies with the worst-case assumptions on the inputs.

Nasipuri et al. (1999) proposed a novel carrier-sense multiple access (CSMA) protocol for multihop wireless networks. The CSMA protocol separates the available bandwidth into numerous channels and chooses an idle channel at random for packet transmission. Besides, it as well utilizes a notion of "soft" channel reservation as it provides preference to the channel that was utilized for the most recent successful transmission. With the aid of simulations they illustrate that the multichannel CSMA protocol offers higher throughput in contrast with its single channel counterpart by reducing the packet loss as a result of collisions.

Two novel methods to build a power efficient planar structure with much less communication costs and lower node degree bounds were presented by Wen-Zhan Song et al. (2004). A geographical adaptive fidelity (GAF) algorithm that reduces energy consumption in ad hoc wireless networks was presented by Ya Xu et al. (2001). GAF identifies nodes that are equivalent from a routing perspective and then turns off the unnecessary nodes, keeping a constant level of routing fidelity, hence conserving energy.

Ware et al. (2000) deal with the performance issues of IEEE 802.11, when used in the ad hoc mode in the presence of hidden terminals. The results demonstrating strong dependence of channel capture behavior on the SNR observed on challenging hidden connections are provided. It demonstrates that in a hidden terminal scenario, the connection possessing the strongest SNR is capable of capturing the channel, in spite of the use of the RTS-CTS-DATA-ACK 4-way handshake designed to address this issue. It is specified that the near-far SNR problem may have a considerable effect on the performance of an ad hoc 802.11 network. Chiung-Ying Wang et al. (2005) projected an efficient power saving protocol for multi-hop MANETs, known as p-MANET. The principal characteristics of the p-MANET protocol comprise of reduced power consumption and transmission latency, and power saving.

The effects of unfairness performance of the IEEE 802.11 standard under a pre-computed routing scheme for two network configurations and a fixed traffic patterns was demonstrated by Yu Yan Ming et al. (2001). The send rate, a quantity similar to the local throughput, for very simple chain and lattice networks with regular traffic patterns, was investigated and subsequently the performance of both types of fairness improvement schemes on this rate was examined.

THE PROPOSED ENERGY

Efficient and High Throughput (EEHT) MAC Protocol

We propose to add a set of new features to the IEEEE 802.11 MAC layer while preserving backward compatibility. The mechanism for every node to select and detect an assistant node is the most important component of the proposed EEHT protocol. A node depends on the likelihood of minimizing the transmission time for the packet in hand to choose an assistant node from a list of potential assistants for its utilization at the time of packet transmission.

The IEEE 802.11 MAC Protocol

For the distributed coordination function (DCF) mode where each station (STA) can initiate a data transmission by itself, the IEEE 802.11 protocol utilizes carrier sense multiple access with collision avoidance (CSMA/CA) as its medium access protocol. In order to evade collisions, it is crucial to sense the channel before packet transmission. A station having data packet to be sent initially senses the channel to ensure the clarity of the channel before initiating the actual transmission. Although the channel is sensed to be free, a collision may occur as not all stations can hear each other. Therefore virtual carrier sensing is employed as well to reserve channel time for the transmitting stations, using the Request to Send (RTS) and Clear to Send (CTS) frames. The channel reservation information is broadcast to the entire network by these two control frames. Any station will be able to hear at least one of these control packets and calculate the time needed for the data packet transmission using the control packets heard. All the stations discover the time for which the channel is going to be free, by employing a Network Allocation Vector (NAV) (1999).

Assistant Table List (ATL)

In our proposed protocol, an Assistant Table List (ATL), the list of potential assistants that can be used for assistance during packet transmission will be maintained by every node in the network. Every node in an 802.11 network passively listens to all ongoing transmissions while checking the packet header of all the packets it receives in order to create and update the ATL. Every time the assistant node changes its position, the ATL can be updated.

Assistant Node Selection

When a source node N_S wants to transmit data to a destination node N_d it selects an assistant node N_a, which lies in between them depending on the following inequality

$$(((Dsa*Dad)/(Dsa+Dad)*wt1)+(Ets+Era)*wt2)<Dsa*wt1+(Ets+Era)*wt2 \quad (1)$$

where,

$D_{sa} \rightarrow$ *Data* Rate from N_S to N_a
$D_{ad} \rightarrow$ *Data* Rate from N_a to N_d
$D_{sd} \rightarrow$ *Data* Rate from N_S to N_d by N_d
$E_{ts} \rightarrow$ *Power* Consumed to transmit 1 packet from N_S
$E_{ra} \rightarrow$ *Power* Consumed to receive 1 packet by N_a
$E_{rd} \rightarrow$ *Power* Consumed to receive 1 packet by N_d
and w_{t1}, w_{t2} weight constants between 0 and 1

Proof

Wt1 & Wt2 are constants. For simplicity we take their value as 1.

$$(D_{sa}*D_{ad}) / (D_{sa}+D_{ad}) < D_{sd} \quad (2)$$

Taking the inverse on both sides of (2), we get

$$(D_{sa}*D_{ad}) / (D_{sa}+D_{ad}) > 1/ D_{sd} \quad (3)$$

Simplifying (3), we get

$$\frac{1}{D_{sa}} + \frac{1}{D_{ad}} > \frac{1}{D_{sd}} \quad (4)$$

The inequality (4) implies that the combined transmission rates of D_{sa} and D_{ad} should be faster than the transmission rate D_{sd}. Similarly we have

$$(E_{ts}+E_{ra}) < (E_{ts}+E_{rd}) \quad (5)$$

The inequality (5) implies that the sum of the energies E_{ts} and R_{ra} should be less than the sum of the energies E_{ts} and E_{rd}. Based on the (4) and (5), we arrive at the combined equation (1). Hence the Proof.

When a transmission from a station N_a is overheard, a source node N_S measures the received signal strength to estimate the channel condition between the sender of that packet and itself. We can find the data rate D_{sa} by using the information from the 802.11k protocol. The source node N_a overhears the transmission between N_a toN_d in order to get the values D_{ad} and D_{sd} from the Physical Layer Convergence procedure (PLCP). During the simulation, the adjustment of the values of w_{t1} and w_{t2} can be performed.

Now, sort the assistant nodes based on the following formula

$$(((D_{sa}*D_{ad}) / (D_{sa}+D_{ad}) *w_{t1}) + (E_{ts}+E_{ra}) *w_{t2}) \quad (6)$$

The assistant nodes are now stored in the ATL.

Determining the Energy Levels of Nodes

When S_n is null, i.e., when the source node cannot detect any assistant node in its transmission path, it swaps to find the high energy node in its neighborhood in an adaptive manner.

We use the maximum value of the residual energy in the entire network to initially group the network nodes into distinct energy levels. We assume that nodes have location information about their neighbors and destination and a non-uniform energy distribution among nodes (since we have deemed a heterogeneous network).

All the nodes query each other regarding their residual energies, thus, the information about the residual energy of the neighbors is possessed by every node. The total residual energy value is obtained by the exchange of this information among all nodes in the network.

$$L_i = E_i/_{E\max} * \beta \quad (7)$$

Where E_i is the residual energy of node *i*, E_{max} is the maximum value of residual energy in the entire network and $\beta\,(0<\beta<1)$ is an experimental parameter.

Transmission Algorithm:

- Whenever a packet is to be transmitted, N_S searches for an assistant node in the ATL. If one such node is found using (1), N_S sends a RTS message with the assistant ID to specify the assistant being selected. If N_S fails to lookup an assistant node from the list it selects the node with the next higher energy level using (6) and sends a RTS message with its ID.
- Even after 2*SIFS* + *CTS* duration, If N_S fails to receive a CTS_a from N_a or a CTS_d from N_d, or a *CTS* is lost after sending a

CTS_a by N_a, N_S performs a exponential random backoff, as if in the case of a collision.

- If N_S fails to receive any CTS_a message fromN_a, and at the same time receives a CTS_d from N_d, it should then directly transmit the data.
- N_S always must send data to N_a at the rate of D_{sa} when CTS_a and CTS_d messages are received.
- If N_a receives a *RTS* message, then it should verify whether eqn.1 can be satisfied. If yes, it then sends a CTS_a message back to N_S.
- If N_a receives a *CTS* from N_d, it then should wait for the data packet from N_S to arrive. If N_a could not receive either the *CTS* message or the data packet as predicted, it goes back to the initial state assuming that the transmission was cancelled. When N_a receives the data packet to be forwarded, N_a is suppose to forward the data packet to N_d at the rate D_a.
- When N_d receives a *RTS*, it will wait for the related CTS_a message from N_a.
- N_d transmits a *CTS* message back to N_S, when CTS_a is received by N_d from N_a. If N_d does not get the data frame within the time frame, it goes back to the initial state assuming that the transmission was cancelled.
- The exchanges of control and data frames in EEHT are shown in Figure 1 and 2, respectively.

ENERGY LEVEL BASED ROUTING

We now present our energy level routing protocol devised on the basis of the standard on-demand routing protocol AODV (Perkins, Belding-Royer, & Das, 2003).

There are many works available in the literature regarding the Quality of Service (QoS) based routing protocols, which improves the throughput of 802.11 adhoc networks.

The incorporation of QoS into routing was proposed by Lei Chen and Wendi B. Heinzelman (2005). They also introduced the bandwidth estimation by disseminating bandwidth information through "Hello" messages.

An attempt to identify the maximum throughput that can be sustained in an 802.11 multi-hop network was made by Ping Chung Ng and Soung Chang Liew(2007). It can be deemed that this is a foremost article in the literature, in which a quantitative analysis on the fundamental impact of hidden nodes and carrier sensing on system throughput is provided. The AODV with multipath and pre-emptive detection was extended to see its affect on the IETF-specification, network performance and the complexity of the implementation, by Mads Østerby Jespersen et al. (2003).

Yihai Zhang and Aaron Gulliver (2005) have presented QS-AODV to offer QoS assurance in ad hoc networks; and investigated the difference between QS-AODV and AODV. A link availability-based QoS-aware (LABQ) routing protocol for mobile ad hoc networks based on mobility prediction and link quality measurement,

Figure 1. Control frame exchange in EEHT

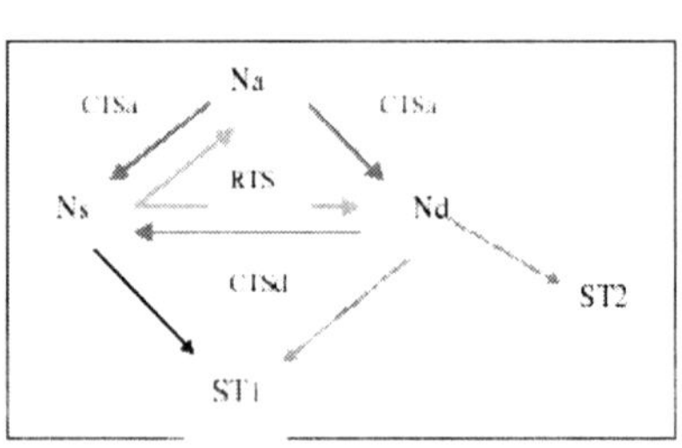

Figure 2. Data frame exchange in EEHT

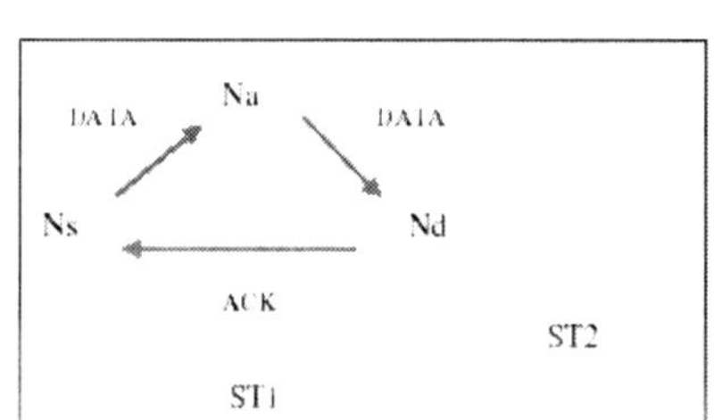

in addition to an energy consumption estimate was presented by Ming Yu et al. (2007). Providing highly reliable and better communication links with energy-efficiency is their objective.

A new technique to estimate the available bandwidth of wireless nodes by the extension of one-hop links in IEEE 802.11-based ad hoc networks was presented by Cheikh Sarr et al. (2005). The fact that a node can estimate the channel occupancy by monitoring its environment is utilized by their technique. A new Quality of Service (QoS) routing protocol for delay sensitive traffic in Mobile Ad Hoc Networks (MANETs) was presented by Xiaojiang Du and Carlos Pomalaza-Ráez (2004).

A novel on-demand routing scheme, namely the Load-Balanced Ad hoc Routing (LBAR) protocol was presented by Hossam Hassanein and Audrey Zhou (2001). In LBAR, routing information on different paths is forwarded to the destination through setup messages. The path with minimum cost in terms of nodal activity is chosen by the destination node. Since packets are transmitted along the least-active path, the congested paths can be avoided by weighing total nodal activity of a path.

AODV Protocol

The AODV routing protocol is a hasty routing algorithm. Until the sources need the established routes, the protocol maintains these established routes. In order to make sure that the routes are fresh, AODV uses sequence numbers.

Route Discovery: Every time a traffic source needs a route to a destination, the route discovery process is initiated. Usually, a network-wide flood of route request (RREQ) packets targeting the destination and waiting for a route reply (RREP) is involved in a route discovery.

- On receiving a RREQ packet for the first time, a reverse path to the source is set up by the node. The node sends a RREP to the source node only if a valid route to the destination is available
- The destination sends a RREP to the source via the reverse path receives, when it receives a RREQ.

Route Maintenance: Route error (RERR) packets are employed for route maintenance. On detecting a link failure, a RERR packet is sent back via separately maintained predecessor links to all sources using the failed link. The RERR erases the routes along its path. On receiving a RERR, a traffic source initiates a new route discovery only if it still requires a route. (2003)

ENERGY LEVEL ROUTING BASED ON AD HOC ON-DEMAND DISTANCE VECTOR ROUTING (ELR-AODV) PROTOCOL

In the proposed protocol, each mobile node decides whether to participate in the selection process of a routing path or not, depending on the local information about the energy levels. An energy-hungry node does not forward data packets on behalf of others in order to conserve its energy. All the relevant nodes share the decision-making process in ELR-AODV. The energy levels of each node can be determined using (6).

Route Discovery: In ELR-AODV, each node depends on its energy level (Li) to determine whether or not to accept and forward the RREQ message. The RREQ is dropped when the energy level is lower than a threshold value Emin ($Li \leq Emin$) and is forwarded in the other case. It is essential that the intermediate nodes along the route have sufficient energy levels, for the destination to receive a route request message.

Route Maintenance: Either when the connections between some nodes on the path are lost due to node mobility or when the energy resources of some nodes on the path are getting depleted too quickly, Route Maintenance is essential. As in

AODV, a new RREQ is sent out and the entry in the route table corresponding to the node that has moved out of range is eliminated, in the first case. However, the node sends a route error RERR back to the source even when the condition $L_i \leq E_{min}$ is satisfied, in the second case. The source is forced to initiate route discovery again, by this route error message. Owing to the fact that it is dependent only on the energy level of the current node, this is a local decision.

Nevertheless, the source may not receive a RREP message even if a route exists between the source and the destination, if this decision is made for every possible route. The source will resend another RREQ message with a higher sequence number, thereby averting such a situation.

In order to allow the packet forwarding to continue, the intermediate node receiving this new request lowers its Emin by Δ. When a node drops a RREQ message, broadcasts an UPDATE message as an alternative. The subsequent nodes closer to the destination lower their threshold values when they come to know that a request message was dropped. Now, the destination receives the second route request message. The destination receiving a RREQ generates a RREP and sends it via the reverse path to the source.

Rate Adaptation: The link loss rate of a node is estimated in ELR-AODV, which serves as the basis for the determination of the new transmission rate. Three factors, congestion, signal interference and fading primarily control the link loss rate.

The congestion condition is signified and weaker links are exposed by precise and real-time link loss rate estimation. The node becomes congested and starts dropping packets once the number of packets reaching a node exceeds its carrying capacity. In order to monitor the congestion status of a node, several metrics are available. We check the congestion status using the following simple method.

A node checks its link-layer buffer available bandwidth in a periodic manner. The difference between the requested bandwidth and the available bandwidth at a node serves as the basis for the determination of the loss rate.

i.e., the loss rate at the node R1 is estimated as

$$L_{R1} = (\mathrm{Re}q.BW - BW_{RI})/F \tag{8}$$

Where $\mathrm{Re}q.BW$ is the requested bandwidth obtained from the RREQ header, BW_{RI} is the available bandwidth at R1 and F is a constant factor.

Then, the host R1 forwards the RREQ packet with it's L_{R1} to the next hop, R2 which then calculates its loss rate L_{R2} similar to (8) and adds it to L_{R1}. Finally, destination node D receives the RREQ with the sum of loss rates.

The destination node D uses the aggregated loss rate to calculate the new transmission rate:

$$N_{rate}\left(A_{rate} - \sum_{i=1}^{n} R_i\right) \tag{9}$$

where A_{Rate} is the arrival rate of packets at node D, which is given by

$$A_{Rate} = NP/T \tag{10}$$

where NP- is the number of packets received and T is the time interval for the packet transmission.

The new rate is sent along with the RREP message to the source. The source adapts its rate to the new rate calculated using (9).

PERFORMANCE EVALUATION AND RESULTS

Simulation Model

We simulated the proposed algorithm using NS2. A channel capacity of 2 Mbps for the mobile hosts is presumed for the simulation. Fifty mobile nodes move for 100 seconds simulation time in a 1000 meter x 1000 meter region. The random waypoint (RWP) model of NS2 is used to obtain the initial locations and movements of the nodes. It is presumed that each node moves with the same average speed in an independent manner. From the physical terrain in this mobility model, a node selects a destination at random. The node moves at a uniformly chosen speed between the minimal speed and maximal speed, in the direction of the destination. When the node reaches the destination, it stays there for a pause time and then moves over. In our simulation, 10 m/s and 10 seconds are the corresponding speed and pause time. The simulated traffic is Constant Bit Rate (CBR). In each case, we conducted ten runs with different random seeds and averaged the results.

Performance Metrics

We compare the performance of the proposed EEHT protocol with the HMAC protocol and the standard IEEE 802.11 MAC protocol. The following metrics are used for the comparison: *Aggregated Throughput:* We measure the aggregated throughput in terms of the number of packets received.

- **Average Energy Consumption:** It is the average energy consumed by the nodes in receiving and sending the packets.
- **Average End-to-end Delay:** The end-to-end-delay is averaged over all surviving data packets from the sources to the destinations.
- **Packet Delivery Fraction:** It is the ratio of the fraction of packets received successfully and the total number of packets sent.
- **Routing Overhead:** It is the number of control packets exchanged by the routing protocol.

SIMULATION RESULTS

Dependence on the Number of Nodes

In the first experiment, we measure the performance of the protocols by varying the number of nodes (25, 50, 75 and 100). From Figure 3, we see that the average end-to-end delay for the EEHT protocol is less in comparison with the other protocols. Figure 4 shows that the average energy consumed by the nodes in receiving and sending the data. Since EEHT makes use of energy efficient nodes, the values are considerably less in EEHT in comparison with HMAC and 8021.11.

When we measure the aggregated throughput for various node sizes, EEHT throughput is high in comparison with HMAC and 802.11, since EEHT makes use of high data rate nodes. Figure 5 shows this result. From Figure 6, we see that the packet delivery fraction (PDF) for EEHT is high in comparison with HMAC and 802.11. The reason is same as in the case of throughput.

Figure 3. Nodes vs. Delay

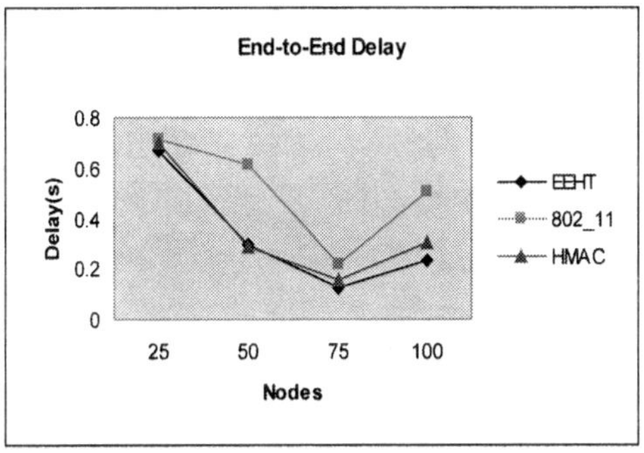

Figure 4. Nodes vs. Energy

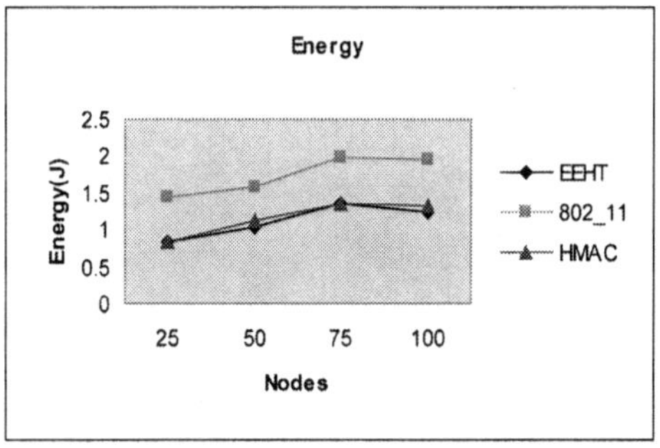

Dependence on Transmission Rate

In the second experiment, we measure the performance of the protocols for varying transmission rates (250, 500, 750 and 1000 Kb). When we measure the aggregated throughput for various rates, EEHT throughput is high in comparison with HMAC and 802.11, since EEHT makes use of high data rate nodes. Figure 7 shows this result.

Figure 8 shows that the average energy consumed by the nodes in receiving and sending the data. Since EEHT makes use of energy efficient nodes, the values are considerably less in EEHT in comparison with HMAC and 8021.11. From Figure 9, we see that the packet delivery fraction (PDF) for EEHT is high in comparison with HMAC and 802.11. The reason is same as in the case of throughput. From Figure 10, we see that

Figure 5. Nodes vs. Throughput

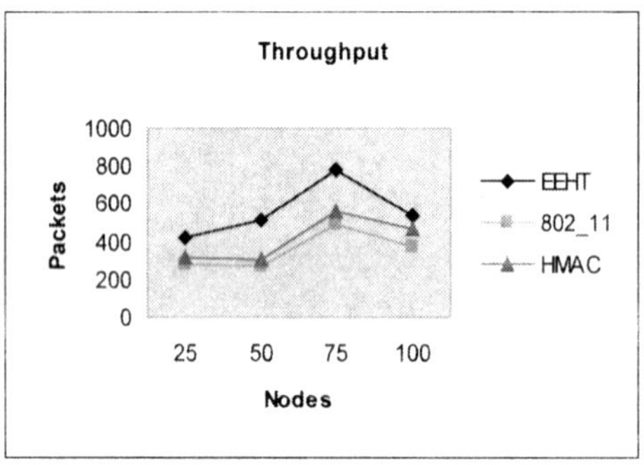

Figure 6. Nodes vs. PDF

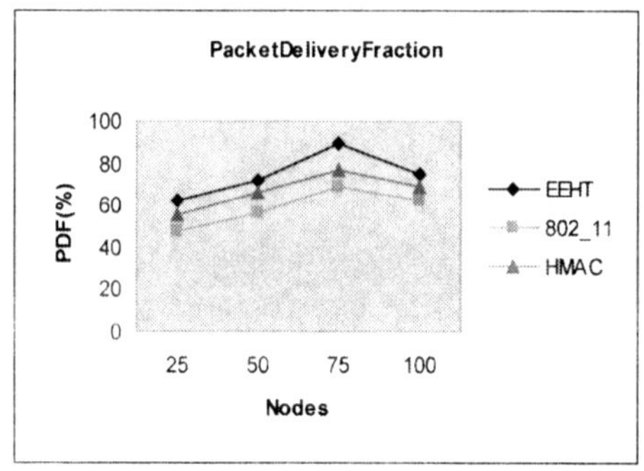

Figure 7. Rate vs.Throughput

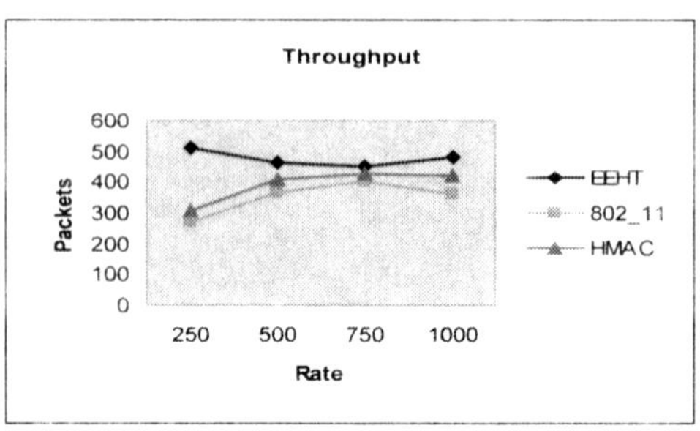

Figure 8. Rate vs. Energy

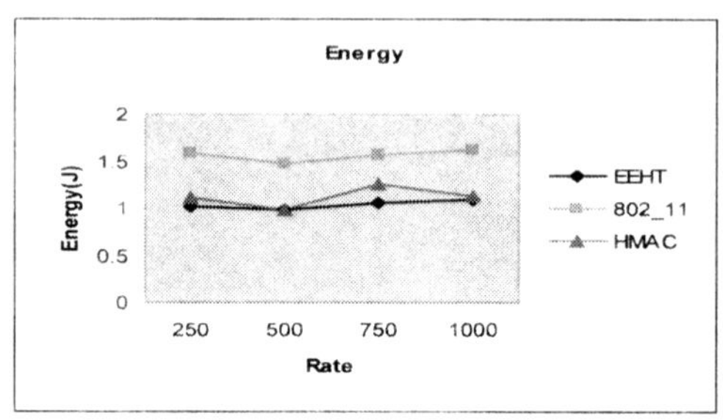

Figure 9. Rate vs. PDF

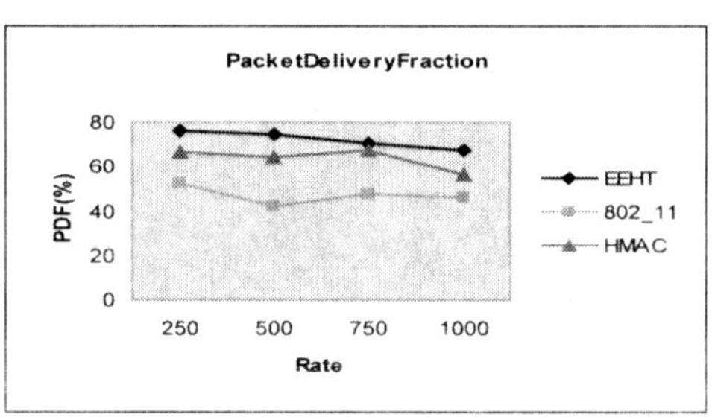

Figure 10. Rate vs. Delay

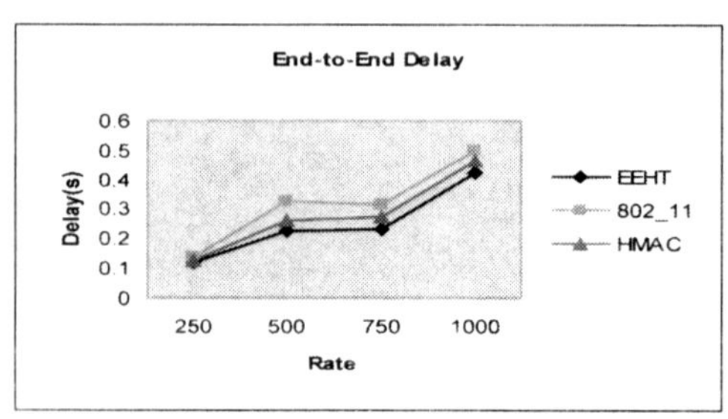

the average end-to-end delay of the proposed EEHT protocol is less in comparison with the other protocols.

Hence our protocol EEHT performs very well in terms of throughput, energy consumption and delay in comparison with HMAC and 802.11 MAC protocols.

Dependence on Speed

In the third experiment, we measure the performance of the ELR-AODV routing protocol by varying the node speed in steps of 5, 10, 15 and 20 m/s. We have considered the AODV routing protocol for the performance comparison. Figure 11 shows that our protocol has less overhead in comparison with AODV.

From Figure 12, we see that the packet delivery fraction (PDF) for ELR-AODV is high in comparison with AODV, since ELR-AODV increases the throughput by minimizing the energy. When we measure the throughput for various speeds also, ELR-AODV throughput is high in comparison with AODV. Figure 13 shows this result. From Figure 14, we see that the average end-to-end delay of the ELR-AODV protocol is less in comparison with the AODV protocol.

Hence our routing protocol ELR-AODV performs very well in terms of throughput, overheads and delay in comparison with the AODV protocol.

CONCLUSION

In this article, we have proposed a new protocol EEHT on the basis of the existing IEEE 802.11, which transmits the data frames through energy efficient nodes or through a list of high data rate assistant nodes, in an adaptive manner. A power efficient high data rate assistant node is detected

Figure 11. Speed vs. Overhead

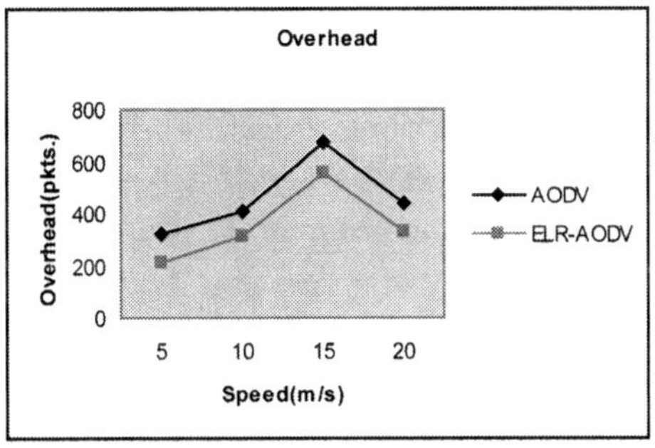

Figure 12. Speed vs. PDF

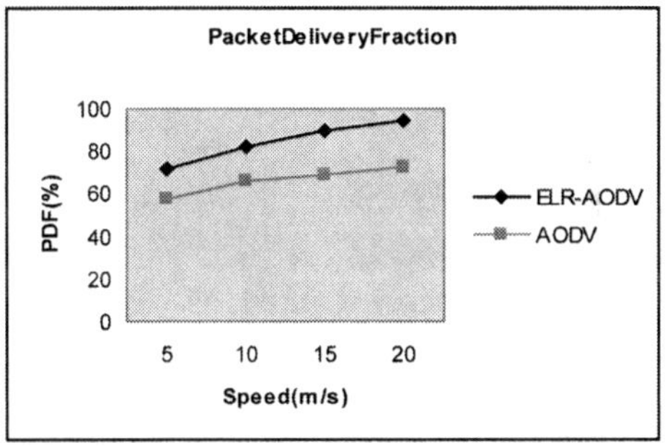

Figure 13. Speed vs. Throughput

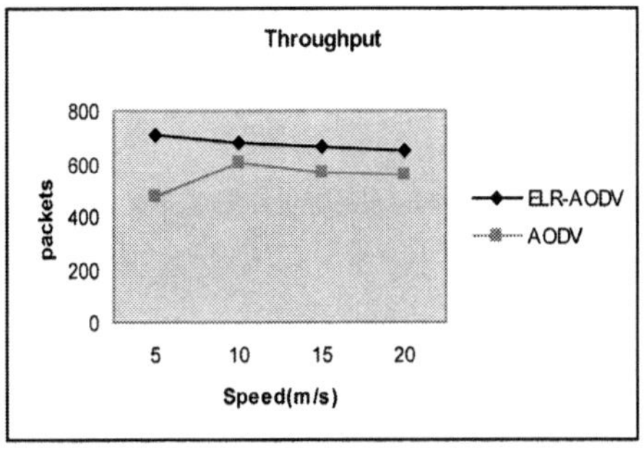

Figure 14. Speed vs. Delay

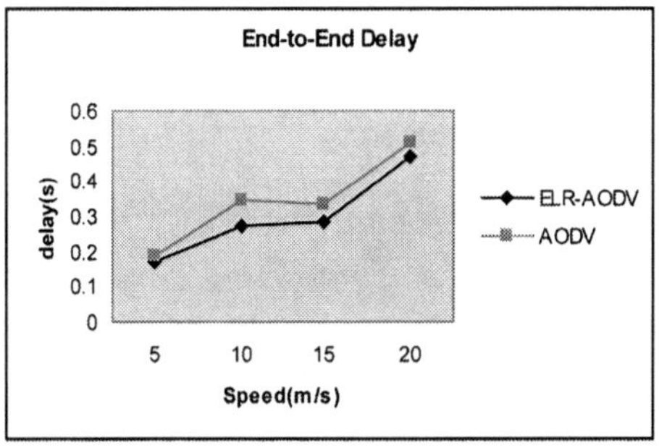

by every node by which an Assistant Table List is maintained. It selects the node with the highest energy level if it does not determine such an assistant node. We have also developed an improved on-demand routing protocol ELR-AODV on the basis of the node energy level, where the congestion level serves as the basis for adjusting the transmission rates.

The analysis and extensive simulation verifies that the performance of EEHT in comparison with HMAC and 802.11 MAC protocols is splendid with respect to throughput, energy consumption and delay. Likewise, the performance of ELR-AODV in comparison with the AODV protocol is splendid with respect to throughput, overheads and delay. The performance of EEHT scheme in a network where node movements are modest is good. Nevertheless, the EEHT is restricted, i.e. the estimation and maintenance of the ATL could be complicated if nodes move at very high speed. Moreover, there is a probable restriction for ELR-AODV, i.e. recurrent modifications in topology may pose challenges in the estimation of energy levels of nodes. Dealing with these limitations will be the intent of our future works.

REFERENCES

Bender, M. A., Farach-Colton, M., Kuszmaul, B. C., & Leiserson, C. E. (2004, April). Adversarial Analyses of Window Backoff Strategie. *Proceedings of the 18th International Parallel and Distributed Processing Symposium, 203*, 26-30.

Bergman, A., & Sidi, M. (2006). Energy efficiency of collision resolution protocols. *Computer Communications, 29*(17), 3397–3415. doi:10.1016/j.comcom.2006.01.016

Bright, C. (2005). *Improving IEEE 802.11 Performance with Power Control and Distance-Based Contention Window Selection*. Master's Thesis, University of Illinois, Urbana-Champaign

Chakeres, I. D., & Royer, E. M. B. (2003, July). Utilizing Resource rich Nodes in Ad hoc Networks. *ACM SIGMOBILE Mobile Computing and Communications Review, 7*(3), 17–18. doi:10.1145/961268.961271

Chen, L., & Heinzelman, W. B. (2005, March). QoS-Aware Routing Based on Bandwidth Estimation for Mobile Ad Hoc Networks. *IEEE Journal on Selected Areas in Communications, 23*(3), 561–572. doi:10.1109/JSAC.2004.842560

Colbourn, C. J., Cui, M., Syrotiuk, V. R., & Lloyd, E. L. (2007). A Carrier Sense Multiple Access Protocol with Power Backoff (CSMA/PB). *Ad Hoc Networks, 5*(8), 1233–1250. doi:10.1016/j.adhoc.2007.02.017

Du, X., & Pomalaza-Ráez, C. (2004). Delay Sensitive QoS Routing For Mobile Ad Hoc Networks. *Proc. of the MILCOM 2004 Conf. Monterrey.*

Fuemmeler, J. A., Vaidya, N. H., & Veeravalli, V. V. (2004). *Selecting Transmit Powers and Carrier Sense Thresholds for CSMA Protocols. Technical report*. University of Illinois at Urbana Champaign.

Hassanein, H., & Zhou, A. (2001). Routing with Load Balancing in Wireless Ad hoc Networks. *International Workshop on Modeling Analysis and Simulation of Wireless and Mobile Systems* (pp. 89-96), Rome, Italy.

Jespersen, M.O., Nielsen, K.D., & Frølund, J. (2003, May 13). *Optimising performance in AOMDV with pre-emptive routing*. University of Aarhus, Computer Science Ny Munkegade DK-8000 Århus C.

LAN MAN Standards Committee of the IEEE Computer Society. (1999). *IEEE Standard 802.11-1999, Part 11: Wireless LAN Medium Access Control (MAC) and Physical Layer (PHY) Specifications*. IEEE.

Ming, Y., Timoney, J., Doyle, L., & O'Mahony, D. (2001, November 27). Evaluation of Channel Fairness Models for Ad-Hoc Networks. *Proceedings of the First Joint IEI/IEE Symposium on Telecommunications Systems Research* (p. 4), Dublin.

Muqattash, A., & Krunz, M. (2003). Power Controlled Dual Channel (PCDC) Medium Access Protocol for Wireless Ad Hoc Networks. Twenty-Second Annual Joint Conference of the IEEE Computer and Communications Societies. *IEEE INFOCOM 2003, 1,* 470-480.

Nasipuri, A., Zhuang, J., & Das, S. R. (1999). A multichannel CSMA MAC protocol for multihop wireless networks. *Proc. IEEE Wireless Comm. and Networking Conf. (WCNC).*

Ng, P. C., & Liew, S. C. (2007, April). Throughput Analysis of IEEE802.11 Multi-hop Ad hoc Networks. [TON]. *IEEE/ACM Transactions on Networking, 15*(2), 309–322. doi:10.1109/TNET.2007.892848

Perkins, C., Belding-Royer, E., & Das, S. (2003, July). Ad hoc On-Demand Distance Vector (AODV) Routing. *Internet Experimental RFC, 3561.*

Sarr, C., Chaudet, C., Chelius, G., & Lassous, I. G. (2005). A node-based available bandwidth evaluation in IEEE 802.11 ad hoc networks. *Proceedings of the 11th International Conference on Parallel and Distributed Systems - Workshops, 2*, 68-72.

Wang, C.-Y., Wu, C.-J., Chen, G. N., & Hwang, R.-H. (2005, July4-7). p-MANET Efficient Power Saving Protocol for Multi-Hop Mobile Ad Hoc. *Third International Conference on Information Technology and Applications, ICITA 2005, 2,* 271-276.

Ware, C., Judge, J., Chicharo, J., & Dutkiewicz, E. (2000). Unfairness and capture behaviour in 802.11 Ad hoc networks. *2000 IEEE International Conference on Communications, 1*, 159-160.

Wen-Zhan, S., Wang, Y., Xiang, Y., & Frieder, L. O. (2004, December). Localized Algorithms for Energy Efficient Topology in Wireless Ad hoc networks. *Mobile Networks and Applications, 10*(6), 911–923.

Xiaojiang, D., & Dapeng, W. (2007, January). Joint Design of Routing and Medium Access Control for Hybrid mobile Ad Hoc Networks. *Mobile Networks and Applications, 12*(1), 57–68. doi:10.1007/s11036-006-0006-9

Xiaojiang, D., Dapeng, W., Wei, L., & Yuguang, F. (2006, January). Multiclass Routing and Medium Access Control for Heterogeneous Mobile Ad Hoc networks. *IEEE Transactions on Vehicular Technology, 55*(1), 270–277. doi:10.1109/TVT.2005.861183

Ya, X., Heidemann, J., & Estrin, D. (2001). Geography informed Energy Conservation for Ad Hoc Routing. *International Conference on Mobile Computing and Networking* (pp. 70-84), Rome, Italy.

Yu, M., Malvankar, A., Su, W., & Foo, S. Y. (2007). A link availability-based QoS-aware routing protocol for mobile ad hoc sensor networks. *Computer Communications, 30*, 3823–3831. doi:10.1016/j.comcom.2007.09.009

Zhang, Y., & Gulliver, T. A. (2005, August 22-24). Quality of Service for Ad hoc On-demand Distance Vector Routing. *IEEE International Conference on Wireless And Mobile Computing, Networking Communications, 3*, 192–196.

This work was previously published in International Journal of Mobile Computing and Multimedia Communications (IJMCMC) 1(2), edited by Ismail Khalil & Edgar Weippl, pp. 48-60, copyright 2009 by IGI Publishing (an imprint of IGI Global).

Chapter 4
Performance Enhancement of Routing Protocols in Mobile Ad Hoc Networks

Kais Mnif
University of Sfax, Tunisia

Michel Kadoch
University of Quebec, Canada

ABSTRACT

This paper proposes to use virtual backbone structure to handle control messages in ad hoc networks. This structure is effective in reducing the overhead of disseminating control information. In the first part, the approach to build the virtual backbone on the setup phase is presented. The construction of backbone is based on the Minimum Connected Dominating Set (MCDS). The novelty is in the way on finding the MCDS. A Linear Programming approach is used to build a Minimum Dominating Set (MDS). Then, a spanning tree algorithm is applied to provide the MCDS. A theoretical analysis based on probabilistic approach is developed to evaluate the size of MCDS. Different techniques of diffusion in ad hoc networks are presented and compared. The flooding technique is simple and efficient, but it is expensive in term of bandwidth consumption and causes broadcast storm problem. Simulation results show that technique using virtual backbone performs flooding and it is compared to MPR (Multipoint Relay). The second part of this paper presents a distributed procedure to maintain the backbone when the mobility of terminals is introduced. A maintenance procedure will be executed by the node which changes its position. This procedure is distributed and guarantees the node connectivity to the backbone. The authors believe that the maintenance of the backbone with small size will be more effective. Simulation results show the performance of this procedure when mobility and scalability are considered.

DOI: 10.4018/978-1-60960-563-6.ch004

INTRODUCTION

Wireless Ad hoc networks are very useful in emergency operations such as search and rescue, crowd control, and commando operations. The major factors that favour ad hoc wireless networks for such tasks are self-configuration of the system with a minimal overhead, independent of fixed or centralized infrastructure, the nature of the terrain of such applications, the freedom and flexibility of mobility, and the unavailability of conventional communication infrastructure.

Communications in wireless ad hoc networks suppose that there is no physical infrastructure. This supposition makes the communication more costly and conduct to a severe problem defined by (Johansson, 1999); the broadcast storm problem. This problem is induced by flooding inherent in on-demand routing protocols. Recently many propositions have been studied, which are inspired by physical backbone to maximize resource utilization and to minimize the number of exchanged messages caused by flooding. Furthermore, backbone can be used to collect topology information for routing, to provide a backup route, to send multicast or broadcast messages.

In Ad hoc environment, the network is distributed, where all network activity including discovering the topology and delivering messages must be executed by the nodes themselves, i.e. routing functionality will be incorporated into mobile nodes. Also, energy efficiency in a multihop network necessitates coordination between the nodes, so that they avoid wasting system resources like energy and bandwidth. While these goals can be met using centralized control, this is not practical in a mobile ad hoc network, or at least not scalable due to the high overhead to monitor and convey the control information throughout the network. A virtual backbone structure is a good solution to significantly reduce the number of nodes which handle control messages on the network. The construction and the maintenance of the virtual backbone impose another control overhead onto the overall communications: the size of the constructed backbone should be as small as possible. And, the role of virtual backbone requires connectivity of nodes. Therefore a minimum connected dominating set can make a good candidate. Nodes belonging to the MCDS set are responsible for relaying messages, while other nodes are not. However, finding the MCDS on a given graph is a NP-complete problem in graph theory.

The study of virtual infrastructures or backbones in wireless ad hoc networks gets more attention in the hope of reducing the communication overhead. But the backbone structure is very vulnerable due to various factors like node mobility and unstable links, and so on.

In most previous propositions (Guha, 1998; Tseng, 2002; Haitao, 2004; Wang, 2005; Ben, 2005; Al-Karaki, 2008), the same idea is used. One algorithm will be charged for the construction and the maintenance of virtual backbone. These propositions differ on the approach used to find the MCDS. They are based on combinatorial technique, graph coloration or marking process approaches. Our approach is different from previous ones; two independent algorithms are developed, the first one is for the construction of the backbone on the setup phase where the whole information of the network is known (number of terminals, capability, position, etc.). This algorithm guarantees a minimal size of the backbone. The second algorithm will be applied to maintain the backbone when mobility is introduced. Each node, which changes its place, applies a distributed maintenance procedure to connect to the backbone.

This paper is organized as follows: section 2 introduced the new approach used to compute the MCDS. Then a performance analysis is presented. Section 3 focuses on comparing the efficiency of the diffusion procedure using virtual backbone with other approaches, such as flooding, MP relay. In section 4, a distributed maintenance procedure will be presented. The main concern of ad hoc network is the mobility of terminal. This procedure guarantees the connectivity of the backbone during the life time of the network. Section 5 will conclude this paper.

VIRTUAL BACKBONE BASED ON MCDS

Notations and Definitions

A simple graph, $G=(V,A)$ is used to represent an ad hoc wireless network, where V represents a set of wireless mobile terminals (nodes) and A represents a set of edges. Each node has a transmission range R. An edge (u, v) indicates that both nodes u and v are within their transmission ranges; $A = \{(u,v) \in V^2 \mid d(u,v) \leq R\}$. Hence, the connections of hosts are based on geographic distances of nodes. Such a graph is also called a unit disk graph (UDG).

Def. 1: *A graph G (V, A) is called connected graph if and only if* $\forall$ $\{u, v\}$ $\subset V$, $\exists$ *a path between u and v.*

Def. 2: *A set S* $\subset$ *V is called a dominating set if every node of G, not in S, has at least one neighbour in S.*

Def. 3: *Based on the concept of domination, each non-dominating node has a dominating neighbour.*

Using the above definitions, the minimum connected dominating set (MCDS) in a given graph can be found as a minimum size subset S of nodes, such that the sub-graph induced by S is connected and S forms a dominating set. Unfortunately, as mentioned in the introduction, finding a MCDS in UDG is known to be NP-complete (Tseng, 2002).

In order to reduce complexity of the MCDS computation, decomposition into two steps is proposed by applying LP approach in each step. The first step finds the MDS in a given graph and the second step computes the spanning tree of the MDS set to get the final solution of the MCDS.

Finding the Minimum Dominating Set (MDS)

Finding the minimum dominating set can be formulated using the integer linear programming approach. A binary variable x_i is defined as a decision variable,

$$x_i = \begin{cases} 1 & \text{if the node } i \text{ is an element in the dominating set, MDS} \\ 0 & \text{otherwhise} \end{cases}$$

The objective function minimizes the number of node of the dominating set:

$$\min \sum_{i \in V} x_i \tag{1}$$

Domination constraint:

$$X + M \times X \geq 1 \tag{2}$$

where $x_i \in \{0,1\}$

$$\begin{bmatrix} m_{11} & \cdots & m_{1n} \\ \vdots & m_{ij} & \vdots \\ m_{n1} & \cdots & m_{nn} \end{bmatrix} \tag{3}$$

Where $X = \left[x_1 \ldots x_n\right]^t$ $M=$ represents the decision vector, and M is $n \times n$ 0/1 adjacent matrix of G, $M=$, $m_{ij} = 1$ iff node i is connected to node j.

With integer linear programming resolution, an optimal solution of the minimum dominating set can be calculated on $O(n)$ running time. According to (Das, 1997), a linear programming problem with d variables and n constraints can be solved in $O(n)$ time as n tends to infinity.

In general, there is no guarantee on the connectivity of the solution $\{x_i\}$, which represents nodes member of the backbone. To get final solution, where backbone nodes are connected, the next step determines the spanning tree of the MDS.

Finding the Connected Set of MDS

To compute the connected set of the MDS, an easy approach consists on finding the minimum spanning tree (MST) of the MDS. The MST problem has been extensively studied for many decades in graph theory and there exist several efficient distributed algorithms described in (Mnif, 2006-a). Most popular algorithms such as *Prim* or *Kruskal* can be applied. The running time of *Prim's* algorithm is $O(m + n.log\ n)$ (Bettstetter, 2001), where n and m denote, respectively, the number of nodes and arcs in the graph. Better than that, the running time can be reduced to $O(m.log\ n)$ if *Sollin's* algorithm is used. *Sollin*'s algorithm is better than *Prim's* algorithm for sparse networks, and is worse for dense network (Bettstetter, 2001).

Lemma

Let $G = (V, A)$, $S \subset V$ is connected dominating set of G if and only if there exist a spanning tree T of G such that V-S is a subset of leaves of T.

Proof

Let $S \subset V$ is connected dominating set of G. Let T_S a spanning tree of the sub graph of G induced by nodes in S. For each node $w \in V - S$, we choose a node $u_w \in S \cap N(w)$, where $N(v) = \{u \in V \mid$ edge $(u,v) \in A\}$. We can see easily that $T = T_S \cup \{u_w \mid w \in V\text{-}S\}$ is a spanning tree of G and U is a subset of leaves of T, then every node in U is attached to at least one node in V-U and the sub graph of G induced by nodes in V-U is connected. Therefore V-U is a connected set of G.

Performance Analysis

In (Johnson, 2000), a comparative study has been presented to compare the efficiency of our approach to others one's (Tseng, 2002; Wang, 2005) on the computation of size of the MCDS. Our algorithm guarantees to provide an optimal size of the MCDS. The choosing set of nodes in step 1 has elements with maximum degree. In this section, an analytical study will be presented to show that our approach has the optimal size MCDS. To do that, the probability, p_{MCDS}, that a node becomes a member of the MCDS, will be determined. So, the expected size of the MCDS, N_{MCDS}, is:

$$N_{MCDS} = p_{MCDS}\, N, \tag{4}$$

where N represents the number of nodes in the graph.

In first step, we will interest on the comparison of the analysis results with results given by our algorithm where the size of network varies, and in the second step, a comparison based on the variation of the transmission range will be presented.

Let's define $\rho = \dfrac{N}{S}$ *the density* of the network, where N is the number of nodes on the network and S the area where nodes are placed. The number of neighbours N_l of a node P with a transmission range R is defined by the number of nodes which are located on the area defined by the circle defined by (center P, radius R), N_l is given by:

$$N_1 = \rho S_1 - 1$$

Where

$$S_1 = \pi R^2$$

Then

$$N_1 = \rho\pi R^2 - 1 \quad (5)$$

Nodes are placed on a plan area with a Poisson distribution, and then the probability to get k nodes in the area S_1 is:

$$p_1(k, N_1) = \frac{N_1^k}{k!}e^{-N_1} \quad (6)$$

The probability that a node will be as MCDS member if and only if the two following probabilities are true:

p_2: the probability for a node to be a dominant; the node determined by the first step of the algorithm, and

p_3: the probability for a node to be an intermediary node; the node determined by the second step of the algorithm

Then the probability for a node to become a member of MCDS is

$$p_{MCDS} = p_2 + (1-p_2)p_3 \quad (7)$$

In the previous section, we did an assumption all nodes have the same transmission range (homogenous); this implies that all nodes have equal probability to be selected, so the probability of a node selecting over k neighbors is $\frac{1}{k+1}$. Furthermore, nodes are placed on the area with Poisson distribution, so the probability that a node will be selected as a dominant is given by

$$p_2 = \sum_{k=1}^{\infty} p_1(k, N_1)\frac{1}{k+1}$$

and we know that

$$\sum_{k=0}^{\infty} \frac{N^k}{k!} = e^N$$

So we get

$$p_2 = \frac{1 - e^{-N_1}(1 - N_1)}{N_1} \quad (8)$$

The summation starts from 1 because the case of k=0 is discarded; the case where the node doesn't have any neighbour on its transmission range.

To evaluate the probability that a node will be an intermediary node, p_3, we consider Figure 1, where *I* and *J* can be two dominants nodes either one node of them is dominant and the other one is an intermediary node of the connected set. Node *K* should be located on the grey area, *S*(*x*), defined by the intersection of the two circles centered on I and *J*. *I* and *J* are two terminals which are not in the transmission range, to communicate to *J*, *I* should transmit to the intermediary node *K*.

$$N(x) = \rho S(x)$$

where

Figure 1. node K should be located on the grey area S(x) to connect nodes I and J and to be an intermediary node

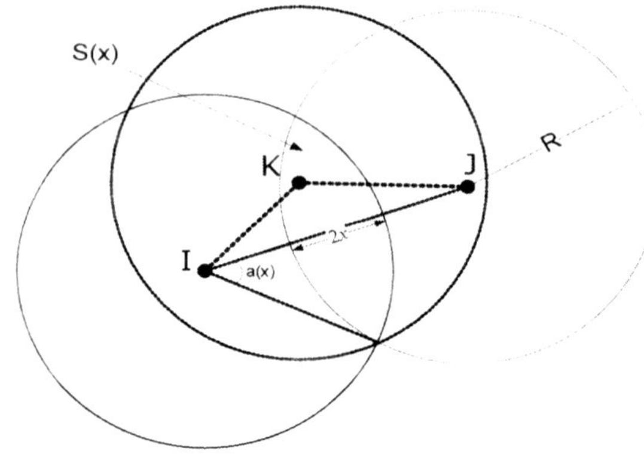

$$S(x) = 2R^2 a\left(x\right)$$

with

$$a(x) = \cos^{-1}\left(\frac{R-x}{R}\right) - \frac{x}{R}\sqrt{1-\left(\frac{R-x}{R}\right)^2}$$

and

$$x = R - \frac{d(i,j)}{2}$$

where *d(i,j)* is the distance between nodes *I* and *J*. Therefore, the probability for a node *K* to be chosen as in intermediary node between nodes *I* and *J* is:

$$p_3 = \sum_{k=1}^{\infty} p_1\left(k, N(x)\right)\frac{1}{k+1}$$

we get

$$p_3 = \frac{1-e^{-N(x)}}{N(x)} - e^{-N(x)} \tag{9}$$

To compare the result given by our algorithm and the one given by equation 4, N nodes are randomly generated using an exceptional distribution and are placing into a square area. In the first simulation, all nodes have the same transmission range and the number of nodes in the network varies from 40 to 320. Figure 2 illustrates simulation results for ρ=0.01 and ρ=0.02 which correspond of a transmission range *R*=20*m* and *R*=40*m* respectively. The size of the MCDS increases linearly to the size of network. Results from simulation are very near, lightly bigger, to results from analysis.

In the second set of simulation, the transmission range is a variable from 10m to 100m. The number of nodes on the network is fixed to *N*=200 which are randomly generated in a square area (100*m*x100*m*). For each value of *R*, multiple runs are executed to get a confident average value of the size of the MCDS. Figure 3 shows that the size of the MCDS drops quickly when the node transmission range increase because node will have larger and larger coverage and therefore the number of neighbor decreases. Results from simulation are closed to results from analysis.

BACKBONE MAINTENANCE

The main feature of ad hoc networks is the mobility of terminals; they are free to move anywhere. In

Figure 2. Size of the MCDS vs. graph size, n

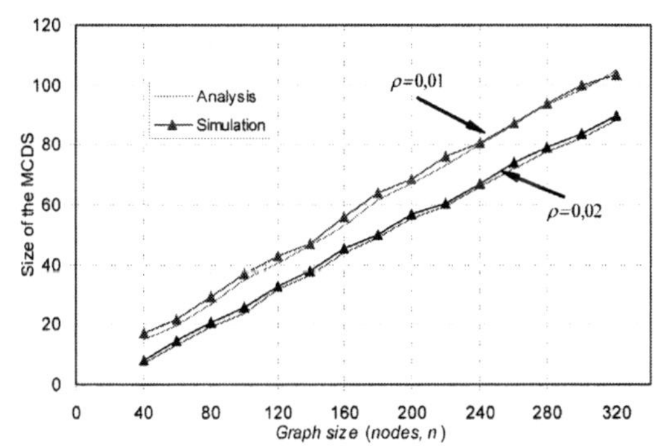

Figure 3. Size of the MCDS vs. the transmission range

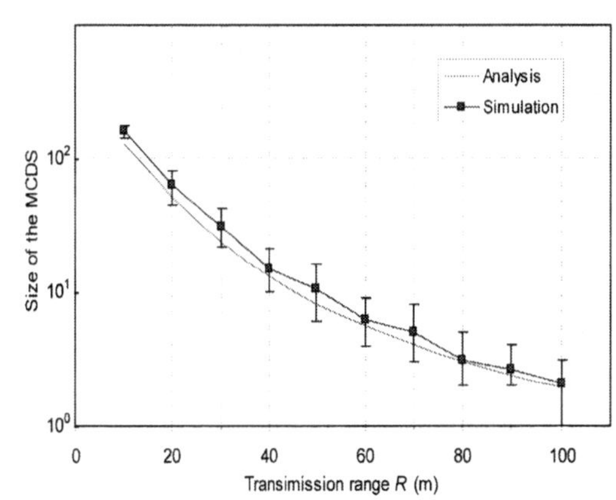

order to maintain the connectivity of the virtual backbone when topology changes. A distributed procedure will be applied the terminal that changes position and tries to connect to the backbone. The maintenance will be executed locally; only on the part of the network where the topology changed; a new terminal comes or a terminal lefts. The following state diagram, Figure 4, presents four states, in any times a terminal can be in one of these four states. It can be:

- Dominant, a member of the backbone,
- Dominate, not a member of the backbone and has at least one neighbour dominant,
- Active, in the process to be dominant or dominate, or
- IDLE, an instantaneously state when terminal changes position.

Routing protocols use hello packets to discover the neighbourhood. Our approach proposes to use the hello packet where some fields have been added, Figure 5.

In order to evaluate the performance of the proposed maintenance procedure, an implementation using Opnet Modeler is proposed. The protocol used in layer 2 is the 802.11b which is already integrated in the simulator. Each terminal is represented by a node, which is modeled by a transmitter and receiver as shown in Figure 6. The node process MCDS is for the construction and the maintenance of the backbone as illustrated in the Figure 7.

For mobility model, we used a modified Random Waypoint RWP proposed by (Bettstetter, 2001) of the initial model described in (Johnson, 1999).

Figure 8 and 9 show the % of connectivity as a function of the number of nodes in the network and the mobility of nodes. Simulations results

Figure 5. Modified format of hello packet

Source (32 bits)			
Destination (32 bits)			
TTL (4 bits)	Id (16 bits)	Type (8 bits)	State (4 bits)
Degree (8 bits)			
dominant (32 bits)			

Figure 4. State diagram

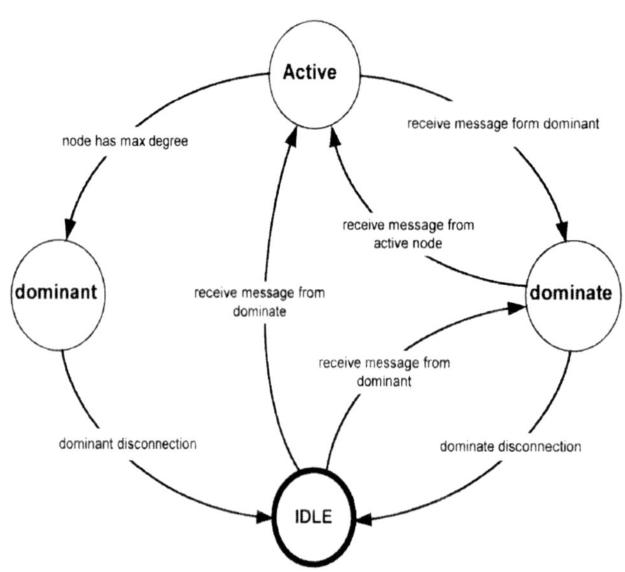

show clearly that the designed procedure is invariant to the network size. Curve in Figure 8 is almost constant when the network size varies. The mobility has a small effect to the performance of our procedure, Figure 9. For a high mobility environment, the connection still over 90%.

Figure 6. Ad hoc terminal node model

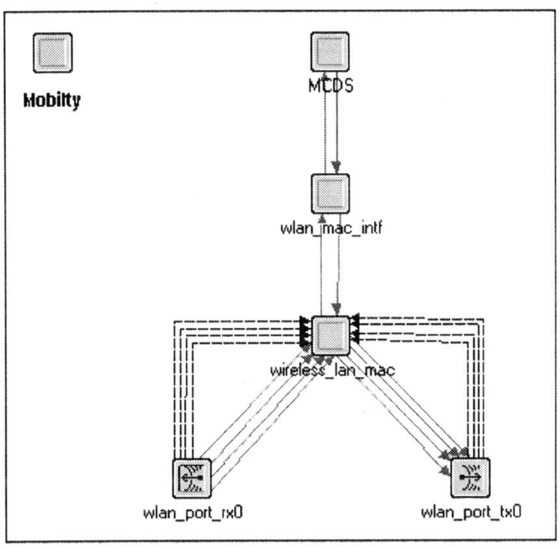

Figure 7. Model of the MCDS process

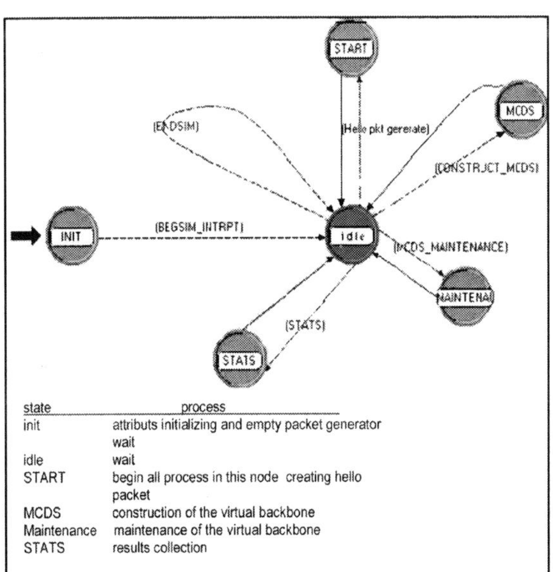

ROUTING PROTOCOLS OVER BACKBONE

A key issue in MANETs in the necessity that the routing protocols must be able to respond rapidly to topology dynamic in the network (Clausen, 2004). Moreover, due to the limited bandwidth available, the quantity of control traffic, used by the routing protocols, should be minimized. With virtual backbone structure, a subset of nodes is selected to be dominants and others as dominates. For example, dominants participate in diffusion of control messages in order to minimize bandwidth consumption and then the energy consumption in

Figure 8. Percentage of connection vs number of nodes

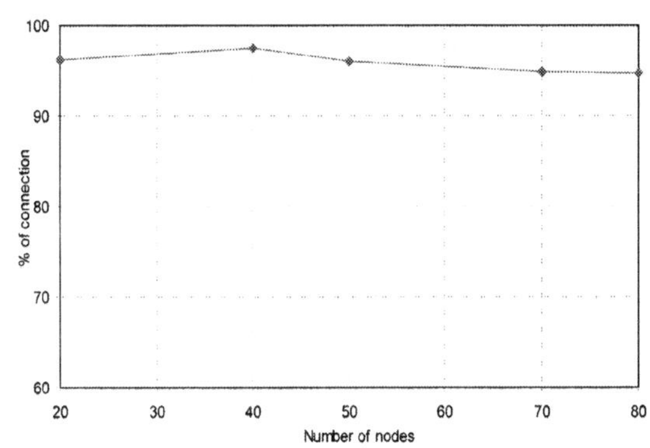

Figure 9. Percentage of connection vs speed of nodes

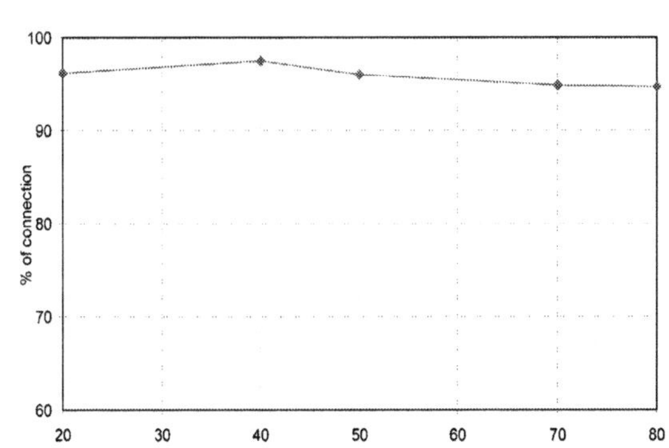

the network. Packet delivery rate and delay are measured for the considered protocols with two levels of load and mobility as variable in a random network scenario.

DSR over Backbone (DSRoB)

DSR is an on demand routing protocol that makes use of routing and an aggressive caching policy. DSR uses network-wide floods as its basic mechanism for computation routes. The use of caching in DSR will effective in limiting the area where the RREQ (Route REQuest) is propagated. In addition to RREQ, DSR uses RREP (Route REPlies) and RERR (Route ERRor). DSRoB retains the above feature of DSR and bringing in the advantage of the backbone structure. DSRoB uses RREQ, RREP and RERR for route request, route replies and route error. However, the route query mechanism is based on the backbone broadcast, rather than the conventional flooding of the RREQ messages. Figure 10 shows the basic idea to diffuse a control message in the presence of the backbone. In this example, the backbone is formed by nodes 3, 5 and 8. Only these nodes have to relay a control message when is received.

AODV over Backbone (AODVoB)

In AODV, nodes in the network maintain distance vector tables to facilitate routing. AODV has an effective feature, it uses a lower byte overheads in relatively static networks (not like DSR, it has to stamp source route on every data packet). Like DSR, AODV uses network-wide floods as its basic mechanism for computation routes. In AODV, the problem is more pronounced because intermediate nodes can respond to a RREQ message only if they have an entry in their distance vector table that particular destination and a node will have an entry in its table only if a flow that originates from or is destined for the destination traverses the node. The AODV protocol includes three components:

- initiation and propagation of RREQ messages,
- initiation and propagation of RREP messages, and
- maintenance of the distance vector table.

AODVoB uses the same messages but the way to flood it is different. Like DSRoB, only nodes from the backbone have to relay the RREQ message. However, the propagation of the RREP message is the same as in AODV. When the RREQ message reaches a domain in which one of the nodes has a route to the destination, the intermediate node replies with the RREP messages as in AODV. Using the backbone can be beneficial to AODV since a significant part of the overhead can be removed from the RREQ messages.

Figure 10. Messages propagation in DSR and DSRoB

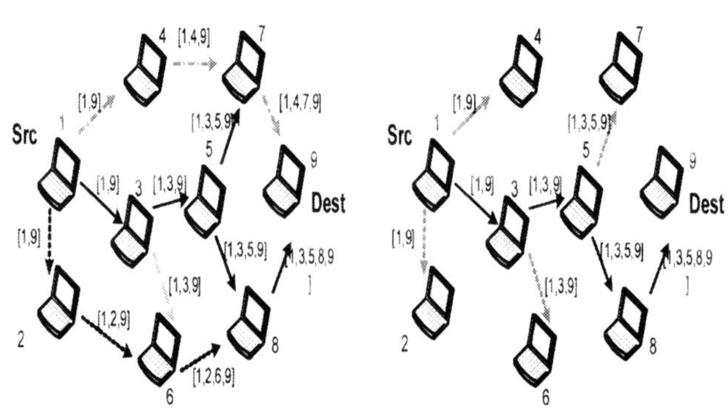

a- RREQ propagation (original DSR)   b- RREQ propagation (DSRoB)

TORA over Backbone (TORAoB)

TORA protocol is loop-free and distributed routing algorithm. It's based on the concept of link reversal. It is source initiated and provides multiple routes for any desired source/destination pair. There are three basic functions of this protocol, namely route creation, route maintenance and route erasure. This protocol uses query (QRY), update (UPD) and clear (CLR) for route creation, route maintenance and route erasure. In the presence of backbone, TORAoB can reduce reaction and communication overhead and thus conserves available bandwidth for data traffic and increases adaptability. Also, TORAoB can reduce the route length where dominants are optimally connected and the route should follow up using the backbone. Recall that TORA doesn't have a mechanism to find the short path between source and destination.

Simulation Results

In this section, we compare the performance of the modified protocols DSRoB, AODVoB and TORAoB to their original versions through simulation using Opnet simulator (Mnif, 2006-b). The parameters of our simulation environment are as follows: 50 nodes randomly distributed, topology dimension is 1500x1500 m^2, each source has CBR traffic with packet size of 512 bytes, the average speed of each nodes varies from 0to 20 m/s and two levels load have been considered: 40 simultaneous sessions for medium load and 80 simultaneous sessions for heavy load). Four parameters have been considered to evaluate the performance of these protocols with and without backbone. These parameters are:

- The delivered packet rate: defined as the rate (in percent) of the total number of data packet delivered to destinations by the number of total number of packets generated by the source,
- The number of retransmission: defined as the number of data packet transmitted divided by the number of data delivered. The number of data packet transmitted takes into consideration each data transmission for each node. It includes packets that are leaved and retransmitted by intermediary nodes,
- The overhead: it includes the overhead used by data packets (for example, the route sequence in DSR),
- The end to end delay; the average delay is the time from where the source generates the data packet until it arrives o the destination. It includes the processing time, queuing delay and the propagation delay.

Figure 11 shows the gap in percent when AODV, DSR and TORA protocols are used in their original version compared to those in presence of backbone. These results show clearly that protocols over backbone outperform those without backbone especially when the level of mobility is high and when the load in the network is heavy. For low mobility and medium charge results show that the presence of backbone doesn't bring any advantage in terms of overhead and delay. Indeed, overhead and delay are more important when protocols are used in presence of the backbone. This can be explained by the presence of backbone add an additional control traffic and additional time for construction.

CONCLUSION

In this paper, a new approach to construct and maintain a virtual backbone is proposed. The construction algorithm ensures a minimal size of the backbone and the maintenance procedure guarantees high level node connectivity even in high mobility environment. Routing protocols such as AODV, DSR and TORA involve all the nodes in the network for routing process and use

Figure 11. Performance comparison of routing protocols

Low mobility

	meduim load (40 sessions)				heavy load (80 sessions)			
	delivered packet rate	retransmission	overhead	delay	delivered packet rate	retransmission	overhead	delay
AODV	+12%	-5%	+18%	+15%	+10%	-10%	+8%	-3%
DSR	+8%	-4%	+25%	+8%	+16%	-12%	-10%	-4%
TORA	+4%	-2%	+7%	+4%	+12%	-6%	-12%	-4%

High mobility

	meduim load (40 sessions)				heavy load (80 sessions)			
	delivered packet rate	retransmission	overhead	delay	delivered packet rate	retransmission	overhead	delay
AODV	+25%	-37%	-23%	-10%	+21%	-32%	-11%	-8%
DSR	+28%	-2%	-18%	-5%	+36%	-12%	-15%	-4%
TORA	+17%	-22%	-28%	-12%	+12%	+25%	-18%	-10%

flooding to diffuse control messages. Most of existing protocols don't perform well in dynamic environment as the static one. Simulations results show that routing protocols enhance their performance in presence of backbone. A significant increase of the packet data delivery rate has been noticed in high mobility environment. Also, the end to end delay has been reduced for high mobility and for medium and high load. This work is a first of a set of simulation studies in mobile ad hoc networks. Studies will include additional performance evaluation of other proposed protocols (including multicast) over backbone and other types of traffic such as TCP. A TCP data flow requires data flows in both directions between source and destination.

REFERENCES

Al-Karaki, J., & Kamal, A. E. (2008, February). Efficient Virtual-Backbone Routing in Mobile Ad Hoc Networks. *Elsevier Computer Networks*, *52*(2), 327–350. doi:10.1016/j.comnet.2007.09.007

Ben, L., & Zygmunt, J. H. (2006). Hybrid Routing in Ad Hoc Networks with a Dynamic Virtual Backbone. *IEEE Transactions on Wireless Communications*, *5*(6), 454–463.

Bettstetter, C., Resta, G., & Santi, P. (2001). The Node Distribution of the Random Waypoint Mobility Model for Wireless Ad hoc Networks. *IEEE Transactions on Mobile Computing*, *2*(3), 257–269. doi:10.1109/TMC.2003.1233531

Clausen, T. (2004). Comparative study of routing protocols mobile Ad hoc Networks. *INRIA Report Research* RR-5135.

Das, B., & Bhagghavan, V. (1997). Routing in Ad Hoc Networks Using Minimum Connected Dominating Sets. *International Conference on Communications*, Montréal (pp. 376–380).

Guha, S., & Khuller, S. (1998). Approximation Algorithms for Connected Dominating Sets. *Algorithmica*, *20*, 374–387. doi:10.1007/PL00009201

Haitao, L., & Rajiv, G. (2004, October 24-27). Selective Backbone Construction for Topology Control in Ad Hoc Networks. *Proceedings of MASS*, Florida, USA, (pp. 41-50).

Johansson, P., Larson, T., Hedman, N., Mielczarek, B., & Degermark, M. (1999). Scenario-Based Performance Analysis of Routing Protocols for Mobile Ad hoc Networks. *The ACM/IEEE MOBICOM'99*, Seattle, WA (pp. 195-206).

Johnson, D., & Maltz, D. A. (1999). Dynamic Source Routing in Ad hoc [On Mobile Computing.]. *Networks*, 153–181.

Mnif, K., & Kadoch, M. (2006). Construction and Maintenance of Backbone for Routing Protocols Enhancement in Mobile Ad hoc Networks. *10th IEEE International Conference on Communication Systems 2006, ICCS'06*, Singapore (pp. 239-243).

Mnif, K., & Kadoch, M. (2006, August, 28-30). Routing Protocols Performance Enhancement using Virtual Backbone in Mobile Ad hoc Networks. *OPNETWORKS'06*, Washington, USA.

Shuhui, Y., Jie, W., & Fei, D. (2008). Efficient Directional Network Backbone Construction in Mobile Ad Hoc Networks. *IEEE Transactions on Parallel and Distributed Systems*, *19*(12), 1601–1613. doi:10.1109/TPDS.2008.43

Tseng, Y.-C., NI, S.-Y., CHEN Y.-S., & SHEU J.-P. (2002). The Broadcast Storm Problem in Mobile Ad Hoc Networks. *Wireless Networks*, *8*, 153–167. doi:10.1023/A:1013763825347

Wan, P.-J., Alzoubi, K. M., & Frieder, O. (2004, April). *Distributed Construction of Connected Domination Set in Wireless Ad hoc and communications, 9*(2), 141-149.

Wang, Y., Wang, W., & Li, X.-Y. (2005) *Distributed Low-Cost Backbone formation for Wireless Ad hoc Networks"6thACMsymposium(Mobihoc'05)* USA, (pp 2-13)

This work was previously published in International Journal of Mobile Computing and Multimedia Communications (IJMCMC) 1(3), edited by Ismail Khalil & Edgar Weippl, pp. 27-39, copyright 2009 by IGI Publishing (an imprint of IGI Global).

Chapter 5
Realization of Route Reconstructing Scheme for Mobile Ad Hoc Network

Qin Danyang
Harbin Institute of Technology, P. R. China

Ma Lin
Harbin Institute of Technology, P. R. China

Sha Xuejun
Harbin Institute of Technology, P. R. China

Xu Yubin
Harbin Institute of Technology, P. R. China

ABSTRACT

Mobile Ad Hoc Network (MANET) is a centerless packet radio network without fixed infrastructure. In recent years tremendous attentions have been received because of capabilities of self-configuration and self-maintenance. However, attenuation and interference caused by node mobility and wireless channels sharing weaken the stability of communication links especially in ubiquitous MANET. A mathematical exploring model for next-hop node has been established. The negative impact of wireless routes discontinuity on pervasive communication is alleviated by a novel route reconstructed scheme proposed in this paper based on restricting the route requirement zone into a pie slice region on intermediate nodes according the solution of the exploring equation. The scheme is an effective approach to increase survivability and reduce average end-to-end delay during route maintenance as well as allowing continuous packet forwarding for fault resilience so as to support mobile multimedia communication. The ns-2 based simulation results show remarkable packets successful delivery rate and end-to-end delay improvements of source-initiated routing protocol with route reconstructing scheme, and especially in the case of high dynamic environments with heavy traffic loads, more robust and scalable performance will be obtained.

DOI: 10.4018/978-1-60960-563-6.ch005

Mobile ad hoc network, always shorted for MANET, is a wireless self-organized network composed of heterogeneous mobile nodes, which can form any provisional topology freely and dynamically to make interconnect and interworking of any people and equipment available in environment without any infrastructure located in advance or those once existing have been destroyed. Since the nodes in a network of this kind can serve as routers and hosts, they can forward packets on behalf of other nodes and run user applications (Frodigh, Johansson, & Larsson, 2000).

With technology of embedded handhold terminal fast mature, MANET will be applied to fighter aircraft, battlewagon, tank, a single soldier, and cars, portable computers, PDA, cell phones *et al* as well (Zhang, CI, & Yang, 2005). As a peer-to-peer network, nodes in MANET are nomadic to create and demolish relationships with others, while being a data communication network, it should be set up and respond to topology changes rapidly enough to support multimedia communication. Different mobile styles and radio wave propagational conditions changing with time and position could cause the connection off and on between adjacent nodes in MANET, the result of which is making MANET a time-varying network (Rickenbach, Schmid, Wattenhofer, & Zollinger, 2005). One of the greatest differences of MANET from other ad hoc networks is the rapidly changing topology, impacted by scalability of network and mobility of nodes (Hean, K. O., Hean, L. O., Tee, & Sureswaran, 2006). There is a large span and hundreds or thousands of nodes are included. Great changes can exist inside the network, such as the nodes motion velocity, direction and so on. Compared to a lot of wireless networks, nontrivial challenges in MANET design and operation are derived from facts of lack of central administration, possibility of nodes moving and all communications lying on wireless medium (Forde, & Doyle, 2006). As wireless communication seems vulnerable in front of propagative damnification, there is seldom any stability in connection. Hops are adopted in MANET due to transmitting scale of nodes is far smaller than that of the whole network. The information in routing tables is continuously upgraded. Such incessant network reconfigurations will cause a lot of control information exchanging frequently to reflect the current state of the network, which will make finite bandwidth a great waste by too much overhead (Cheikh, Claude, Guillaume, & Guerine, 2008). An exploring model for next hop is presented in this paper for the condition that when wireless link is unavailable. Based on the solution of the optimal exploring equation, a route reconstructing scheme is proposed to restrict requirement zone into a pie slice region on intermediate nodes, that is to say, when route from source to the destination is broken up, the routing discovery for optimal path to substitute initials from the upstream node rather than from the current one, the purpose of which is to improve survivability, keep data transmitting when a route is invalid, at the same time reduce overhead and delay in route reconfiguration, thereby realize route reconstructing performance. Actually, the scheme focuses on reducing average end-to-end delay on premise of maintaining connectivity to shorten the time for recovery.

OPTIMAL EXPLORING MODEL

Nodes mobility in MANET will cause network topology changing dynamically for sure, which is, to a great extent, a random rapid and unpredictable change. In wireless link environment, the necessary procedure to ensure robustness, reliability and effectiveness of network is to realize dependable communications between nodes.

The Optimal Exploration for Route

The optimal exploring theory is on how to find a target already existed, which is called exploring target, in an optimal way (Dimitrakakis, 2006).

The probabilities of distribution function of target location and moving path, detection function and constraint condition are main parameters of this theory (Groot, 1970). Detection function and target position function can help calculate probability of the target being found successfully in every distributive scheme correspondingly to each area of exploring space (Ohsumi, 1986). Therefore, the solution of optimal exploring problem is to find an optimal distributive scheme on exploring time to maximize possibility of searching the target successfully or minimize the expectation value of cost needed (Li, 2001).

For optimal exploring model being set up better, some terms should be defined first as follows (Zhu, 2005).

Assume $X(t) \in R^n$ is the position of target node at t, and $S(t) \in R^n$ is the position of reconstructing node at t. Joint probability density $f(x, t, S)$ can be defined as:

$$f(s,t,S)\mathrm{d}x = \mathrm{Prob}[x \le X(t) \le x + \mathrm{d}x \mid \textit{unsuccessful exploration along } S(\tau)(0 \le \tau \le t)]$$

$f(x, t, S)$ is the joint probability density of position of target node and exploring process. It is obvious that $f(x, t, S)$ tends to initial probability density $f_0(x)$ when t approaches to 0. The detection probability at t can be acquired as (1) according to joint probability density.

$$b_t = 1 - \int_D f(x,t,S)\mathrm{d}x \tag{1}$$

Where integral region D is the area target node lying in, in other words, D is the subset of exploring space R^n and the probability of target lying in it is greater than 0. In two dimensional conditions, joint probability density function satisfies(2), which includes information on target node moving model and detection model. $f_0(x, y)$ is an initial density function of target node position and satisfied(3).

$$\begin{aligned}\frac{df}{dt} &= \frac{1}{2}\left[\frac{\mathrm{d}^2}{\mathrm{d}x^2}\left(a_{11}\left(x,y,t\right)f\right) + 2\frac{\mathrm{d}^2}{\mathrm{d}x\mathrm{d}y}\left(a_{12}\left(x,y,t\right)f\right)\right. \\ &\left. + \frac{\mathrm{d}^2}{\mathrm{d}y^2}\left(a_{22}\left(x,y,t\right)f\right)\right] - \frac{\mathrm{d}}{\mathrm{d}x}\left(b_1\left(x,y,t\right)f\right) \\ &- \frac{\mathrm{d}}{\mathrm{d}y}\left(b_2\left(x,y,t\right)f\right) - \psi\left(x,y,t,S\right)f \\ f\left(x,y,0,S\right) &= f_0\left(x,y\right)\end{aligned} \tag{2}$$

$$f_0\left(x,y\right)\mathrm{d}x = \mathrm{Prob}\begin{bmatrix} x \le X\left(0\right) \le x + \mathrm{d}x, \\ y \le Y\left(0\right) \le y + \mathrm{d}y. \end{bmatrix} \tag{3}$$

Conditional probability density in unsuccessfully exploring case can be obtained by *Bayes* formula (Wang, & Zeng, 2007) once $f(x, t, S)$ is known. Generally, it is easy to get $f(x, t, S)$, and $\rho(x\ t|S)$ can be got by(4).

$$\rho\left(x,t \mid S\right) = \frac{f\left(x,t,S\right)}{\int_D f\left(x,t,S\right)\mathrm{d}x} \tag{4}$$

Probability of survival $u(x, t, T, Z)$ (Xi'an University of Electronic Science and Technology, 2006) on the interval of time $[t, T]$ can be defined as:

$$u\left(x,t,T,Z\right) = \mathrm{Prob}\left[\textit{unexplored on } [t,T] \mid X\left(t\right) = x\left(t \le \tau \le T\right) \textit{ given}\right]$$

Having noted that $f(x, t, Z)$ depends on current position while $u(x, t, T, Z)$ lies on the initial one, and when $t \to T$, there is always $u(x, t, T, Z) \to 1$. Denominator in (4) denotes possibility of failed exploration until t, thereby $P[t;Z]$ can be defined as(5). In this way, $P[t;Z]$ is successful exploring probability along path Z, which can be rewritten as(6).

$$P[t,Z] = 1 - \int_D f(x,t,Z)\,\mathrm{d}x \tag{5}$$

$$P[t;Z] = 1 - \int_D f_0(x)\,u(x,0,T,Z)\,\mathrm{d}x \tag{6}$$

Other useful information can be educed from these basic probability characteristic metric. T is the time taken to find the target node, as a positive random variable, the expectation can be calculated by (Shi, Deng, & Qi, 2007):

$$\begin{aligned} E[T] &= \int_0^\infty \mathrm{Prob}[T > t]\,\mathrm{d}t \\ &= \int_0^\infty \mathrm{Prob}\left[\mathit{undetected} \quad in \quad [0,t)\right]\mathrm{d}t \\ &= \int_0^\infty \int_D f(x,t,S)\,\mathrm{d}x\mathrm{d}t \end{aligned}$$

The Optimal Exploring State Equations

Exploring equations for static target. Assume exploration until t and that after t are two independent events as(7), which means that the reconstructing node is not able to learn experience from past failure on how to exploring better. $b(x, t, z)$ in(7) is detection function.

$$f(x,t+\Delta t,Z) = f(x,t,Z)\left(1 - b(x,t,z)\Delta t\right) \tag{7}$$

Probability of survival $u(x, t, T, Z)$ can be obtained by a similar deduction.

$$u(x,t,T,Z) = \left(1 - b(x,t,z)\Delta t\right)u(x,t+\Delta t,T,Z) \tag{8}$$

Exploring equations for target of deterministic movement. If the position of target at time t is x, it is located at $x+a(x, t)$ at time $t+\Delta t$. Here $a(x, t)$ is a known function.

As $t \to T$, there is $u(x, t, T, Z) \to 1$.

With the help of conveying function $q(\xi, t, \Delta t, x-\xi)$ and independence assumption, joint probability density function and probability of survival can be obtained. Exploring problem of static target transforming to target of deterministic movement, exploring equations have been transformed from ordinary differential equations to partial differential ones. If an operator of L_x^d is defined as(10), and its adjoint operator is L_x^{d*}, the exploring equations for target of deterministic movement is shown in(11).

$$\begin{aligned} u(x,t,T,Z) &= \left(-b(x,t,z)\Delta t\right)u\left(x + a(x,t)\Delta t, t+\Delta t, T, Z\right) \\ &= \left(-b(x,t,z)\Delta t\right)\Big[\sum_i u_{x_i}(x,t,T,X)\,a_i(x,t)\Delta t + \\ &\quad + u(x,t,T,Z) + u_t(x,t,T,Z)\Delta t\Big] + o(\Delta t) \end{aligned} \tag{9}$$

$$L_x^d = \sum_i a_i(x,t)\frac{\partial}{\partial x_i} \tag{10}$$

$$\begin{aligned} &u_t + L_x^d u = b(x,t,z)u \\ &f_t = L_x^{d*} f - b(x,t,z) f \end{aligned} \tag{11}$$

Exploring equations for target of random movement. Letbe $\xi = \Delta X$ is variable of target position, and assume moments of ξ possesses forms as follows.

$$\begin{aligned} &\int \xi_i q(\xi,t,\Delta t,x)\,\mathrm{d}\xi = a_i(x,t)\Delta t + o(\Delta t) \\ &\int \xi_i \xi_j q(\xi,t,\Delta t,x)\,\mathrm{d}\xi = a_{ij}(x,t)\Delta t + o(\Delta t) \\ &\int \xi_{i_1}\xi_{i_2}\cdots\xi_{i_n} q(\xi,t,\Delta t,x)\,\mathrm{d}\xi = o(\Delta t), \qquad n \ge 3 \end{aligned} \tag{12}$$

Based on the Taylor expansion (He, 2005) and independence assumption, a forward equation and a backward one can be obtained (Li, & Zhou, 2004). By defining operator L_x as(13), exploring equations could be written simply as(14).

$$\mathrm{L}_x = \sum b_i(x,t)\frac{\partial}{\partial x_i} + \frac{1}{2}\sum c_{ij}(x,t)\frac{\partial^2}{\partial x_i \partial x_j} \tag{13}$$

$$\begin{aligned} &u_t + \mathrm{L}_x u = b(x,t,z)u \\ &f_t = \mathrm{L}_x^* f - b(x,t,z)f \end{aligned} \tag{14}$$

Solutions of Exploring Equations

Solutions for static target. The solutions of(7) and (8) are quite simple that we only need to solve ordinary differential equations as in(15).

$$\begin{aligned} &f(x,t,z) = f_0 \exp\left[-\int_0^t b(x,s,Z(s))\,\mathrm{d}s\right] \\ &u(x,t,T,Z) = \exp\left[-\int_t^T b(x,s,Z(s))\,\mathrm{d}s\right] \end{aligned} \tag{15}$$

Solutions for target of deterministic movement. In this case, exploring equations with form of(16) could be solved by classical radial method (Courant, & Hilbert, 1962).

$$\begin{aligned} &f_t(x,t,Z) = -\sum \frac{\partial}{\partial x_i}\left(a_i(x,t) f(x,t,Z)\right) - b(x,t,z) f \\ &u_t(x,t,T,Z) + \sum a_i(x,t)\frac{\partial u}{\partial x_i} = b(x,t,z)u \end{aligned} \tag{16}$$

Considering initial conditions based on differential chain rule, a new equation can be obtained as(17).

$$\begin{aligned} &f_t + \sum a_i(x,t) f_{x_i} = -f\left[\nabla \cdot a + b(x,t,z)\right] \\ &f(x,t,0) = f_0(x) \end{aligned} \tag{17}$$

According to radial method (Huang, & Zhao, 2002), characteristic curve of (17) is the solution of(18).

$$\frac{\mathrm{d}x_i}{\mathrm{d}t} = a_i(x,t), \qquad x_i(0) = x_{0_i} \tag{18}$$

Where x_0 is a node in exploring scale and satisfied $f_0(x_0) \neq 0$.

Solution of (17) is a curve in space, the projection of which in x-space is called *a radial from* x_0, and (17) turned to be(19).

$$\begin{aligned} &\frac{\mathrm{d}f}{\mathrm{d}\tau} = -f\left[\nabla \cdot a + b(x(t), t(\tau), Z(\tau))\right], \\ &f(0) = f_0(x_0) \end{aligned} \tag{19}$$

Radials are disjoint for pointing at different sub-spaces, as shown in Figure *1*. Then solution of (17) can be written as(20), where integral denotes radial starts at x_0 from 0 to t.

Figure 1. Pointing to different numbers of sub-spaces based on radial method

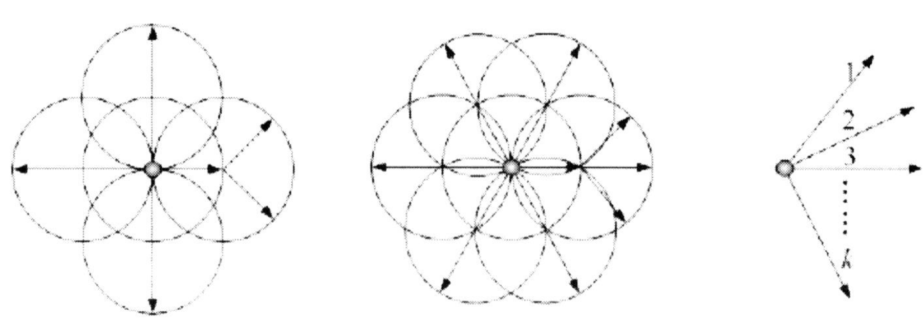

(a) Point to 4 sub-spaces (b) Point to 6 sub-spaces (c) Point to k sub-spaces

$$f\left(x\left(t,x_0\right),t,Z\right)=f_0\left(x_0\right)\exp\left[-\int\left(b\left(x\left(s\right),s,Z\left(s\right)\right)+\nabla\cdot a\left(x\left(s\right),s\right)\right)\mathrm{d}s\right] \quad (20)$$

Therefore algorithm descriptions of exploring equation solved by radial method are as follows:

1. appointing a node x corresponding to t;
2. find x_0 to make radial started at x_0 pass x at t;
3. if $f_0(x_0)=0$, then $f(x, t, Z)=0$, end the algorithm and quit;
4. Or else, calculate $f(x, t, Z)=0$ by solution of (18).

Approximate solution of exploring equations. In the most general condition can be deduced as (21) and (22).

$$u_t+\frac{1}{2}\sum c_{ij}\frac{\partial^2 u}{\partial x_i\partial x_j}+\sum a_i\frac{\partial u}{\partial x_i}=b\left(x,t,z\right)u \quad (21)$$

$$f_t=\frac{1}{2}\sum\frac{\partial^2}{\partial x_i\partial x_j}\left(c_{ij}f\right)-\sum\frac{\partial}{\partial x_i}\left(a_if\right)-b\left(x,t,z\right)f \quad (22)$$

By rescaling functions introduced (Zhu, Hong, & Wang, 1999), time length, velocity of target node and covariance matrix eigenvalue are represented by T_c L_c b_c *and* a_c respectively.

T_c = total available time for exploring,
L_c = the largest linear dimension
$b_c = \max_x|a(x, t)|$,
$a_c = \max_x\|c_{ij}(x, t)\|$.

A rescaling transformation about variable and function is defined according to a set of characteristic quantity selected from above as.

$$t=\hat{t}T_c,\quad x=\hat{x}L_c,\quad z=\hat{z}L_c,\quad a\left(x,t\right)=\hat{a}\left(\hat{x},\hat{t}\right)b_c,$$
$$c_{ij}\left(x,t\right)=\hat{c}_{ij}\left(\hat{x},\hat{t}\right)a_c,\quad b\left(x,t,z\right)=\frac{\hat{b}\left(\hat{x},\hat{t},\hat{z}\right)}{T_c}.$$

In the first case, assume

$$\frac{a_c}{L_c^2/T_c}\equiv\varepsilon\ll 1,\qquad \frac{b_c}{L_c/T_c}\sim O\left(1\right),$$

The exploring equations after simplified will be as(23).

$$f_t=\frac{\varepsilon}{2}\sum\frac{\partial^2}{\partial x_i\partial x_j}c_{ij}f-\sum\frac{\partial}{\partial x_i}a_if-bf$$
$$u_t+\frac{\varepsilon}{2}\sum_{ij}c_{ij}f\frac{\partial^2 u}{\partial x_i\partial x_j}+\sum_i a_i\frac{\partial u}{\partial x_i}=bu \quad (23)$$

In the second case, assume

$$\frac{b_c}{L_c/T_c}\equiv\varepsilon\ll 1,\quad \frac{a_c}{L_c^2/T_c}\sim O\left(\varepsilon^2\right)$$

Exploring equations will be as (24)

$$f_t=\frac{\varepsilon^2}{2}\sum\frac{\partial^2}{\partial x_i\partial x_j}c_{ij}f-\varepsilon\sum\frac{\partial}{\partial x_i}a_if-bf$$
$$u_t+\frac{\varepsilon^2}{2}\sum_{ij}c_{ij}f\frac{\partial^2 u}{\partial x_i\partial x_j}+\varepsilon\sum_i a_i\frac{\partial u}{\partial x_i}=bf \quad (24)$$

If ε is set to be 0, (23)will be simplified to be exploring equations for target of deterministic movement, and (24)will be for static target. Thus, solutions of target with deterministic movement can be regarded as approximate results of general situation.

Asymptotic analysis on exploring equations for target of random movement. The idea of radial method for solving parabolic partial differential equations is to find an ε small enough to obtain approximate results for exploring equations above (Harbin Intitute of Technology, 2006). Suppose the initial value of the first equation in (23) and (24) appears as the form shown in(25).

$$f_0(x) = e^{-\phi(x)/\varepsilon} \sum_{k=0} \varepsilon^k h_k(x) v \tag{25}$$

Then solution of this equation has the form of (26)

$$f(x,t,Z) = \left[\sum_{n=0} \varepsilon^n g_n(x,t,Z)\right] e^{-\phi(x,y,Z)/\varepsilon} \tag{26}$$

$\varphi(x, y, Z)$ and $g_n(x, t, Z)$, $n = 0,1,2$, are unknown functions that are uncertain. All-order derivative or differential expressions of $f(x, t, Z)$ obtained from forms above, will be substituted into(22), and arranged by index of ε, there is (27), shown in Box 1.

Coefficients of every ε^m (m=-1, 0, 1...) are set to be 0 for each, different equations will be obtained (Grindrod, 1991). Firstly, coefficient of ε^{-1} is set to be 0, and so called Eikonal Equation (Lu, Zhao, Zhu, & Bao, 1981) will be obtained as(28).

$$\frac{\partial \phi}{\partial t} + \sum_i a_i \frac{\partial \phi}{\partial x_i} + \frac{1}{2}\sum_{i,j} c_{ij} \frac{\partial \phi}{\partial x_i}\frac{\partial \phi}{\partial x_j} = 0 \tag{28}$$

Then coefficients of ε^m(m=0,1,2,...) are set to be 0, m-order propagation equation is obtained as(29). Eikonal Equation and propagation equation are quite similar with asymptotic solution of wave equation in electromagnetic field and wave (Fornberg, & Whitham, 1978), which can be solved by *radial method*.

$$\begin{aligned}\frac{\partial g_m}{\partial t} = \sum_{i,j} \Bigg[& \frac{c_{ij}}{2}\left(\frac{\partial^2 g_{m-1}}{\partial x_i \partial x_j} - 2\frac{\partial g_m}{\partial x_i}\frac{\partial \phi}{\partial x_j} - g_m \frac{\partial^2 \phi}{\partial x_i \partial x_j}\right) \\ & -2\frac{\partial c_{ij}}{\partial x_i}\frac{\partial \phi}{\partial x_j} - 2\frac{\partial c_{ij}}{\partial x_j}\frac{\partial \phi}{\partial x_i} g_m - \frac{\partial^2 c_{ij}}{\partial x_i \partial x_j} g_{m-1}\Bigg] \\ & + \sum_i \frac{\partial g_m}{\partial x_i} - \left(b + \sum_i \frac{\partial a_i}{\partial x_i}\right) g_m\end{aligned} \tag{29}$$

Considering Eikonal Equation to be a first-order nonlinear partial differential equation (Maxell, Seamon, & Verstraete, 2007), Hamiltonian function is defined as(30).

$$H(x,p,t) = \sum_i a_i p_i + \sum_{i,j} \frac{1}{2} c_{ij} p_i p_j \tag{30}$$

Then (28) can be written as:

$$\frac{\partial \phi}{\partial t} + H\left(x, \frac{\partial \phi}{\partial x}, t\right) = 0$$

Differential equations (31) can be regarded as a radial sent out from (x_0, p_0) in space (x, p), and if being satisfied by $x(t)$ and $p(t)$, (32) will be deduced from (28), the solution of which is at least a local solution of (28), and the initial conditions

Box 1.

$$\begin{aligned} e^{-\phi/\varepsilon}\left(\sum_{n=0}^{\infty} \varepsilon^n \frac{\partial g_n}{\partial t} - \varepsilon^{n-1}\frac{\partial \phi}{\partial t} g_n\right) = e^{-\phi/\varepsilon} & \left\{\sum_{ij} \frac{c_{ij}}{2} \sum_{n=0}^{\infty}\left[\varepsilon^{n-1}\frac{\partial^2 g_n}{\partial x_i \partial x_j} - \varepsilon^n\left(2\frac{\partial g_n}{\partial x_i}\frac{\partial \phi}{\partial x_i} + g_n \frac{\partial^2 \phi}{\partial x_i \partial x_j}\right) + \varepsilon^{n-1} g_n \frac{\partial \phi}{\partial x_i}\frac{\partial \phi}{\partial x_j}\right]\right\} \\ & - e^{-\phi/\varepsilon}\left[\sum_i\left(a_i - \varepsilon\sum_j \frac{\partial c_{ij}}{\partial x_j}\right)\sum_{n=0}^{\infty}\left(\varepsilon^n \frac{\partial g_n}{\partial x_i} - \varepsilon^{n-1}\frac{\partial \phi}{\partial x_i} g_n\right)\right] - e^{-\phi/\varepsilon}\left(b + \sum_i \frac{\partial a_i}{\partial x_i} - \frac{\varepsilon}{2}\sum_j \frac{\partial^2 c_{ij}}{\partial x_i \partial x_j}\right)\sum_{n=0}^{\infty}\varepsilon^n g_n \end{aligned} \tag{27}$$

can be selected as (33). Along the direction of the radial generated in(31), (28)will transform to an ordinary differential equation as(34).

$$\frac{dx_i}{dt} = \frac{\partial H}{\partial p_i}, \qquad x_i(0) = x_{i_0}$$
$$\frac{dp_i}{dt} = -\frac{\partial H}{\partial x_i}, \qquad p_{i_0}(0) = p_{i_0} \tag{31}$$

$$\frac{d\phi}{dt} = -H(x,p,t) + \sum_i p_i \frac{dx_i}{dt} = \frac{1}{2}\sum_{i,j} c_{ij} p_i p_j, \phi(0) = \phi_0 \tag{32}$$

$$x(0) = x_0, p_i(0) = \left.\frac{\partial \phi}{\partial x_i}\right|_{x_0}, \phi(0) = \phi(x_0) \tag{33}$$

$$\frac{dg_m}{dt} = -g_m\left[b(x,t,z) + \Gamma(t)\right] + G_{m-1}(t), g_m(0) = h_m(x_0) \tag{34}$$

Where

$$\Gamma(t) = \sum_i \frac{\partial a_i}{\partial x_i} + \sum_{i,j}\left(\frac{\partial c_{ij}}{\partial x_j} p_j + \frac{1}{2} c_{ij} \frac{\partial^2 \phi}{\partial x_i x_j}\right)$$
$$G_{m-1}(t) = \sum_{i,j} \frac{c_{ij}}{2}\left(\frac{\partial^2 g_{m-1}}{\partial x_i \partial x_j} - 2\frac{\partial c_{ij}}{\partial x_j}\frac{\partial g_{m-1}}{\partial x_i} - \frac{\partial^2 c_{ij}}{\partial x_i \partial x_j} g_{m-1}\right)$$

Solution of the original (28) can be obtained by solving this ordinary differential equation.

HOW TO REALIZE THE OPTIMAL EXPLORATION IN MANET

First there is an assumption that when a path to destination node is destroyed or invalided, the new reconstructed route should not be too different from the former one. In view of background of this work, AODV (Perkins, Royer, & Das, 2003) is adopted as the original routing protocol for MANET. The route reconstructing scheme is designed to repair the local part of unavailable path without informing other nodes that the current link has been invalided so as to minimize negative impact of overhead on other affairs.

Route Discovery

Route discovery protocol will be out of function as long as there is an effective path making endpoints of a communicating link be connected. When a route is needed to set up in order to reach a new destination node, the node has to broadcast RREQ (Route REQuest) control message to discover a path reaching this destination, or reaching an intermediate node which possesses a path new enough to get to the destination, and a route now can be confirmed by transmitting RREP (Route REPly) control message back to the source node sending RREQ. Such end-to-end route can be established by the routing tables which are generated when RREP arrives at source node as AODV, or by the routing tables which are generated when the first data packet reaches destination from source node. After route setting up, data packets can be transmitted according to routing tables (Wisitpongphan, & Tonquz, 2006).

Route reconstructing process of invalided route is transparent to other nodes in network, especially the source node, which will be reflected directly in routing tables. Since transmitting path of data packets is decided by routing tables, difference between the original information carried in data packet head file and that in current routing tables will not affect data transmitting at all (Lee, S., Youn, Jung, Lee, J., & Kang, 2005). Accordingly, there is no need to report route error or alteration to up stream node after accomplish local route reconstruction.

Route Reconstructing Model

According to the mathematics model established above, radial method is the key to solve minimax problems, based on which the route reconstructing scheme is proposed, and extending to a pie slice. In actual communication, directions of receiving and transmitting information are measurable (Radzievskij, & Ufaev, 2003). It is assumed that information transmitting directions are knowable to each other after route being established. Geographic orientation can be used to set up as reference in network, as shown in Figure 2, θ_i is vector angle between the *i*th hop node and its up stream one, where the subscript can be obtained by protocol forwarding searchipcn hop count.

Vector angle γ_i between any node *i* and destination node D could be obtained as(35).

$$\gamma_i = \sum_{j=i}^{N} \theta_j \tag{35}$$

Figure 3 shows the design philosophy of route reconstructing process. When the signal strength on link F-G reduced caused by the mobility of node G or current worsening environment, node F detected the critical case $(LSV(S\_MAC_{packet}) < LSV_{threshold})$ firstly, and transmitted RRREQ (Route Repair REQuest) to upstream node A. After receiving RRREQ and before the link disconnecting completely, node A restored the unstable route S→......→A→F→G→D to be a currently available one S→......→A→B→C→D. During this

Figure 2. Diagram of vector angleθ_i

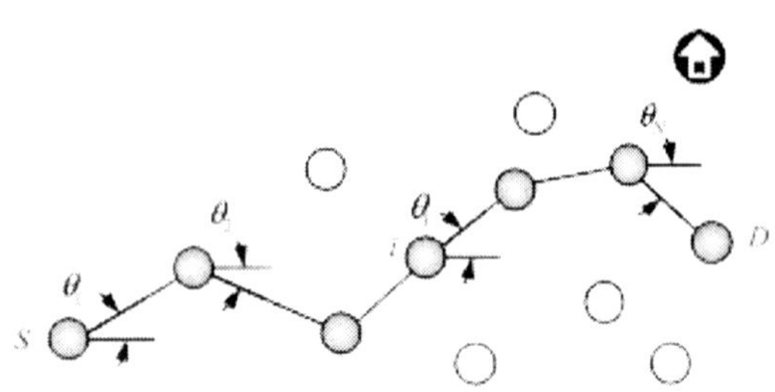

restoration, flood is restricted locally, and the communication will not be kept ongoing. Preemptive routing algorithm is adopted in this scheme to inspect link state, which is already a comparatively mature method. When node A detected the link A→F→G→D will be invalided, a RRREQ with vector angle information on destination node will be broadcasted. In this paper, reconstructing area of node A is designed to be a pie slice with node A as its vertex, and line connection between node A and destination node D as medial axis. Field angle α_0 of the sector can be adjusted according to the request on practical communicating demand; larger angle means larger scale of reconstructing request covering, while overhead introduced will increase correspondingly.

Any nodes lying in this pie slice region, if there is no valid route to destination node D in local cache, will broadcast request again, which will follow the same orientation principle. After node B received a RRREQ, a pie slice reconstructing scale with B as vertex and line connection between B and D as medial axis had been formed. A new path would be established only when an intermediate node with valid route to D stored or D, in

Figure 3. Route reconstructing process

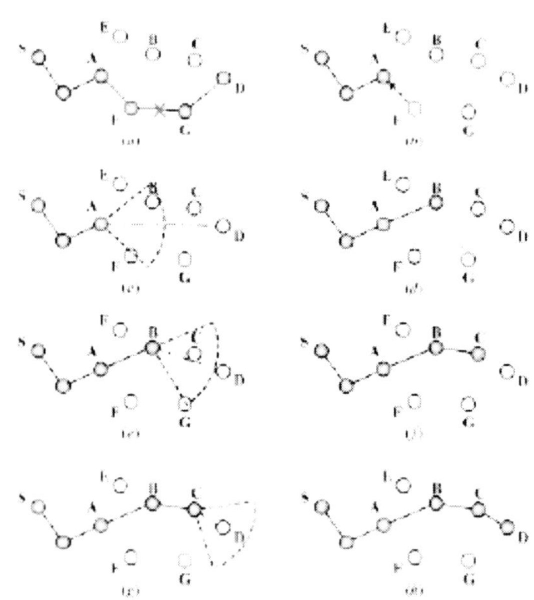

itself, received RRREQ and then transmitted RRREP (Route Repair REPly) back. Data packets that failed to pass on and the subsequent will be stored in local cache to avoid route reconstructing process being observed. A timeout is set to describe the longest waiting time so as to avoid buffer over running simultaneously.

This local route reconstructing process can be epitomized as follows:

1. When a node failed to transmit data to next hop (like F in Figure 3), the up stream node (like A in Figure 3) would broadcast RRREQ with reconstructing area information to explore the route reconstruction and the timer would be started simultaneously;
2. All nodes received RRREQ would detect if there is valid route; if there is, RRREP will be sent back and a new route will be established to restore the destroyed one; or else, go to next step;
3. Every node will calculate whether it is lying in effective reconstructing scale of RRREQ, if it is, the node will forward RRREQ; or else, discard it;
4. If a node on receives RRREP before timeout, it will transmit data packets in local buffer immediately along the new route, or else, it should report route error to source node in order to initiate another route discovery.

Determination of Forwarding Domain

In the process of route reconstruction being established, initial node will broadcast RRREQ with information on current vector angle position, destination position and reconstructing angle. Node B gets to know about node A node D and α_0 from RRREQ derived from A, as shown in Figure 4. The node received RRREQ will calculate α by (36) to decide whether needs to forward RRREQ.

Figure 4. Field angle of route reconstructing request

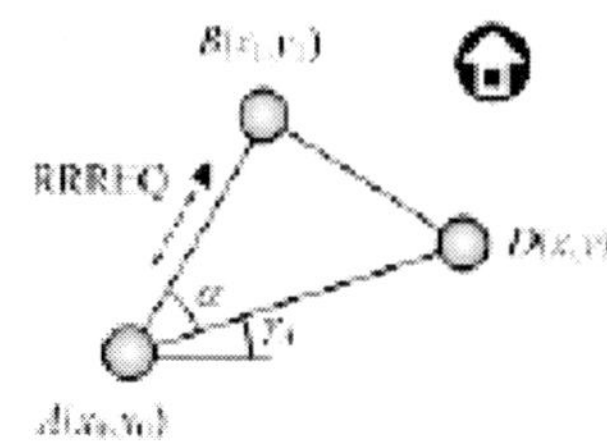

$$k_1 = \frac{y - y_0}{x - x_0}, \qquad k_2 = \frac{y_1 - y_0}{x_1 - x_0}$$

$$\tan\alpha = \frac{k_2 - k_1}{1 + k_1 k_2} = \frac{\frac{y_1 - y_0}{x_1 - x_0} - \frac{y - y_0}{x - x_0}}{1 + \frac{y_1 - y_0}{x_1 - x_0}\frac{y - y_0}{x - x_0}}$$

$$= \frac{(y_1 - y_0)(x - x_0) - (y - y_0)(x_1 - x_0)}{(y_1 - y_0)(y - y_0) + (x_1 - x_0)(x - x_0)}$$

$$\alpha = \arctan\frac{(y_1 - y_0)(x - x_0) - (y - y_0)(x_1 - x_0)}{(y_1 - y_0)(y - y_0) + (x_1 - x_0)(x - x_0)} \tag{36}$$

Assume node broadcasting RRREQ is located at (x_0, y_0), node receiving RRREQ is at (x_1, y_1), destination node is at (x, y), the field angle of reconstructing domain is α_0, and α is result by calculating. When $\alpha \leq \alpha_0$, the node receiving RRREQ is inside reconstructing domain and needs to forward new RRREQ; or else, discard RRREQ.

What is more, when a link disconnected in an active route and distance from disconnected spot to destination is not more than max repairing length *MAX_REPAIR_TTL*, up stream node is able to select local repair to restore this link by adding 1 to destination serial number and broadcasting a RREQ, initial value of TTL in which can be set as:

max(*MIN_REPAIR_TTL*,0.5×*#hops*)+*LOCAL_ADD_TTL*

where *#hops* denotes number of hops of reaching sending nodes that failed to deliver packets. Therefore, local repair operation is always transparent and always satisfies(37).

$$TTL \geq MIN\_PEPAIR\_TTL + LOCAL\_ADD\_TTL \quad (37)$$

Data Structure and Algorithm Description

Upstream node initializes directional flood RREQ (*Next hop*, *hop_count* = 0) to downstream node to discover route. Intermediate node receiving RREQ will start calculating vector angle, then directionally forward RREQ only when there is $LSV(S\_MAC_{RREQ}) > LSV_{threshold}$, or else, discard it. When RREQ gets to destination, a unidirectional temporary route from reconstructing node to destination node has been established. Destination node will reply RREP along temporary route and confirm the path as bidirectional formal path. Accordingly, when RREP arrived at reconstructing node, the new route has already been established.

In this process, every node monitors stability of upstream link continuously. When a data frame is received, an SNR is received simultaneously. The node first refreshes *LSV*, then calculates $LSV(S\_MAC_{frame})$, when there is $LSV(S\_MAC_{frame}) < LSV_{threshold}$, it will send RRREQ to upstream node to request it restoring unstable route to destination. Downstream is able to predict link disconnection by stability monitoring, and request to restore the unstable path before the link breaking through completely so as to realize route reconstruction.

When upstream node on unstable link receives RRREQ, it will initialize local restoring process. The reconstructing node will start a timer with timeout set to be T_{repair}, then broadcast RREQ (*next node*, $hop\_count = hop\_count_{local}$) for *TTL* being TTL_{repair}. The smaller TTL_{repair} will restrict flooding scale into local pie slice in RREQ. *hop_count* should be initialized as hop_count_{local} in local buffer to avoid loop. The downstream node will compare flooding angle when receiving RREQ, if lying in pie slice region, send a RREP back, or else, reduce *TTL* by 1, increase *hop_count* by 1, and forward RREQ in the same pie slice region until *TTL*=0. Before time is out, if reconstructing node receives correspondent RREP, route reconstruction is successful, and it will inform source node that the route has been restored once by RRN (Route Repair Notice). In route reconstructing, RREQ should be checked by pie slice flooding as well to ensure the stability. For keeping current communication, temporary route set by RREQ will be stored in *NEXT_HOP_in_repaired*, and the communication will not achieving on this unstable link, until RREP comes to confirm it to be formal one (*NEXT_HOP=NEXT_HOP _in_repaired*), to replace the unstable path in order to maintain communication in order to realize unstable route seamless recovery. After completing reconstruction, route length may be longer than that of before, in other words, the total hops maybe increase, distance vector in buffer of nodes out of forwarding domain may incorrect, which has a negative impact on communication being on the rails and repair work later, therefore certain mechanism should be introduced to refresh the route to be consistent. When receiving RRN, source node gets to know that route has already been repaired once, and it will transmit RUREP (Route Update REQuset), RREREQ (*hop_count*=0), and destination node will reply *RUREP* (*hop_count*=0). RUREQ and RUREP will propagate along reconstructed path, where each node will update their correspondent distance vector according to them.

PERFORMANCE EVALUATION

Performance of AODV with route reconstructing scheme, shorted for AODV-RRS, will be stimulated on NS-2 (The Network Simulator, *n.d.*) and the reconstructing performances, which means that some certain number of links will be broken out artificially selected randomly from all current connections, are compared with traditional AODV, shorted for AODV-TRA.

Simulation Model

Performance of MANET will be simulated in LINUX environment based on NS-2 with UDP (User Datagram Protocol) in transport layer, CBR (Constant Bit Rate) dataflow to generate data, and 802.11b DCF (Distributed Control Function), which transmits data packets to neighbors based on RTS and CTS control packets, adopted in MAC layer (Scanchez, 2006). RTS/CTS packets interaction is achieved before data delivering to realize virtual carrier sensing and channel reservation, so as to alleviate impact of hidden terminal. Parameters considered in this simulation are set as in Table 1.

Simulation Results

PSDR (Packets successful delivery rate), total control overhead and average end-to-end delay are concerned in this paper, for which some certain

Table 1. Simulation Parameters

Parameters	Value
Scenario size [m^2]	1000×1000
Mobile model	RWM[a]
Routing protocols	AODV-RRS, AODV-TRA
Wireless interface bandwith [Mbps]	2
Power radius [m]	250
Packet size [Byte]	512
Simulation duration [s]	600
Link disconnected rate [link(s)/s]	1[b], 2, 5, 10
Number of nodes	5, 10, 20, 30, 40
Number of source nodes	1, 2, 5, 10, 20
Velocity [m/s]	0, 1, 2, 5, 10, 15, 20
PDR[c] [packet(s)/s]	1, 2, 5, 10, 15, 20

a. RWM is short for Random Waypoint Model
b. 1 link/s is a default setting for link disconnected rate
c. PDR is short for packet deliver rate

number of links connected currently are selected randomly to be broken out by setting program properly. Impact on each metric will be compared by nodes moving velocity, number of source nodes, number of nodes, packets delivery rate and link disconnected rate. For simple express, PSDR and PDR denote packets successful delivery rate and packets delivery rate respectively.

Impact on PSDR. Figure 5 to Figure 7 show impacts on PSDR by different parameters. It is obvious that PSDR will increase with moving velocity rising and number of source nodes reducing as in Figure 5 Stat. result shows when the velocity is 20m/s and there are 20 source nodes transmitting data, PSDR with AODV-RRS will be increased by 15.82% and 15.17% respectively. The increase of number of nodes will reduce PSDR. From Figure 6, it can been seen that when PDR is not large, both route protocols will basically ensure data packets transmitting successfully, while PDR is over 10 packets/s, PSDR will toboggan, in such condition both these two protocols show quite similar performance. Link disconnected artificially to verify route reconstructing performance in adverse communication environment also in Figure 6. With the rise of link disconnection rate, PSDR drops sharply; however, the one with AODV-RRS is better. When there are 5 links disconnected, PSDR with traditional AODV drops to

Figure 5. Impact on PSDR by velocity & number of source nodes

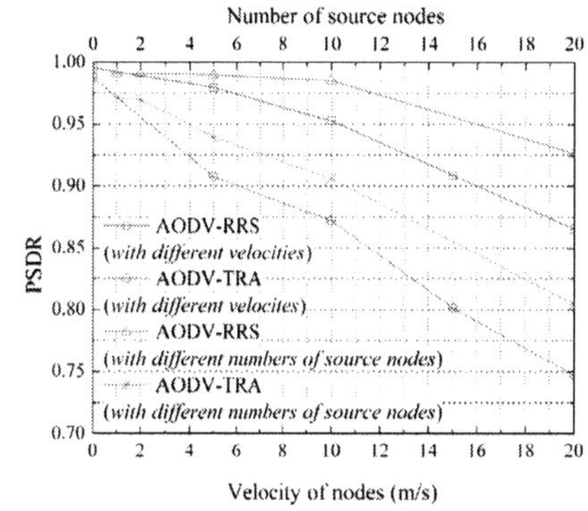

Figure 6. Impact on PSDR by PDR & disconnection rate

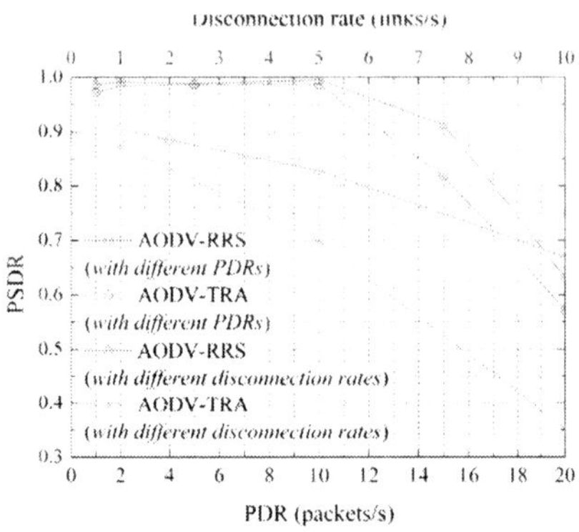

70% which means that communication is unavailable, while that with AODV-RRS can maintain above 80%. Therefore, AODV-RRS proposed in this paper is able to survive in diverse network environment so as to improve PSDR effectively. More nodes suggest more dense distribution, due to local route pie slice directional exploration, the situations with AODV-RRS can be alleviated as shown in Figure 7.

Impact on total communication overhead. In AODV-RRS, each node has to calculate vector angle, which will increase control overhead obviously, and after that, the vector angle has to be transmitted together with data, which will increase transmitting overhead simultaneously. Despite rise of transmitting overhead and control overhead, AODV-RRS reduces times of full-network broadcasting and request after failing to discover new route. Figure 8 to Figure 10 show comparisons of impact with AODV-RRS and traditional AODV by some key parameters on total overhead.

The communication total overhead will increase with moving more quickly and rise of number of source nodes, no matter AODV-RRS or AODV-TRA adopted, however, the gentler curve of AODV-RRS denotes that it will serve better in mobile conditions with more source nodes. When the velocity is 20m/s, communication total overhead with traditional AODV exceeds

Figure 7. Impact on PSDR by number of nodes

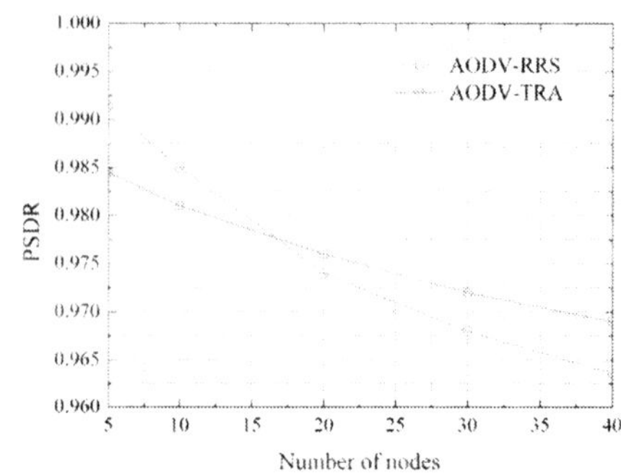

Figure 8. Impact on total overhead by velocity & number of source nodes

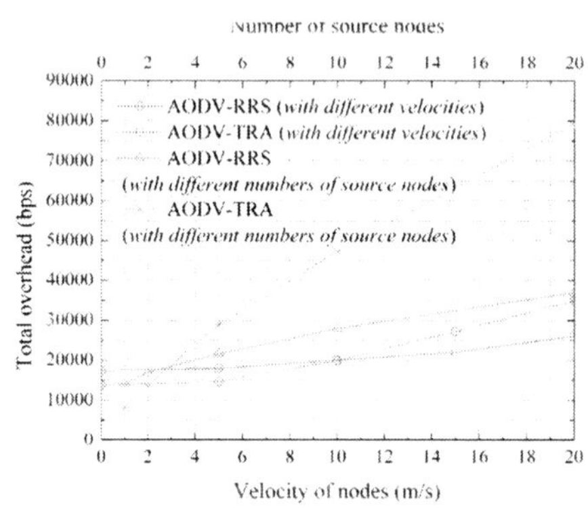

25%, and when there are 20 source nodes, it will exceed as much as 57%.

Impact on average end-to-end delay. In pervasive communication, voice service is quite common, which makes average end-to-end delay be concerned greatly. A series of similar results are obtained shown in Figure 11 to Figure 13. Comparison in Figure 11 denotes AODV-RRS is beneficial to develop real-time business. From Figure 12 we can see that when PDR is smaller than 20 packets/s, the average end-to-end delay with AODV-RRS is larger than that with AODV-TRA; and they are nearly the same when PDR is 20~25 packets/s; only when PDR is more than 25 packets/s, superiority of AODV-RRS can be embodied. These illuminate completely that AODV-

Figure 9. Impact on total overhead by PDR & disconnection rate

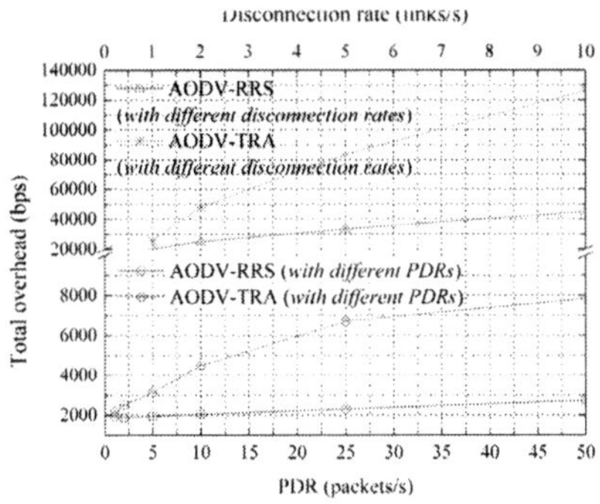

Figure 10. Impact on total overhead by number of nodes

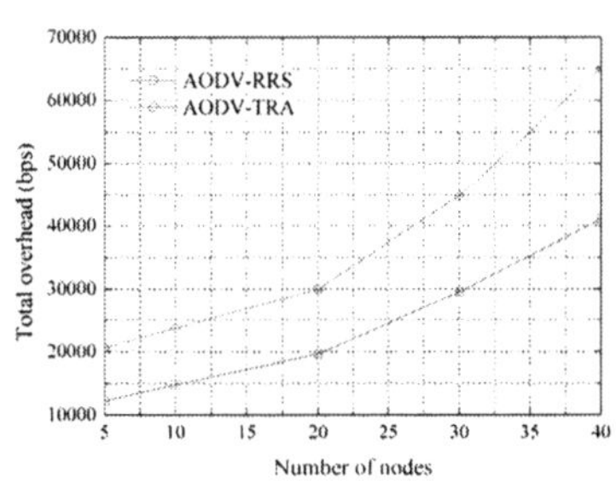

RRS possesses a better adaptability for network environment with high dynamic, high density, more source nodes, large data and more frequent disconnections. In view of Figure 13, the greater the density of nodes in network are, the heavier flooding the exploration with AODV-TRA will cause, and the larger the control overhead will be, which will cause information transmitting delay increasing. But for AODV-RRS, rise of number of nodes is not a serious factor for worsening the performance, because more nodes in pie slice region will make it easier for new route being established.

Figure 11. Impact on average end-to-end delay by velocity of nodes & number of source nodes

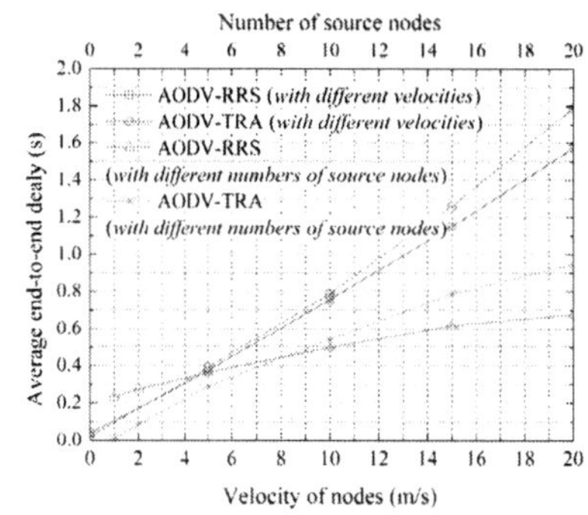

Figure 12. Impact on average end-to-end delay by PDR & disconnection rate

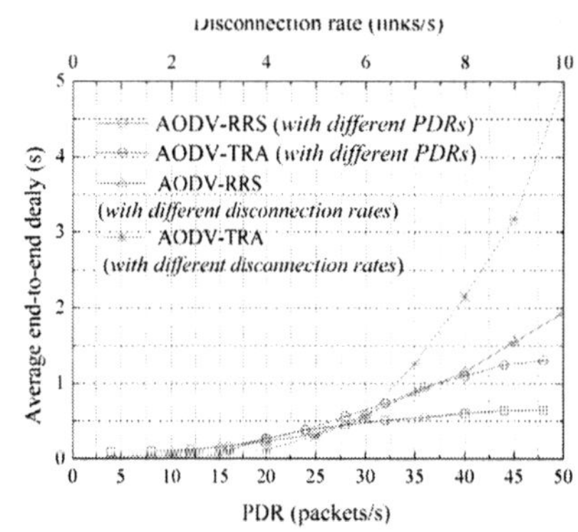

Analysis for Simulation Results

Impact on network communicating when current links disconnected artificially by AODV-RRS proposed in this paper has been strived to verified according to series of simulations, the results of which indicated a better adaptability to network performance with high dynamic, high density, multi source nodes, huge data quantity and frequent disconnections.

High dynamic means the network topology changing frequently. In simulation, the highest velocity is up to 20m/s, that is 48km/h, which is normal for urban traffics and also the maximum value of M tanks of USA. Therefore, the simula-

Figure 13. Impact on average end-to-end delay by number of nodes

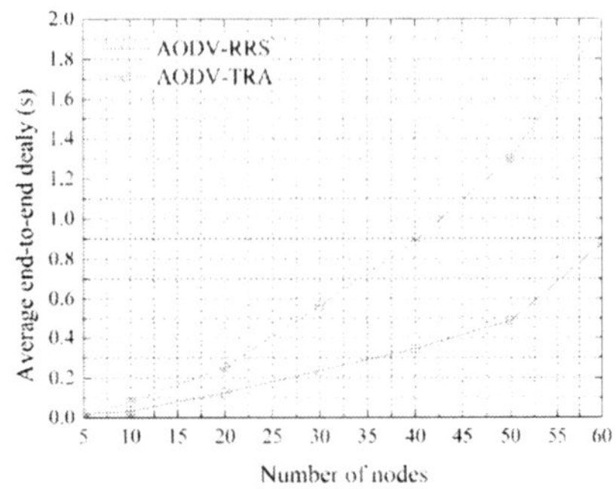

tion has practical significance no matter in MAN or battlefield communication environment. It has been shown that in such speed, transmitting data packets successfully delivery rate has been increased by 15.82%, and the total overhead and average end-to-end delay have been reduced by 25% and 13.8% respectively based on AODV-RRS.

High density indicates many more nodes. Compared with AODV-TRA, more nodes there are in network, more bandwidth will be taken up by flood broadcasting in route reconstruction, which will cause network performance descending rapidly. Statistics results from simulation denote that when there are more than 20 nodes in an area of 1000m×1000m, PSDR with AODV-RRS is higher than that of AODV-TRA. The total overhead reduced by 41.17% in the highest density situation, and the average end-to-end delay reduced by as mush as 55.71%, which widens the road for developing real-time traffic.

Multi source nodes expresses that many nodes are allowed to initiate data transmitting request simultaneously and could be responded, that is to say, once having burst data to send, any node in MANET will have a much larger probability to find the destination and achieve a successful transmission. The results show a much more gently changing tendency in curves of AODV-RRS than that of AODV-TRA, therefore, more benefits will be obtained in the occasion with more source nodes or potential ones.

Huge data quantity describes a condition that information or traffic need be sent is huge, which will cause congestion obviously, and affect network performance severely. However, a nicer performance has been presented based on AODV-RRS even in such condition with huge data quantity. With packets delivery rate rising up, most part of control overhead comes from RREQ for AODV-TRA, while adopting local directional exploring strategy, AODV-RRS limits too much RREQ flood so as to alleviate congestion. Total overhead increasing tendency of AODV-RRS is much gentler with PDR as 65.14% lower than that of AODV-TRA. When PDR is larger than 10packets/s, PSDR has increased by 10.31%, and larger than 20packets/s, average end-to-end has reduced by 50.92%.

Frequent disconnection means that the current link is broke off by factors such as noise, environmental interference et al, so as to make link invalid. In actual communication case, links disconnection rate is not as high as in simulation. The reason it has been set so is to verify performance of AODV-RRS in unstable network by artificially breaking off. The results show that with AODV-RRS PSDR has been increased by 47.43%, total overhead and average end-to-end delay have been reduced by 47.43% and 64.39% respectively.

CONCLUSION

As a distributed network, MANET does not depend on any infrastructure. Mobility of nodes, interference of channel and energy consumption make it great challenge for route protocol design. Aimed at negative impact of unstable route on communicating quality, an effective route reconstructing model is established in this paper in order to avoid network structural damage caused by mobile nodes, change of RF transmitting condition and nodes being destroyed *et al*, reducing energy consumption of frequent route discovery to realize survivability of MANET and usability of the route. The model based on motion state and exploring process is an optimal searching model according to linear programming in nature. Considering different states of nodes in practical network, the optimal exploring equations have been set up for static target nodes, determined mobile nodes and randomly mobile nodes. By defining Hamiltonian, the original equations are transformed into a parabolic partial differential equation and a first-order nonlinear partial differential equation, which can be solved by radial method. To realize optimal exploration in practical communication, a reconstructing model is proposed in this paper by restricting exploring scale into a pie slice pointing to destination nodes. When disconnection happens, a reconstructing node starts local route repair by broadcasting RREQ containing a field angle, rather than informing source node as usual. Node that received RREQ would calculate to determine whether it lies in forward area, if it does, goes on forwarding, or else, discards. An optimization procedure is designed to guide selection of the final route against the condition that more than one links existing in route reconstructed scale.

Network communicating performance has been simulated on NS-2 to verify reconstructing capability by disconnecting certain number of links artificially. Compared with traditional AODV in packets successfully delivering rate, total overhead and average end-to-end delay, an obvious predominance is obtained by AODV with route reconstructing scheme, especially in environment with high moving velocity, large number of source nodes, high nodes density, huge data quantity and frequent link disconnections.

There is a close relationship between MANET performance and the way of reconstruction after links broke out, which has already been proved in simulation. The scheme of restricting route request into a pie slice scale is a realization of optimal exploration for dynamical nodes, which is not proposed only for AODV. Therefore, universality of this route reconstructing algorithm, the exact energy consumption brought in by calculating, and how to obtain a suitable field angle for flooding in practical communication as well will be focused on in next phase work.

REFERENCES

Athans, M., & Falb, P. L. (1966). *Optimal Control*. McGraw-Hill, Inc.

Cheikh, S., Cluade, C., Guillaume, C., & Guerin, L. I. (2008). Bandwidth estimation for IEEE 802.11-based ad hoc networks. *IEEE Transactions on Mobile Computing*, *7*(10), 1228–1241.. doi:10.1109/TMC.2008.41

Courant, R., & Hilbert, D. (1962). *Method of Mathematical Physics, 11*, 167-180. New York: Wiley.

Dimitrakakis, C. (2006). Nearly optimal exploration- exploitation decision thresholds. *IDIAP Research Report* (IDIAP-RR-06-12). *Artificial Intelligence*, 1–10.

Esa, J., & Virtamo, J. (2007). Random waypoint mobility model in cellular networks. *Wireless Networks*, *13*(2), 177–188..doi:10.1007/s11276-006-4600-3

Forde, L., & Doyle, L. (2006). Ad hoc Innovation: Distributed Decision Making in Ad Hoc Networks. *IEEE Communications Magazine*, *44*(4), 131–137. doi:10.1109/MCOM.2006.1632660

Fornber, B., & Whitham, G. B. (1978). A numerical and theoretical study of certain non-linear wave phenomena. *Phil. Trans. R. Soc London, Series A*, *289*, 373–404.

Frodigh, M., Johansson, P., & Larsson, P. (2000). Wireless *ad hoc* networking—The art of networking without a network. *Ericsson Review*, *4*, 248–263.

Grindrod, P. (1991). *Patterns and Waves: The Theory and Applications of Reaction-Diffusion Equations* (p. 177). Oxford: Clarendon Press.

Groot, M. H. (1970). *Optimal Statistical Decisions*. New York: McGraw-Hill.

Harbin Institute of Technology (2006). *Almost Periodic Type and Remotely Almost Periodic Type Solutions to Parabolic Inverse Problems* [Doctor]. Yang, F. L.

Hasib, A., & Fapojuwo, A. O. (2008). A mobility model for heterogeneous wireless networks. *2008 IEEE Radio and Wireless Symposium,* pp. 815-818. doi:10.1109/RWS.2008. 4463617

He, Y. N. (2005). Taylor Expansion Method for Nonlinear Evolution Equations. *Applied Mathematics and Mechanics, 26*(4), 522–529. doi:10.1007/BF02465392

Hean, K. O., Hean, L. O., Tee, E. R., & Sureswaran, R. (2006). Scalability study of ad-hoc wireless mobile network routing protocol in sparse and dense networks. *Proc. of DFMA Int. Conf. on Distributed Frameworks for Multimedia Applications* (pp. 29-36). doi:10.1109/DFMA.2006.296891

Huang, D. S., & Zhao, W. B. (2002). *A Comprehensive Understanding for Radial Basis Probabilistic Neural Networks* (pp. 1237–1241).

Kwan, Julian, S. H., & Sun, H. W. (2006). An improved landslide mobility model. *Canadian Geotechnical Journal, 43*(5), 531–539..doi:10.1139/T06-010

Lee, S. J., Youn, J. S., & Jung, S. H. Lee, J., & Kang, C. H. (2005). Selective route-request scheme for on-demand routing protocols in ad hoc networks. *Lecture Notes in Computer Science, v 3738 LNCS, Ad-Hoc, Mobile, and Wireless Networks-4th International Conference, ADHOC-NOW* (pp. 153-163).

Li, C. M. (2001). Random exploration and optimal exploration. *Military Operations Research and Systems Engineering, 2*(4), 10–20.

Li, L. L., & Zhou, D. H. (2004). Robust Fault Diagnosis of Nonlinear System Based on Analytical Models: a Surevey. *Information and Control, 33*(4), 451–456.

Lu, S. Q., Zhu, X. Q., Zhao, W. Q., & Bao, C. G. (1981). Eikonal-Born Expansion And Its Application In Large-Angle Scattering. *Physica Energiae Fortis Et Physica Nuclearis, 5*(1), 121–126.

Maxell, P., Seamone, B., & Verstraete, J. (2007). Independent dominating sets and Hamiltonian cycles. *Journal of Graph Theory, 54*(3), 233–244. doi:10.1002/jgt.20205

Ohsumi, A. (1986). Optimal Searching for a Markovian Target and Relation to Optimal Stochastic Control. *Theory and Applications of Nonlinear Control Systems,* CByrnes, A Lindquist North Holland, Amsterdam (pp. 569-583).

Perkins, C. E., Royer, E. B., & Das, S. (2003). *Ad hoc On-Demand Distance Vector (AODV) Network Working Group, Routing. Draft-ietf-manet-aodv-13.txt*. The Internet Society.

Radzievskij, V. G., & Ufaev, V. A. (2003). Primary signal processing in digit6al panoramic detectors-direction finders. *Radiotekhnika, 2*, 26–31.

Rickenbacn, P. V., Schmid, S., Wattenhofer, R., & Zollinger, A. (2005). A robust interference model for wireless ad-hoc networks. *Proceedings of the 19th IEEE International Parallel and Distributed Processing Symposium* (pp. 1-8). doi:10.1109/IPDPS.2005.65

Sanchez, M. (2006). Mobility Models. http://www.dis-ca.upv.es/misan/mobmaodel.htm

Shi, H. L., Deng, W., & Qi, L. (2007). Properties of maximal mathematical expectation. *Journal of Shandong University. Natural Science, 42*(2), 92–94.

The Network Simulator-ns-2. (*n. d.*) Retrieved June 20, 2006, from http://www.isi.edu/nsnam/ns/

Wang, W., & Zeng, G. S. (2007). Trusted dynamic level scheduling based on Bayes trust model. *Science in China (Ser F-:Inf Sci), 50*(3), 456-469

Wisitpongphan, N., & Tonquz, O. K. (2006). Scalable broadcast strategies for ad hoc routing protocols. *2006 1st International Symposium on Wireless Pervasive Computing,* pp. 1-6.

Xi'an University of Electronic Science and Technology. (2006). *Schema Theory and Convergence Theory for Genetic Algorithms* [Doctor]. Xi'an: Wang, Y. P.

Zhang, Z., CI, Y. W. & Yang, X. Z. (2005). Design of wearable computer software platform applying to multimedia information collection based on ad hoc network. *Computer Engineering, 31*(21), 173–175.

Zhu, Q. X. (2005). Optimal Search Theory and Some Applications. *World Sci-tech R & D, 27*(4), 39-49.

Zhu, X. G., Hong, B. R., & Wang, D. M. (1999). Implementation of Time\Scale Transformation Based on Continuous Wavelet Theory. *Systems Engineering and Electronics, 21*(12), 11–13.

This work was previously published in International Journal of Mobile Computing and Multimedia Communications (IJMCMC) 1(2), edited by Ismail Khalil & Edgar Weippl, pp. 55-77, copyright 2009 by IGI Publishing (an imprint of IGI Global).

Chapter 6
Buffer Management in Cellular IP Network using PSO

Mohammad Anbar
Jawaharlal Nehru University, New Delhi

Deo Prakash Vidyarthi
Jawaharlal Nehru University, New Delhi

ABSTRACT

Cellular IP networks deal with the concepts of micro-mobility. Buffer management in Cellular IP networks is very crucial as its proper usage not only increases the throughput of the network but also results in the reduction of the call drops. This article proposes a model for buffer management in Cellular IP network using Particle Swarm Optimization (PSO), an evolutionary computational method often used to solve hard problems. The model considers two kinds of buffers; Gateway buffer and Base Station buffer. In the proposed two-tier model, the first tier applies a prioritization algorithm for prioritizing real-time packets in the buffer. In the second tier PSO algorithm is used on a swarm of cells in the network. PSO is applied for a given time slot, called window. In each window period the swarm can store number of packets depending on the window size and the total number of packets. The effect of various parameters e.g. number of packets, size of packets, window size, and a threshold value on buffer utilization has been studied by conducting the simulation experiments.

INTRODUCTION

Cellular IP protocol is a solution for mobility management at micro level, where it intends to provide handoff support besides the micro mobility management. Being designed for micro level of mobility, Cellular IP supports frequently moving mobile hosts. It can also efficiently serve rarely moving or even stationary mobile hosts. A Cellular IP network consists of interconnected nodes. These nodes accomplice two tasks. They route IP packets inside the Cellular IP network, on the other hand they communicate with mobile hosts via wireless interface. Components of Cellular

DOI: 10.4018/978-1-60960-563-6.ch006

IP network are Cellular IP Base Station which controls and is responsible for nodes in a cell, Cellular IP Gateway that connects Cellular IP network to Internet and stores the packets intended to be sent through Internet in its buffer, and the Cellular IP Mobile Node which implements the Cellular IP protocol. Many operations take place in Cellular IP network such as routing, paging, and handoff etc. Quality of Service (QoS) in Cellular IP network depends on many factors and is provided according to the resources available in the network. Buffer, as an example, is one of the important resources in Cellular IP networks which must be managed and utilized in an effective manner so that a Cellular IP network gives better performance.

Considering two types of traffic, real-time and non real-time, there should be good schemes for buffer management by which least packets loss is guaranteed. In wireless communications packets are prone to be dropped, lost, or delayed which results in connection disturbance. Proposed schemes for buffer management in wireless communication networks especially in Cellular IP networks must be as robust as possible, and give good results in terms of QoS.

One of the tools that can be used to optimize the performance of the buffer is Particle Swarm Optimization (PSO) algorithm. PSO is a swarm intelligence based algorithm and is often used to find solution to an optimization problem in a search space. This algorithm is based on the sociological behavior associated with birds flocking (Nedjah & Mourelle, 2006). PSO has been applied in many optimization applications and has given good results so far. As an example PSO algorithm is used for bandwidth allocation in cellular networks (Huang, Chuang, Lai, Sun, & Guan, 2007). PSO algorithm can be summarized as follows (Seo, Im, Heo, Kim, Jung, & Lee, 2006).

```
PSO ()
{   Initialize random swarm location
and velocity.
   Repeat until the optimum value is
obtained
   {     Update velocity.
         Update positions.
         Update Pbest, (best position
for each particle)
         Update Gbest (the best posi-
tion for the swarm so far)
   }
```

For the buffer management problem of Cellular IP network, a two tier model is being proposed. In the first part it prioritizes the real-time packets in the buffer and arranges them in the front portion of the buffer. It uses an algorithm to search and sort these packets. After that the packets are served from the front end of the buffer till a specified threshold is met. Rest of the packets will not be dropped straightforward but they will be transferred back to the swarm consisting of seven cells in the Cellular IP network. The swarm will search for the best solution to store all the returned packets from the Gateway buffer. Eventually, the packets will be transferred to the Gateway buffer with higher priority. Most of the returned packets to the swarm are normally (most probably) non real-time packets. It is due to keeping the real-time packets in the front side of the buffer and serving them. Therefore the non real-time packets, considered to be less important, might be delayed.

The rest of the article is organized as follows. In the next section related models have been described. In section 3 Particle Swarm Optimization (PSO) is explained step by step. Section 4 briefly explains buffer types in Cellular IP networks. In section 5, the proposed model is elaborated. Section 6 contains the simulation experiments for the performance of the model and its comparison with the buffer management for Intelligent Gateway Gadget. Some observations have been made on the results obtained and the comparison between the two models in section 7. Section 8 concludes the proposed work.

RELATED WORK

Quite a few works has been done for buffer management in Cellular IP network. (James, Hou, & Ho, 2007) proposed an application-driven MAC-layer buffer management framework. It is based on active dropping (AD) mechanism for real-time video streaming in IEEE 802.16 point-to-multipoint (PMP) networks. In this scheme the basic idea is that MAC-layer protocol data units (MPDUs) of video streams could be actively dropped at the Base Station (BS) if the corresponding frame is not with a sufficient confidence to be successfully delivered to the recipient within its application layer delay bound.

(Lim & Qiu, 2001) mentioned a predictive scheme for Buffer Management in ATM networks based on Fuzzy Logic. This scheme is referred to as Fuzzy Logic Selective Cell Drop (FSCD). The basic idea of this scheme is its ability to accept or drop new incoming packets dynamically based on the predicted future buffer condition in the switch. In this scheme Fuzzy Logic prediction had been used for the production of a drop factor and then based on this factor, the packet dropping decision is taken.

A scheme for Buffer Management based on RED algorithm for Intelligent Gateway in wireless LAN is presented by (Noh, & Bae, 2007). This scheme is applicable in MAC layer of Gateway Gadget and depends on Early Random Discard (RED) algorithm. Simulation for this scheme has been done and the results show the dropping rate with different number of thresholds.

Buffer Management for real-time traffic over wireless networks based on Random Early Expiration Detection is studied by (Chen, & Lemin, 2005). The basic idea behind this scheme is to predict whether new arrived packet can reach to the receiver under the limit of maximal delay bound, and randomly discard the packet based on estimate result so as to prevent invalid transmission and at the meantime to minimize queue delay and to reduce expiration probability of following packets.

(Ryu, & Koh, 1996) discussed dynamic Buffer Management Technique for minimizing the necessary buffer space in continuous media server depending on window principle where the service duration of clients is divided into several small time intervals called windows and reallocates the buffer space in every window period. In this scheme the effect of varying the window size on average necessary buffer size was studied and analyzed and also the effect of varying the window size on average buffer utilization.

PARTICLE SWARM OPTIMIZATION (PSO)

PSO is an innovative computational way to solve hard problems. It is inspired by the behavior of insects or fish or birds such as fish schooling and bird flocking evident from Figure 1. In this method, if one of the particles (fish or bird) discovers a good path to food, the rest of the swarm follows it even if they are far away in the swarm. The basic idea of PSO algorithm can be explained this way. Suppose that swarm of birds is randomly searching food in an area. There is only one piece of food in the area being searched. All the birds don't know where the food is, but they know how far the food is. So the best strategy to find the food is to follow the bird which is nearest to the food. It means that this bird is the best solution for this problem among all the birds in the swarm. There

Figure 1. An example of swarm (birds flocking)

are two types of best values in PSO algorithm: *Pbest* which is the best position for each particle in the swarm and must be updated depending on the fitness value for each particle. The second best value is *Gbest* which is the global best value for the swarm in general. This value must be checked with each iteration and be changed by the best Pbest if this Pbest for the current iteration is better than the best Pbest for the previous iteration. Various steps of PSO algorithm can be listed as follows (Seo, Im, Heo, Kim, Jung, & Lee, 2006).

Step 1: Initialize Random Swarm Location and Velocity

Initial locations and velocities for all particles are generated randomly. Current position for each particle is considered as Pbest. Gbest is selected as the best Pbest among all the Pbests in the swarm.

Step 2: Update Velocity

For each particle we must update the particle velocity as in equation 1.

$$V_j^{k+1} = w.V_j^k + c1.r1.(P_{best_j}^k - X_j^k) + c2.r2.(G_{best}^k - X_j^k) \quad (1)$$

Where

- w is inertia weight and is varied from 0.9 till 0.4.
- r1, r2 are random numbers between 0 and 1
- c1, c2 are acceleration factors that determine the relative pull for each particle toward Pbest and Gbest and usually c1, c2 = 2.
- V_j^{k+1} is the velocity of particle (j) in iteration k+1.
- V_j^k is the velocity of particle (j) in iteration k.
- $P_{best_j}^k$ is the best position of particle(j)so far.
- X_j^k is the current position of particle (j) in iteration k.
- G_{best}^k is the global best position of the swarm so far.

Step 3: Update the Position

Each particle modifies its position according to the equation 2.

$$X_j^{k+1} = X_j^k + V_j^k.\Delta t \quad (2)$$

Where,

- V_j^k is the velocity of particle j in iteration k.
- Δt is a time step and usually 1.

Step 4: Update Pbests

The fitness function must be evaluated for each particle and if the fitness function for a particle (j) at iteration (k) is better than fitness function for this particle in the previous best position then the best position for this particle must be the current position for it.

If (fitness value (Pbest) < fitness value (P)) then Pbest =P.

Step 5: Update Gbest

For all the particles of the swarm if the fitness value at any Pbest of any particle is better than this value at previous Gbest, at this time Gbest must be updated and it will be Pbest of the particle which has the best fitness value at its Pbest.

If(fitness value($P_{best\,j}^{\,k}$)> fitness value($G_{best}^{\,k-1}$) then $G_{best}^{\,k-1} = P_{best\,j}^{\,k}$.

Step 6: Loop Iteration

Repeat steps from 2 till 5 until convergence.

Fitness function in PSO algorithm is a function against which the fitness of each individual in the swarm is tested. It is designed based on the problem and the parameter to be optimized.

PSO has many applications in many fields and can be applied to solve many optimization problems (Nedjah & Mourelle, 2006).

BUFFER IN CELLULAR IP NETWORKS

As mentioned earlier, there are three main components in Cellular IP networks. Based on the components, there are three kinds of buffers also, that are mobile node buffer, base station buffer and gateway buffer. Mobile node buffer is one in which the packets are arranged to be sent to the Base Station of the cell in which the Mobile Node is lying. Each Base Station of the network has its own buffer which is considered as a part of the total buffer of the swarm to which this Base Station belongs. Third kind of buffer is the buffer of the Gateway that, as mentioned before, connects Cellular IP network to Internet and through which the packets are forwarded from Cellular IP network to correspondent nodes in another network. Gateway buffer has importance in Cellular IP network where all the packets will be stored before further transmission through Internet. Buffer Management can be done at two levels which are Base Station's buffer and Gateway's buffer. Base Station's buffer can be managed by grouping the cells into swarms and apply Particle Swarm Optimization algorithm in order that the swarm will be able to avail buffer space for storing the incoming packets. Gateway's buffer can be managed by applying prioritization algorithm on the stored packets in this buffer. This algorithm gives more priority to the real-time packets rather than non-real time packets.

THE MODEL

The proposed model, in this article, considers that there is a Cellular IP network connecting to Internet through Gateway. Cells in Cellular IP networks are grouped into swarms and each swarm consists of seven cells as shown in Figure 2.

Packets are sent from Mobile Nodes in the network to Base Stations and later forwarded to the Gateway to be sent through Internet to the correspondent nodes. The model also considers that there are two types of traffic and therefore, two types of packets: real-time packets and non real-time packets. The model is divided in two tiers.

Tier One (Prioritization of Real-Time Packets in Gateway Buffer)

This part considers the arrived packets to the Gateway buffer where the packets are stored and eventually served from the buffer. The model refers to the real-time packets as (1) when stored in the buffer and to the non real-time packets as (0). The Gateway buffer is shown in Figure 3.

Figure 2. The system model

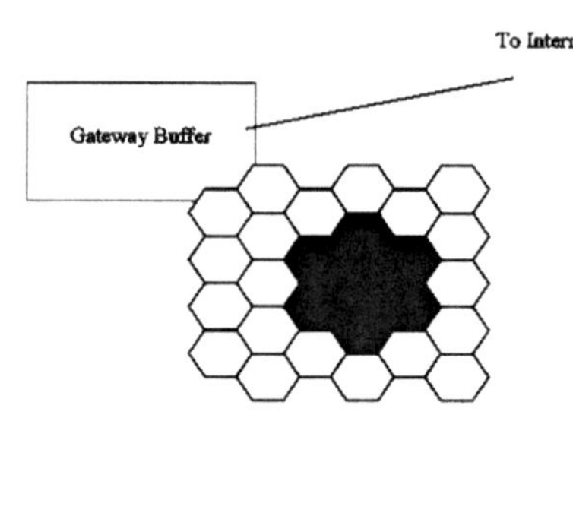

Real-time packets and non real-time packets arrive randomly to the buffer and so kept randomly as shown in Figure 4.

In tier one a prioritization algorithm is applied in which the buffer is searched for the real-time packets. These real-time packets are arranged in the front positions of the buffer as shown in Figure 5.

After this, packets' departure from the buffer will take place until a specific threshold is met. The packets after this threshold will not depart from the buffer as discussed in the example shown in Figure 6. The threshold is randomly generated.

Part one of this model shows the difference between Connection Completion Probability (CCP) for real-time packets and non real-time packets before and after prioritization. Connection Completion Probability (CCP), in general, is defined as follows.

$$CCP = \frac{Number\ of\ served\ packets}{Total\ number\ of\ packets}$$

The packets that haven't been served from the buffer won't be dropped from the buffer as doing so will cause connection termination for the connections to which these packets belong to. These packets will be returned back to the Cellular IP network.

Figure 3. Gateway buffer as considered in the model

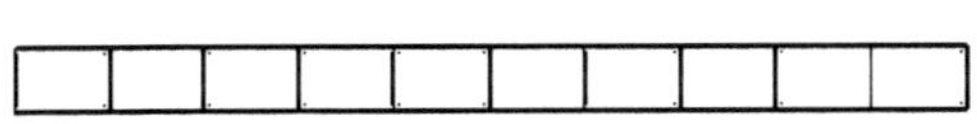

Figure 4. Mixed packet arrangement in the buffer

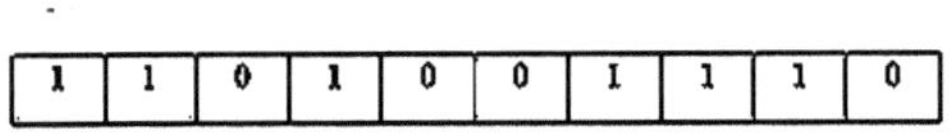

Tier Two (PSO Algorithm)

Packets that have been returned to the network will arrive to a swarm, consisting of seven cells, in Cellular IP network. Swarm will execute PSO algorithm to avail buffer spaces in Base Stations buffers to store the coming packets and later on they will be returned to the buffer of the Gateway with high priority.

PSO algorithm is executed for three cases: in first case the swarm will be initialized with zero buffer spaces for all particles (base stations) i.e. there is no available buffer space in the swarm. In the second case the swarm will be initialized with some of the buffer spaces to be zero and the others are non-zero. In the third case the swarm will be initialized with non-zero buffer space for all particles (Base Stations). PSO algorithm is executed in every time period called window. Window size is varied to study the effect of the window size on the number of dropped packets. Number of windows will depend on window size and the number of returned packets.

EXPERIMENT

Experiments have been conducted to study the effect of various related parameters on Connec-

Figure 5. Arrangements of packets after prioritization

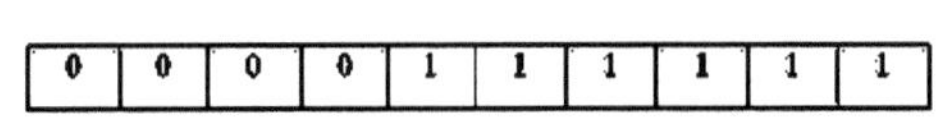

Figure 6. Serving packets till the specified threshold TH

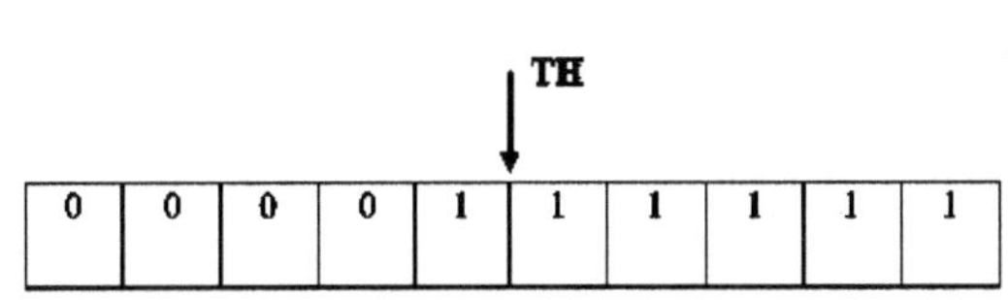

tion Completion Probability (CCP), dropped packets, buffer utilization and starvation. These experiments have been presented in the following subsections in the graphs.

Effect of the Number of Packets with Fixed Threshold in the Buffer

In this experiment number of packets is being changed from 30 till 80 packets while the threshold (TH) is fixed and is equal to 15 (i.e. packets will be served from the buffer till packet number 15). The effect of number of packets on Connection Completion Probability (CCP) is observed before and after prioritization of packets as mentioned in tier one. The graph is shown in Figure 7.

$$CCP = \frac{Number\ of\ served\ real-time\ packets}{Total\ Number\ of\ real-time\ packets} \quad (3)$$

Effect of Various Thresholds on CCP

In this experiment the threshold values have been varied from 20 to 70. The number of packets is 100 for each experiment. This is also done before and after prioritization. The graph is shown in Figure 8.

CCP for Different Number of Packets with Varying Threshold

For different number of packets with varying threshold CCP has been observed. This is done before prioritization of the packets. The graph is shown in Figure 9.

CCP Before and After Prioritization for Three Different Numbers of Packets

Figures 10, 11 and 12 show the CCP before and after prioritization with number of packets to be 100, 150 and 200 respectively. X-axis represents the various threshold values.

Figure 7. CCP with fixed threshold

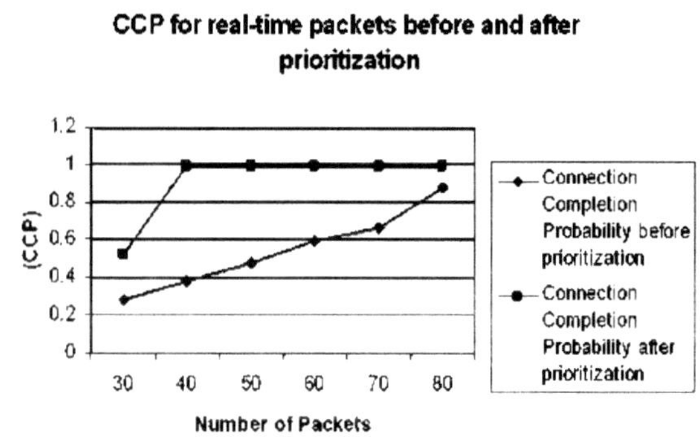

Figure 8. Effect of threshold on CCP

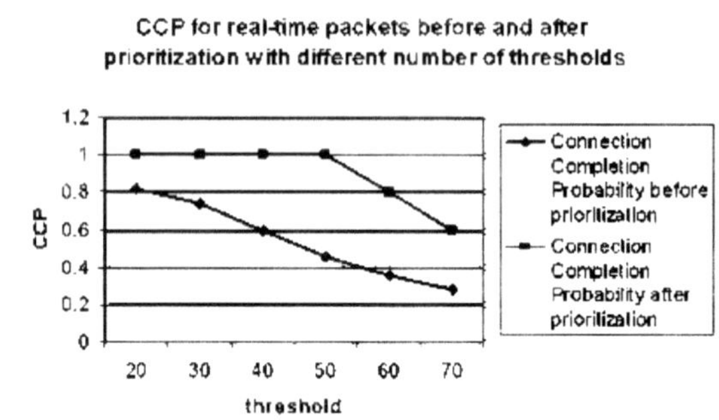

Figure 9. CCP before prioritization for different number of packets

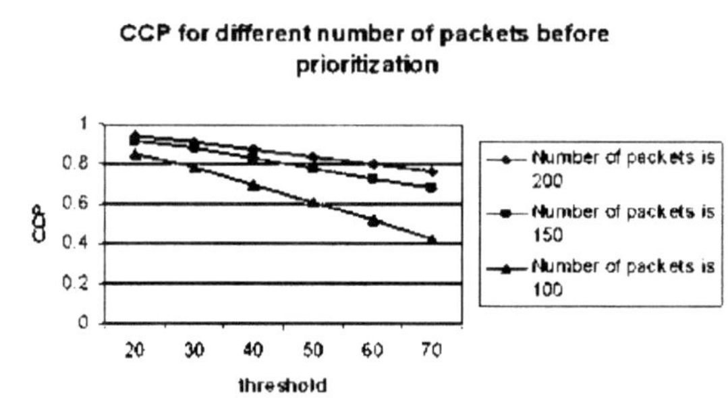

Figure 10. CCP before and after prioritization for 100 packets

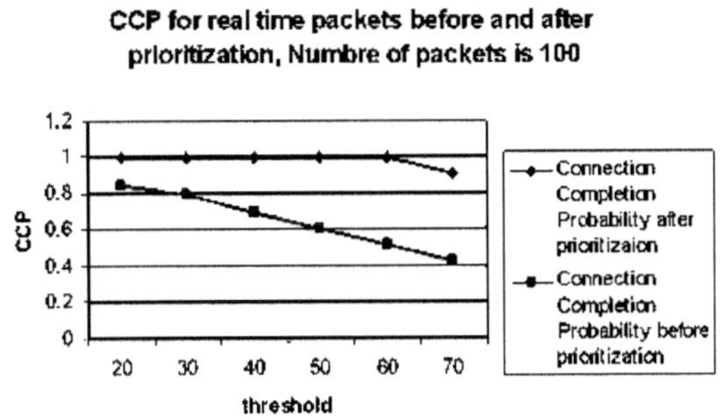

Figure 11. CCP before and after prioritization for 150 packets

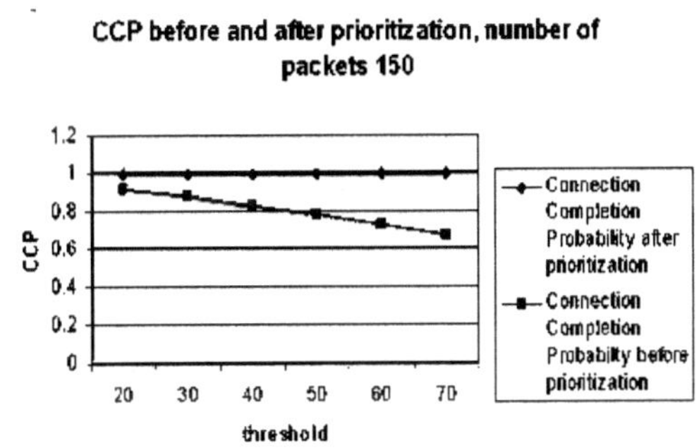

Figure 12. CCP before and after prioritization for 200 packets

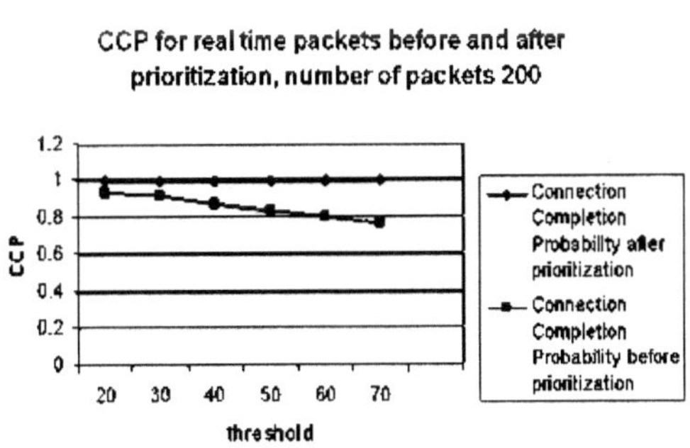

Effect of Packet Size on the Dropped Packets

This experiment is carried out for different number of packets returned form the Gateway buffer to the network and with three different sizes of packet. Window size for the whole experiment is 10 packets. The graphs in Figure 13 show comparison between these three cases. The x-axis represents the number of returned packets to the network.

Figure 13. Dropped packets for different packet sizes

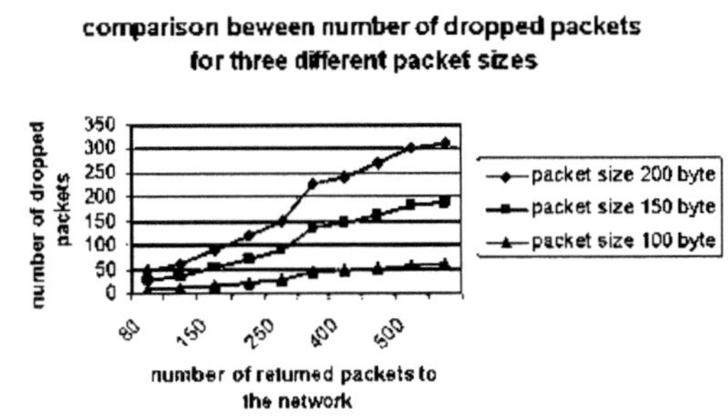

Effect of Window Size on the Dropped Packets

The experiment is conducted to study the effect of window size on the dropped packets. For this number of packets are generated randomly. In this experiment the number of packets generated is 86, 173, 259 with the packet size of 300 byte. Figure 14 shows comparison between these three cases.

Study of Overhead with Varying Window Size

Experiment is conducted to study the overhead from increasing window size. This experiment is conducted when the swarm has been initialized with zero buffer space. The number of generated packets is 200 for all the experiments here.

The observation is done on the overhead involved in sense of buffer starvation as evident from Figure 15. Buffer starvation is defined as follows.

$$Buffer\ Starvation\ Ratio = \frac{\text{Required } buffer\ size}{Available\ buffer\ size}$$

Figure 14. Dropped packets for different number of packets

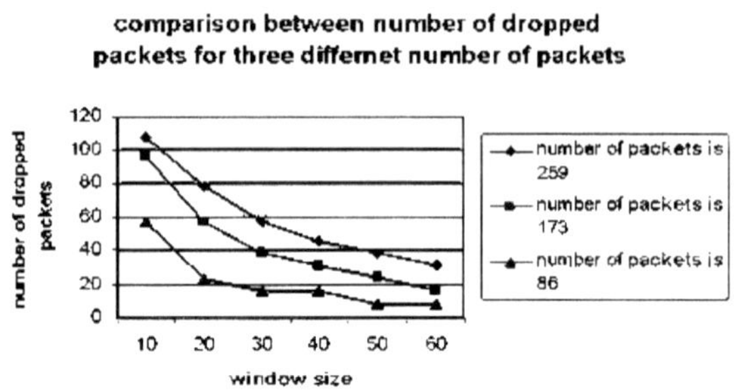

Figure 15. Overhead of buffer starvation from varying window size

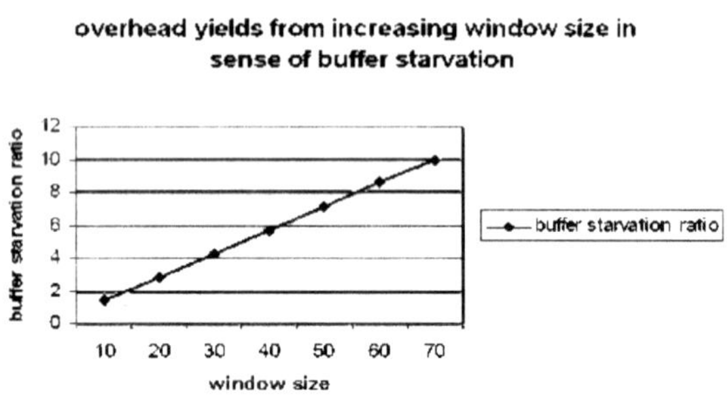

Figure 16. Overhead of dropped packets from varying window size

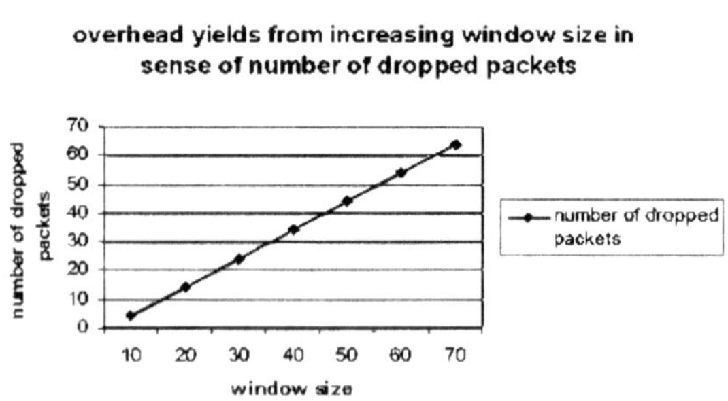

b. Again the overhead involved in sense of number of dropped packets has been studied as shown in Figure 16.

Buffer Utilization

After the swarm could avail these amounts of buffer space to store the sent packets from the Gateway buffer, another issue is how to utilize this available buffer space. For that, these experiments show the change of buffer utilization with the available buffer space and packet size. We observed the buffer utilization also by the experiment. Buffer Utilization is defined as follows.

$$Buffer\ Utilization(BU) = \frac{Used\ amount\ of\ buffer}{Available\ buffer\ space} \quad (5)$$

The experiment is conducted for three different packets sizes 100, 150 and 200 bytes. Figure 17 shows a comparison between all these three cases.

Comparison with RED model

Experiment is conducted to compare the proposed model with that of a QoS RED Buffer Management Scheme of Intelligent Gateway Gadget based on Wireless LAN. The result is shown in Figure 18. The inputs are in conformity with the RED model.

OBSERVATIONS

The following observations have been derived from the experiments conducted on the model.

1. As the packets arrive to the buffer of the Gateway randomly, it is observed when the number of packets is less, the Connection Completion Probability (CCP) is less and when the number of packets increases, the CCP also increases. Further, as the real-time packets being served are more there is an increase in CCP. After prioritization, CCP increases and reaches 100% due to serving all real-time packets from the front positions of the buffer. From this experiment,

Figure 17. Buffer utilization for three different packet sizes

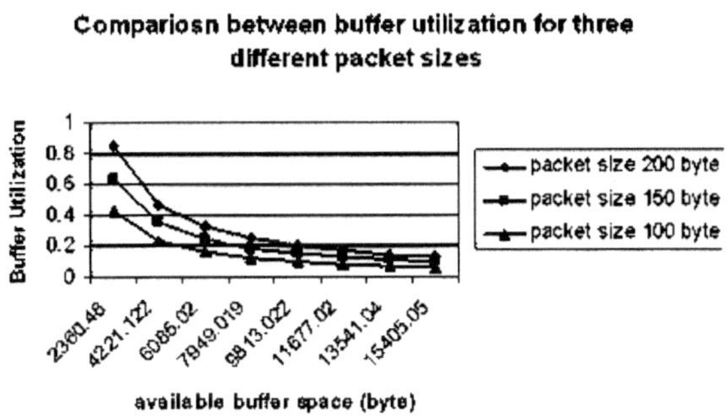

Figure 18. Comparison between the proposed and the RED model

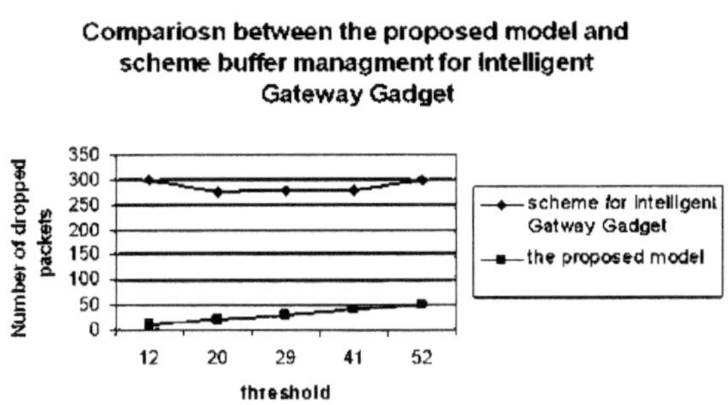

the conclusion is derived that prioritization ensures better CCP for any number of packets especially for bigger number of packets.

2. When the number of packets is fixed and threshold is varying, it is observed that when the threshold is small the number of served packets are more as evident from Figure 7; therefore, we conclude that bigger the threshold is, the less the number of served packets and the probability of completing the connection is.
3. When comparing CCP for three different numbers of packets it is clear that when number of packets is less then the number of served real-time packets may be less resulting in small CCP.
4. Comparing between CCP after and before prioritization for different number of packets (100,150,200) and different thresholds shows that CCP is (1) after prioritization when number of packets are 150 and 200 for the same distribution of real-time and non real-time packets in the buffer, this is with all thresholds, it refers that prioritization algorithm, when arranging the real-time packets in the front positions of the buffer, is helping in serving them out and improving CCP. Before prioritization the packets are randomly being served from the front position of the buffer; therefore, there is no priority for delay-sensitive packet (i.e. real-time packets) and then CCP is less.
5. Experiment of number of dropped packets was performed for different number of packets returned from the gateway buffer. By increasing packet size from 100 byte to 200 byte during window period of size 10 packets, there will be dropped packets due to less capacity within this specific window period.
6. Comparing number of dropped packets in cases: 100 byte, 150 byte, 200 byte in size shows that the bigger the packet sizes, the bigger the number of dropped packets is. This is due to unavailability of buffer space during a window period.
7. Generating number of packets randomly 86 packets, 173 packets and 259 packets with different window sizes to check the effect of window size on number of dropped packets shows that number of dropped packets decreases while increasing the window size. This is because of higher capacity to store more packets during window period. Increasing window size will result in an overhead as discussed. Comparing these three cases of the varying number of packets with different window sizes, the conclusion is that when the window size is big enough; the number of dropped packets is less.
8. Increasing window size for reducing number of dropped packets can't be done for the

desired size because it incurs an overhead in two respects; the first one in sense of number of dropped packets and the second is in sense of buffer starvation. Suppose number of packets is 100 for window period 1 and considering the first case of swarm initialization with packet size 10, the number of dropped packets is 4 within this window period. By increasing window size to 20 packets the number of dropped packets for the same window period is 14 packets. Moreover, the buffer space needed will be more. Therefore considering the buffer starvation ratio for window size 10 packets the ratio is 1.4299 and number of dropped packets is 4. For window size 20 packets the buffer starvation ratio is 2.8599 and number of dropped packets is 14. As mentioned before, this is during one window period.

9. Applying PSO algorithm, when the swarm avails the buffer space for the returned packets from the gateway buffer, after that lot of this space is used to store the coming packets; therefore, the buffer utilization is better. When packet size is bigger the buffer utilization is also better.
10. Comparison between the three different cases proves that bigger the packet size is better the buffer utilization is.
11. Comparison is done between the proposed model and another related model which is the QoS RED Buffer Management scheme of Intelligent Gateway Gadget based on a wireless LAN. This scheme (RED model) calculates the dropping rate with different number of thresholds. To be comparable with the proposed model, the dropping rate has been converted into the number of dropped packets. The comparison shows that the proposed model using PSO drops less number of returned packets in comparison to the RED model.

CONCLUSION

A model for Gateway Buffer Management in Cellular IP network has been proposed in this article. Buffer Management in this model is done in two tiers, the first is prioritization of real-time packets arriving to the buffer of the gateway and then calculating Connection Completion Probability (CCP) after and before the prioritization. It has been observed that with prioritization, CCP could be improved. Remaining packets in the buffer, as per the threshold value, are not dropped from the buffer; rather they are returned back to a swarm of cells in the Cellular IP network which executes PSO algorithm in order to avail buffer space for the returned packets. This algorithm has been performed in three cases with three different initialization of the swarm, in the second and the third cases the swarm could avail more buffer space for the returned packets; therefore, the buffer utilization has been improved. By comparing number of dropped packets with the same threshold between the proposed model and the QoS RED model, it has been observed that for the given packet size, the proposed model performs better in sense of number of dropped packets.

REFERENCES

Boukerche, A., & Jing, F. (2006, November 14-16). Buffer Management for 3D Image-based Rendering over Wireless Network with QoS Adaptation. *Proceedings of 31st IEEE Conference on Local Computer Networks* (pp. 55-62). Tampa, Florida, USA: IEEE Press.

Chen, Y., & Lemin, L. (2005, September 21-23). A random early expiration detection based buffer management algorithm for real-time traffic over wireless networks. *Proceedings of the IEEE International Conference on Computer and Information Technology (CIT 2005) (pp. 507–511). Shanghai, China: IEEE Press.*

Datta, A., Mukherjee, S., & Viguier, I. R. (1999). Buffer management in real-time active database systems. *IEEE Transactions on, Systems, Man and Cybernetics. Part A, 29*(2), 216–224.

Dianhui, W., Wong, A. K. Y., & Dillon, T. S. (2006, December 2-5). Heuristic rule based neuro-fuzzy approach for adaptive buffer management for Internet-based computing. *Proceedings of 31st IEEE Conference on Local Computer Networks, 1*, 296 – 299. Melbourne, Victoria, Australia: IEEE Press.

Harai, H., & Murata, M. (2006). High-speed buffer management for 40 GB/s-based photonic packet switches. *IEEE/ACM Transactions on Networking, 14*(1), 191–204. doi:10.1109/TNET.2005.863450

Hou, Y. T., Dapeng, W., Yao, J., & Takafumi, C. (2000, November 8-10). A core-stateless buffer management mechanism for differentiated services Internet. *Proceedings of 25th Annual IEEE Conference on Local Computer Networks (LCN 2000)*, (pp. 168 – 176). Tampa, Florida, USA: IEEE Press.

Huang, C. J., Chuang, Y. T., Lai, W. K., Sun, Y. H., & Guan, C. T. (2007). Adaptive resource reservation schemes for proportional DiffServ enabled fourth-generation mobile communications system. *Computer Communications Journal, 30*(7), 1613–1623. doi:10.1016/j.comcom.2007.01.015

Ippoliti, D., Xiaobo, Z., & Liqiang, Z. (2007, August 13-16). Packet Scheduling with Buffer Management for Fair Bandwidth Sharing and Delay Differentiation. *Proceedings of 16th International Conference on Computer Communications and Networks (ICCCN 2007), (pp. 569 - 574). Honolulu, Hawaii, USA: IEEE Press.*

James, S., Hou, F., & Ho, P. H. (2007, May, 21-23). An Application-Driven MAC-layer Buffer Management with Active Dropping for Real-time Video Streaming in 802.16 Networks. Proceedings of the IEEE International Conference on Advanced Information Networking and Applications (AINA '07), *(pp 451–458). Niagara Falls, Ontario, Canada: IEEE Press.*

Jardosh, S., Zunnun, N., Ranjan, P., & Srivastava, S. (2008, December 15-18). Effect of network coding on buffer management in wireless sensor network. *Proceedings of International Conference on Intelligent Sensors, Sensor Networks and Information Processing (ISSNIP 2008)*, (pp.157 - 162). Sydney, NSW, Australia: IEEE Press.

Junhua, T., Gang, F., Chee-Kheong, S., & Liren, Z. (2008). Providing Differentiated Services Over Shared Wireless Downlink Through Buffer Management. *IEEE Transactions on Vehicular Technology, 57*(1), 548–555. doi:10.1109/TVT.2007.905246

Kalil, M. A., Al-Mahdi, H., & Mitschele-Thiel, A. (2008, July 27-August 1). Performance Evaluation of Hop-Aware Buffer Management Scheme in Multihop Ad Hoc Networks. *Proceedings of the Fourth International Conference on Wireless and Mobile Communications (ICWMC '08)*, (pp. 31–36). Athens, Greece: IEEE Press.

Krachodnok, P., & Benjapolakul, W. (2001). Buffer management for TCP over GFR service in an ATM network. In *Proceedings of Ninth IEEE International Conference on Networks,* (pp. 302 - 307). Washington, DC, USA: IEEE Press.

Krifa, A., Baraka, C., & Spyropoulos, T. (2008). Optimal Buffer Management Policies for Delay Tolerant Networks. *Proceedings of 5th Annual IEEE Communications Society Conference on Sensor, Mesh and Ad Hoc Communications and Networks (SECON '08)*, (pp. 260-268). San Francisco, CA, USA: IEEE Press.

Li, M. (1993).Determination of optimal buffer sizes in static buffer management. *Proceedings of IEEE Region 10 Conference on Computer, Communication, Control and Power Engineering (TENCON '93)*, (pp. 479–486). Beijing, China: IEEE Press.

Lihua, S., Xiaotong, Z., Qin, W., & Yanfei, G. (2007). MYBUF: A High Performance Buffer Management Mechanism. *Proceedings of International Conference on Wireless Communications, Networking and Mobile Computing (WiCom 2007)*, (pp. 3007 - 3010). Shanghai, China: IEEE Press.

Lim, H. H., & Qiu, B. (2001). Predictive fuzzy logic buffer management for TCP/IP over ATM-UBR and ATM-ABR. *Proceedings of the IEEE International Conference on Global Telecommunications (GLOBECOM '01)*, (Vol. 4) (pp. 2326–2330). San Antonio, Texas, USA: IEEE Press.

Mishra, P., & Saksena, M. (1998). Designing buffer management policies at an IP/ATM gateway. In *Proceedings of IEEE ATM Workshop*, (pp.144–153). Fairfax, Virginia, USA: IEEE Press.

Nedjah, N., & Mourelle, L. D. M. (2006). *Swarm Intelligent Systems*. Berlin, Heidelberg: Springer-Verlag. doi:10.1007/978-3-540-33869-7

Noh, K. J., & Bae, C. S. (2007). A QoS RED Buffer Management Scheme of Intelligent Gateway Gadget based on a Wireless LAN. *Proceedings of the IEEE International Conference on Advanced Communication Technology (ICACT '07)*, (Vol.1) (pp. 286-291). Gangwon-Do, South Korea: IEEE Press.

Razouqi, Q., Seong-Soon, J., & Ghosh, S. (2000). Performance analysis of fuzzy thresholding-based buffer management for a large-scale cell-switching network. *IEEE Transactions on Fuzzy Systems*, *8*(4), 425–441. doi:10.1109/91.868949

Ryu, Y. S., & Koh, K. (1996). A dynamic buffer management technique for minimizing the necessary buffer space in a continuous media server. In *Proceedings of the third IEEE International Conference on Multimedia Computing and Systems (MMCS 1996)*, (pp. 181-185). Hiroshima, Japan: IEEE Press.

Seo, J. H., Im, C. H., Heo, C. G., Kim, J. K., Jung, H. K., & Lee, C. G. (2006). Multimodal function optimization based on particle swarm optimization. *IEEE Transactions on Magnetics*, *42*(4), 1095–1098. doi:10.1109/TMAG.2006.871568

Yerima, S. Y., & Al-Begain, K. (2007). Buffer Management for Multimedia QoS Control over HSDPA Downlink. *Proceedings of 21st International Conference on Advanced Information Networking and Applications Workshops (AINAW '07)*, (Vol.1) (pp. 912 – 917). Niagara Falls, Ontario, Canada: IEEE Press.

Yousefi'zadeh, H., & Jonckheere, E. A. (2005). Dynamic neural-based buffer management for queuing systems with self-similar characteristics. *IEEE Transactions on Neural Networks*, *16*(5), 1163–1173. doi:10.1109/TNN.2005.853417

This work was previously published in International Journal of Mobile Computing and Multimedia Communications (IJMCMC) 1(3), edited by Ismail Khalil & Edgar Weippl, pp. 78-93, copyright 2009 by IGI Publishing (an imprint of IGI Global).

Chapter 7
Throughput Optimization of Cooperative Teleoperated UGV Network

Ibrahim Y. Abualhaol
Broadcom Corporation, USA

Mustafa M. Matalgah
The University of Mississippi, USA

ABSTRACT

Cooperative communications among group of teleoperated unmanned ground vehicles (UGVs) allows to exploit spatial diversity in wireless fading channels by relaying signals between each other. Due to the high speed of the UGVs, the nature of the channel environments and the possible co-channel interference, the effect of multipath propagation and the Doppler spread are more pronounced. In this article, we proposed a low complexity dynamic channel assignment (DCA) technique with adaptive modulation and coding (AMC) strategy to allocate the available bandwidth over a number of communications links in a cooperative UGV network. In many processing algorithms and transmission protocols reported in the literature, performance improvement in terms of system throughput and reliability has been demonstrated. The proposed DCA with AMC in a cooperative UGV network has two objectives. First, to maximize the overall throughput of the cooperative UGV network and second, to significantly reduce the probability of outage in the system. In this article, the outage is defined as the percentage of time the links are incapable of supporting a minimum required transmission rate which is determined by the application. The DCA approach is formulated in terms of a binary optimization problem that is solved using the branch-and-bound method. The authors assum the links in the network to be Rayleigh faded and we used a finite state Markov chain (FSMC) for their modeling. Using Monte Carlo simulation, we showed that the proposed DCA approach in a cooperative UGVs provides significant gain in the overall throughput and reduction in the outage probability compared to the static channel assignment (SCA).

DOI: 10.4018/978-1-60960-563-6.ch007

INTRODUCTION

Nowadays, traffic transportation systems faces a lot of challenges, such as advanced traffic management, vehicle control, safety control, and networking and information services for users on the road. One suggested solution is developing cooperative unmanned ground vehicles (UGVs) to improve the safety, security and efficiency of the transportation systems, and to enable new mobile services and applications for the traveling public. The main challenge that a cooperative UGV network faces is the harsh nature of the communication links. The links in a network of UGVs suffer from many problems like power fluctuation of the received signal due to multipath propagation and Doppler spread that becomes more severe at high speeds and high carrier frequencies. Besides, the limited weight of the UGV imposes a restriction on its mission times. The relaying cooperation problem appeared in the information theory community (Cover, 1979), and were inspired by the concurrent development of the ALOHA system at the University of Hawaii. The relay channel model is comprised of three terminals: a source that transmits information, a destination that receives information, and a relay that both receives and transmits information in order to enhance the communication between the source and the destination. Recently multiple relays have been examined in (Kramer, 2005). Combination of relaying and cooperation are also possible, and are often referred to as cooperative communications and most of these models fall within the broader class of generalized feedback wireless channels (Cover, 1981). Unfortunately, the fundamental performance limits, in terms of Shannon capacity are not known in general till now. Although, some useful bound in capacity have been obtained. The application of cooperative diversity in communication system has been proved to have significant performance improvement in terms of various performance metrics, including, capacity as in (Kramer, 2005), improved reliability as in (Laneman, 2004), diversity multiplexing-tradeoff in (Azarian, 2005), and bit/symbol error probability in (Sendonaris, 2003).

In (Edrich, 2002), Edrich & Schmalenberger proposed the use of combined direct sequence spread spectrum (DSSS) and frequency hopping spread spectrum (FHSS) technique to reduce the effect of interference in the unmanned airbone vehicle (UAV) wireless links. This technique has the advantage of averaging the effect of interference but not avoiding it. Using dynamic channel assignment (DCA) with adaptive modulation and coding (AMC) (Ye, 2002), it is possible to significantly reduce the effect of interference by assigning sub-channels to the links with better conditions (since these sub-channels experience a better signal-to-interference plus noise ratio (*SINR*) over that link). Two recent contributions (Song, 2005; Song, 2005) showed that using the inherent frequency diversity of OFDM can optimize the network throughput by using AMC according to the channel state information (CSI) of each sub-channel. The quality of service (QoS) and the utility-oriented bandwidth allocation was studied in (Cao, 2002).

To the best of our knowledge no such performance analysis and optimization have been performed for teleoperated cooperative UGV network taking into consideration the complexity of the cooperative UGVs relay channel. In this article we propose a DCA with AMC approach in a cooperative UGV network to optimize the performance in terms of UGV network throughput and communication link reliability using relay transmission with optimal resource allocation. We adopted a finite state Markov chain (FSMC), previously introduced in (Wang, 1995) to model the Rayleigh fading in the wireless links. Each UGV is assumed to have a certain QoS requirement, which depends on the application the UGV is assigned for. Our objective is to maximize the network throughput and to reduce the system outage (i.e, increase reliability) through DCA of a group of sub-channels while satisfying the minimum rate requirement of each link. We formulated the DCA with AMC in a cooperative relay UGV network to maximize the aggregate throughput and to satisfy the required QoS as a binary optimization problem that can be solved using the branch-and-bound method.

The rest of this article is organized as follows. In Section 2, the network model of cooperative UGVs is discussed. Then in Section 3, the finite state Markov chain model (FSMC) of the Rayleigh fading channel is presented. In Section 4, the optimization problem is formulated. In Section 5, the simulation results are presented before the article is finally concluded in Section 6.

THE NETWORK MODEL

Network Topology

Consider a network of cooperative teleoperated UGVs as shown in Figures 1 and 2 for the reverse link and forward link, respectively. In this configuration, each UGV (such as V_1, V_2,..., V_N) transmits its data to the most suitable UGV to be a relay (*M* in Figure 1), which in turn relays this information to the control unit (CU). This topology has the advantage of consuming less power than direct transmission from the UGVs to the CU. Each UGV requires one forward channel through which the UGV mission control data and the control information of the on-board sensor payload can be transmitted. In the reverse link direction, two channels are needed, one provides the position of the UGV, its mission path and navigation data as well as the internal state of the UGV and the sensor payload. The second channel is responsible for providing real time transmission of the captured video data. Based on the collected data from the UGVs, the CU makes the following decisions:

1. Choice of the relay UGV according to the UGVs position information.
2. Antenna directivity according to UGV position and traveling path.
3. The sub-channel assignment for each link.
4. The modulation and coding assignment for each sub-channel.
5. The power transmission assignment for each link.

These control information are sent directly to all UGVs instantaneously after the decision is made by the CU.

Channel State Information

Assuming that we have *N* links $L_1, L_2, \ldots, L_N$ and that the available bandwidth (BW) is divided into *K* sub-channels $B_1, B_2, \ldots, B_K$, where $N < K$. The information of each link can be transmitted over any sub-channel. We assume that the channel state information (CSI) are known for all link/sub-channel combination in terms of *SINR* and are mapped into certain rates by using propriate modulation and coding scheme. According to CSI, the available rate for each sub-channel is one of *m* values $R_1, R_2, \ldots, R_m$, where each rate is determined by the type of the modulation/coding scheme. Mathematically, the CSI matrix can be written as

$$\mathbf{CSI}_{K\times N} = \begin{bmatrix} SINR_{11} & SINR_{12} & \ldots & SINR_{1N} \\ SINR_{21} & SINR_{22} & \ldots & SINR_{2N} \\ .. & . & . & . \\ . & . & . & . \\ .. & . & . & . \\ . & . & . & . \\ SIR_{K1} & SIR_{K2} & \ldots & SIR_{KN} \end{bmatrix}, \quad (1)$$

where $SINR_{ij}$ is the received *SINR* of sub-channel B_i over link L_j. The data rate matrix, $\mathbf{R}_{K\times N}$, can be written as

Figure 1. Reverse link transmission for teleoperated cooperative UGV network

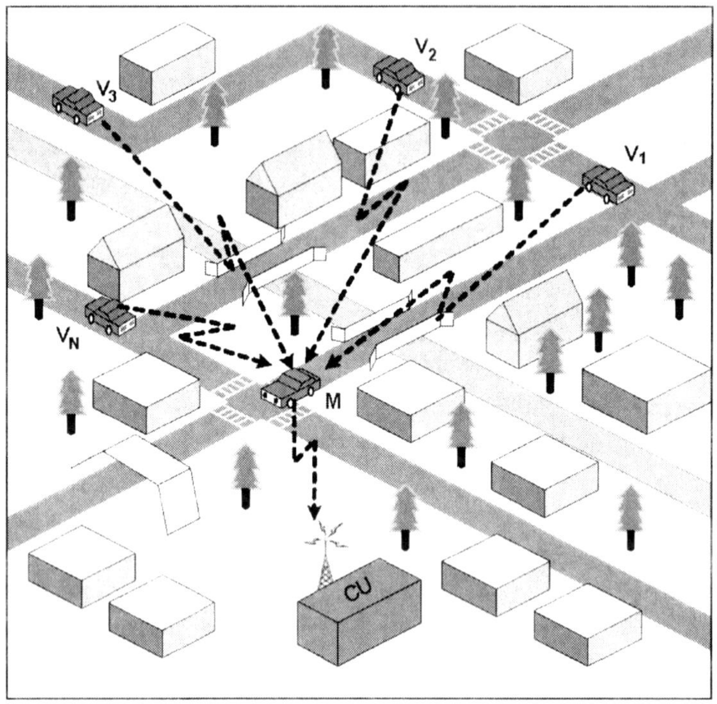

Figure 2. Forward link transmission for teleoperated cooperative UGVs network

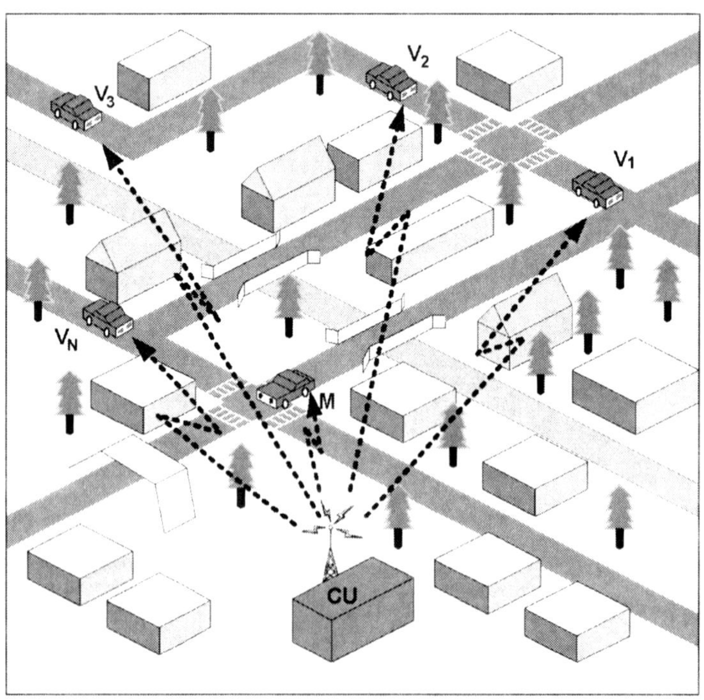

$$\mathbf{R}_{K\times N} = \begin{bmatrix} r_{11} & r_{12} & \cdots & r_{1N} \\ r_{21} & r_{22} & \cdots & r_{2N} \\ \cdot\cdot & \cdot & \cdot & \cdot \\ \cdot & \cdot & \cdot & \cdot \\ \cdot\cdot & \cdot & \cdot & \cdot \\ \cdot & \cdot & \cdot & \cdot \\ r_{K1} & r_{K2} & \cdots & r_{KN} \end{bmatrix}, \qquad (2)$$

where r_{ij} is the achievable rate when using sub-channel B_i over link L_j and $r_{ij} \in \{R_1, R_2, \ldots, R_m\}$. The mapping between $\mathbf{CSI}_{K\times N}$ and $\mathbf{R}_{K\times N}$ can be impelmented using AMC (Nanda, 2000; Goldsmith, 1997), *where* a higher order modulation and coding is used over a sub-channel with better *SINR* while not exceeding a target bit error rate (*BER*) value. This choice will improve the rate for this sub-channel, and consequently, increase the overall aggregate throughput of the cooperative UGV network. Now, if continuous rate adaptation (Qiu, 1999) is used, r_{ij} can be expressed as

$$r_{ij} = \log_2(1 + \beta \times SINR_{ij}) \text{ [bits/sec/Hz]}, \qquad (3)$$

where β is calculated from

$$\beta = \frac{-1.5}{\ln(5 \times \mathrm{BER}_{\text{target}})}, \qquad (4)$$

and $\mathrm{BER}_{\text{target}}$ is the required minimum *BER*.

Channel Model

The wireless channel used in UGVs communication suffers from scattering and reflection that results in multipath propagation. Each path will be associated with a specific propagation delay and attenuation factor depending on the path conditions. The received signal envelope fluctuations can be modeled by a Rayleigh distribution (Proakis, 1995). Moreover, the motion of UGVs will result in variations in the received signal level, which is known as Doppler frequency shift.

Based on the fact that the received *SINR* is proportional to the square of the received signal envelope, which is Rayleigh distributed, *SINR* has the following exponential probability density function (pdf) (Proakis, 1995):

$$P(\gamma) = \frac{1}{\rho}\exp\left(\frac{-\gamma}{\rho}\right), \qquad (5)$$

where ρ is the average value of the received *SINR* which includes the effect of large scale fading, co-channel interference, and, the additive noise. The worst case value (maximum value) of the Doppler spread, f_m in any communication link can be taken as

$$f_m = \frac{v}{\lambda} = \frac{v \times f_c}{c}, \qquad (6)$$

where v is the relative velocity, f_c is the carrier frequency and $c = 3 \times 10^8$ m/s is the speed of light. From (Wang, 1995), the expected number of times per second that the received *SINR* passes downward across a given *SINR* threshold γ_i is given by

$$N_i = \sqrt{\frac{2\pi \times \gamma_i}{\rho}} \times f_m \times \exp\left(\frac{-\gamma_i}{\rho}\right). \qquad (7)$$

We choose to model the cooperative UGVs sub-channels using a FSMC model as explained in (Wang, 1995). Towards that end, we need to find the state transition matrix **T** and the steady state probability vector **P** of the FSMC model. We assume that the *SINR* range is divided into D regions with *SINR* thresholds $\gamma_0 = 0 < \gamma_1 < \gamma_2 < \ldots < \gamma_{D-1} = \infty$. The Rayleigh fading channel is considered to be in state S_k, $k = 0, 1, \ldots, D-1$, if the received *SINR* is in the interval $[\gamma_k, \gamma_{i+1}]$. Recall that according to (3)

Figure 3. D-States FSMC modeling of a Rayleigh fading channel

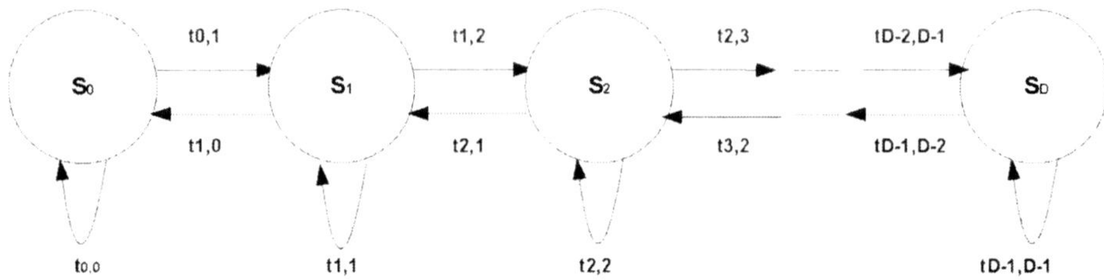

and using the lower limit, γ_k, of each *SINR* bin (as a worst case value of the *SINR* over the whole range), a certain rate can be achieved while being in state S_k. The FSMC of the Rayleigh fading channel with *D* states is given in Figure 3. It is opvious that the only possible transition from any state is to either the same state or to its adjacent neighbors, which mathematically can be written as

$$t_{i,j} = 0, \quad \forall \quad |i-j| > 1, \tag{8}$$

where t_{ij} is the transition probability from state *i* to state *j*.

In our system we propose the use of sub-channels to transmit information over links between UGVs. Given that R_b is the available rate per sub-channel, then on the average, the rate achieved by this sub-channel while being in state *k* is $R_b^{(k)} = R_b P_k$, where P_k is the steady state probability of being in state *k*,

$$P_k = \exp\left(\frac{-\gamma_k}{\rho}\right) - \exp\left(\frac{-\gamma_{k+1}}{\rho}\right). \tag{9}$$

The Markov transition probabilities can be approximated as given in (Wang, 1995)

$$t_{i,i+1} \approx \frac{N_{i+1}}{R_b^{(i)}}, \quad i = 0,1,2,\ldots,D-2, \tag{10}$$

$$t_{i,i-1} \approx \frac{N_i}{R_b^{(i)}}, \quad i = 1,2,3,\ldots,D-1, \tag{11}$$

$$t_{0,0} \approx 1 - t_{0,1}, \tag{12}$$

$$t_{D-1,D-1} = 1 - t_{D-1,D-2}, \tag{13}$$

$$t_{i,i} = 1 - t_{i,i-1} - t_{i,i+1}, \quad i = 1,2,\ldots,D-2. \tag{14}$$

Consequently, the steady state probability vector **P** and the transition probability matrix **T** can be written as

$$\mathbf{P} = \begin{bmatrix} P_0 \\ P_1 \\ \vdots \\ P_{D-1} \end{bmatrix}, \tag{15}$$

$$\mathbf{T} = \begin{bmatrix} t_{0,0} & t_{0,1} & 0 & \cdots & 0 \\ t_{1,0} & t_{1,1} & t_{1,2} & 0 & \cdots \\ \vdots & \vdots & \vdots & \vdots & \vdots \\ 0 & \cdots & t_{D-2,D-3} & t_{D-2,D-2} & t_{D-2,D-1} \\ 0 & \cdots & 0 & t_{D-1,D-2} & t_{D-1,D-1} \end{bmatrix}. \tag{16}$$

Transitioning to the adjacent states is based on the assumption of slow fading where the envelope of *SINR* changes slowly enough to stay in the same state or jump up or down to the adjacent state as it is clear in (16).

Optimization Problem

When the CU obtains the channel state matrix $\mathbf{CSI}_{K\times N}$, it maps it into the rate matrix $\mathbf{R}_{K\times N}$ for a certain acceptable BER by using (3). Obtaining the CSI is beyond the scope of this article. We define an assignment matrix, $\mathbf{X}_{K\times N}$, which is given by

$$\mathbf{X}_{K\times N} = \begin{bmatrix} x_{11} & x_{12} & \cdots & x_{1N} \\ x_{21} & x_{22} & \cdots & x_{2N} \\ .. & . & . & . \\ . & . & . & . \\ .. & . & . & . \\ . & . & . & . \\ x_{K1} & x_{K2} & \cdots & x_{KN} \end{bmatrix}, \tag{17}$$

where $x_{kn} = 1$ indicates that the sub-channel k is assigned to link n and $x_{kn} = 0$ otherwise. Now, the aggregate throughput of the teleoperated cooperative UGV network THR(**X**), can be formulated as

$$\mathrm{THR}(\mathbf{X}) = \sum_{k=1}^{K}\sum_{n=1}^{N} r_{kn} \times x_{kn}, \tag{18}$$

where any sub-channel is used by only one link at any time to avoid any co-channel interference. This requirement can be formulated in the form of the following constraint:

$$\sum_{k=1}^{K} x_{kn} \le 1, \quad \forall \quad n = 1,2,\ldots,N. \tag{19}$$

The inequality in (19) indicates that at most the sub-channel can be assigned to one link and if the sub-channel experiences low *SINR* over all links, that spcific sub-channel will not be assigned to any link. The reverse link topology shown in Figure 1, in addition to the nature of the application for which the system of UGVs is used, enforce an extra set of constraints as follows. First, the application might require a minimum, $R_{n,\min}$, and a maximum value, $R_{n,\max}$, for the data rate R_n over link l_n. The minimum data rate is determined according to the nature of the application and is selected to satisfy a specific QoS metric. On the other hand, the maximum rate represents the maximum utility that can be achieved for this application where any increase of the rate over this maximum value will result in a waste of the bandwidth and will provide no extra benefit for the application that this UGV is operated for. Hence, to get maximum utility of our available bandwidth the following constraints should be added to all links except the Relay-CU link (link j),

$$\begin{aligned} R_n &\ge R_{n,\min}, \quad \forall \quad n \ne j, \\ R_n &\le R_{n,\max}, \quad \forall \quad n \ne j. \end{aligned} \tag{20}$$

Second, the use of a of Relay-CU link as a backbone of the network in the reverse transmission dictates the following constraint

$$\sum_{\substack{n=1 \\ n\ne j}}^{N} R_n + R_{j,min} \le R_j. \tag{21}$$

It can be seen from (21) that the extra available rate when all non Relay-CU links reach their maximum rate can be assigned to the relay-CU link. The optimization problem can now be summarized as follows:

$$\mathit{Maximize}\ \ \mathrm{THR}(\mathbf{X}) = \sum_{k=1}^{K}\sum_{n=1}^{N} r_{kn} x_{kn}$$

$\mathit{Subject\ to}:$

$$(1)\sum_{k=1}^{K} x_{kn} \leq 1, \quad \forall \quad n = 1,2,\ldots,N$$

$$(2) x_{kn} \in [0,1], \quad \forall \quad n = 1,2,\ldots,N, k = 1,2,\ldots,K$$

$$(3) R_{n,min} \leq \sum_{k=1}^{K} r_{kn} x_{kn} \leq R_{n,max}, \quad \forall \quad n \neq j$$

$$(4)\sum_{k=1}^{K}\sum_{\substack{n=1\\ n\neq j}}^{N} r_{kn} x_{kn} + R_{j,min} \leq \sum_{k=1}^{K} r_{kj} x_{kj}. \quad (22)$$

In summary, the suggested topology and the formulation of the optimization problem, to be used in teleoperated cooperative UGV network, has two main goals. The first on is the reduction of the system outage time (increasing reliability), which is a very critical issue in UGVs communications. The outage of any UGV means that the assigned bandwidth to this UGV is not enough to send the required minimum rate. This will produce some delay in acquiring the sensors and video data, so using DCA of the available bandwidth will prevent and decrese the outage probability. The second goal is maximizing the total aggregate throughput of the system which, in turn, increases the amount of collected data. This dynamic scheme will be compared with the static channel assignment (SCA) of the available bandwidth over all links for the same link conditions.

Figure 4. Three reverse links example of teleoperated cooperative UGV network

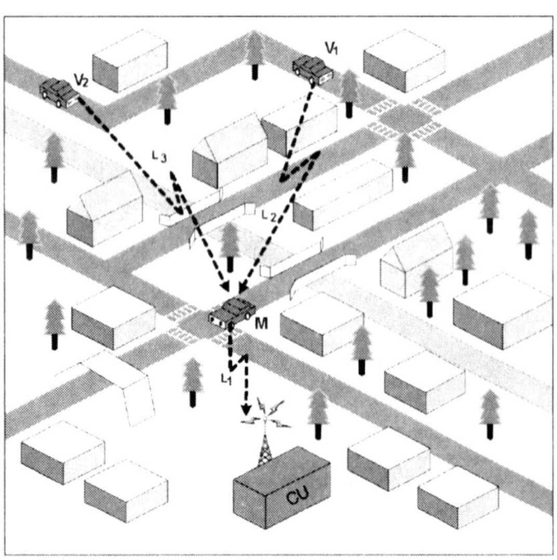

SIMULATION RESULTS

As an example, we assume a three-links cooperative UGVs topology as shown in Figure 4. In our simulations, the three-links L_1, L_2 and L_3 are assumed to experience Rayleigh fading with fading parameter $\rho = 15$. We assume a 16 states FSMC model with *SINR* thresholds diven in dB as [0, 2, 4, 6, 8, 10, 12, 14, 16, 18, 20, 22, 24, 26, 28, 30]. The *SINR* thresholds related to the FSMC model and the mapping of these states into rates are given in Table 1. The rates shown in Table 1 are chosen roughly just for demonstration and might not be physically realizable. In other words, it might be difficult to find a modulation/coding scheme that exactly provides such rates. The system parameters used in our simulations as well as the link parameters are summarized in Tables 2 and 3, respectively.

In Table 3, the velocity entry corresponding to L_1 is the velocity of the relay (i.e., UGV V_1) relative to the CU, while the entries corresponding to L_2 and L_3 are the velocities of UGV (V_2) and UGV (V_3), respectively, relative to UGV ($V1$). The transition probability matrix and the steady state probability vector for L_3 were calculated and are given in Table 4. We conducted 25000 simulation runs and the DCA optimization was invoked in each run. We assumed that the minimum required rates over L_1, L_2 and L_3 are $R_{1,min}$ = 25 Mbps, $R_{2,min}$ = 25 Mbps and $R_{3,min}$ = 10 Mbps, respectively. Also, the maximum rate constraint was relaxed in the example we are presenting. Figures 5, 6, and 7 depict the changes in the achievable rate over links L_1, L_2 and L_3, respectively, vs. the

Table 1. Thresholds & state-rate mapping

SINR Range	State	Rate(Mbps)
0-2	1	0
2-4	2	3.054
4-5	3	5.461
5-6	4	7.461
6-8	5	9.161
8-10	6	10.064
10-12	7	11.955
12-14	8	13.134
14-16	9	14.203
16-18	10	15.182
18-20	11	16.084
20-22	12	16.921
22-24	13	17.701
24-26	14	18.431
26-28	15	19.119
28- ∞	16	19.767

Table 2. System parameters

Carrier Frequency	2.4 GHz
Total Bandwidth	80 MHz
Sub-channel Bandwidth	8 MHz
Number of Sub-channels	10
Number of links	3

Table 3. Links parameters

Link Number	Velocity[m/s]	Average SINR (dB)
1	15	15
2	25	15
3	40	15

simulation run with both DCA and SCA. We considered that the system in outage if any of the

Figure 5. Achievable rate of L_1 using DCA and SCA

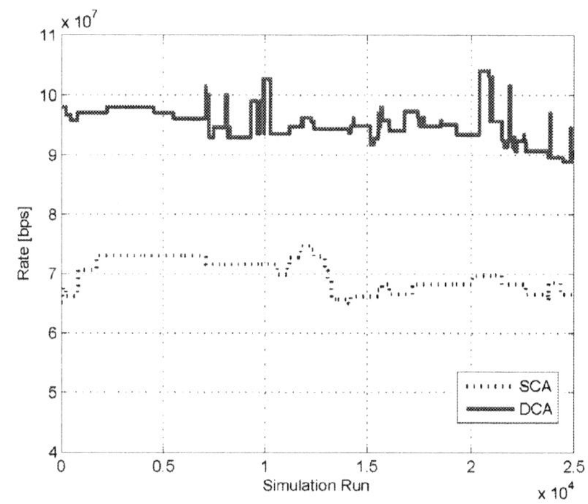

links fail to support their minimum rate constraint. Also, it is shown that the relay-CU link (L_1 in our example) needs to support its minimum requirement as well as the achievable rates over the other tow links in the network. From these figures, and based on the minimum rate requirements on each link mentioned earlier we can arrive at the interesting observation that the outage probability using DCA is assured by the optimization constraints which clearly proves the superiority of the proposed DCA scheme.

In Figure 8, the overall aggregate throughput of the teleoperated UGV network is shown. The average throughput and the average aheiveable rates over the three links were calculated and are given in Table 5. The use of DCA instead SCA gave 26% improvement in the overall average throughput. We expect to obtain even more improvement in the overall average throughout as we increase the number of sub-channels within the same bandwidth. It should also be mentioned here that the mechanism of getting the CSI (i.e, the required rate in reverse link to transmit such information), in addition to the required rate in forward link needed to provide control information about the assigned sub-channels, will reduce the expected 26% throughput gain. This reduction can be further investigated and is beyond the interest of this article.

Table 4. transition matrix & steady state probability vector

k	$t_{k,k-1}$	$t_{k,k1}$	$t_{k,k+1}$	p_k
1	-	0.9993	0.0007	0.1248
2	0.0008	0.9983	0.0009	0.1092
3	0.0011	0.9978	0.0011	0.0956
4	0.0013	0.9974	0.0013	0.0837
5	0.0015	0.9970	0.0015	0.0732
6	0.0017	0.9967	0.0016	0.0641
7	0.0018	0.9964	0.0017	0.0561
8	0.0020	0.9961	0.0019	0.0491
9	0.0021	0.9959	0.0020	0.0430
10	0.0023	0.9956	0.0021	0.0376
11	0.0024	0.9954	0.0022	0.0329
12	0.0025	0.9952	0.0023	0.0288
13	0.0025	0.9952	0.0023	0.0252
14	0.0027	0.9948	0.0025	0.0221
15	0.0028	0.9946	0.0026	0.0193
16	0.0004	0.9996	-	0.1353

Figure 6. Achievable rate of L_2 using DCA and SCA

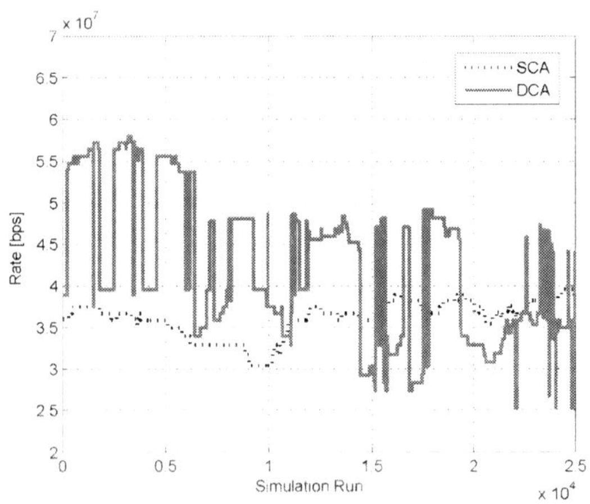

Figure 7. Achievable rate of L_3 using DCA and SCA

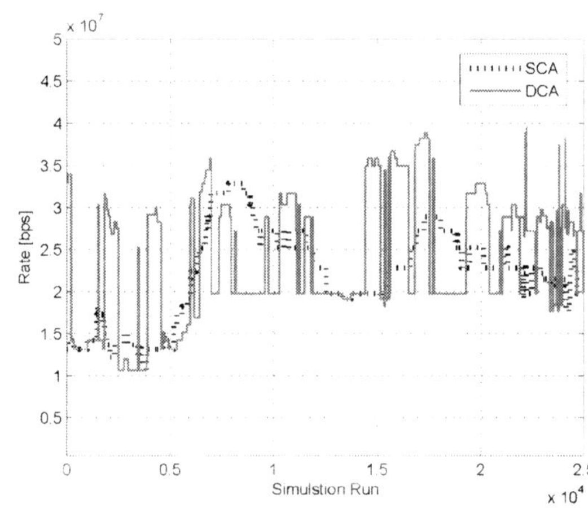

CONCLUSION

Cooperative teleoperated UGV network is a possible solution to achieve advanced traffic management, vehicle control, safety control, and information services for users on the road. Due to the high speed of the UGVs, the nature of the channel environments where they are usually deployed and the possible co-channel interference in addition to limited power capabilities of the UGVs, those make a huge challenge for communication system designers to optimize their solutions. In

Figure 8. Overall network throughput using DCA and SCA

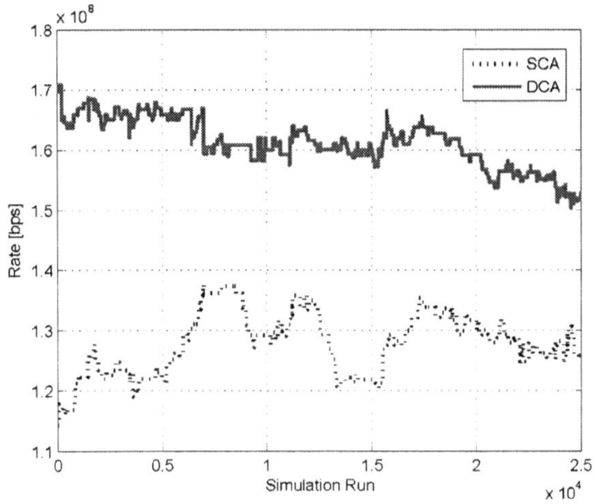

Table 5. Average throughput with ACA & SCA

Link Number	ACA [Mbps]	SCA [Mbps]
1	95.222	69.909
2	42.122	36.056
3	23.869	22.011
All	161.213	127.976

this article, DCA with AMC in a relay UGVs cooperative network was suggested to maximize the network throughput and to reduce the outage probability of the system. The use of DCA helps to avoid interference, which can cause the loss of valuable information that might affect critical decisions if a UGV goes out of the CU control range. Using Monte Carlo simulations, we showed how our proposed relay based cooperative scheme with DCA and AMC can result in significant gains in the achieved throughput and reduction in the overall system outage.

REFERENCES

Azarian, K., Gamal, H. E., & Schniter, P. (2005). On the achievable diversity-multiplexing tradeoff in half-duplex cooperative channels. *IEEE Transactions on Information Theory*, *51*(12), 4152–4172. doi:10.1109/TIT.2005.858920

Cao, Y., & Li, V. O. K. (2002). Utility-oriented adaptive QoS and bandwidth allocation in wireless networks. *IEEE International Conference on Communications* (pp. 3071-3075). Anchorage, AK.

Cover, T., & Gamal, A. E. (1979). Capacity theorems for the relay channel. *IEEE Transactions on Information Theory*, *25*(5), 572–584. doi:10.1109/TIT.1979.1056084

Cover, T., & Leung, C. (1981). An achievable rate region for the multiple-access channel with feedback. *IEEE Transactions on Information Theory*, *27*(3), 292–298. doi:10.1109/TIT.1981.1056357

Edrich, M., & Schmalenberger, R. (2002). Combined DSSS/FHSS approach to interference rejection and navigation support in UAV communications and control. *IEEE Seventh International Symposium on Spread Spectrum Techniques and Applications*, *3*, 687-691. Prague, Czech Republic.

Goldsmith, A. J., & Chua, S. G. (1997). Variable-rate variable-power MQAM for fading channel. *IEEE Transactions on Communications*, *45*, 1218–1230. doi:10.1109/26.634685

Kramer, G., Gastpar, M., & Gupta, P. (2005). Cooperative strategies and capacity theorems for relay networks. *IEEE Transactions on Information Theory*, *51*(9), 3037–3063. doi:10.1109/TIT.2005.853304

Laneman, J. N., Tse, D. N. C., & Wornell, G. W. (2004). Cooperative diversity in wireless networks: Efficient protocols and outage behavior. *IEEE Transactions on Information Theory*, *50*(12), 3062–3080. doi:10.1109/TIT.2004.838089

Nanda, S., Balachandran, K., & Kumar, S. (2000). Adaptation techniques in wireless packet data services. *IEEE Communications Magazine*, *38*, 54–64. doi:10.1109/35.815453

Proakis, J. G. (1995). *Digital Communications*. New York: McGraw-Hill.

Qiu, X., & Chawla, K. (1999). On the performance of adaptive modulation in cellular systems. *IEEE Transactions on Communications*, *47*, 884–895. doi:10.1109/26.771345

Sendonaris, A., Erkip, E., & Aazhang, B. (2003). User cooperation diversity-Part II, Implementation aspects and performance analysis. *IEEE Transactions on Communications*, *51*(11), 1939–1948. doi:10.1109/TCOMM.2003.819238

Song, G., & Li, Y. (2005). Cross-layer optimization for OFDM wireless networks-part I: theoretical framework. *IEEE Transactions on Communications*, *4*, 614–624. doi:10.1109/TWC.2004.843065

Song, G., & Li, Y. (2005). Cross-layer optimization for OFDM wireless networks-part II: algorithm development. *IEEE Transactions on Communications*, *4*, 625–634. doi:10.1109/TWC.2004.843067

Wang, H., & Moayeri, N. (1995). Finite-state Markov channel-a useful model for radio communication channels. *IEEE Transactions on Vehicular Technology*, *44*, 163–171. doi:10.1109/25.350282

Ye, S., Blum, R. S., & Cimini, L. J. (2002, Spring). Adaptive modulation for variable-rate OFDM systems with imperfect channel information. *IEEE Veh. Technol. Conf.* [Birmingham, AL.]. *VTC*, *2*, 767–771.

This work was previously published in International Journal of Mobile Computing and Multimedia Communications (IJMCMC) 1(4), edited by Ismail Khalil & Edgar Weippl, pp. 32-46, copyright 2009 by IGI Publishing (an imprint of IGI Global).

Section 2
Innovations in Mobility Engineering, Performance and Optimization

Chapter 8
The Accuracy of Location Prediction Algorithms Based on Markovian Mobility Models

Péter Fülöp
Budapest University of Technology and Economics, Hungary

Sándor Imre
Budapest University of Technology and Economics, Hungary

Sándor Szabó
Budapest University of Technology and Economics, Hungary

Tamás Szálka
Budapest University of Technology and Economics, Hungary

ABSTRACT

The efficient dimensioning of cellular wireless access networks depends highly on the accuracy of the underlying mathematical models of user distribution and traffic estimations. The optimal placement/deployment of e.g. UMTS, IEEE 802.16 WiMAX base stations or IEEE 802.11 WLAN access points is based on user distribution and traffic characteristics in the service area. In this paper we focus on the tradeoff between the accuracy and the complexity of the mathematical models used to describe user movements in the network. We propose a novel Markov chain based model capable of utilizing user's movement history thus providing more accurate results than other models in the literature. The new model is applicable in real-life scenarios, because it relies on information effectively available in cellular networks (e.g. handover history). The complexity of the proposed model is analyzed, and the accuracy is justified by means of simulation.

1 INTRODUCTION

Random Walk Mobility model is often used in network planning and in analyzing network algorithms, because of its simplicity (Jardosh, Belding-Royer, Almeroth, & Suri, 2003; Zonoozi, & Dassanayake, 1997). On the other hand this very simple model presumes unrealistic conditions, like uniform user distribution in the mobile network.

DOI: 10.4018/978-1-60960-563-6.ch008

However, in real-life networks, geographical characteristics, such as streets and parks influence the cell residence time (dwell time) and movement directions of users in the network, and result in a non-uniform user density. While these models are appropriate for mathematical analysis, easy to use in simulations and for trace-generation, they fail to capture important characteristics of mobility patterns in specific environments, e.g. time variance, location dependence, unique speed and dwell-time distributions, etc. (Yoon, Noble, Liu, & Kim, 2006; Camp, Boleng, & Davies, 2002; Wong, & Leung, 2000).

User movements in a cellular network can be described as a time-series of radio cells the user visited (Chellappa, Jennings, & Shenoy, 2003). The handover event of active connections (e.g. cell boundary crossing) is recorded in the network management system's logs, thus the information can be extracted from the management system of cellular mobile networks, such as GSM/GPRS/UMTS networks. The users movements are described by the dwell-time and outgoing probabilities (the probability of a user leaving for each neighboring cell). These parameters can be calculated for each cell based on the time-series of visited cells of the users. However, in some cases, these two parameters – dwell-time and outgoing probabilities – are not enough to capture all the information in the time-series of user movements. In many situations, the outgoing probabilities are correlated with the incoming direction of the users, so the movement contains memory.

We introduce a precise Markov mobility model based on these information, to capture the additional information contained in the user location traces. Our goal is to provide a synthetic model capable of capturing the unique properties of specific locations, e.g. urban areas, such as crowded parks, one-way streets etc.

The results are applicable for location based services, network dimensioning and more effective Call Admission Control algorithms can be applied in order to ensure user's satisfaction and optimal resource usage in cellular wireless mobile networks (Fongen, Larsen, Ghinea, Taylor, & Tacha Serif, 2003; Michaelis, & Wietfeld, 2006).

In this paper we introduce a Markov chain based mobility model with enhanced memory states, and compare the performance with a memoryless traditional Random Walk model. The models are able to capture the typical movement and cell dwell time patterns in arbitrary shaped mobile cell clusters.

2 RANDOM WALK MODEL IMPROVEMENTS

Our work is based on the utilization of time-series of mobile users' movement patterns in cellular mobile networks. In our work we assume that there is a given trace of a mobile service provider's network history. This dataset consists of all signals that were transferred in the network in the examined time interval. Beside many other network parameters and properties, two main information sets can be recovered:

- the cell-path that each user visited before
- the time intervals users have spent in each cell

The series of visited cells is crucial to analyze the similarities in the users' motion. Based on the motion patterns of the terminals amongst the cells, we can describe some drifts of the users' motion in a given cell or point. A drift may be caused by geographical or infrastructural objects (like highways, etc.) or some time-dependent circumstances (like mass events, concert, football matches, etc.) (Jardosh, Belding-Royer, Almeroth, & Suri, 2003; McDonald, & Znati, 2000).

From a mathematical point of view, the drift of the motion can be interpreted as different transition probabilities from one cell to another. A probability-vector can be defined for each cell that describes the probabilities of moving from the source cell to an adjacent one. This vector is called Handover Vector (HOV) (Liao, Tie, & Du, 2006). The HOV is a discrete distribution, it contains transition probabilities to neighboring cells on the condition that the user moves out from the given cell (Jayasuriya, 2001). The HOV might have time dependent components (e.g. the morning or the afternoon rush hours). We describe the usage of the HOV in Section 2.2.

The cell dwell time is an important parameter that helps to estimate the velocity of the motion. Velocity can vary in a wide range since the users frequently change their speed or motion direction (Wang, & Li, 2002).

Motion dynamics depend on the speed and the drift, from the network operator's point of view the accurate prediction of these components are useful to estimate the users' position in the near future, especially in case of handoffs. The analytically precise model that is able to predict both, helps the network operator in dynamic resource planning for short intervals based on actual network state. A long-term resource planning scheme (i.e. day-night scheme) can be supported by a short-term prediction that can follow dynamic escalations of call-initiations or degradation of the quality of the radio channel (Chlamtac, & Lin, 2003).

In this paper we observe methods for modeling motion drift and velocity with direction and cell dwell time prediction. All methods have in common that the prediction is made based on cell parameters. In the first part of this paper we focus on the accurate estimation of cell parameters, individual user patterns and history are not taken into account. Viewed from the user's side, the prediction is memoryless since all users in a given cell is considered the same passive entity, their speed and direction depend on the cell dwell time and the transition probabilities. In Section 2.3 we observe a mobility model extended with memory that defines different user-groups according to previous handover events and movement directions.

2.1 Random Walk

In the literature several mobility models can be found which describe aggregate or individual movement behavior (Michaelis, & Wietfeld, 2003; Wong, & Leung, 2000).

The most general mobility model is the traditional Random Walk (RW) model. The traditional discrete time random walk model is a series of steps, each taken in a uniformly distributed random direction. In the one-dimensional case, the random walk is the movement of an individual on a straight line. A decision is made in each point of time, the model steps forward with probability p and backwards with $(1-p)$. In the two-dimensional RW model the movement space is an infinite square grid where four possible directions are available in each state. The model can be extended in the similar way to higher dimensions.

Although it is easy to be used, the traditional random walk model has disadvantages in user motion modeling:

- Without any improvements, the RW model is not capable of simulating the mobility in an environment where geographical or infrastructural objects determine the motion behavior. Beside this, the RW model simulates the user distribution in the network in a uniform way which is clearly not applicable in real-life situations.
- The model steps in each time slot. The previously visited states or the origin state are recurrent in one and two dimensions (Pólya, 1921), but the model does not allow the same state in two following time slots. Thus the time spent in each state is not taken into account, each time tick means a transition to another cell that can be interpreted as a constant velocity user-motion.
- The model does not include the user's motion history, the states that were visited in the past. The uniformly distributed successor directions in a given state are not

precise enough in a real-life application. It can be seen that the possibility of moving forward in a user drift is higher than stepping backward. The sophisticated weighted possibilities of successor directions can be constructed by knowing the previously visited cells (Michaelis et al., 2006) or by mapping the geographical or infrastructural circumstances of the area covered by the given cell.

2.2 Random Walk Model Extension

We decide to use the RW model since it can be applied in a wide range of motion patterns (from pedestrian walk to highways) and can be specialized for uncommon scenarios also. The focus of our work is the elimination of the drawbacks outlined in Section 2.1. With a few extensions the RW model becomes able to accurately estimate future user distribution in a network or cell cluster.

Since the RW model assumes a motion with constant absolute value of velocity it cannot simulate different movement speeds. The best method to implement this feature is ensuring the possibility of staying in the same RW state for arbitrary amount of time. This can be achieved by two different approaches that mostly give similar results.

In the basic RW model the user stays in the actual cell until the end of the actual time slot. The lengths of slots are equal, since it is a discrete time system. We propose two methods to replace the constant time slot-lengths with a distribution that converges better to real-life motion patterns.

The handoff trace can be used to calculate several dwell times for each cell. Based on these derived data-series, we modeled each cell by a phase-type (PH) system, which produces a distribution of the cell dwell time with the appropriate absorption time. The phase-type system is defined in Equation (1) with a transient matrix ($\underline{\mathbf{A}}$), a vector that contains the transition intensities to the absorbing state (**a**) and the initializing vector ($\boldsymbol{\alpha}$).

$$Q = \begin{bmatrix} A & a \\ 0 & 0 \end{bmatrix} pdf_{PH} = f(A, a) \qquad (1)$$

The Cell Phase Type System is defined based on the trace, a unique distribution can be created for each cell. It is started when a user registers in the given cell, and the model unregisters the user when the PH system absorbs. Immediately after the absorption the user registers into one of the adjacent cells.

In Figure 1 the simplest PH system can be seen, combined with the uniform successor distribution. The PH contains one transient state which means that the absorption time is exponentially distributed.

Based on the handoff trace another dwell time simulation method can be derived. We provided the possibility of staying in the same cell for the next fixed length time slot as an extension of the RW model. There is an additional transition, the

Figure 1. Exponentially distributed (exp(λ)) dwell time simulator with HOV

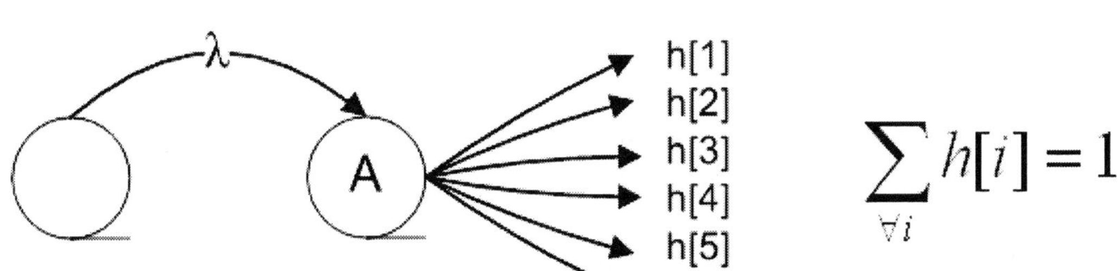

Figure 2. Looping transition in the modified RW structure

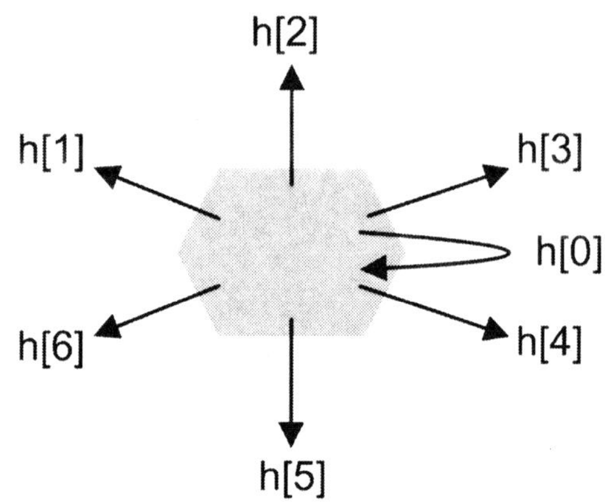

loopback direction. The simple RW model does not allow the user to stay in the same state, so our proposition enables the model to simulate different cell dwell times. Formally that means the probability of stepping into the same cell at the end of the slot, such as a looping step shown in Figure 2.

With the correct tuning of the loopback value, arbitrary distributions of cell dwell times can be simulated if the time slot length is appropriately chosen.

$$\mathbf{h}' = \left[\mathbf{h}'[0], \mathbf{h}'[1], \mathbf{h}'[2], \mathbf{h}'[3], \mathbf{h}'[4], \mathbf{h}'[5], \mathbf{h}'[6]\right]$$
$$\mathbf{h}'[0] = p_{stay}$$
$$\mathbf{h}'[i] = (1 - p_{stay}) \cdot \mathbf{h}[i] \qquad (2)$$

Equation (2) shows that the outgoing transition probabilities have to be weighted in order to keep them as a distribution. This model is the Extended Random Walk (ExtRW).

The remaining drawback of the Extended Random Walk model is the lack of handling different motion directions. Our extension is substitutes the uniformly distributed random successor direction in ExtRW to a special distribution that is valid only in the given cell. This distribution is represented by the Handover Vector (HOV). The HOV can be determined from the trace of handoffs made in an arbitrary network (See section 2.3.2). The given data of the trace provides measures of the relative frequency of handoffs between each adjacent cell-pairs. Based on the relative frequency, the probabilities of the HOV can be easily calculated. The HOV is cell-specific, each cell has its own geographical properties that affects the users' motion drift and the transition probabilities into the adjacent cells.

The Extended Random Walk model combined with the cell-dependent transition probabilities gives the best estimation of future user distribution. In following sections the simulation results prove that the Handover Vector can simulate the constant or time-dependent changing density of the users distributed in the different cells of the network.

3 MARKOVIAN APPROACH

The Random Walk model, as it was analyzed in the previous section, is the most commonly used mobility model to describe individual movement behavior. The extended RW and HOV eliminate the disadvantages of the RW model with variable dwell time and weighted transition probabilities calculated from known parameters of the system in the past. Based on these extensions the distribution of user population in the cluster can be estimated with a limited accuracy.

The limits of accuracy even in the most sophisticated HOV model is caused by the lack of memory, since the RW-like models do not possess the knowledge of the actual distribution of recent moments. Beside the trends of user movement flows of a longer period in the past, the application of the recent user locations have a capital importance in variable, directional user motion. Neglecting the actual transition series of a user in the cluster, i. e. the model is memoryless, the estimation works with a significantly higher error rate. Figure 3 shows a simple example which represents the effect of memory in the model. If we consider two roads shown is Figure 3.b, the accuracy of transition probability estimations are better when the model knows where the users come from than the RW-like estimation which cannot differentiate the users on the two roads (Figure 3.a.).

1. model without memory, unknown where is the users come from
2. model with memory, the previous steps of the users taken in account

To clarify the error rate of a memoryless HOV model compared to an algorithm that calculates with memory, we use the cell shown on Figure 3. Let us define the following parameters:

- the number of incoming users on the upper road at timeslot t is $in1_t$,
- the number of incoming users on the lower road at timeslot t is $in2_t$,
- similarly the number of users leaving on the upper road at timeslot t is $out1_t$,
- the number of users leaving on the lower road at timeslot t is $out2_t$,
- and the user movement directions with a simple transition matrix $\mathbf{P} = \begin{bmatrix} p_{11} & p_{12} \\ p_{21} & p_{22} \end{bmatrix}$

The HOV model in Figure 1.a calculates the number of leaving users (*out1, out2*) with a historical estimation of the **P** matrix, and with the sum of *in1* and *in2* (the total number of users in the observed cell), but without the knowledge of *in1*, *in2*.

In our proposed algorithm, the use of *in1*, *in2* means that the model calculates based on the information of the incoming users. The incoming users can be considered uniform or different (marked users). Since the latter is more accurate, in this comparison we use marked users.

Assume that the historical estimation of the HOV model's **P** matrix is based on the previous

Figure 3. User prediction methods

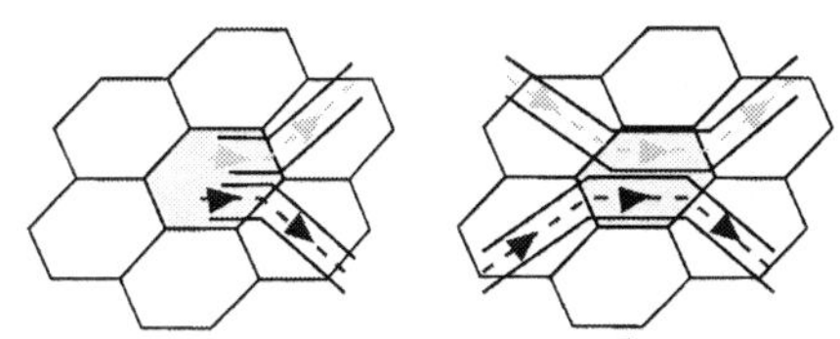

timeslot. That is the P($out1_{t+1}$) and P($out2_{t+1}$) probabilities are $out1_t/(out1_t+out2_t)$ and $out2_t/(out1_t+out2_t)$, respectively. Applying the same assumption on the algorithm with memory, the number of leaving users can be calculated with the **P** matrix itself, that is $out1_{t+1} = in1_t{*}p_{11} + in2_t{*}p_{21}$ and $out2_{t+1} = in1_t{*}p_{12} + in2_t{*}p_{22}$. At a given and constant **P** matrix let us assume that the incoming user distribution varies, that is the $in1_t/in2_t$ ratio (Incoming Distribution - ID) changes. Figure 4. shows the error of HOV compared the estimation using memory.

The HOV model works with error if ID_{t-1} is different from ID_t which is caused by the fact that the HOV historical **P**-estimation in this special case equals the number of leaving users of the previous timeslot. That is it does not include the actual ID_t value. Contrarily the memory-model calculates with the actual number of incoming users and the **P** matrix itself, which results the

Figure 4. HOV prediction error in percents

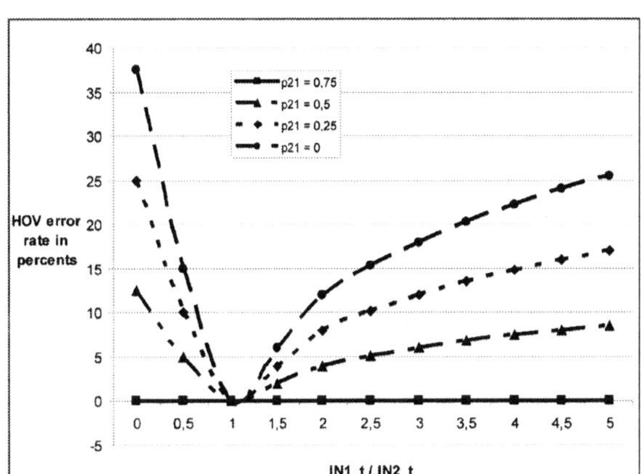

exact probabilities of the leaving users distribution. The error rate caused by the lack of memory increases as the variance of ID is increased (that is $in1_t/in2_t$ ratio is changed).

Using memory cannot enhance furthermore the accuracy of the estimation if $p_{21} = p_{11}$ and Figure 4. shows a constant zero error rate ($p_{11} = p_{21} = 0.75$). In this case the leaving direction of each user is independent of the incoming direction and the memory is useless, since users arriving from each direction are leaving towards a given direction with the same probabilites.

The results show that our proposition of using memory in a mobility model significantly increases the accuracy of the model in cases where the ID distribution in an arbitrary cell has high variance, or has periodicities without stationary distribution.

To increase the accuracy with the use of memory in the mobility model, we introduce our Discrete Time Markov model [5] with memory (M_n, where n denotes the number of Markov states). The states of the model show and store where the users came from. The memory can be interpreted in two meanings. Time level memory shows the number of timeslots in the past that the model considers. Thus a model with m time levels in time t can calculate the next transtion based on the user position in $(t-m, \ldots\ t-2,\ t-1)$. Direction level memory shows the number of directions that the model can differentiate. In general a cell cluster consists of hexagonal cells. The direction level of a model on this cluster is maximum 6. If transition from the central cell to two adjacent cells are not differentiated then the direction level is decreased by 1.

Obviously with increasing the levels of time and direction memory exponentially increases the complexity of the model, contrarily the logarithmical increase of accuracy. Our work was motivated to find an optimal size of estimation parameter space with the highest accuracy beside tolerable complexity. In the following subsections we will give example models with different time and direction level. We will show the coherence of the complexity and accuracy in a model with a given statespace.

3.1 Markov Modell

In our method the direction of a user is identically distributed between 0 and 2π. The user's speed is between 0 and V_{max}. After moving in a direction with a randomly chosen speed for a given Δt time, the user changes its direction and speed. N_{cell} denotes the number of cells in the network.

In a simple Markov-chain based model a user can be located in three different states during each time slot, the stay state (S) and the left-move state (L) and the right-move state (R). This classification of cells can be seen in Figure 5.

Let us define $X(t)$ random variable, which represents the movement state of a given terminal during time slot t. We assume that $\{X(t), t=0,1,2,\ldots\}$ is a Markov chain with transition probabilities p, q, v.

We assume that a user is in cell i at the beginning of a timeslot. If the user is in state S, similarly to the previous model, it remains in the given cell in the next slot. If the user is in state R, it moves to one of the cells on the right-hand side, if in state L, it moves to the left-hand side of the dividing line (direction level is 3).

Since the transition propensities are not symmetric, the left-move state and the right-move state

Figure 5. Neighbour cells separated into two groups

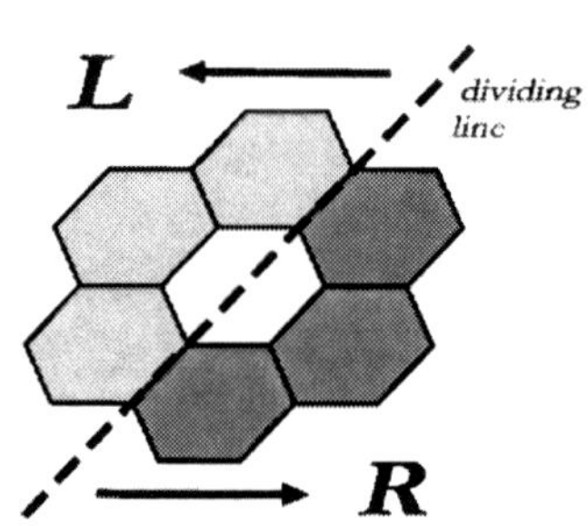

have different probabilities. Figure 6. depicts the Markov chain and transition (Π) matrix.

$$\begin{aligned} &P_S \cdot (p_1 + p_2) = P_L \cdot (1 - q_1 - v_2) + P_R \cdot (1 - q_2 - v_1) \\ &P_L \cdot (1 - q_1) = P_S \cdot p_1 + P_R \cdot v_1 \\ &P_R \cdot (1 - q_2) = P_S \cdot p_2 + P_L \cdot v_2 \end{aligned} \tag{3}$$

We also know that $P_S + P_L + P_R = 1$ thus the steady state probabilities can be calculated.

With knowledge of the result we can predict the number of mobile terminals for time slot $t+1$ for each cell, using Equation (4), where $S^i_{adj(r)}$ is the set of right neighbours of cell I (The cell numbering we used for models is shown by Figure 7.).

$$\begin{aligned} &N_i(t+1) = N_i(t) \cdot P_S(i) + \\ &+ \frac{1}{3} \sum_{l, C_l \in S^i_{adj(l)}} N_l(t) \cdot P_R(l) + \frac{1}{3} \sum_{r, C_r \in S^i_{adj(r)}} N_r(t) \cdot P_M(r) \end{aligned} \tag{4}$$

As we mentioned earlier this model performs well when the user's movement has only one typical direction, because in this case the handover intensities of the right-move cells does not differ significantly. It is the same in the aspect of the left-move cells.

Figure 6. State diagram and Π matrix of 3-state M-model(M_3)

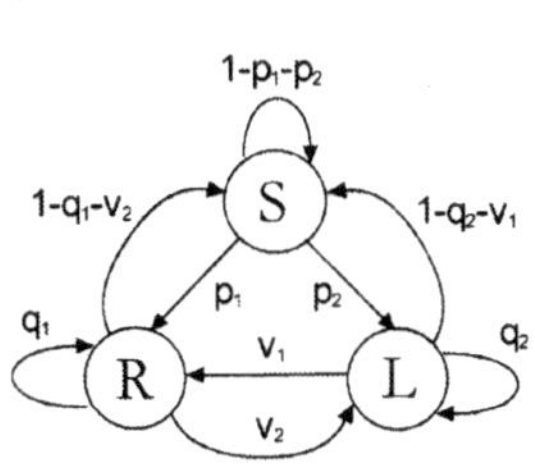

$$\Pi = \begin{bmatrix} 1 - p_1 - p_2 & p_1 & p_2 \\ 1 - q_1 - v_2 & q_1 & v_2 \\ 1 - q_2 - v_1 & v_1 & q_2 \end{bmatrix}$$

If we try to predict the user's distribution in a city having irregular, dense road system, or in a big park where people are able to move around then the handover intensities could differ thus the calculations above could produce errors. From this point of view the best way is if we represent all of the neighbour cells as a separated Markov state, so we create 7 states, because we assume 7 elements in a cluster (direction level is 7):

- stationary state (S)
- move to neighbour 1..6 state (M_{N1}.. M_{N6})

The Markov chain and transition matrix are more complex as it can be seen in Figure 7. As in the previous cases the steady state probabilities can be evaluated.

To calculate the number of users in C_i for time slot t, then Equation (5) is to be used, where the neighbour cells of cell i are indexed from 1 to 6.

$$N_i(t+1) = N_i(t) \cdot P_S(i) + \sum_{j, C_j \in S^i_{adj}} N_j(t) \cdot P_{M_{Nj+3 \bmod 6}}(l) \tag{5}$$

Figure 7. State diagram and Π matrix of seven-state Markov model

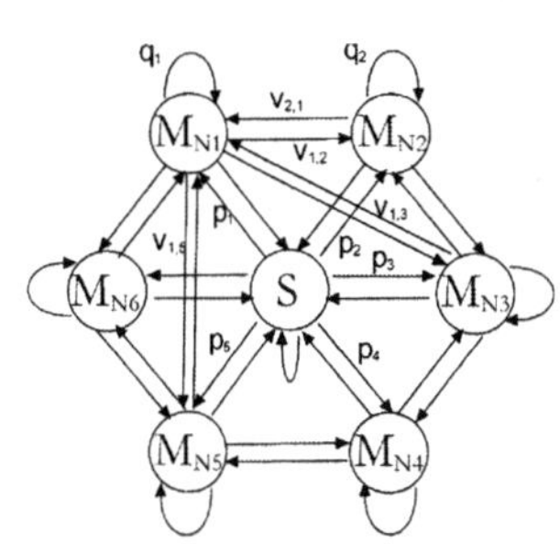

$$\Pi = \begin{bmatrix} 1 - \sum_{m=1}^{6} p_m & p_1 & p_2 & p_3 & p_4 & p_5 & p_6 \\ 1 - q_1 - v_{1,2} - v_{1,6} & q_1 & v_{1,2} & v_{1,3} & v_{1,4} & v_{1,6} & v_{1,6} \\ 1 - q_2 - v_{2,3} - v_{2,1} & v_{2,1} & q_2 & v_{2,3} & v_{2,4} & v_{2,5} & v_{2,6} \\ 1 - q_3 - v_{3,4} - v_{3,2} & v_{3,1} & v_{3,2} & q_3 & v_{3,4} & v_{3,5} & v_{3,6} \\ 1 - q_4 - v_{4,5} - v_{4,3} & v_{4,1} & v_{4,2} & v_{4,3} & q_4 & v_{4,5} & v_{4,6} \\ 1 - q_5 - v_{5,6} - v_{5,4} & v_{5,1} & v_{5,2} & v_{5,3} & v_{5,4} & q_5 & v_{5,6} \\ 1 - q_6 - v_{6,1} - v_{6,5} & v_{6,1} & v_{6,2} & v_{6,3} & v_{6,4} & v_{6,5} & q_6 \end{bmatrix}$$

It is to be taken into account that in the real networks a cell does not always have six neighbours depending on the coverage. This model has to be generalized for a common case when a cell has n neighbouring cells. We expanded our previous mentioned model to an n-neighbour case (Figure 8), when all the n neighbours are represented with a Markov state:

- stationary state (S)
- move to neighbour 1..n state ($M_{N1}..M_{Nn}$)

The steady state probabilities can be calculated as in the previous cases (Equation 6).

$$P_S \cdot \sum_{i=1}^{n} p_i = \sum_{j=1}^{n} (1 - q_i - \sum_{l \neq j}^{n} v_{j,l}) P_{Ni}$$

$$\ldots$$

$$P_{Nk}(1 - q_k) = P_S \cdot p_n + \sum_{i \neq k} P_{Ni} \cdot v_{i,k} \qquad 1 \leq k \leq n$$

$$\ldots$$

$$\sum_{k=1}^{n} P_{Nk} = 1 - P_S$$

(6)

Using the result the predicted number of users in the next time slot is given in Equation (7), where $P_{M_i}(j)$ denotes the steady state probabilities of moving to the cell i from cell j.

$$N_i(t+1) = N_i(t) \cdot P_S(i) + \sum_{j, C_j \in S^i_{adj}} N_j(t) \cdot P_{M_i}(j) \tag{7}$$

3.2 Model Parameters and the Error of the Module

In this chapter we present the transition probabilities and its determination of our Markov model. We introduce a method to process the network traces. In this section we show the error of the module based on the parameter determination.

The logs from the network provider contain the handover information and current time. Namely the handover information consist of the user ID and a source and destination cell. We split the examined interval into Δt parts. In every timeslot a cell is appended to every user. This results the input for the determination algorithm (Figure 9.).

First of all, the transition probabilities, used in previous section, are defined as:

p_n = the probability of moving from stationary state (S) to M_n
q_n = the probability of staying in a move state (M_n)
$v_{n,m}$ the probability of moving from a move state (M_n) to another one (M_m)

Figure 8. State diagram and Π matrix of n+1-state Markov model (M_{n+1})

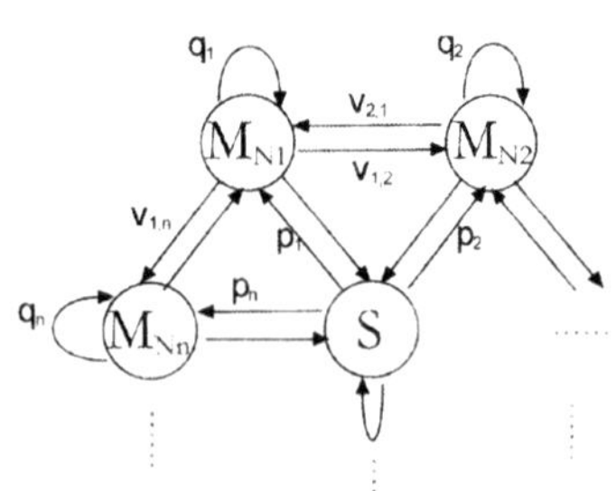

$$\Pi = \begin{bmatrix} 1-\sum_{m=1}^{n} p_m & p_1 & p_2 & \ldots & \ldots & \ldots & p_n \\ 1-q_1-v_{1,2}-v_{1,n} & q_1 & v_{1,2} & \ldots & \ldots & \ldots & v_{1,n} \\ \ldots & \ldots & \ldots & \ldots & \ldots & \ldots & \ldots \\ \ldots & \ldots & \ldots & \ldots & \ldots & \ldots & \ldots \\ 1-q_i-v_{i,i+1}-v_{i,i-1} & \ldots & \ldots & v_{i,i-1} & q_i & v_{i,i+1} & \ldots \\ \ldots & \ldots & \ldots & \ldots & \ldots & \ldots & \ldots \\ 1-q_n-v_{n,1}-v_{n,n-1} & v_{n,1} & \ldots & \ldots & \ldots & v_{n,n-1} & q_n \end{bmatrix}$$

These values are calculated based on above mentioned result trace. Let us introduce the following terminologies to process the traces and to determine the probabilities mentioned above:

- Pa(X,Y)
- A two-step move pattern from C_i to C_j, in other way a user, who is in C_i at time t, and in C_j at time t+1. This trace could be used for ExtRW and HOV parameter determination also, in fact this is the "memory less trace"; we don't know, where the user came from, we only know is where it is heading to: $Pa(C_{i,}\ C_j)$.
- *Pa(X,Y,Z)*
- A three-step move pattern from C_h to C_j, across C_i, in another way a user, who is in C_h at time t-1, in C_i at time at t, and in C_j at time t+1. This trace could be called a "memory trace", because during we are in C_i, we know, that the user came from C_h): $Pa(C_{h,}\ C_i,\ C_j)$.

Figure 9. The format of the result trace (U_x means the user x)

	U_1	U_2	U_3	...
t_1	C_i	C_j	C_j	...
t_2	C_k	C_k	C_j	...
t_3	C_k	C_l	C_j	...
t_4	C_j	...	...	...

Figure 10. Example for the move patterns

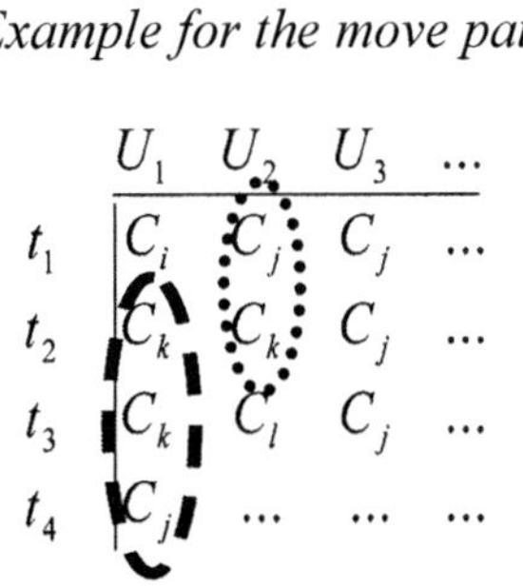

Dashed line: three-step move pattern, Dotted line: two-step move pattern.

- For all users and for all timeslots we search the $Pa(C_{a,}\ C_b)$ pattern in the trace. The number of the found patterns:
- $Num(Pa(C_{a,}\ C_b))$.
- The index of a cell, which is in *z* direction from the current C_j : *Ind(z,i)*.

Using these determinations the probabilities belonging to C_j can be given by the following way from real network traces:

$$\hat{p}_{(j)\ i} = \frac{Num(Pa(C_j, C_j, C_{Ind(j,i)}))}{\sum_{l=0}^{n}\sum_{k=0}^{n} Num(Pa(C_{ind(j,l)}, C_j, C_{ind(j,k)})) = Sum[j]} \quad (8)$$

$$\hat{q}_{(j)\ i} = \frac{Num(Pa(C_{ind(j,i)}, C_j, C_{Ind(j,i)}))}{Sum[j]} \quad (9)$$

$$\hat{v}_{(j)\ i,k} = \frac{Num(Pa(C_{ind(j,i)}, C_j, C_{Ind(j,i)}))}{Sum[j]}, \quad (10)$$

Based on these equations a prediction can be given for the users distribution. The question is the accuracy of the predicting model.

It can be seen that these $\hat{p}_i, \hat{q}_i, \hat{v}_{i,k}$ values were defined based on relative frequency in the network traces. Accuracy of this calculation depends mainly on the number of samples. That is a finite set of samples could not be sufficient, there is always a minimal error which is the difference between the real model ($p_i, q_i, v_{i,k}$) and the calculated ($\hat{p}_i, \hat{q}_i, \hat{v}_{i,k}$) values. Let us define the ε_p, $\varepsilon_{q,}\ \varepsilon_v$ as the error of $p_i, q_i, v_{i,k}$ determination of the parameters of the Markov model (As ε_p in Equation 11, $\varepsilon_{q,}\ \varepsilon_v$ can be determined similarly).

The mean value of calculated parameters (for example $E(\hat{p}_i)$) is the real value because of the law of averages.

$$\varepsilon_p = \left|\hat{p}_i - p_i\right| = \left|\hat{p}_i - E(\hat{p}_i)\right| \rightarrow p_i = \hat{p}_i + \varepsilon_p \tag{11}$$

As we defined earlier the n-state Markov model (see chapter 3.1) satisfies the general matrix equation $P = P\Pi$, which can be solved. But we are not able to determine exactly the Π matrix. Calculating with the relative frequencies we get a $\hat{\Pi}$ matrix, which estimates the Π matrix with calculation errors. When we solve the matrix equation $P = P\Pi$ using $\hat{\Pi}$ instead of Π, we get a $\hat{P}$ equals P with the addition of the model error. Summarizing these coherences:

- theoretical solution: $P = P\Pi$
- practical solution: $\hat{P} = \hat{P}\hat{\Pi}$

$$\hat{P} = P + H \tag{12}$$

$$\hat{P} = \hat{P}\hat{\Pi} \rightarrow P + H = (P + H)(\Pi + \mathrm{E}) \tag{13}$$

where H denotes the error of the steady state probability comes from the parameter calculation error, and E means the matrix derived from the ε_p, $\varepsilon_{q,}$ ε_v values. Our aim by the following to calculate the dependency of the model error on the network parameters.

Let us start from the Equation 6. described with the calculated parameters instead of the real parameters (Equation 14).

$$\begin{aligned}
&\hat{P}_S \cdot \sum_{i=1}^{n} \hat{p}_i = \sum_{j=1}^{n} (1 - \hat{q}_i - \sum_{l \neq j}^{n} \hat{v}_{j,l})\hat{P}_{Ni} \\
&.... \\
&\hat{P}_{Nk}(1 - \hat{q}_k) = \hat{P}_S \cdot \hat{p}_n + \sum_{i \neq k} \hat{P}_{Ni} \cdot \hat{v}_{i,k} \qquad 1 \leq k \leq n \\
&...
\end{aligned} \tag{14}$$

Replace the variables by Equation 12 and 11 to get the following equation (Equation 15).

$$\begin{aligned}
&(P_S + H_S) \cdot \sum_{i=1}^{n} (p_i + \varepsilon) = \sum_{j=1}^{n} (1 - (q_i + \varepsilon) - \sum_{l \neq j}^{n} (v_{j,l} + \varepsilon))(P_{Ni} + H_{Ni}) \\
&... \\
&(P_{Nk} + H_{Nk})(1 - (q_k + \varepsilon)) = (P_S + H_S) \cdot (p_n + \varepsilon) + \sum_{i \neq k} (P_{Ni} + H_{Ni}) \cdot (v_{i,k} + \varepsilon) \\
&... \\
&1 \leq k \leq n
\end{aligned} \tag{15}$$

As next step we denote H_S, H_{Nk} using the terms from Equation 6. and the coherence $\varepsilon = max(\varepsilon_{p,}$ $\varepsilon_{q,}$ $\varepsilon_v)$:

$$\begin{aligned}
&H_S = \sum_{j=1}^{n} (1 - q_i - \sum_{l \neq j}^{n} v_{j,l})H_{Ni} + (1 - \sum_{i=1}^{n} p_i)H_S - n\varepsilon(nH_{Ni} + 1 + H_S) \\
&... \\
&H_{Nk} = H_S p_m + \sum_{i \neq k} H_{Ni} \cdot v_{i,k} + H_{Nk} q_k + H_S \varepsilon + (nH_{Nk} + 1)\varepsilon \\
&... \\
&1 \leq k \leq n
\end{aligned} \tag{16}$$

We have an equation system with $n+1$ equation (Equation 16). Let us do some simplifying in the following way:

$$\begin{aligned}
p &= avg(p_i) && \forall i \\
q &= avg(q_i) && \forall i \\
v &= avg(v_{i,j}) && \forall i, \forall j
\end{aligned} \tag{17}$$

These three general probability parameters can describe typical user movements in the current cell.

Parameter p means the probability of the moving after a stop, q is for describing how the user can hold its moving direction, and at last v, the opposite of q, namely how often changed the moving direction of the users changed.

Using these general parameters yield an upper estimation with an equation system in 2 variables:

$$\begin{aligned}
&H_S = n(1 - q - (n-1)v)H_N + (1 - np)H_S - n\varepsilon(nH_N + 1 + H_S) \\
&H_N = H_S p_m + (n-1)H_N v + H_N q + H_S \varepsilon + (nH_N + 1)\varepsilon
\end{aligned} \tag{18}$$

If we examine the ε error rate, we find out that an upper estimation can be applied for it as well using the Chebyshev inequality. The variance of relative frequency is well known:

$$\sigma(\hat{p}) = \sqrt{\frac{p(1-p)}{m}} \tag{19}$$

where $p = E(\hat{p}_i)$, and m are the number of samples in the trace.

Let us make an upper estimation on Equation 19 to eliminate variable p:

$$\sqrt{\frac{p(1-p)}{m}} \le \sqrt{\frac{1}{4m}} = \frac{1}{2\sqrt{m}} \tag{20}$$

Using this result in the Chebyshev inequality:

$$P(|\hat{p} - E(\hat{p})| \ge \varepsilon) \le \frac{1}{2\sqrt{m}\varepsilon^2} \tag{21}$$

If we assume that $|\hat{p} - E(\hat{p})|$ is not greater than ε with 99% probability, and that $m=10000$, we reach an upper estimation for ε:

$$0.01 \le \frac{1}{2\sqrt{m}\varepsilon^2}$$

$$\ldots \tag{22}$$

$$-\frac{1}{\sqrt{2}} \le \varepsilon \le \frac{1}{\sqrt{2}} \to \varepsilon = \frac{1}{\sqrt{2}}$$

In order to get an equation system that depends only on $p,q,v,$ the ε is replaced in Equation 18.

$$H_S = n(1 - q - (n-1)v)H_N + (1 - np)H_S - n\frac{1}{\sqrt{2}}(nH_N + 1 + H_S)$$

$$H_N = H_S p_m + (n-1)H_N v + H_N q + H_S \frac{1}{\sqrt{2}} + (nH_N + 1)\frac{1}{\sqrt{2}} \tag{23}$$

H_S, H_N, the error of the steady state probability can be calculated from the equation system, but due to the limitation of this paper we do not present it in detail.

As the last step we define H_{Model}, the model error from H_S, H_N: $H_{Model} = H_S + nH_N$

Using the plotting abilities of *Mathematica* we examined this model error depending on p,q,v,n. The next figures depicts the results. The x axis shows the number of states (n), the y axis represents the error of the module.

We analyzed four different scenarios considered as typical user movements. The result of the first combination is depicted on Figure 11. In this case the all of the general transition parameters

Figure 11. Model error, p=0.1, q=0.1, v=0.1. Slow motion in current cell

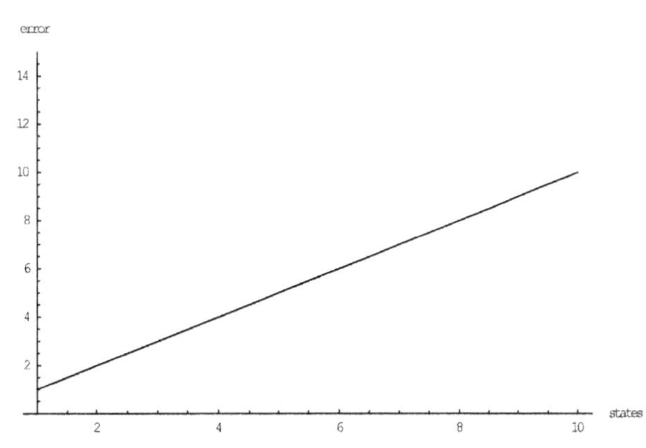

(p,q,v) take low values, meaning that the users in the current cell move slowly, stop several times. The function of model error increases linearly, as it is depicted.

The second scenario shows the opposite of the previous case, namely fast movement in all direction, none of the users stop (Figure 12).

Third case simulates a fast, direction-changing movement environment. The function on Figure 13 has a local maximum in x=3, after that it is decreasing logarithmical.

In the last situation the users move fast, and hold their direction. The model error increasing slowly (Figure 14).

In this section we determined the previously introduced Markov model parameters from the network traces, defined that the model error derived from the parameter calculations, and examined it dependently on the user movement behaviors.

3.3 Complexity and Accuracy in the Markovian Approach

In the previous section we introduced a Markov model generator method, an optional model can be derived depending on the resource requirements (complexity) and the demanded precision (accuracy).

Figure 12. Model error, p=0.2, q=0.9, v=0.9. Fast motion with direction changes and direction holding as well

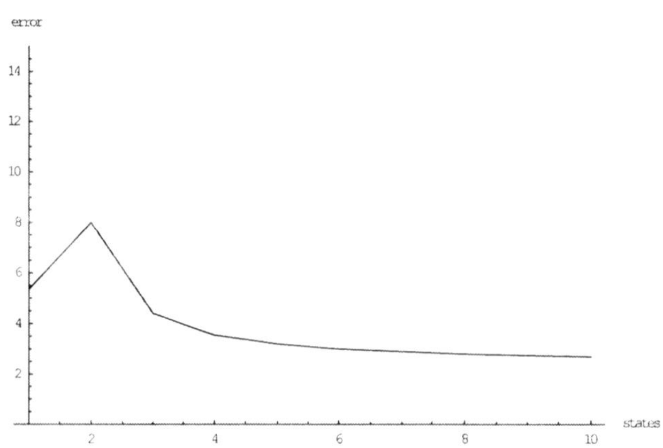

Figure 13. Model error, p=0.2, q=0.1, v=0.9. Fast motion with direction changes

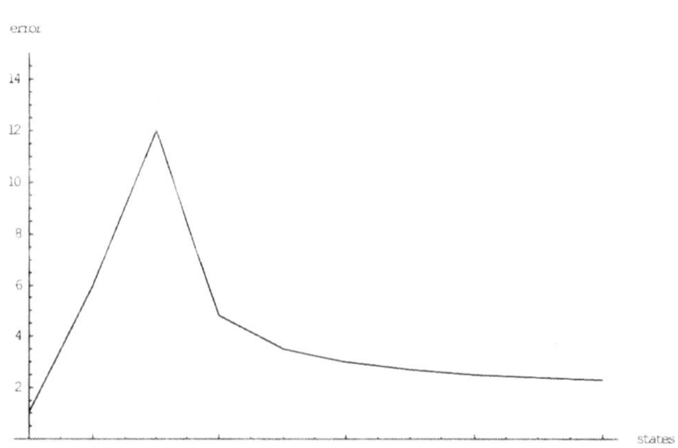

The accuracy of the model is increasing in the function of the number of states. The number of states is increasing when the memory (time level) is increased, or when the number of direction (direction level) is increased. Time level rises the states exponentially, direction level rises it linearly. With the state-space increasing, the computational complexity of the Markov steady state calculations are also follow a rising curve. The question is the characteristic of these functions and the existence of a theoretical or practical optimum point.

It is assumed that each cell has N neighbors and the 3-state (stay, left and right-move state) model is used to determine user movement. It is also assumed that N/2 cells belong to both left and right Markov-states, and the users are uniformly distributed between cells. A theoretical error can be derived from this assumption since in most cases the user motion pattern does not results a uniform distribution in the N/2 cells. In the worst case the users move with 1 probability into one of the neighbouring cells. In our estimation calculations the error can be measured with the difference between the uniform distribution and the worst case. This difference in mathematical equation is the following:

$$1-\frac{1}{N/M}+\frac{1}{N/M}*((N/M)-1), \qquad (24)$$

where N means the neighbour numbers, and M means the direction numbers in the model.

We measured the computation complexity also as the function of state number. This enables us to compare complexity and prediction error in an easy way. Based on the 1..M-state model the prediction computation is calculated with the costs of Markov steady state mathematical operations and other procedures necessary for transition probabilities. The complexity can be estimated with $o(M_{states}^{\ 3} + M_{states} + 1/M_{states})$.

Figure 15 shows the complexity and error characteristic. In the given model calculations the optimal point of operation is around 5 states where error is minimal at this level of complexity.

In the previous comparison the direction level sizing is used only. If we want to look back for the estimation, than memory has to be included into the model. In fact this means that every state in the Markov chain has to be changed with M states. This causes exponential state number explosion that can be seen on the Figure 16.

As we have extended our model step by step in time and direction level, its precision increased as well as the complexity of the model. In order to decrease it, we developed a simple algorithm, which is able to minimize the number of states based on merging adjacent cells. Because of the limitation of the paper we will only show the

Figure 14. Model error, p=0.3, q=0.9, v=0.1. Fast motion with direction holding

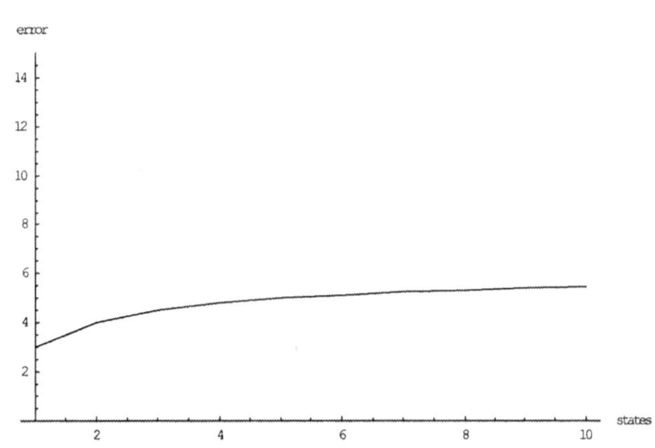

Figure 15. Complexity (black) and accuracy of the calculation (red)

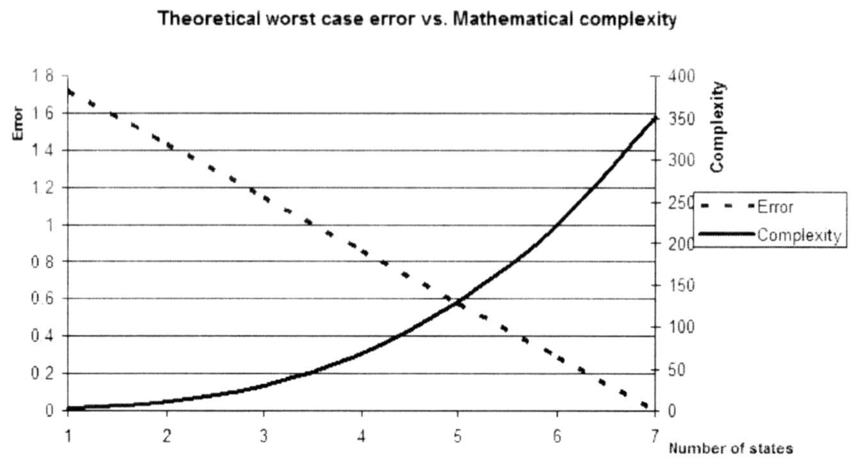

Figure 16. The increase of number of states in case of different direction levels (3,4,5,6,7 states), and different time levels (1,2,3,4)

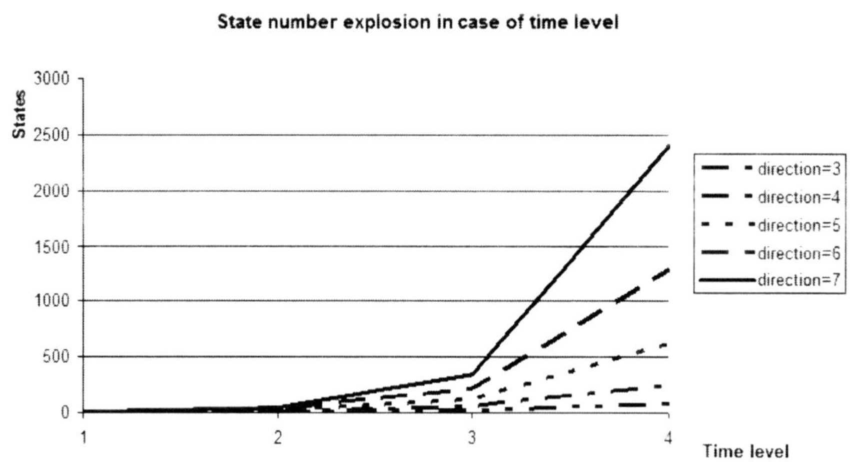

algorithm for direction level. The input parameters are the following:

- an acceptable *Err* rate of error caused by the merging
- h_{kl} elements of the handover matrix (where h_{kl} is the handover probability from C_l to C_k, $\forall C_k, \forall C_l \in S'_{adj}, k, l = 1 \ldots N_{cell}$). Q_{Hi} $(n_i+1 \text{ x } n_i+1)$ matrix is defined (where n_i is the number of neighbour cells of cell i). The elements in the first row and column are the handover probabilities of the examined cell i.

Based on the Q_{Hi} matrix (elements denoted by h'_{kl}) and the error rate a group of cells is to be

Box 1.

```
original_Q = QHi;
minimal_grouping_Q= QHi;
minimal_grouping_num= ni;
// For all neighbour cell of cell i
For x=2 to ni +1 {
     // Clear the container of the grouped cells
     grouping.clear();
     nullstations.clear();
     QHi = original_Q;
     // Try to group some cells into each other
     For j=x-1 to ni+x-1 {
          k=(j mod 7)+1;
          If ((k!=x) && (k!=i) && (!grouping.member(k))) {
               // Get the k. row from the h_i matrix
               group_row=h_i[k]
                group_column=trans(h_i[k]);
               // Create a null vector
               temp_g_row=0;
               temp_g_column=trans(0);
               // Create a new k group
               group[k].create();
               // Add element k to new group
               group[k].add(k);
               // Try to find a cell to k, which behaves similar like k
               For l=2 to ni  {
                    if (!grouping.member(l)) {
                    greater_than_Err=false;
                     For m=1 to ni{
                           if((!nullstation.member(m)) && (m!=l)){
                        if((|hikm- hilm |>Err) || (|himk- himl |>Err)){
                          greater_than_Err=true;
                          break;
                         } else{
                          temp g row[m]=(|group_row[m]+ hilm |/2);
                          temp_g_column[m]=(|group_column[m]+ himl |/2);
     } } }
     // If the calculated difference (Err) is greather
     // than the limit, than add the cell l to cell k.
   // From this cell l and cell k acts like one cell, a cell group
     if (!greater_than_Err){
          nullstations.add(m);
          group[k].add(l);

          group_row=temp_g_row;
          group_column=temp_g_column;
     } } }
     QHi[k,_]=group_row;
     QHi[_,k]=trans(group_column);
     }}
     if(grouping.length<minimal_grouping_num) {
     minimal_grouping_Q= QHi;

     minimal_grouping_num=grouping.length;
}}
```

found, where the absolute difference between h^i_{ki} and h^i_{li}, and between h^i_{ik} and h^i_{il} ($\forall C_l$, $\forall C_k$ in a group) is less than *Err*.

The algorithm is given with its pseudo code: (see Box 1).

As a result of the algorithm the simplified equation is Equation (25), where *K* denotes a

Figure 17. Cell cluster with streets and park

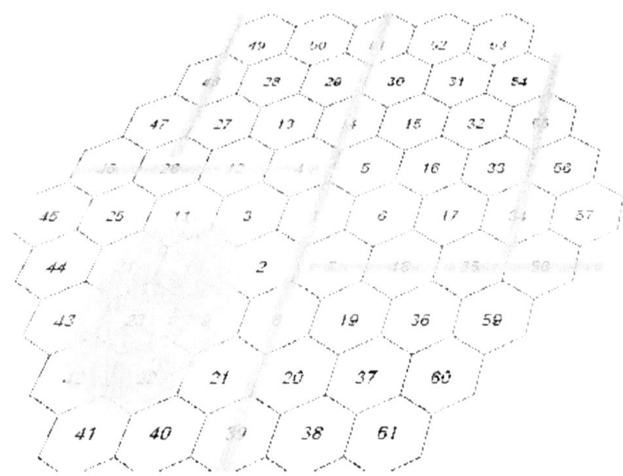

group of cells, m_K^j the number of elements in group K, and $P_{M_K}(j)$ is the steady state probability of moving to group K from cell j.

$$N_i(t+1) = N_i(t) \cdot P_S(i) + \sum_{j:C_j \in S_{adj}^i} N_j(t) \cdot \frac{1}{m_K^j} P_{M_K}(j)$$

$$C_i \in K \tag{25}$$

In this section we examined complexity and theoretical errors. We exhibited the state number explosion at time level increasing. The next chapter contains measurements with simulations, which confirm our mathematical results.

4 ACCURACY MEASUREMENTS WITH SIMULATION

The inaccuracy of the Random Walk-based mobility models depend on the properties of the transition probabilities. The RW model is only capable of accurate prediction of user movements in case of uniform movement distributions (e.g. all elements in the transition probability vector are equal to 1/6).

According to the calculations and simulations, the Markovian approach of user movement modeling and the HOV produces better estimation of the users' distribution in a cell cluster.

The estimation procedure was validated by a simulation environment of a cell cluster shown on Figure 17. The cluster consisted of 61 named cells, the simulation environment included geographical data that is interpreted as streets on the cluster area. The drift of the movement is heading to the streets from neutral areas.

The simulation used 610 mobile terminal (10 for each cell), in the initial state uniformly distributed in the cluster. The average motion velocity of the users is parameterized with a simple PH cell dwell time simulator (reciprocal of exponentially distributed values).

The simulation consists of two parts. The reference simulation is the series of the transitions that the mobiles have initiated between cells. It produces a time-trace that contains the actual location data for each mobile terminal in the network. We have used this reference simulation as if it was a provider's real network trace.

The estimation procedure uses the past and the current reference simulation results to estimate future number of users in each cell. The estimation error is interpreted as the measure of accuracy of each mobility model in this paper.

The prediction starts 100 timeslots after the reference simulation initiation. During the warm-up process the reference simulation produces enough sample data for the correct estimation which uses the previous reference results as an input to estimate the future user distribution. Each user-transition in the 100-timeslot reference period is used to derive transition probabilities, motion speed and patterns in the simulation cell-space. These patterns serve as an input for the simulation threads of each mobility model. The models have the same input throughout the simulation process so that the results are comparable [10].

The estimation errors of the models in the simulation are measured in three different ways. The first approach calculates the average error of the cells in each timeslot. It produces a time-dependent relative error value (TREV) in each timeslot for the cell-cluster. TREV shows the average error compared to the actual user number in the cells. It can be seen on Figure 18. that TREV depends on the dynamics of the motion, basically on the cell dwell time. The generic lambda (λ) parameter effects the motion velocity of the simulated users, the higher value means longer dwell time thus slower motion.

The relative performance of the models can be seen on the Figure 19. The execution time of the methods of the model are plotted in each timeslot on logarithmical scale.

5 CONCLUSION

In this paper we proposed extensions for the Random Walk model that enabled it to simulate real-life user movement accurately. We proposed an alternative Markov-chain based method which is also comparable with the HOV method. The simulation results proved the analytical properties of the proposed mobility models.

The simple RW model works with a significant error rate at all time. Since the user movement pat-

Figure 18. TREV values in RW, ExtRW, M_3, M_7, HOV models with λ = (1, 4)

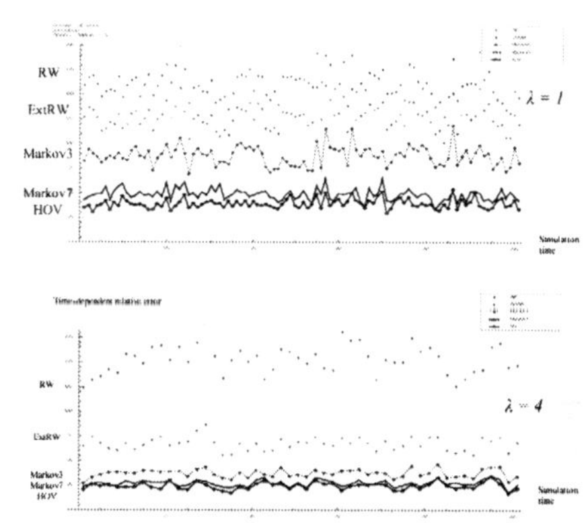

Figure 19. CPU usage of the models on logarithmical scale

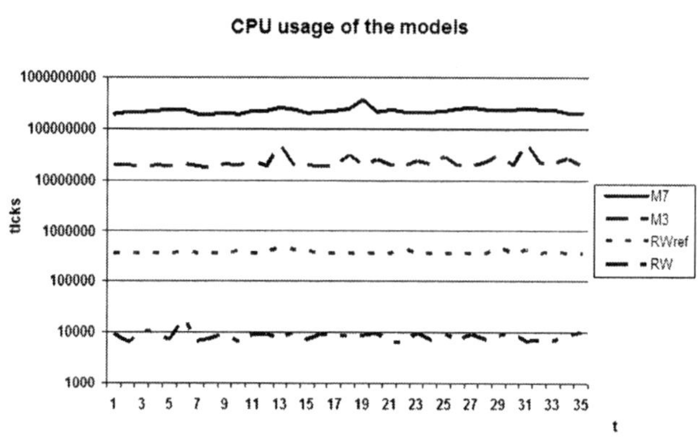

terns in the simulation are not completely random due to the streets and geographical circumstances, the uniformly distributed Random Walk pattern cannot model it.

The HOV model produces better results. By the extension the model is able to precisely simulate cell dwell time, and the six elements of the outgoing transition probabilities are correctly weighted based on the warm-up data trace. The accurate dwell time and movement direction estimation makes the Handover Vector model the best of the memoryless approaches.

The Markov mobility model is the most accurate in the estimation process since it has the ability to calculate with motion direction, speed and the recent handover event (user history) also. The three-state model focuses on cell dwell time since it differentiates only two motion directions which cannot follow general drifts. The seven-state model is sophisticated in both cell dwell time and motion direction since it is capable of following six different drifts in the cell-cluster.

The network operator may use the HOV or the seven-state Markov model to make predictions on the future distribution and location of users among radio cells to justify CAC or other QoS decisions.

6 ACKNOWLEDGMENT

This work was supported by the Mobile Innovation Center, Hungary.

7 REFERENCES

Camp, T., Boleng, J., & Davies, V. (2002). A Survey of Mobility Models for Ad Hoc Network Research. Wireless Communications and Mobile Computing (WCMC). *Special issue on Mobile Ad Hoc Networking: Research, Trends and Applications, 2*, No.5.

Chellappa, R., Jennings, A., & Shenoy, N. (2003, May). A Comparative Study of Mobility Prediction in Cellular and Ad Hoc Wireless Networks. *Proceedings of the IEEE Int'l Conference on Communications 2003 (ICC2003)*, Alaska, USA.

Chlamtac, Y. F., & Lin, I. Y.-B. (2003). *Portable movement modeling for PCS networks*. IEEE Vehicular Technology.

Fongen, A., Larsen, C., & Ghinea, G., Taylor, & Tacha Serif, S.J.E. (2006). Location based mobile computing - A tuplespace perspective. *Mobile Information Systems, 2*(2-3), 135–149.

Hong, X., Kwon, T., Gerla, M., Gu, D., & Pei, G. (2001, January). A Mobility Framework for Ad Hoc Wireless Networks. *Proceedings of ACM 2nd International Conference on Mobile Data Management, (MDM 2001)*, Hong Kong.

Jardosh, A., Belding-Royer, E. M., Almeroth, K. C., & Suri, S. (2003, September). Towards Realistic Mobility Models for Mobile Ad hoc Networks. *Proceedings of ACM MobiCom* (pp. 217-229).

Jayasuriya, A. U. (2001). *Improved Handover Performance Through Mobility Prediction*, PhD thesis, University of South Australia.

Liang, B., & Haas, Z. (1999, March). Predictive distance-based mobility management for PCS networks. *Proceedings of the Joint Conference of the IEEE Computer and Communications Societies (INFOCOM)*.

Liao, H., Tie, L., & Du, Z. (2006, April 24). A Vertical Handover Decision Algorithm Based on Fuzzy Control Theory. *Computer and Computational Sciences, 2006. IMSCCS '06. First International Multi-Symposiums, 2*, 309–313.

Mathematica Documentation. Wolfram Research Inc. from http://www.wolfram.com/.

McDonald, A., & Znati, T. (2000). Predicting Node Proximity in Ad Hoc Networks: A Least Overhead Adaptive Model for Selecting Stable Routes. Proceedings of *First Annual Workshop on Mobile Ad Hoc Networking and Computing, (MobiHoc 2000)*, Boston.

Michaelis, S., & Wietfeld, C. (2006). *Evaluation and comparison of prediction stability for user movement pattern detection algorithms*. Athens: European Wireless.

Michaelis, S., & Wietfeld, C. (2006, May). Comparison of User Mobility Pattern Prediction Algorithms to increase Handover Trigger Accuracy. *IEEE Vehicular Technology Conference*, Melbourne.

Wang, K., & Li, B. (2002, April). Group Mobility and Partition Prediction in Wireless Ad-Hoc Networks. *IEEE International Conference on Communications, (ICC 2002),* New York.

Wong, V. W.-S., & Leung, V. C. M. (2000, September/October). Location Management for Next-Generation Personal Communication Network. *IEEE Network*.

Yoon, J., Noble, B. D., Liu, M., & Kim, M. (2006, June). Building realistic mobility model from coarse-grained wireless trace. *Proceedings of the Fourth International Conference on Mobile Systems, Applications, and Services (MobiSys)*, Uppsala, Sweden.

Zonoozi, M. M., & Dassanayake, P. (1997, September). User mobility modelling and characterization of mobility patterns. *IEEE Journal on Selected Areas in Communications*, *15*(7). doi:10.1109/49.622908

This work was previously published in International Journal of Mobile Computing and Multimedia Communications (IJMCMC) 1(2), edited by Ismail Khalil & Edgar Weippl, pp. 1-21, copyright 2009 by IGI Publishing (an imprint of IGI Global).

Chapter 9
Frequency Domain Equalization and Adaptive OFDM vs. Single Carrier Modulation

Inderjeet Kaur
Ajay Kumar Garg Engineering College, India

ABSTRACT

In the present article an attempt is made to compare multi-carrier and single carrier modulation schemes for wireless communication systems with the utilization of fast Fourier transform (FFT) and its inverse in both cases. With the assumption that in OFDM (orthogonal frequency division multiplexing), the inverse FFT transforms the complex amplitudes of the individual sub-carriers at the transmitter into time domain, the inverse operation is carried out at the receiver. In case of single carrier modulation, the FFT and its inverse are used at the input and output of the frequency domain equalizer in the receiver. Different single carrier and multi-carrier transmission systems are simulated with time-variant transfer functions measured with a wideband channel sounder. In case of OFDM, the individual sub-carriers are modulated with fixed and adaptive signal alphabets. Furthermore, a frequency-independent as well as the optimum power distribution are used. Single carrier modulation uses a single carrier, instead of the hundreds or thousands typically used in OFDM, so the peak-to-average transmitted power ratio for single carrier modulated signals is smaller. This in turn means that a SC system requires a smaller linear range to support a given average power. This enables the use of cheaper power amplifier as compared to OFDM system.

DOI: 10.4018/978-1-60960-563-6.ch009

INTRODUCTION

The basic recipe followed in this article is as follows:

1. For investigating the transmission of digital signals we have used wideband frequency-selective radio channels.
   It has been observed that the frequency-selective fading caused by multipath time delay spread degrades the performance of digital communication channels by causing inter-symbol interference, thus results in an irreducible BER and imposes an upper limit on the data symbol rate.
2. We have compared the performance of single carrier and multi-carrier modulation schemes for a frequency-selective fading channel considering un-coded modulation scheme.
   Our analysis shows that the un-coded OFDM loses all frequency diversity present in the channel which results in a dip in the channel. As a result of this, the information data on the subcarriers, affected by the dip, is erased. Further, this erased information cannot be recovered from the other carriers. Consequently, it results in a poor Bit Error Rate (BER) performance. However, we can recover frequency diversity and improve the BER performance by adding sufficiently strong coding which spreads the information over multiple subcarriers.

Alternatively, the performance of OFDM can also be improved significantly by using different modulation schemes for the individual sub-carriers. In this scenario, the modulation schemes have to be adapted to the prevailing channel transfer function. Moreover, each modulation scheme provides a trade off between spectral efficiency and the bit error rate. The spectral efficiency can be maximized by choosing the highest modulation scheme that will give an acceptable (BER). In a multipath radio channel, frequency selective fading can result in large variation in the received power of each carrier.

The article is organized as follows: In section II the fixed and adaptive OFDM transmitters are described. A brief description of a single carrier system with frequency domain equalization is given in section III. Section IV deals with the simulations results. Finally, the main conclusions are drawn in section V.

ADAPTIVE OFDM TRANSMISSION

Figure 1 shows the block diagram of the OFDM transmitter used. As can be seen, Binary data is, first, fed to a modulator which generates complex symbols on its output. The modulator either uses a fixed signal alphabet (QAM) or adapts the signal alphabets of the individual OFDM sub-carriers. Both, signal alphabets and power distribution can be optimized corresponding to the channel transfer function. Because of the slow variation of transfer function with time (as shown by the

Figure 1. Block diagram of a) an OFDM and b) a single carrier transmission system with frequency domain equalization

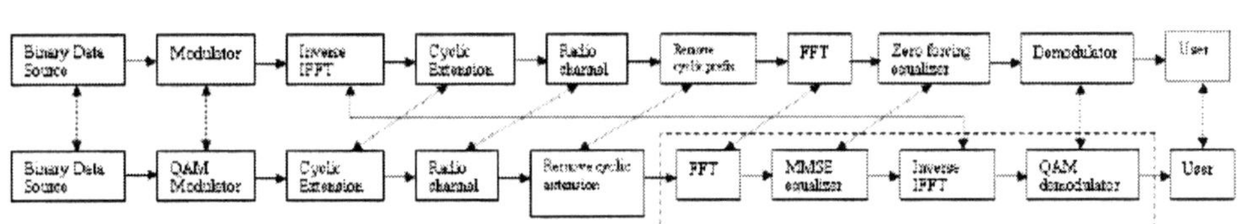

Figure 2. Simulation results for a line-of-sight (LOS) radio channel with a) a fixed (measurement 1) and b) a mobile (measurement 2) user terminal antenna

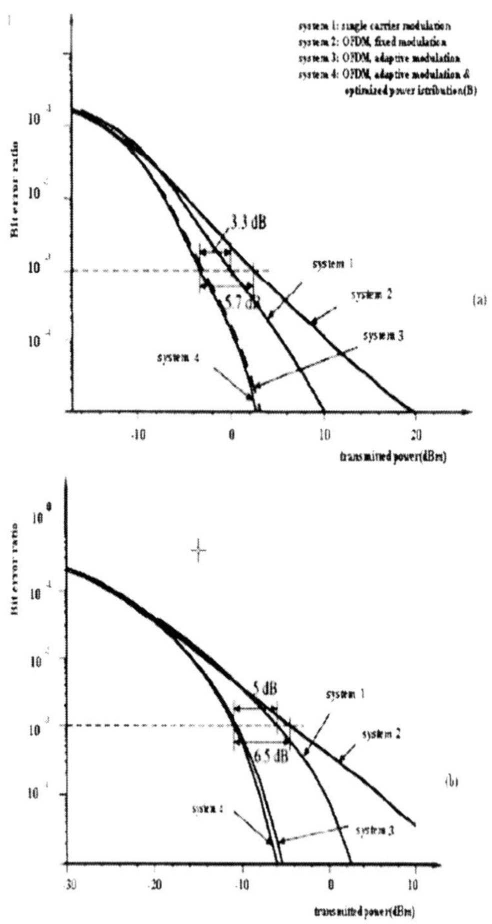

propagation measurements of radio channels with fixed antennas), it is safe to assume that the instantaneous channel transfer function can be estimated at the receiver and can be communicated back to the transmitter. The third block transforms the symbols into time-domain using inverse fast Fourier transform (FFT) at the transmitter. The next block inserts the guard interval. The output signal is transmitted over the radio channel. At the receiver, the cyclic extension is removed and the signal is transformed back into frequency domain with an FFT. Prior to demodulation, the signal is equalized in frequency domain with the inverse of the transfer function of the radio channel corresponding to a zero-forcing equalizer.

In this article, we have considered two different adaptive modulator/demodulator pairs A and B. In modulator A, the distribution of bits on the individual sub-carriers is adapted to the shape of the transfer function of the radio channel. Modulator B optimizes simultaneously both, the distribution of bits and the distribution of signal power with respect to frequency.

The algorithms for the distribution of bits and power are described in (Czylwik, 1997). The adaptive modulators select from different QAM modulation formats: no modulation, 2-PSK, 4-PSK, 8-QAM, 16-QAM, 32-QAM, 64-QAM, 128-QAM, and 256-QAM. This means that 0, 1, 2, 3... 8 bit per sub-carrier and FFT block can be transmitted. In order to get a minimum overall error probability, the error probabilities for all used sub-carriers should be approximately equal.

In case of modulator A, the distribution of bits is carried out in an optimum way so that the overall error probability becomes minimum. The algorithm for modulator A maximizes the minimum (with respect to all sub-carriers) SNR margin (difference between actual and desired SNR for a given error probability).

Modulator B optimizes the power spectrum and distribution of bits simultaneously. The result of modulator B is that the same SNR margin is achieved for all sub-carriers. The obtained SNR margin is the maximum possible so that the error probability becomes minimum. Therefore, modulator B calculates the optimum distribution of power and bits.

The results of the optimization processes of both modulator A and modulator B are shown in Figure 3. For comparison, the upper diagram gives the absolute value of the transfer function. For the specific example presented in Figure 3, both modulators yield the same distribution of bits.

Figure 3. Simulation results for a non-line-of-sight (NLOS) radio channel with a) a fixed (measurement 3) and b) a mobile (measurement 4) user

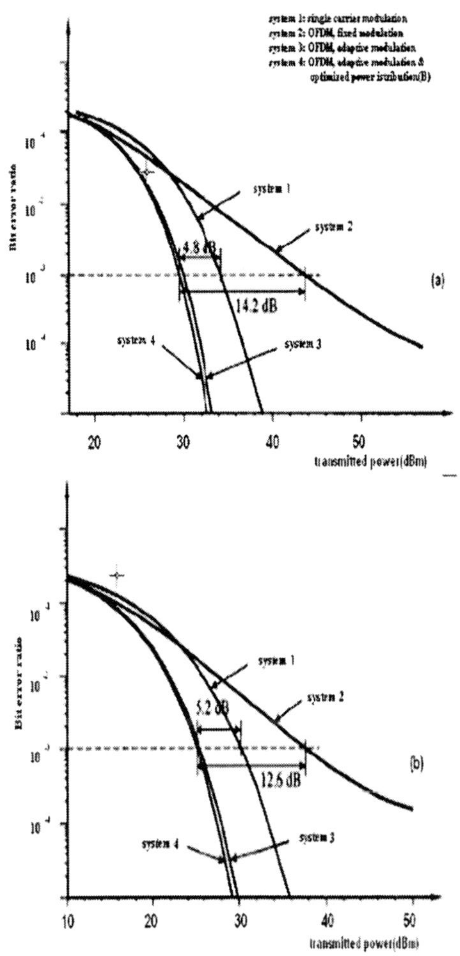

Furthermore, the power distribution and SNR is shown for both modulators.

SINGLE CARRIER TRANSMISSION WITH FREQUENCY DOMAIN EQUALIZATION

The lower part of the block diagrams in Figure 1 shows the considered single carrier transmission system. The figure shows that the basic concepts for single carrier modulation with frequency domain equalization and OFDM transmission are almost similar. The main difference, as shown by Sari et al. (1994) is that the block "inverse FFT" is moved from the transmitter to the receiver. Therefore, both, single carrier modulation and OFDM without adaptation exhibit the same complexity. Moreover, since in case of single carrier modulation too the FFT algorithm is used, a block-wise signal transmission has to be carried out. Similarly in an OFDM system, a periodic extension (guard interval) is required in order to mitigate inter-block interference.

In order to realize a constant bit rate transmission, a fixed symbol alphabet is used, for single carrier modulation in contrast to adaptive OFDM. There is however, a basic difference between the single and multi-carrier modulation schemes: In case of the single carrier system, the decision is carried out in time domain, whereas in case of the multi-carrier system the decision is carried out in frequency domain. In case of the single carrier system, an inverse FFT operation is located between equalization and decision. This inverse FFT operation spreads the noise contributions of all the indivi.dua l sub-carriers on l all the samples in time domain. Since the noise contributions of highly attenuated

Sub-carriers can be rather large; a zero-j-forcing equalizer shows a poor noise performance. Because of this reason, a minimum mean square error (MMSE) equalizer is used for the single carrier system. The transfer function of the equalizer *H,(w,t)* depends on the SNR of the respective sub-carriers

S/Nlr(w, t) at the input of the receiver:

$$H_e(w,t) = \frac{1}{H(w,t)} \cdot \frac{S/N|_r (w,t)}{S/N|_r (w,t)+1}$$

H(w, t) denotes the time-variant transfer function of the radio channel. For large SNRs the MMSE equalizer turns into the zero-forcing equalizers which multiplies with the inverse transfer function.

The main advantage of single carrier modulation compared with multi-carrier modulation is the fact that the energy of individual symbols is distributed over the whole available frequency range. Therefore, narrowband notches in the transfer function have only a small impact on error probability. Furthermore, the output signal of a single carrier transmitter shows a small crest factor whereas an OFDM signal exhibits a Gaussian distribution.

SIMULATION RESULTS

In the present article, by measuring transfer function of the channels, following systems are compared:

1. Single carrier modulation with minimum mean square error (MMSE) frequency domain equalizer
2. OFDM with fixed modulation
3. Sub-carriers and frequency-independent power distribution, OFDM with optimized modulation schemes and frequency-independent power distribution (modulator A)
4. OFDM with optimized modulation schemes and optimized power distribution (modulator B).
5. For all transmission systems

a. a complex base band simulation is carried out with ideal channel estimation and synchronization.

b. No over sampling was used since only linear components (except the detectors) are assumed in the transmission systems.

c. The temporal location of the FFT interval with respect to the cyclic extension at the receiver (i.e. the time synchronization of the OFDM blocks) is optimized so that the bit error ratio becomes minimum.

d. For both, single carrier and multi-carrier modulation, QAM schemes with different bandwidth efficiencies are used.

Simulation results for four typical radio channels at a carrier frequency of 1.8 GHz are presented. Table 1 summarizes the parameters for all measurements. In case of the mobile scenarios (measurements 2 and 4), the user terminal antenna was moved over a distance of 1 m with a low velocity. Examples of the simulation results are presented in Figure 2 and 3. The figures show the bit error ratio as a function of the average transmitted power. In all the examples shown, 16- QAM (bandwidth efficiency: 4 bit/symbol) is used for single carrier modulation and fixed OFDM (systems 1 and 2). In case of adaptive modulation, the average bandwidth efficiency is the same as in case of fixed modulation. Therefore, only transmission systems with the same average data rate are compared. The main parameters of the simulations are shown in Table 2.

The results show that an enormous improvement in performance (12 to 14 dB) is obtained from OFDM with adaptive modulation. Adaptive OFDM also shows a significant gain compared with single carrier modulation. But only a gain of less than 0.5 dB is achieved using an optimized power spectrum for OFDM instead of a frequency-independent. Because of this small difference, it is recommended to use a constant power spectrum in order to save computational or signaling effort.

For the LOS measurements also, a significant gain (5 to 6 dB) is obtained from adaptation, but the gain is smaller than in the NLOS case. This results from a higher coherence bandwidth of the LOS radio channel transfer function. Particularly in the NLOS case with single carrier modulation, a high gain (7 to 9 dB) compared with fixed OFDM is obtained. In case of the LOS channels single carrier modulation yields only a signal gain of 1 to 2 dB.

Additional simulations show that the gain from adaptive modulation increases when higher-

Table 1. Parameters of radio channel propagation measurements

Measurement	1	2	3	4
Distance of antennas	100m	100m	250m	250m
Propagation conditions	LOS	LOS	NLOS	NLOS
Base station antenna	Omni directional fixed	Omni directional fixed	Sectional fixed	Sectional fixed
User terminal antenna	Omni directional fixed	Omni directional fixed	Omni directional fixed	Omni directional fixed
Carrier frequency	1.8 GHz			
Average attenuation	77.4 dB	66.1 dB	112 dB	105.5 dB
Delay spread	0.41 ms	0.34 s	1.43 ms	0.74 ms

Table 2. Simulation parameters

Length of FFT interval	**256 samples**
Length of guard interval	50 samples
RF bandwidth	5MHz
Average data rate	16.8 Mbps
Noise figure of the receiver	6 dB
Number of transmitted bits	$2\ 10^5$

level modulation schemes are used. Furthermore, adaptive OFDM is less sensitive to inter-block interference due to an insufficient long guard interval than fixed OFDM and single carrier modulation Czylwik (1997).. This can be explained by the fact that in the adaptive system, bad channels are not used or only used with small signal alphabets so that a small amount of inter-block interference is not critical. But adaptive OFDM exhibits also some disadvantages: The calculation of the distribution of modulation schemes causes a high computational effort. Additionally, the channel must not vary too fast because of the required channel estimation. A rapidly varying channel causes also a high amount of signaling information with the effect that the data rate for the communication decreases. Furthermore, an OFDM signal exhibits a Gaussian distribution with a very high crest factor. Therefore, linear power amplifiers with high power consumption have to be used.

If channel coding is included in the transmission system also, it has been shown in Sari, Karama, & Jeanclaudea (1994) that OFDM with fixed modulation schemes shows approximately the same performance as single carrier modulation with frequency domain equalization.

The better performance of adaptive OFDM compared with single carrier modulation results due to the capability of adaptive OFDM to adapt the modulation schemes to sub-channels with very different SNRs in an optimum way. In order to improve the performance of single carrier modulation, the latter can be combined with antenna diversity using maximum ratio combining (Kadel, 1996).

CONCLUSION

On the basis of the simulations / analysis done in this article, the following conclusions can be drawn:

1. By using adaptive modulation schemes for the individual sub-carriers in an OFDM transmission system, the required signal

power can be reduced dramatically compared with fixed modulation.
2. Simulations show that for a bit error ratio of a gain of 5 to 14 dB can be achieved depending on the radio propagation scenario.
3. Significantly better performance is obtained with single carrier modulation than with OFDM with fixed modulation schemes.
4. Adaptive OFDM outperforms single carrier modulation by 3 to 5 dB.
5. In addition to the modulation schemes (bit distribution) also the power distribution of adaptive OFDM can be optimized.
6. simulations reveal that from the optimum power distribution only a small gain of less than 0.5 dB is obtained. Therefore, it is recommended to refrain from optimizing the power distribution since either additional computation or additional signaling for the synchronization is needed.

Finally, it is to be noted that with adaptive OFDM and single carrier modulation, higher gains - compared with conventional OFDM-are obtained for NLOS channels than for LOS channels. Since NLOS radio channels exhibit usually higher attenuation, this property is of particular advantage. Furthermore, the simulation results yield no significant differences between radio channels with fixed and mobile user triennial antennas.

REFERENCES

Chowj, P. S., Cioffi, M., & Binghama, J. A. C. (1995). Practical discrete multitone transceiver loading algorithm for data transmission over spectrally shaped channels. *IEEE Transactions on Communications*, *43*, 773–775. doi:10.1109/26.380108

Czylwik, C. (1996). Comparison of the channel capacity of wideband radio channels with achievable data rates using adaptive OFDM. *Proceedings of the 5th European Conference on Fixed Radio Systems and Networks ECRR '96, Bologna,* (pp. 238-243).

Czylwik, C. (1997). Adaptive (OFDM for wideband radio channels. *Proceedings of the GLOBECOM '96, London,* (pp. 713-718).

Hirosakai, B. Yoshida, Tanakas, Hasegawak, Inoue, & Watanabea, K. (1985). 191.2 kbps voice band data modem based on orthogonally multiplexed QAM techniques. *IEEE International Conference on Communications* (pp. 661-665).

Hughes-Hartogs, D. (1987). *Ensemble modem structure for imperfect transmission media.* U. S. Patent 4,679,227

Kadel, G. (1996). Diversity and equalization in frequency domain - a robust and flexible receiver technology for broadband mobile communication systems. In *Proceedings of the IEEE Vehicular Technology Conference '97, Phoenix*

Sari, H., Karama, G., & Jeanclaudea, I. (1994). An analysis of orthogonal frequency-division multiplexing for mobile radio applications. *Proceedings of the VTC '94 in Stockholm,* (pp. 1635-1639).

Sari, H., Karama, G., & Jeanclaudea, I. (1994). Frequency-domain equalization of mobile radio and terrestrial broadcast channels. *Proceedings of the Globecom '94 in San Francisco* (pp. 1-5).

This work was previously published in International Journal of Mobile Computing and Multimedia Communications (IJMCMC) 1(3), edited by Ismail Khalil & Edgar Weippl, pp.1-7, copyright 2009 by IGI Publishing (an imprint of IGI Global).

Chapter 10
Definition and Analysis of a Fixed Mobile Convergent Architecture for Enterprise VoIP Services

Joel Penhoat
France Telecom Orange Labs, France

Olivier Le Grand
France Telecom Orange Labs, France

Mikael Salaun
France Telecom Orange Labs, France

Tayeb Lemlouma
Université de Rennes 1 IRISA, France

ABSTRACT

Fixed Mobile Convergence is an important challenge for telecommunication operators given the heterogeneity of access networks technologies and the variety of terminals. Fixed Mobile Convergence, which introduces the concept 'Being always best connected', is considered to be the next step in the evolution of telecommunication networks and should increase the operators' revenues. In order to enforce the concept 'Being always best connected', this paper presents and analyzes an architecture for enterprises, named 'Business Zone'. After defining the Business Zone, we present its architecture and we analyze its main components while limiting our study to the transport of VoIP flows. Then we present two methods that we have patented: the first method authorizes a VoIP flow to be transmitted according to the available resources in the Business Zone; the second method enhances the decision process during a handover.

DOI: 10.4018/978-1-60960-563-6.ch010

INTRODUCTION

Fixed Mobile Convergence (FMC) is an important challenge for telecommunication operators because of the heterogeneity of access networks technologies and the diversity of terminals. From a technical point of view, FMC is considered to be the next step in the evolution of telecommunication networks. Bihannic *and al* (BIHANNIC, 2009) provide an overview of the different technologies and the different network elements involved in defining the operator convergence strategy. From a commercial point of view, FMC enables new services for residential users and enterprises and should increase the operators' revenues (Katsianis, 2007).

In this proliferation context, the concept 'Being always connected' becomes 'Being always best connected'. In order to enforce the concept 'Being always best connected', this paper presents and analyzes a FMC architecture for enterprises, named 'Business Zone' (BZ).

In the BZ, a user can use a fixed and/or a mobile terminal. A mobile terminal compliant with the Third Generation Partnership Project Unlicensed Mobile Access (3GPP UMA) specification (3GPP TS 43.318, 2007), is named 'UMA terminal'. To facilitate handovers between heterogeneous networks, the BZ implements a mechanism compliant with the MIH (Media Independent Handover) Services Standard as defined in (IEEE 802.21, 2008). During a handover, a mobile terminal selects a network from a network list located in a database named 'Network Base'. This database provides the topology of the networks constituting the BZ.

The remainder of this paper is organized as follows. After defining the Business Zone, we present the BZ architecture and analyze the main BZ components while limiting the scope of our study to the transport of VoIP flows. Then, we describe a method which authorizes a UMA terminal to transmit or receive a VoIP flow according to the available resources in the BZ. We have patented this method. After that, we define a method to automatically update the WiFI network topology in the Network Base. This method, that we have patented and which can be extended to different radio access network technologies, enhances the decision process during a handover. Then we conclude the paper and introduce the foreseen future studies.

DEFINITION OF THE BUSINESS ZONE

The BZ is a logical network that enables a small enterprise to access to voice, video and data services. It is composed of four network technology types, Ethernet, WiFI, GSM, UMTS, and is connected to an IP Multimedia Subsystem (IMS) architecture. Camarillo (2008) describes an overview of an IMS architecture. The BZ implements the 3GPP UMA specification (3GPP TS 43.318, 2007), the IEEE 802.21 Standard (IEEE 802.21, 2008), the IEEE 802.11e Standard (IEEE 802.11, 2007), the IEEE 802.11i Standard (IEEE 802.11, 2007), the IEEE 802.11k Standard (IEEE 802.11k, 2008), the IEEE 802.11r Standard (IEEE 802.11r, 2008).

Fixed terminals can access to Voice, Video and Data services while the UMA terminals can access to voice and data services. A UMA terminal has an AMR (Adaptive Multi-Rate) codec compliant with the 3GPP TS 26.071 specification (3GPP TS 26.071, 2007). A packet sent by an Adaptive Multi-Rate codec is named 'AMR' packet.

The BZ transports voice, video, data and UMA signalling flows. These flows are classified into 'FIXED' flows and 'UMA' flows. A 'FIXED' flow is a flow sent or received by a fixed terminal. A 'UMA' flow is a flow sent or received by a UMA terminal. A UMA signalling flow is composed of the flows used to set up and maintain an IPsec connection in ESP UDP-Encapsulated-Tunnel mode between a UMA terminal and the SEGW (Secure Gateway). In this paper, a 'FIXED' voice flow is named 'FIXED Voice' flow, and a 'UMA' voice flow is named 'UMA Voice' flow. A 'FIXED' video and/or data flow is named 'FIXED Video+Data' flow and a 'UMA' data flow is named 'UMA Data' flow. A UMA signalling flow is named 'UMA Sign' flow.

THE ARCHITECTURE OF THE BUSINESS ZONE

Figure 1 presents the architecture of the BZ. In this section, we analyze the major components necessary to transport the VoIP flows. In order to facilitate the proper understanding of the architecture, the analysis looks separately at the IP level and at the Link level.

Analysis of the Architecture at the IP Level

At the IP level, the key points to be analysed are:

- The mechanism to supply an IPv4 address to a UMA terminal;
- The establishment of the IPsec connection between a UMA terminal and the SEGW;
- The use of QoS mechanisms.

Regarding the mechanisms to supply an IPv4 address to a UMA terminal, the assumption is that the BZ has non routable IPv4 addresses compliant with the RFC 1918 (RFC 1918, 1996). A DHCP (Dynamic Host Configuration Protocol) server located in the BZ grants non routable IPv4 addresses to the BZ equipments. This DHCP server is compliant with the RFC 3361 (RFC 3361, 2002) to enable a terminal to obtain the IPv4 address of the P-CSCF (Proxy-Call Session Control Function). The P-CSCF analyzes SIP (Session Initiation Protocol) messages received from the terminals before sending them to further entities of the IMS architecture (Camarillo, 2008). To communicate with the equipments located in the public networks, a NAT (Network Address Translation) function is implemented on the gateway interface connecting the BZ to the fixed access network.

Concerning the establishment of the IPsec connection between a UMA terminal and the SEGW

Figure 1. Architecture of the Business Zone

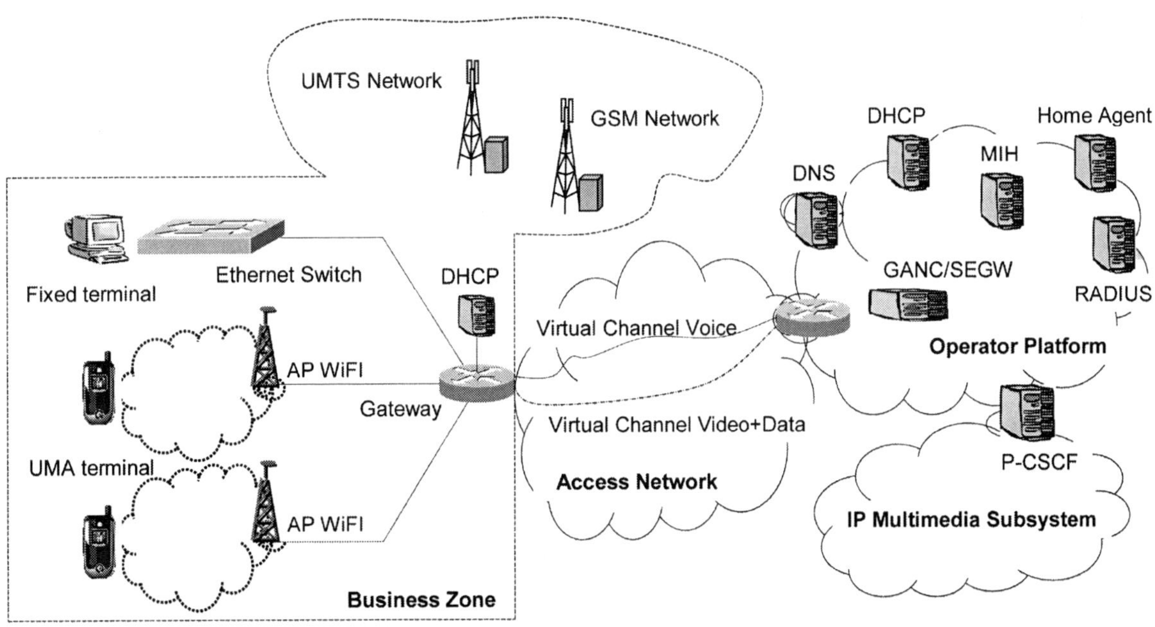

(3GPP TS 43.318, 2007), a UMA terminal connected to a WiFI network makes use of the ESP UDP-Encapsulated-Tunnel mode when setting up an IPsec connection with the SEGW located in the GANC (Generic Access Network Controller). During IPsec connection establishment, the IKEv2 protocol (RFC 4306, 2005) creates two unidirectional security associations called 'IPsec Security Association'.

During the IPsec connection establishment process, three main operations are carried out:

- The detection of a NAT function on the gateway interface connecting the BZ to the fixed Access Network in accordance with the RFC 3947 (RFC 3947, 2005) and the RFC 3948 (RFC 3948, 2005);
- The authentication of the UMA terminal using the EAP-SIM (RFC 4186, 2006) or EAP-AKA (RFC 4187, 2006) mechanisms. The RADIUS server (see Figure 1) queries the HLR (Home Location Register) using the UMA terminal IMSI (International Mobile Subscriber Identity) to obtain authentication vectors;
- The allocation of a routable IPv4 address to the UMA terminal by a DHCP server. The DHCP server is located in the operator platform (see Figure 1) and manages the routable IPv4 addresses of all the BZs.

At the end of the IPsec connection establishment process, the UMA terminal owns two IPv4 addresses: a non routable IPv4 address allocated by the BZ DHCP server, and a routable IPv4 address allocated by the DHCP managing the routable IPv4 addresses of all the BZs. Figure 2 illustrates the format of an IPsec packet in ESP UDP-Encapsulated-Tunnel mode and points out the authenticated fields and the ciphered fields. The non routable IPv4 address is located in the IPsec header, while the routable IPv4 address is located in the IP header. The UDP header is necessary because of the presence of the NAT function.

The Mobile IPv4 protocol (RFC 3344, 2002) is implemented in the UMA terminals. After the IPsec connection establishment process, the UMA terminal sends its routable IPv4 address to its Home Agent.

When there is no traffic on an IPsec connection, a mechanism for keeping alive this connection in the gateway NAT table is indispensable. In accordance with the RFC 3947 (RFC 3947, 2005) and the RFC 3948 (RFC 3948, 2005), a 'NAT-keepalive' packet is sent every twenty seconds by the UMA terminal when its IPsec connection is in the idle state. The SEGW ignores these packets.

Finally, with regard to QoS mechanisms: since the resources necessary to transmit the flows in the BZ are limited, flows are classified according to transmission priority: the highest transmission priority is granted to the voice flows, then to the video and data flows. The video flows and the data flows have the same priority. The classification is based on the DSCP (Differentiated Service Code Point) field contained in the header of each IP packet (RFC 3260, 2002). The classification rules are:

Figure 2. Format of an IPsec packet in ESP UDP-Encapsulated-Tunnel mode

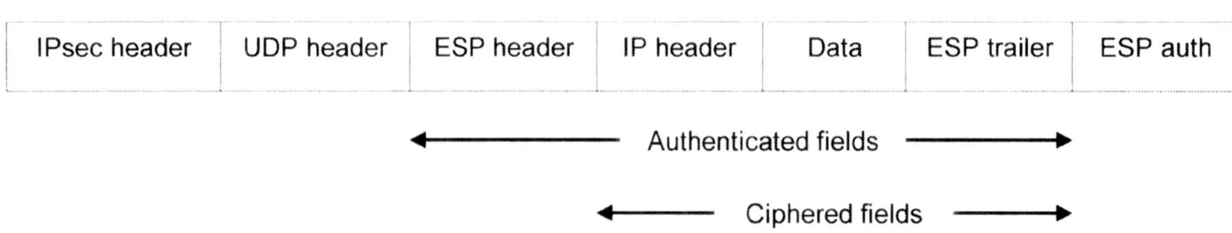

- The 'FIXED Voice' flows have a DSCP value equal to DSCP_1;
- The 'UMA Voice' flows have a DSCP value equal to DSCP_2;
- The 'FIXED Video+Data' flows have a DSCP value equal to DSCP_3;
- The 'UMA Data' flows have a DSCP value equal to DSCP_3;
- The 'UMA Sign' flows have a DSCP value equal to DSCP_3;
- The values DSCP_1, DSCP_2, DSCP_3 are such as: DSCP_1 > DSCP_2 > DSCP_3.

By referring to the Figure 2, we notice that the IP header is ciphered. Therefore, to classify the flows, the UMA terminal and the SEGW must re-write the DSCP value from the IP header to the IPsec header before ciphering the IP header.

Analysis of the Architecture at the Link Layer

At the Link Layer, the key points are:

- QoS mechanisms in the WiFI networks;
- Protection of the flows in the WiFI networks;
- Network selection during a handover;
- Re-authentication during a WiFI-WiFI handover;
- Implementation of two Virtual Channels on the ATM interface of the gateway.

The QoS in the WiFI networks: the WiFI networks implement a QoS mechanism compliant with the IEEE 802.11e Standard (IEEE 802.11, 2007). In accordance to this standard, the WiFI frames are classified in four classes: AC_BK, AC_BE, AC_VI, AC_VO. The classification is based on the User Priority field contained in the QoS Control field in each MPDU (Mac Protocol Data Unit) header frame. To achieve consistency between the classification rules at the IP level and at the Link Level, the UMA terminals and the gateway implement a mapping between the DSCP field and the User Priority field (see Table 1). In accordance with the Home Gateway Initiative recommendations (HGI, 2008), the User Priority value is obtained by taking the three most significant bits of the DSCP value.

The protection of the flows in the WiFI networks: in each WiFI network, we implement the CCMP (Counter Mode with CBC-MAC Protocol) protocol (IEEE 802.11, 2007) to protect the flows. Since the protection applies to all flows (ciphered or not with the IPsec protocol), an IPsec connection is also ciphered with the CCMP protocol. The format of a PLCP (Physical Layer Convergence Protocol) frame carrying one AMR packet is compliant with the one shown in Figure 3. In this figure, a PLCP frame carries one IPsec

Table 1. Mapping between the DSCP field and the User Priority field

DSCP	User Priority	IEEE 802.11e Access Category	IEEE 802.11e designation
0x08 (DSCP_3)	1	AC_BK	Background
0x10 (DSCP_3)	2	AC_BK	Background
0x00 (DSCP_3)	0	AC_BE	Best Effort
0X18 (DSCP_3)	3	AC_BE	Best Effort
0x20 (DSCP_3)	4	AC_VI	Video
0x28 (DSCP_3)	5	AC_VI	Video
0x30 (DSCP_2)	6	AC_VO	Voice
0x38 (DSCP_1)	7	AC_VO	Voice

Figure 3. Format of a PLCP frame carrying one AMR packet

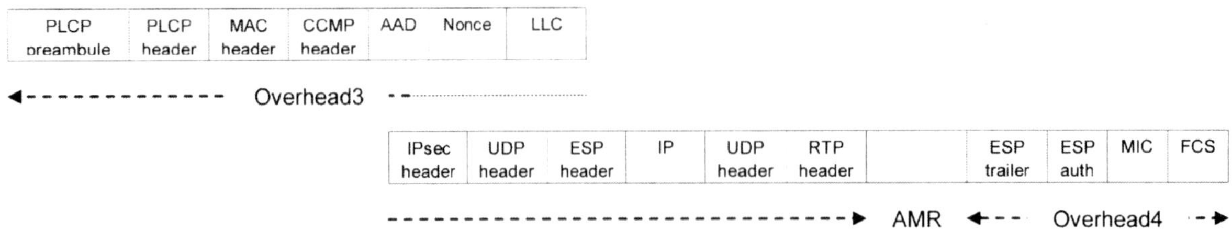

packet, and the IPsec packet carries one AMR packet.

The network selection during a handover: the BZ implements a network selection process compliant with the MIH Services Standard (IEEE 802.21, 2008). A MIH server, located in the operator platform (see Figure 1), contains a network list in the Network Base. The messages 'MIH_Get_Information request' and 'MIH_Get_Information response' (IEEE 802.21, 2008) exchanged between a mobile terminal and the MIH server enable the terminal to learn this list. The mobile terminal stores the list in a base named 'Local Network Base'. Figure 4 shows the protocol stack in a UMA terminal compliant with the MIH Services Standard (IEEE 802.21, 2008).

During a handover, the UMA terminal Network Selector Entity (see Figure 4) selects a network in its Local Network Base based on four criteria: the network list in its Local Network Base, the constraints of the application to transmit, the user's preferences, the characteristics of the networks present in its neighbourhood. The IEEE

Figure 4. Protocol stack in a UMA terminal compliant with the MIH Services Standard

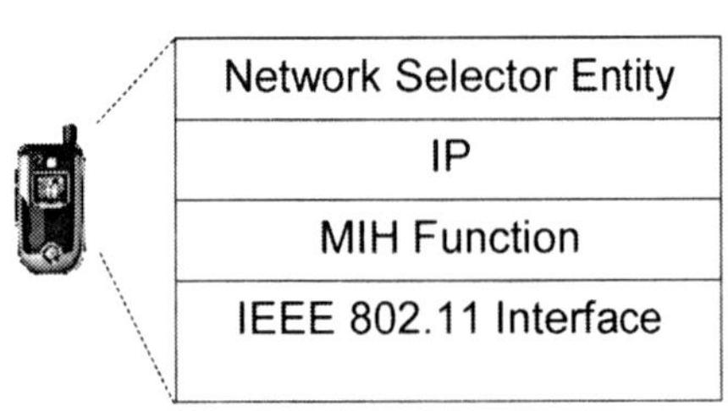

802.11k Standard (IEEE 802.11k, 2008) enables the UMA terminal to obtain the characteristics of the WiFI networks present around it.

The re-authentication during a WiFI-WiFI handover: after selecting a WiFI network, a UMA terminal must re-authenticate itself to the new Access Point. During the re-authentication process, we use the IEEE 802.11r Standard (IEEE 802.11r, 2008) to decrease the number of exchanged messages: four messages are exchanged between the terminal and the Access Point (802.11 Authentication Request, 802.11 Authentication Response, Reassociation Request, Reassociation Response) instead of twenty three messages exchanged between the terminal, the Access Point, the RADIUS server and the HLR (Home Location Register) with the EAP-SIM mechanism. Since the re-authentication delay is reduced, the QoE (Quality of Experience) is improved during a WiFI-WiFI handover.

The implementation of two Virtual Channels on the ATM interface of the gateway: the gateway has an ATM interface connecting the BZ to the fixed Access Network (see Figure 1). Two Virtual Channels are implemented on this interface:

- A Virtual Channel 'Voice' carrying the 'FIXED Voice' flows and the 'UMA Voice' flows. Each Ethernet frame carrying an IPsec packet (RFC 2684, 1999) is encapsulated in an AAL5 CPCS frame. Then, each AAL5 CPCS frame is conveyed in the ATM Virtual Channel 'Voice'. Figure 5-a presents the format of an AAL5 CPCS frame carrying an AMR packet. Figure 5-b presents an AAL5 CPCS frame conveyed over several ATM cells;

- A Virtual Channel 'Video+Data' carrying the 'FIXED Video + Data' flows, the 'UMA Data' flows and the 'UMA Sign' flows.

Admission of a 'UMA Voice' Flow in the Business Zone

In this section we describe the admission control of 'UMA Voice' flows taking into account the available resources in the BZ. Then we briefly describe the admission protocol which authorizes or denies the admission of a new 'UMA Voice' flow based on the available resources in the BZ.

The Conditions of Admission of a 'UMA Voice' Flow in the Business Zone

Since the resources in the BZ are limited, it is essential to define the conditions of admission of a 'UMA Voice' flow. We define these conditions in three steps. In the first step, we calculate, at time t, the available rate for conveying the 'UMA Voice' flows over the Virtual Channel 'Voice'. In the second step, we calculate, at time t, the available rate for conveying the 'UMA Voice' flows over a WiFI network. In the third step, we authorize or deny the admission of a new 'UMA Voice' flow based on the available rates calculated in the first two steps.

The first two steps use the notations below:

- The number of packets per second sent by an Adaptive Multi-Rate codec is named PPS;
- A packet sent by an Adaptive Multi-Rate codec is named AMR;
- The function that calculates the number of bytes in a packet is named length(packet);
- A flow sent by a terminal is named uplink flow;
- A flow received by a terminal is named downlink flow.

Calculation of the Available Rate for Conveying the 'UMA Voice' Flows over the Virtual Channel 'Voice'

The calculation of the available rate for conveying the 'UMA Voice' flows over the Virtual Channel 'Voice' uses the notations below:

- The uplink rate of the Virtual Channel 'Voice' is named Rate_ATM_VC_Voice_up. This Virtual Channel offers a CBR (Constant Bit Rate) service;
- The total uplink rate of the 'FIXED Voice' flows is named Rate_ATM_Total_Fixed_Voice_up. This connection offers a CBR service;
- The total uplink rate of the 'UMA Voice' flows is named Rate_ATM_Total_UMA_Voice_up. It is equal to the difference (Rate_ATM_VC_Voice_up – Rate_ATM_Total_Fixed_Voice_up);
- The uplink rate of a 'UMA Voice' flow is named Rate_ATM_UMA_Voice_up;
- The downlink rate of the Virtual Channel 'Voice' is named Rate_ATM_VC_Voice_down. This Virtual Channel offers a CBR service;
- The total downlink rate of the 'FIXED Voice' flows is named Rate_ATM_Total_Fixed_Voice_down. This connection offers a CBR service;
- The total downlink rate of the 'UMA Voice' flows is named Rate_ATM_Total_UMA_Voice_down. It is equal to the difference (Rate_ATM_VC_Voice_down – Rate_ATM_Total_Fixed_Voice_down);
- The downlink rate of a 'UMA Voice' flow is named Rate_ATM_UMA_Voice_down.

At time t, (m-1) 'UMA Voice' flows are conveyed over the Virtual Channel 'Voice'. The available uplink rate for conveying new 'UMA Voice' flows is given by the equation (1):

Rate_ATM_Available_UMA_Voice_up(t) = Rate_ATM_Total_UMA_Voice_up –

$$\sum_{i=1}^{i=m-1} (Rate\_ATM\_UMA\_Voice\_up)i \qquad (1)$$

The available downlink rate for conveying new 'UMA Voice' flows is given by the equation EQUATION1-bis:

Rate_ATM_Available_UMA_Voice_down(t) = Rate_ATM_Total_UMA_Voice_down –

$$\sum_{i=1}^{i=m-1} (Rate\_ATM\_UMA\_Voice\_down)i \qquad (1\text{-bis})$$

In the equations (1) and (1-bis), the rate of the 'UMA Voice' flow number *i* is equal to the number of packets per second sent by the Adaptive Multi-Rate codec multiplied by the length of a packet conveyed over the Virtual Channel 'Voice'. The calculation of the length of a packet conveyed over the Virtual Channel 'Voice' requires the knowledge of the different encapsulations used for this packet. By referring to Figures 5-a and 5-b, the uplink or downlink rate of the 'UMA Voice' flow number *i* is given by the equation (2). In this equation, length(ATM header) = 5 bytes, and length(ATM payload) = 48 bytes.

Rate_ATM_UMA_Voice_(up or down) = PPS*(length(Overhead1 + AMR + Overhead2) +

length(ATM header)*(length(Overhead1 + AMR + Overhead2)/length(ATM payload))) (2)

Calculation of the available rate for conveying the 'UMA Voice' flows over a WiFI network

The BZ has several WiFI networks. The calculation of the available rate for conveying the 'UMA Voice' flows over a WiFI network uses the notations below:

- The rate of a WiFI network is named Rate_WLAN_Network;
- The uplink rate of a 'UMA Voice' flow is named Rate_WLAN_UMA_Voice_up;
- The downlink rate of a 'UMA Voice' flow is named Rate_WLAN_UMA_Voice_down;
- The uplink or downlink rate of a 'UMA Data' flow is named Rate_WLAN_UMA_Data;
- The uplink or downlink rate of a 'UMA Sign' flow is named Rate_WLAN_UMA_Sign;
- The uplink or downlink rate of a flow other than those previously listed is named Rate_WLAN_OTHER;

Figure 5.

Figure 5-a: Format of an AAL5 CPCS frame carrying an AMR packet

LLC-SNAP OUI	PID PAD	Ethernet header	IPsec header	UDP header	ESP header	IP header	UDP header	RTP header		ESP trailer	ESP auth	PAD	CPCS-PDU trailer

◄------ Overhead1 ------► AMR ◄------ Overhead2 ------►

Figure 5-b: An AAL5 CPCS frame conveyed over several ATM cells

ATM header	AAL5 CPCS	ATM header	AAL5 CPCS	ATM header	AAL5 CPCS	ATM header	AAL5 CPCS	ATM header	AAL5 CPCS

- The radio channel access delay is named Taccess;
- The transmission delay of a PLCP frame is named Tframe;
- The transmission delay of an acknowledgment is named Tack.

At time t, (n-1) 'UMA Voice' flows and other flows are conveyed over a WiFI network. The rate necessary to transmit these flows is given by the equation (3):

$$Rate\_WLAN\_Network(t) \geq \sum_{i=1}^{i=n-1}((Rate\_WLAN\_UMA\_Voice\_up)i + (Rate\_WLAN\_UMA\_Voice\_down)i) + \sum_{j}(Rate\_WLAN\_UMA\_Data)j + \sum_{k}(Rate\_WLAN\_UMA\_Sign)k + \sum_{1}(Rate\_WLAN\_OTHER)1 \quad (3)$$

At time t, the available rate for conveying new 'UMA Voice' flows over the WiFI network is given by the equation (4):

$$Rate\_WLAN\_Available\_UMA\_Voice(t) = Rate\_WLAN\_Network(t) \sum_{i=1}^{i=n-1}((Rate\_WLAN\_UMA\_Voice\_up)i + (Rate\_WLAN\_UMA\_Voice\_down)i) \quad (4)$$

In the equation (4), the calculation of the rate of the 'UMA Voice' flow number *i* requires the knowledge of the format of one PLCP frame carrying one AMR packet and the time necessary to transmit the PPS PLCP frames. Figure 3 presents the format of a PLCP frame carrying one AMR packet. Figure 6 presents the method to transmit the PPS PLCP frames (IEEE 802.11, 2007). When the HCCA (Hybrid coordination function Controlled Channel Access) mode of the IEEE 802.11e Standard (IEEE 802.11, 2007) is used to transmit the frames, the value of the parameter Taccess is equal to zero. When the EDCA (Enhanced Distributed Channel Access) mode of the IEEE 802.11e Standard (IEEE 802.11, 2007) is used, this parameter has a non-zero value. In that case, Hole and Tobagi (HOLE, 2004) describe a method to estimate the value of the parameter.

By referring to Figures 3 and 6, the uplink or downlink rate of the 'UMA Voice' flow number *i* is given by the equation (5):

Rate_WLAN_UMA_Voice_(up or down) = PPS*(length(Overhead3 + AMR + Overhead4)) /(Taccess + (Tframe + Tack)*PPS + (2*PPS - 1)*SIFS) (5)

The Conditions of Admission of a 'UMA Voice' Flow

When a UMA terminal, connected to a WiFI network, wants to send or receive a new 'UMA Voice' flow, the BZ checks that the available rate in the Virtual Channel 'Voice' and the available rate in the WiFI network permit the transmission of the new flow. The equations (6), (6-bis) and

Figure 6. Method to transmit the PPS PLCP frames

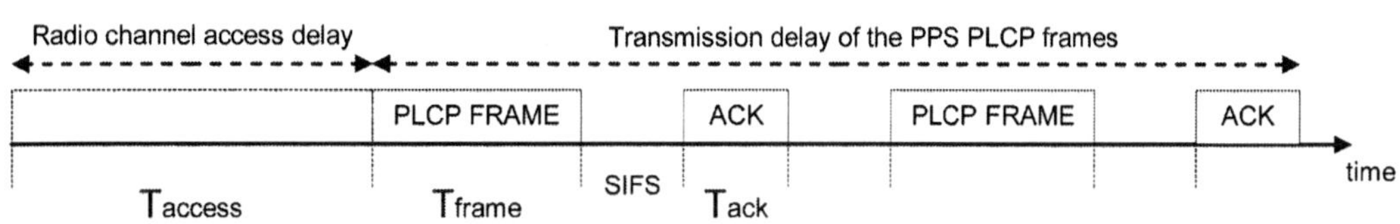

(7), defined below, convey these two conditions of admission.

At time tk, the Virtual Channel 'Voice' transmits (m-1) 'UMA Voice' flows. The admission of the flow number *m* is possible if the equations (6) and (6-bis) are verified:

Rate_ATM_Available_UMA_Voice_up(tk) > (Rate_ATM_UMA_Voice_up)m (6)

Rate_ATM_Available_UMA_Voice_down(tk) > (Rate_ATM_UMA_Voice_down)m (6-bis)

At time tk, the WiFI network, on which the UMA terminal is connected, transmits (n-1) 'UMA Voice' flows. The admission of the flow number *n* is possible if the equation (7) is verified:

Rate_WLAN_Available_UMA_Voice(tk) > (Rate_WLAN_UMA_Voice_up)n +

(Rate_WLAN_UMA_Voice_down)n (7)

The Admission Protocol of a 'UMA Voice' Flow

The equations (6), (6-bis) and (7) show that the admission of a new 'UMA Voice' flow is based on a calculation of a rate made by the BZ gateway (equations (6), (6-bis)) and by the Access Point on which the UMA terminal is connected (equation (7)). Since the checking of the equations (6), (6-bis) and (7) is done by different equipments (the gateway and the Access Points), we define a protocol allowing a dialogue between these equipments. In this chapter, we give an overview of this protocol.

At time t, when a UMA terminal wants to send a call, it transmits an admission request towards its Access Point by sending an ADDTS Request frame (IEEE 802.11, 2007). On receipt of the request, the Access Point checks the equation (7).

- If the equation (7) is not verified, the Access Point transmits towards the terminal an ADDTS Response frame (IEEE 802.11, 2007) to tell it that it can't send the call;
- If the equation (7) is verified, the Access Point transmits an admission request towards the gateway;
  - If the equation (6) or the equation (6-bis) is not verified, the gateway informs the Access Point that it can't transmit the call. On receipt of this information, the Access Point transmits towards the terminal an ADDTS Response frame (IEEE 802.11, 2007) to tell it that it can't send the call;
  - If the equations (6) and (6-bis) are verified, the gateway informs the Access Point that it can transmit the call. On receipt of this information, the Access Point transmits towards the terminal an ADDTS Response frame (IEEE 802.11, 2007) to tell it that it can send the call.

At time t', when the gateway receives a call from the Access Network (see figure 1), it transmits an admission request towards the Access Point on which the terminal is connected. On receipt of the request, the Access Point checks the equation (7).

- If the equation (7) is not verified, the Access Point informs the gateway that it can't receive the call;
- If the equation (7) is verified, the Access Point informs the gateway that it can receive the call.

Method to Automatically Update the WiFI Network Topology Contained in the MIH Server Network Base and in the Mobile Terminal Local Network Base

Since there is no mechanism for a mobile terminal to update the Network Base, we define a method to automatically update the WiFI network topology in the MIH server Network Base and in the mobile terminal Local Network Base. This method, which can be extended to different radio access network technologies, enhances the handover decisions.

During the roaming of a user, his mobile terminal can know the WiFI network topology in its neighbourhood by scanning the networks with the IEEE 802.11k Standard (IEEE 802.11k, 2008). The information collected by the terminal enables an automatic update of the WiFI network topology in the Local Network Base. This information is important for the terminal and for the MIH server since contributing to the decision-making process during a handover. Therefore, after a scanning, the terminal must be able to transmit to the collected information to the MIH server thus automatically updating the MIH server Network Base.

The MIH server can also automatically inform the terminal about the modifications of the WiFI network topology contained in its Network Base, e.g. when the operator managing the MIH server modifies the Network Base. The term 'modification' means adding or removing one or several WiFI Access Point(s) in the Network Base.

The automatic update of the WiFI network topology in the Local Network Base and in the Network Base requires the definition of a protocol to exchange information between a mobile terminal and a MIH server. Since the IEEE 802.21 Standard (IEEE 802.21, 2008) does not support such protocol, we have specified a peer-to-peer protocol between a mobile terminal and a MIH server. This protocol enables a MIH server to validate the information submitted by a mobile terminal; it provides a reliable transmission of the updates; it avoids the flooding of the MIH server and the flooding of the mobile terminals; it also avoids bursts of packets in the networks.

Before presenting this protocol, it is necessary to examine the structure of a WiFI entry of the mobile terminal Local Network Base and of the MIH server Network Base. A WiFI entry identifies the characteristics of a single Access Point.

Structure of a WiFI Entry of the Mobile Terminal Local Network Base

A WiFI entry in the mobile terminal Local Network Base is made up of eight fields:

- ACCESS-TYPE: a two bytes length field containing the type of the network. The syntax is compliant with (IANA, 2009);
- BSSID: a six bytes length field containing the BSSID (Basic Service Set Identifier) of the WiFI network;
- AUTHORIZE: a one byte length field indicating if the MIH server authorizes the terminal to connect to the Access Point during a handover. Only bit 0 is used. Bit 0 is equal to one if the MIH server authorizes the terminal to connect to the Access Point during a handover. Bit 0 is equal to zero if the MIH server doesn't authorize the terminal to connect to the Access Point during a handover;
- LOCATION: a variable length field containing the location of the Access Point. The syntax is compliant with (ABOBA, 2008);
- RATE: a variable length field containing the different rates of the Access Point. The syntax is compliant with (IEEE 802.11, 2007);
- CAPABILITIES: a four bytes length field containing the capabilities of the Access Point. The syntax is compliant with (IEEE 802.21, 2008);
- SECURITY: a variable length field containing the authentication and ciphering methods of the Access Point. The syntax is compliant with (IEEE 802.11, 2007);
- OPERATOR: a variable length field containing the operator's name managing the Access Point. The syntax is compliant with (ABOBA, 2008).

Structure of a WiFI Entry of the MIH server Network Base

A WiFI entry in the MIH server Network Base consists of seven fields. The fields named 'ACCESS TYPE', 'BSSID', 'LOCATION', 'RATE', 'CAPABILITIES', 'SECURITY', 'OPERATOR', are identical to the description provided above.

Validation of the Information Provided by a Mobile Terminal

After a scanning, a mobile terminal adds a WiFI entry in its Local Network Base for each newly detected WiFI network. The value of the AUTHORIZE field of a new WiFI entry is equal to zero. To set up to one the value of the AUTHORIZE field, the terminal sends a request towards its MIH server to validate the corresponding WiFI entry. The validation criteria are based on the policy of the operator managing the MIH server.

Reliability of Update Transmissions

The transmission of updates of the WiFI network topology between a mobile terminal and a MIH server must be reliable. Reliability of the transmission is obtained by:

- Authenticating the exchanged messages between a mobile terminal and its MIH server. The authentication method chosen is HMAC-SHA1 (FIPS, 2002);
- Retransmitting the lost messages in the networks.

If the messages exchanged between a mobile terminal and its MIH server are not lost in the networks, the number of WiFI entries in the mobile terminal Local Network Base is equal to the number of WiFI entries in the MIH server Network Base. We use UDP (User Datagram Protocol) to transport the messages. UDP has been preferred to TCP (Transmission Control Protocol) because of the time needed to open a TCP connection. Since UDP is not a reliable protocol, we have defined an acknowledgment mechanism, a protection mechanism, and a retransmission mechanism.

The acknowledgment mechanism: when receiving an update of the WiFI network topology, the receiver sends a message towards the sender in order to acknowledge the receipt of the update message. We have defined three update scenarios and two acknowledgment messages that we present below.

- Scenario 1 (Figure 7-a): a mobile terminal sends a 'Validation request' message towards its MIH server to validate one or more WiFI entries. On receipt of the request, the MIH server validates some of the WiFI entries. The validation criteria are based on the policy of the operator managing the server. Then, the server sends the validated WiFI entries to the terminal in a 'Validation response' message. Upon receipt of the validated WiFI entries, the terminal sets up the AUTHORIZE field to 1 for each validated WiFI entry;
- Scenario 2 (Figure 7-b): a MIH server sends a 'Modification' message towards a mobile terminal to indicate a modification of one or more WiFI entries. Upon receipt of this message, the terminal sends an 'Acknowledgment' message back to the server;
- Scenario 3 (Figure 7-c): a mobile terminal sends a 'Validation request' message towards the MIH server to validate one or more WiFI entries. On receipt of the request, the MIH server validates some of the WiFI entries and also can inform the terminal about a modification of one or more WiFI entries. The server sends the validated WiFI entries to the terminal and also sends in the same message the modified WiFI entries ('Validation response+Modification' message in Figure

7-c). On receipt of this message, the terminal sets up the AUTHORIZE field to 1 for each validated WiFI entry, and adds or removes WiFI entries based on the modified WiFI entries received. Then, the terminal sends an 'Acknowledgment' message towards the MIH server.

A 'Validation request' message contains the characteristics of each Access Point that the mobile terminal wishes to include in the Local Network Base. The eight fields, defined in the paragraph 'Structure of a WiFI entry of the mobile terminal Local Network Base', make up each characteristic. A UDP datagram carries the request. A twenty bytes length Message Authentication Code (FIPS, 2002) is appended to the UDP datagram to authenticate the request. Its calculation does not include the IP/UDP headers because network equipments can modify these headers.

A 'Modification' message contains the characteristics of each Access Point that the MIH server wants to add in or remove from the mobile terminal Local Network Base. A UDP datagram carries this message. A Message Authentication Code is appended to the UDP datagram to authenticate the message. As above, its calculation does not include the IP/UDP headers.

The protection mechanism: this mechanism is based on Raptor Codes (SHOKROLLAHI, 2006).

The retransmission mechanism: when a message is lost, retransmission is performed by the sender. On the MIH server side, a message is retransmitted if a timer expires before an 'Acknowledgment' message is received. The initial value of the timer is equal to the RTT (Round Trip Time) between the MIH server and the mobile terminal. Then it is multiplied by two after each retransmission of a message. On the terminal side, a message is retransmitted if a timer expires before a 'Validation response' message is received. The initial value of the timer is equal to the RTT between the mobile terminal and the MIH server. Then it is multiplied by two after each retransmission of a message.

The retransmission mechanism has two retransmission parameters, named 'R_TERM' and 'R_MIH'. The R_TERM parameter is related to the retransmission number of the mobile terminal. The R_MIH parameter is related to the retransmission number of the MIH server. We present below how these two parameters are used depending on the three scenarios previously defined.

- Scenario 1 (Figure 7-a): if, after R_TERM retransmissions, the mobile terminal has not received a 'Validation response' message, the terminal must remove in the Local Network Base the WiFI entries sent in the 'Validation request' message. As a consequence these entries will not be used during a handover. In that case, the number of WiFI entries in the Local Network Base is equal to the number of entries in the Network Base;
- Scenario 2 (Figure 7-b): if, after R_MIH retransmissions, the MIH server has not received an 'Acknowledgment' message, the number of WiFI entries in the Local Network Base should be different from the number of WiFI entries in the Network Base;
- Scenario 3 (Figure 7-c): this case is the combination of the two previous cases.

The analysis of the scenarios 2 and 3 shows that the WiFI network topology known by a mobile terminal can be different from the topology known by the MIH server. To avoid this situation, it is necessary to efficiently protect the 'Modification' and 'Validation response+Modification' messages against error transmissions. These messages are outlined in bold lines in the Figure 7.

Avoiding the Flooding of the MIH Server and Flooding of the Mobile Terminals

To avoid the flooding of the MIH server, the maximum number of 'Validation request' and 'Acknowledgment' messages sent per hour by a mobile terminal is limited. Likewise, to avoid the flooding of a mobile terminal, the maximum number of 'Modification' and 'Validation response+Modification' messages sent per hour by the MIH server is limited.

Avoiding Bursts of Packets in the Networks

To avoid bursts of packets in the networks, the mobile terminals randomly send their 'Validation request' messages. On the server side, a Token Bucket filter as defined in (WANG, 2001) provides a smooth sending rate of the IP packets.

CONCLUSION AND FUTURE WORKS

In this paper we have presented the architecture of the Business Zone which implements the concept 'Being always best connected'. We have limited our study to the transport of VoIP flows.

First we have described a patented protocol between the BZ gateway and the BZ Access Points which authorizes a UMA terminal to transmit a VoIP flow according to the available resources in the BZ.

Secondly, since there was no method for a mobile terminal to update the Local Network Base and the Network Base, we have defined and patented a method to automatically update the WiFI network topology in the Local Network Base and in the Network base. This method, which can be extended to different radio access network technologies, enhances the handover decisions.

In the future, our study will be extended to cover the transport of Video and Data flows. The first area of study aims to develop a method of admission for VoIP flows and for Video flows in the BZ. A dialogue between our admission protocol and the IMS Policy decision Function (3GPP TS 23.207, 2008) may be considered as a possible way forward.

The second area of study concerns the use of the scalable video flows to improve the QoE. A video flow is called 'scalable' when parts of the bit stream can be removed in a way that the resulting substream forms another valid bit stream for some target decoder, and the substream represents the source content with a reconstruction quality that is less than that of the complete original bit stream but is high when considering the lower quantity of remaining data. The SVC (Scalable Video Coding) extension of a H.264/AVC codec produces a scalable video bit stream made up of one or more substreams named 'layers' (SCHWARZ, 2007). The Base layer, which is H.264/AVC compliant, provides a basic video quality with a low bit rate.

Figure 7. Update scenarios

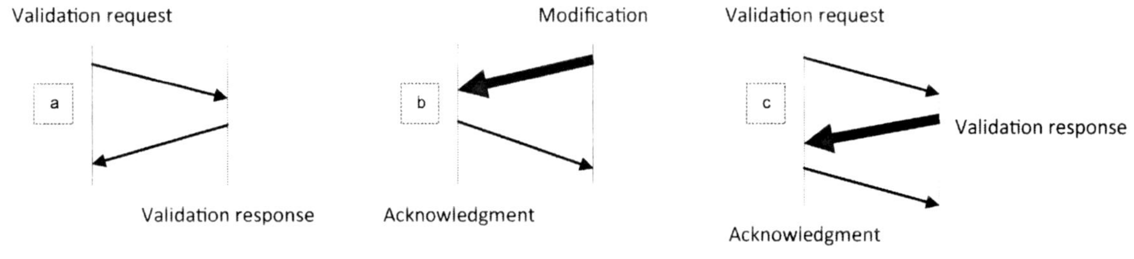

Enhancement layers are used to refine the Base layer (MARPE, 2007). When a video content is conveyed over a radio network, Grüneberg *and al* (GRUNEBERG, 2007) have shown that the selective removal of RTP (Real time Transport Protocol) packets and the implementation of a protection mechanism against transmission errors can improve the QoE in case of network congestion.

REFERENCES

Aboba, B., Adrangi, F., Jones, M., Lior, A., & Tschofenig, H. (2008, November). *IETF Geopriv: Carrying Location Objects in RADIUS and Diameter draft-ietf-geopriv-radius-lo-20.txt.*

Arkko, J., & Haverinen, H. (2006). *IETF Network Working Group: RFC 4187 Extensible Authentication Protocol Method for 3rd Generation Authentication and Key Agreement*. EAP-AKA.

Bihannic, N., Boutaba, R., Javaid, U., Meddour, D-E., & Rasheed, T. (Manuscript submitted for publication). Completing the Convergence Puzzle- from the Survey to a tentative Roadmap. *IEEE Wireless Communications Magazine.*

Camarillo, G., & Garcia-Martin, M. A. (2008). *The 3G IP Multimedia Subsystem (IMS): Merging the Internet and the Cellular Worlds* (3rd ed.). Chippenham, England: John Wiley & Sons, Ltd. doi:10.1002/9780470695135

de Groot, G. J., Karrenberg, D., Lear, E., Moskowitz, B., & Rekhter, Y. (1996). *IETF Network Working Group: RFC 1918 Address Allocation for Private Internets.*

DiBurro, L., Huttunen, A., Swander, B., Stenberg, M., & Volpe, V. (2005). *IETF Network Working Group: RFC 3948 UDP Encapsulation of IPsec ESP Packets*.

3GPP TS 23.207 V8.0.0. (2008). *3rd Generation Partnership Project; Technical Specification Group Services and System Aspects; End-to-end Quality of Service (QoS) concept and architecture* (Release 8).

3GPP TS 26.071 V7.0.1. (2007). *3rd Generation Partnership Project; Technical Specification Group Services and System Aspects; Mandatory speech CODEC speech processing functions; AMR speech CODEC; General description* (Release 7).

3GPP TS 43.318 V7.2.0. (2007). *3rd Generation Partnership Project; Technical Specification Group GSM/EDGE Radio Access Network; Generic access to the A/Gb interface; Stage 2* (Release 7).

Grossman, D. (2002). *IETF Network Working Group: RFC 3260 New Terminology and Clarifications for Diffserv.*

Grossman, D., & Heinanen, J. (1999). *IETF Network Working Group: RFC 2684 Multiprotocol Encapsulation over ATM Adaptation Layer 5.*

Grüneberg, K., Hellge, C., Mirta, S., Schierl, T., & Wiegand, T. (2007). Using H.264/AVC-based Scalable Video Coding (SVC) for Real Time Streaming in Wireless IP Networks. *IEEE International Symposium on Circuits and Systems*, ISCAS 2007.(pp. 3455-3458).

Haverinen, H., & Salowey, J. (2006). *IETF Network Working Group: RFC 4186 Extensible Authentication Protocol Method for Global System for Mobile Communications (GSM) Subscriber Identity Modules*. EAP-SIM.

Hole, D. P., & Tobagi, F. A. (2004). Capacity of an IEEE 802.11b Wireless LAN supporting VoIP. *2004 IEEE International Conference on Communications, 1,* 196-201.

Home Gateway Technical Requirements: Residential Profile Version 1.0. (2008, April). *Home Gateway Initiative.*

Huttunen, A., Kivinen, T., Swander, B., & Volpe, V. (2005). *IETF Network Working Group: RFC 3947 Negotiation of NAT-Traversal in the IKE.*

IEEE P802.21/D9.0. (2008). Draft Standard for Local and Metropolitan Area Networks: Media Independent Handover Services.

IEEE Standard for Information technology. (2007). *Telecommunications and information exchange between systems- Local and metropolitan area networks- Specific requirements- Part 11: Wireless LAN Medium Access Control (MAC) and Physical Layer (PHY) Specifications.*

IEEE Standard for Information technology. (2008). Telecommunications and information exchange between systems-Local and metropolitan area networks-Specific requirements- Part 11: Wireless LAN Medium Access Control (MAC) and Physical Layer (PHY) Specifications Amendment 1: Radio Resource Measurement of Wireless LANs IEEE 802.11k.

IEEE Standard for Information technology. (2008). Telecommunications and information exchange between systems-Local and metropolitan area networks-Specific requirements- Part 11: Wireless LAN Medium Access Control (MAC) and Physical Layer (PHY) Specifications Amendment 2: Fast Basic Service Set (BSS) Transition IEEE Std 802.11r.

Katsianis, D., Rokkas, T., Sphicopoulos, T., & Varoutas, D. (2007). Fixed Mobile Convergence For an Integrated Operator: A Techno-Economic Study. *18th Annual IEEE International Symposium on Personal, Indoor and Mobile Radio Communications* (pp. 1-5).

Kaufman, C. (2005). *IETF Network Working Group: RFC 4306 Internet Key Exchange (IKEv2).* Protocol.

Marpe, D., Schwarz, H., & Wiegand, T. (2007). Overview of the Scalable Video Coding Extension of the H.264 / AVC Standard. *IEEE Transactions on Circuits and Systems for Video Technology, 17*(9), 1103–1120. doi:10.1109/TCSVT.2007.905532

National Institute of Standards and Technology Information Technology Laboratory. (2002, March). *The Keyed-Hash Message Authentication Code (HMAC).* Federal Information Processing Standards Publication.

Online Internet Assigned Numbers Authority. (2009). *Radius Types* [WWW page]. URL http://www.iana.org/assignments/radius-types

Perkins, C. (2002). *IETF Network Working Group: RFC 3344 IP Mobility Support for IPv4.*

Schulzrinne, H. (2002). *IETF Network Working Group: RFC 3361 Dynamic Host Configuration Protocol (DHCP-for-IPv4) Option for Session Initiation Protocol (SIP).* Servers.

Schwarz, H., Sullivan, G., Wiegand, T., & Wien, M. (2007). *Text of ISO/IEC 14496-10:200X/FDIS Advanced Video Coding* (4th ed). International Organisation For Standardisation ISO/IEC JTC1/SC29/WG11 Coding of Moving Pictures and Audio.

Shokrollahi, A. (2006). Raptor Codes. *IEEE Transactions on Information Theory, 52*(6), 2551–2567. doi:10.1109/TIT.2006.874390

Wang, Z. (2001). *Internet QoS Architectures and Mechanisms for Quality of Service.* San Francisco: Morgan Kaufmann Publishers.

This work was previously published in International Journal of Mobile Computing and Multimedia Communications (IJMCMC) 1(3, edited by Ismail Khalil & Edgar Weippl, pp. 40-56, copyright 2009 by IGI Publishing (an imprint of IGI Global).

Chapter 11
A Qualitative Resource Utilization Benchmarking for Mobile Applications

Reza Rawassizadeh
Vienna University of Technology, Austria

Amin Anjomshoaa
Vienna University of Technology, Austria

A Min Tjoa
Vienna University of Technology, Austria

ABSTRACT

There are many mobile applications currently available on the market, which have been developed specifically for smart phones. The operating system of these smart phones is flexible enough to facilitate the high level application development. Similar to other pervasive devices, mobile phones suffer from limited amount of resources. These resources vary from the power (battery) consumption to the network bandwidth consumption. In this research the mobile resources are identified and classified. Furthermore, a monitoring approach to measure resource utilization is proposed. This monitoring tool generates traces about the resource usage which is followed by a benchmarking model which studies monitoring traces and enables users to extract qualitative information about the application from quantitative trace of resource usage.

INTRODUCTION

According to Compass Intelligence studies, U.S. companies will spend $11.6 billion on mobile applications by 2012 (Burney, 2009). This indicates a significant increase in the number of mobile applications from both quantity and quality perspectives. Consequently many duplicate applications with similar functionalities and features will enter into the market.

Mobile phones as wide accepted pervasive devices have finite energy sources (Satyanarayanan, 1996). Pervasive devices also suffer from client thickness (Satyanarayanan, 2001) which implies

DOI: 10.4018/978-1-60960-563-6.ch011

that there will always be the challenge of willingness to increase the quality of an application while dealing with the shortage of existing resources.

Studies revealed that users prefer buying mobile phones with more features (Thompson, Hamilton, & Rust, 2005). Adding more features increases the resource utilization and fosters more powerful devices. Using less resources and providing the desired functionalities increase the application efficiency.

Developers and researchers try to handle resource shortages of the pervasive devices in different ways, such as studying the context and adapting the device to the current context (Alia et al., 2007), optimizing energy usages with a CPU scheduler (Yuan & Nahrstedt, 2003) and so forth.

Such research results suggest that the resource usage of an application is an important factor for mobile devices, which could affect the application's quality. In addition to other performance metrics such as application response time, throughput, reliability and availability (Jain, 1991), resource utilization is an important performance metric also.

For instance consider a scenario where a user intends to purchase an audio player for his smart phone. There are many different choices on the market, "Music player X" is an audio player which plays user desired audio formats such as MP3 and gets some information about the current music track from the Internet. "Music player Y" is another audio player which not only play the desired audio format and gets the information from the Internet, but also sends the audio track name to the user's micro-blog account (such as Twitter or Friendfeed accounts). These features might be attractive for some users, but wireless network bandwidth might be limited and expensive. A quality operator can study which application performs the assigned task with less network activities as a capability indicator and decides upon choosing the appropriate application. Large scale industrial mobile device producers, who are interested in purchasing applications from third parties and embed them into their devices, can benefit from studying resource usage of the applications. In this context, small amount of disk space or memory allocation plays an important role.

This paper focuses on measuring resource usage of applications via a monitoring tool, and benchmarking capability of the target application from the resource consumption point of view. Qualitative features such as user interface design or application features are not within the scope of this paper. The proposed monitoring tool tracks resource usage of the device during the runtime of the target applications. It generates a trace from the resource usage which can be used to study capabilities of the application or for studying QoS issues. Our monitoring approach which resides on the same device, does not require any information about the target application and monitors resources during the application execution. In order to be flexible and scalable, the proposed approach is designed independent of the target application.

The remainder of this paper is organized as follows. Next section describes the resource classification; afterward the related works will be introduced. Then we discuss controversies and restrictions. Afterward we introduce the benchmarking model,and utility function and finally resource monitoring methods will be described. Then the experiment evaluation will be described and last section concludes the paper.

RESOURCE CLASSIFICATIONS

As the first step, the resources of the mobile devices, which are worth measuring and affect the capability of the application must be identified. Each mobile application consumes five type of resources namely *CPU*, *memory*, *battery* and probably *disk* and *network* activities.

These resources can influence each other, for example CPU utilization affects battery consumption, but we intend to study them separately regardless of their interdependencies. Currently there is no standard model to prioritize mobile resources. According to (Kravets & Krishnan, 1998) battery, which is responsible for the device power, is known as the most important resource for mobile phones.

RELATED WORKS

To the best of our knowledge, there is no mobile platform benchmarking suite for evaluating applications based on their resource consumption. Most of the existing benchmarking tools merely compare the CPU utilization of an application or other hardware capabilities of mobile devices. Unlike these approaches, our intention is to benchmark application resource usage and not the hardware capabilities of the device. The related work, described below, is not necessarily used to benchmark resources, but they have common ideas to track resource usages, benchmarking models, etc.

The JBenchmark is a Java ME (Mobile Edition) based mobile application, which measures processor and clock frequency of Java enabled mobile phones (JBenchmark, 2004). It provides a test suite, which contains a set of Java ME applications that perform tests like 3D graphics, CPU intensive processes, etc. They are installed on a Java ME capable mobile phone and are meant to benchmark the CPU performance of these phones.

Some other approaches such as Dulan and Tabirca, provide a test bed to analyze coding techniques via fractal image generation, which requires significant mathematical processing and consumes enormous amount of CPU time (Doolan & Tabirca, 2007). This research proved that inlining all calculations is more efficient than using an object oriented approach for mobile devices. We interpret this as another effort to monitor CPU activity. PennBench (Chen, Vijaykrishnan, & Irwin, 2005) is another benchmarking suite for Java enabled mobile devices. It benchmarks Java ME application memory usage characteristics based on the memory size constraints and heap footprints. Similar to JBenchmark, PennBench uses a test suite which contains a set of Java ME applications. Keijzers et al. (Keijzers, Ouden, & Lu, 2008) provide an approach for smart phone usability benchmarking based on the effectiveness, efficiency and satisfaction of using smart phones. The benchmarking is based on the user surveys. Powerscope (Flinn & Satyanarayanan, 1999) profiles the energy usage of mobile applications. It tries to map energy consumption to the program structure. Paternò et al. (Paternò, Russino, & Santoro, 2007) proposed an approach to remotely evaluate usability on mobile applications based on logging audio channel, battery consumption, position of the user and etc. SYSMark 2007 (SYSmark2007 Preview, 2007) is used to benchmark four types of applications (E-learning, video creation, office productivity and 3D modeling) on personal computers with a predefined set of actions. Then it compares the results of the applications with similar functionalities. SPEC (SPEC, 2006) provides standard benchmarking approaches for CPU, Web servers, SIP, power usage, etc. Some of SPEC benchmarking approaches like SIP or transaction benchmarking focused on a specific application type.

CONTROVERSIES AND RESTRICTIONS

Any monitoring approach can be used to dynamically manage QoS (Chalmers & Sloman, 1999). In this research a resource utilization monitoring prototype is proposed to track the resource usage of mobile applications. A quality operator or application evaluator can assign weights to resources for prioritizing the importance of the measured resources, and subsequently evaluates application resource utilization via a *utility function*. A utility function supports the quality operator in comparing different applications with the same functionality by profiling the resource usages of each application. This will be explained in more detail in the next section. It is notable that evaluating end user quality factors such as usability or response time are not in the scope of this research and only the application resource consumption is measured. Despite the fact that they are worth to be considered, this will not be supported by proposed monitoring approach because it resides separate from the application and will not have any interaction with the target application. Separation of the monitoring tool from the target application makes this tool more flexible and scalable. However it restricts the end user quality factor measurements of the target application. Most end user quality measurement approaches require quality evaluators to delve deep into the usage or the architecture of the target application.

We profile application resource usage by tracking the usage of a specific resource which will be used for the target application or it will be used for the whole device. It could be argued that overall resource usage of the device should be considered during the target application monitoring. Otherwise target application resource consumption is calculated with an overhead, which comes from two sources: one from the general device resource usage overhead, and the second one from the monitoring application overhead. This introduces a systematic measurement error Δ_i for resource i. If the user intends to measure precisely the resource utilization of the target application A without this error, he/she can execute the monitoring tool without starting the target application, and then record the consumption of resource i and compute an estimate for Δ_i. In the next step both monitoring tool and target application will be executed and the total consumption $R_i(A)$ will be derived under the same condition. The true resource consumption $R_i^{true}(A)$ of the application can then be estimated by

$$R_i^{true}(A) = R_i(A) - \Delta_i \quad (1)$$

It is important to note that there is no guarantee that the returned numbers describe precisely the resource utilization under the assumed condition. In each of our experiments, numbers have slightly changed, hence in any calculation a percentage of error must be considered. Running the experiments for a couple of times and taking averages is thus mandatory. A special problem is given by the fact that due to random fluctuations, for a resource that is not used at all by some applications, $R_i^{true}(A)$ could actually be smaller than zero. This problem will be addressed by using max $\{R_i^{true}(A),0\}$ instead of $R_i^{true}(A)$.

BENCHMARKING APPLICATIONS VIA A UTILITY FUNCTION

In the presented work, resource consumption is described by metric data and not categorical data. Furthermore, we ignore the relation between performance indicators, for example network activity is not related to the disk activity. In order to calculate capabilities of applications based on the resource utilization, we need to analyze the resources together and make sure that all of them are using the same format. In order to fulfill this requirement, this approach calculates the sum of the scores for each one of resources. When users intend to compare two applications, the experiment conditions should always remain constant. The conditions are composed of different components such as device state, experiment duration, number of inputs issued to the target application, etc.

Utility Function

According to Walsh et al. (Walsh, Tesauro, Kephart, & Das, 2004), utility functions allow indicating the degree of desirability of a service; therefore we chose a utility function to represent the outcome of our benchmark. A utility function maps the resource usage of the application to a numerical degree of satisfaction. This utility function provides a multidimensional mapping from consumption measurements of n resources (Smith, 1988) to one single utility value. In this mapping, the importance of resource i is represented by a weighting factor W_i. We propose to set W_i to 0 in case the resource is not important. When the resource is considered low priority, W_i is set to 1, and for higher priorities, it is set to 2. A real example of using the utility function execution will be described in the evaluation and verification section of this paper. Other factors such as "execution time of an operation" could also be considered using this utility function, but our monitoring tool does not support them. In the following, $0 \leq U_i(A) \leq 1$ defines the normalized utility of resource $(UR)_i$ under condition c, with systematic errors removed due to (1). Then the utility $U_c(A)$ of application A under condition c is defined by

$$\sum_{i=1}^{i=m-1} (Rate\_ATM\_UMA\_Voice\_up)i \tag{2}$$

The result of the utility function is a value between 0 and 1. The higher the value, the higher is the utility of the application.

Utility of a Resource

One of the major challenges is to make the consumption of different resource types comparable with each other, in order to use them in one formula. Therefore, we seek a $0 \leq U_i(A) \leq 1$, i.e., the normalized utility of resource i for application A, which should be a function of the respective consumption of resource i. We start by mapping the resource consumptions onto the interval [0,1]. The resources battery and CPU consumptions are represented via percentages. These can be mapped to the interval [0,1] in a natural way. The size of memory is always limited. For instance Android emulator SDK 1.1 shows about 94572 KB total memory, without running the target application it uses about 61296 KB of memory. Normalizing the usage of memory is thus done by dividing the memory used by the application by the total amount of memory that could be used by the application, which is about 23296 KB. This value will use the total amount of available memory in our calculation as shown in Table 2. Note that this is the total amount of memory minus the amount of memory used by the operating system, running system tasks, and the monitoring tool. For instance we assume that during the benchmarking process, resource i is sampled K times for application A, resulting in the values $0 \leq R_{ij}(A) \leq 1,\ 1 \leq j \leq K$, and the respective arithmetic mean $\overline{R}_i(A)$. Furthermore let Δ_i be an estimate for the systematic error as described in (1), also being mapped to the interval [0,1]. The normalized utility $U_i(A)$ of resource i for the battery is then defined by

$$U_i(A) = 1 - \max\{\overline{R}_i(A) - \Delta_i, 0\} \tag{3}$$

Resource usage of CPU and Memory can be measured disregard to the monitoring tool overhead. It means Δ is zero for CPU and memory.

On the other hand, the number of disk and network activities potentially could be very high and are only limited by the total network and internal bus bandwidth, and the bandwidth and access times of the disk. In practice, most applications will use only a small fraction of the possible maxima. In order to derive a mapping into the interval [0,1] we specify, that the highest utility, i.e. the value 1, should be achieved in case the resource is not used at all. Likewise, the higher the observed numbers are, the smaller the utility should be. For deriving a stable utility of application A and resource i (including disk and network), we thus propose to run the application several times, and observe the measurement values $R_{ij}(A)$ for each run j, with arithmetic mean $\overline{R}_i(A)$. Again, let Δ_i denote the systematic error, this time in the original value range (not normalized). Then the utility for resource *i* is defined by:

$$U_i(A) = (\max \{ \overline{R}_i(A) - \Delta_i, 0\} + 1)^{-\alpha_i} \quad (4)$$

This definition depends on constants α_i reflecting how fast the utility approaches zero for resource *i*, and is influenced by the possible value range of observations. For instance, it reflects whether the networking activity is reported in bytes, KB, MB etc. This way, the benchmarking results of the applications are normalized, and combine scores from different normalized resources, but on the other hand, the utility of one application does not depend on the measurement results of another application. What should be kept fixed, though, when comparing applications with each other, is the device that is used for measuring. One possible issue may arise by the fact that (3) is a linear function, while (4) decreases with geometric speed. However, we think that if a resource is used in such a high degree that it is almost fully utilized, the resulting utility should be near zero anyway. This is achieved by both functions. The influence of this highly utilized resource then disappears in (2), and it more or less reflects the utility of the other resources only.

Weights

Applications vary from their resource usages based on their functionality. For example a music player does not require a network connection or it has minimal network activity but it consumes a significant amount of battery usage. The reason is that listening to music generally runs significantly longer than for instance checking emails. On the other hand a social-networking application requires heavy network activities but consumes less battery than the music player. These applications are totally different from their resource usage perspective. Assigning priority or weight to the resources can define the level of importance in a numerical value. For instance in case of a music player, battery gets high priority and network activity gets zero or low priority, while weights must be assigned by the user (quality operator) based on the application functionality. As a result in a music player application battery has more weight than the network or in a social-networking the client's battery is less important compared to network bandwidth usage. Users need to classify applications based on their category (music player, social-networking, games, etc.) and assign relevant weights to them.

RESOURCE MONITORING METHODS

Android[1] has been chosen as the implementation platform. It is an open source mobile operating system, which is based on the Linux kernel. We have developed a resource-monitoring tool, which tracks and logs the resource consumption of a given process. The monitoring tool will run in the background parallel to the target application via using Android services. The developed application uses a simple GUI for starting and stoping the monitoring processes on selected resources (shown in Figure 1).

The sampling interval for the battery, the CPU and the memory can be set by the Settings button. Network and disk activities are sampled only twice, once at the start, while the start button is pressed and next time when the stop button is pressed. Unlike CPU and memory, other resources are not measured per process. Battery usage is mainly the hidden resource and it is difficult to be measured.

Android facilitates measuring battery by creating a broadcast receiver and calling it by an intent, which contains ACTION_BATTERY_CHANGED as the action. Other resources have been measured via reading */proc* subfolders.

Sampling is done based on the predefined intervals. The sampling interval depends on the resource type, e.g. sampling interval for battery

Figure 1. Resource monitoring tool

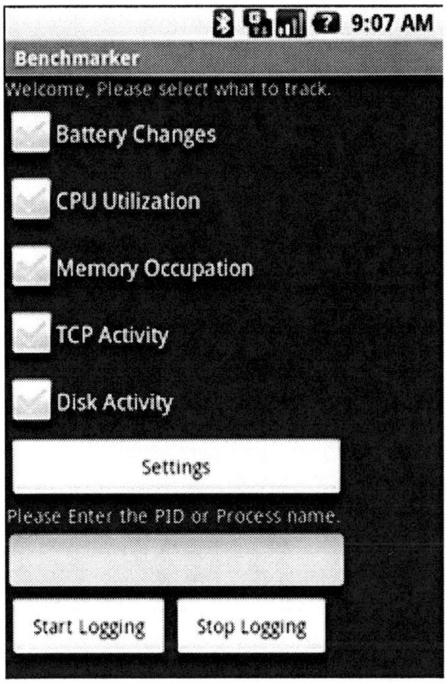

will be done each 10 minutes but CPU utilization could be sampled each second. As explained before, monitoring an application itself consumes resources. This indicates that the sampling result is a mixture of monitoring application and the target application. To increase quality of the monitoring result, the user can monitor the target application under the same condition more than once and calculate the arithmetic mean of the observations.

The monitoring tool runs in the background as a set of services (Android services). Each service is responsible to log one resource. The result of the monitoring is a trace file. Box 1 is an example of such traces:

EVALUATION AND VERIFICATION

To evaluate this model, we monitored resource utilization under similar conditions for two Sudoku games[2] via the proposed monitoring tool. Our test environment is an Apple Mac Book Pro with 2.4 GHz CPU which contains Android emulator SDK 1.1. The tests are repeated four times. At the end the largest dataset has been chosen for the analysis. The arithmetic means of the measurements in each observation do not show significant differences. The dataset is pruned to remove near zero CPU and memory usages before and after Monkey user simulation. Monkey (UI/Application Exerciser Monkey, Android Developer Guide, 2009) is a user simulator feature of the Android platform, that generates pseudo-random streams of the user events like click, tap the screen, etc.

To ensure that there are enough samples, outliers are found and removed. Afterward the normality of data with Quantile-Quantile plot (Boslaugh & Watters, 2008) for each dataset is examined. The result is shown in Figure 2. Shapiro-Wilk test (Shapiro, Wilk, & Chen, 1968) is used to verify the normality of data. These datasets passed the Shapiro-Wilk test for the CPU and not for the memory. The statistical error level for both the CPU and memory of both applications is calculated for 95% confidence interval and they were acceptable. As stop criteria we used an estimate for the standard error of the mean. The result shows that the magnitude of standard errors is much smaller than the mean. The arithmetic mean for CPU and total memory allocated for application (vmsize) has been used to calculate the utility function (Table 1).

The monitoring tool created an application trace from the CPU and memory usage. Both applications have been tested via the Monkey. We named these Sudoku games "Application A" and "Application B". As it has been noted before, testing application under the same condition (same test duration, same device state, etc.) is very important. In order to adhere to the same condition for each test, we used the new Android emulator instance with an empty 1 Gigabyte SDCARD for each experiment and the same experiment duration. First the Monkey simulator is started. The result will be used by the utility function to evaluate these two applications and compare them. Figure 3 shows the CPU usage for approximate-

Box 1.

[29 Mar 2009 16:17:30 GMT][GUI] START LOGGING RESOURCES
[29 Mar 2009 16:17:30 GMT][NET][Receive:487, Send:405]
[29 Mar 2009 16:17:30 GMT][DSK][Read issued:9, Write Completed:1]
[29 Mar 2009 16:17:33 GMT][CPU][PID:217, CPU:1%, #Thread:6]
[29 Mar 2009 16:17:33 GMT][MEM][PID:217,VmSize: 115084KB, VmRSS:22760KB]
[29 Mar 2009 16:17:37 GMT][CPU][PID:217, CPU:2%, #Thread:6]
[29 Mar 2009 16:17:37 GMT][MEM][PID:217,VmSize: 115124KB, VmRSS:22852KB]
...
[29 Mar 2009 16:24:15 GMT][MEM][PID:217,VmSize: 118696KB, VmRSS:24992KB]
[29 Mar 2009 16:24:18 GMT][Battery][The phone's battery is charging, %50]
[29 Mar 2009 16:24:22 GMT][CPU][PID:217, CPU:25%, #Thread:9]
...
[29 Mar 2009 16:27:34 GMT][MEM][PID:217,VmSize: 117360KB, VmRSS:25172KB]
[29 Mar 2009 16:27:36 GMT][NET][Receive:25088, Send:21161]
[29 Mar 2009 16:27:36 GMT][DSK][Read issued:10, Write Completed:86]
[29 Mar 2009 16:27:36 GMT][GUI] LOGGING RESOURCES HAVE BEEN STOPED

Figure 2. Quatile-Quantile Plots for CPUs and VmSizes

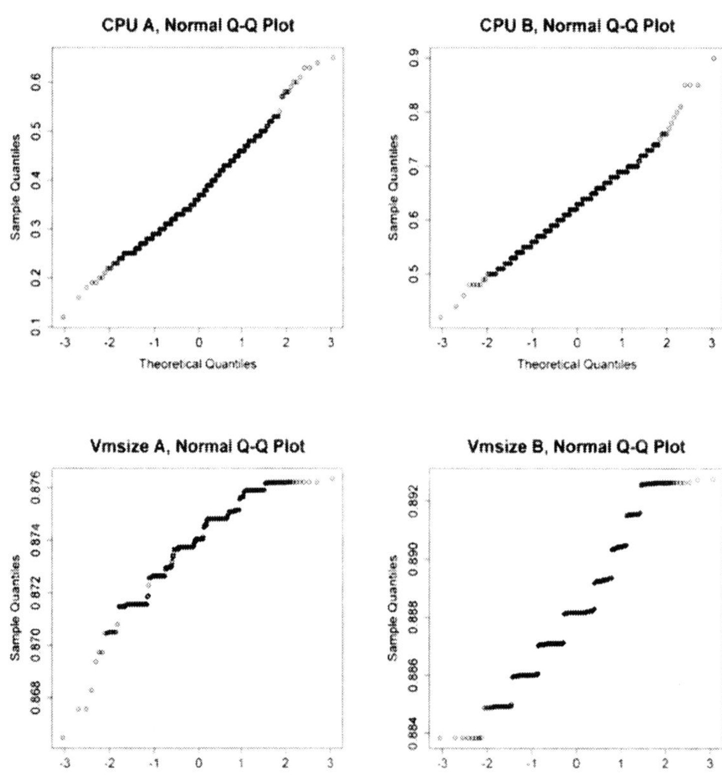

- *Table 1. Values for utility of resource*

Recourse	Weight	Values for application A	Values for application B
CPU	2	62%	37%
Memory	1	vmSize=11919, vmRSS=25001	vmSize=10563, vmRSS=17657
Disk	0	-	-
Network	0	-	-
Battery	0	-	-

Table 2. Utility function (23296 is the available amount of memory for applications, as described before)

	Application A		Application B	
Weight	2 (CPU)	1(Memory)	2 (CPU)	1(Memory)
Utility of Resource	62% → 1–(62÷100)	vmSize=11919 → 1-(11919÷23296)	37% → 1–(37÷100)	vmSize=10563 → 1-(105631÷23296)
Utility Function	$\frac{(2\times 0.38)+(1\times 0.4884)}{2+1}$		$\frac{(2\times 0.63)+(1\times 0.5466)}{2+1}$	
Result	0.4161		0.6022	

ly the same number of samples (about 80) collected from the CPU traces of these applications. In this experiment CPU sampling has been done based on three seconds intervals. There are some zero or near zero CPU usages, at the beginning and at the end of the monitoring dataset. This is because the Monkey user simulator is called after the monitoring process has been started and after stopping the Monkey application the monitoring application is still running.

To measure memory utilization, the monitoring tool logs "total memory size allocated for the target application" (vmsize) and the "resident set size" (vmrss) in a three second interval for each application. vmrss and vmsize for each application are shown in Figure 4. We use only vmsize for the utility function. Table 1 shows Utility of Resources, weights and their associated values. Table 2 shows the result of the calculation and the utility function result for each application. Result of the utility function calculations are shown in Figure 3 and Figure 4.

Results indicate that Application B consumes less CPU and memory than Application A. Application state could also affect utility function's error. In the Android platform application components have different states which represent the application's life cycle (Component Lifecycles, Application Fundamentals, Android Developer Guide, 2009). For instance, if an application is visible to the user and it has focus (start state), then more CPU will be consumed. Likewise when another Android activity is resumed (pause state), then less CPU is assigned. The same approach could be also done for battery, memory, disk and network resources. For instance, if the application is visible to the user, then battery consumption will increase. In case the application is suspended and another application get the GUI focus (pause), then battery consumption will decrease.

CONCLUSION

In these research mobile resources which are being used by different mobile applications, are identified and classified. A software monitoring approach to measure these resources and profile

Figure 3. CPU utilization trace

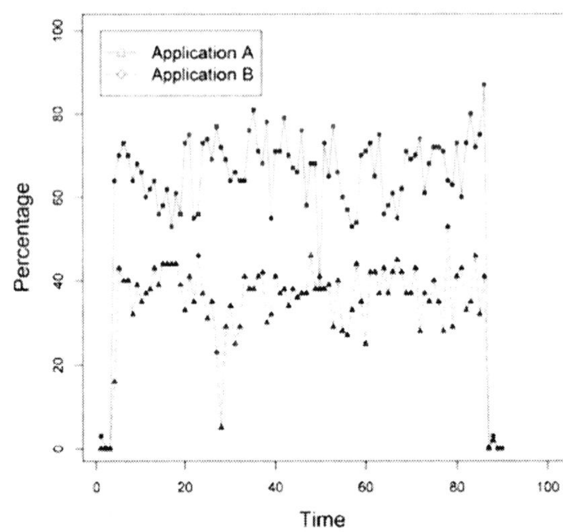

Figure 4. Average memory usage

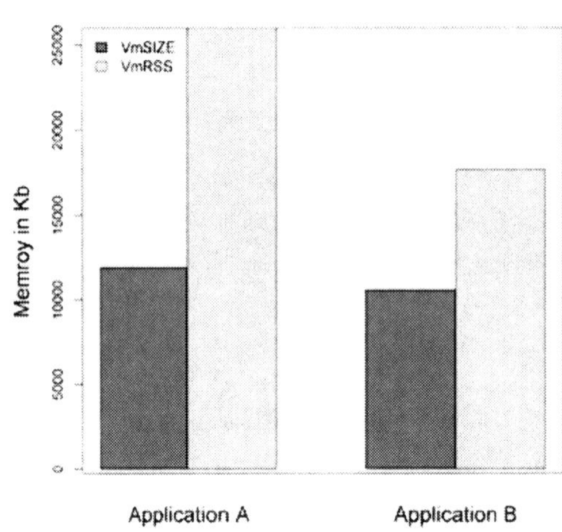

the resource utilization of applications has been proposed. It is a light software monitoring tool, which consumes very few resources by itself. It supports flexibility and modularity by encapsulating the resource monitoring process from the target application. Implementation of a monitoring tool has been done on the Android platform. The Android platform has been chosen because of its ease of access to the lower layers of the operating system. To benchmark applications from their resource usage perspective, a utility function has been proposed. Its activity is based on the weight of the resource and the average resource usage consumption. The received data objects from different resources are heterogeneous and in order to use them together, it is necessary to convert them into a uniform numerical data. This process will be done by utility of resource and resource weight which represents the level of the importance of each resource and will be set by the application evaluator based on the application type. The utility of resource is a normalized arithmetic mean of resource usage during the simulation. As proof of concept two Sudoku applications, based on their CPU and memory consumption, are analyzed and compared.

REFERENCES

Alia, M., Eide, V. S. W., Paspallis, N., Eliassen, F., Hallsteinsen, S. O., & Papadopoulos, G. A. (2007). A Utility-Based Adaptivity Model for Mobile Applications. *International Conference on Advanced Information Networking and Applications Workshops, 2,* (pp. 556-563).

Benchmarks, S. P. E. C. (Standard Performance Evaluation Corporation). (2006). Retrieved from http://spec.org/ benchmarks.html.

Boslaugh, S., & Watters, P. (2008). *Statistics in a Nutshell: A Desktop Quick Reference*. (pp. 118-119). Sebastopol,CA: O'Reilly Media, Inc.

Burney, K. (2009). *New Year's Bustle? Vertical Market Expectations for 2009 ICT Spending in the US- From Crisis to Slow, Long-term Recovery*. Compass Intelligence.

Chalmers, D., & Sloman, M. (1999). A Survey of Quality of Service in Mobile Computing Environments. *IEEE Communications Surveys, 2* (2).

Chen, G., Vijaykrishnan, N., & Irwin, M. J. (2005). PennBench: A benchmark suite for embedded Java. In *The IEEE 5th Annual Workshop on Workload Characterization.*

Doolan, D. C., & Tabirca, S. (2007). The Need for Speed: coding styles for Mobile Devices. In *Elmar* (pp. 193-196).

Flinn, J., & Satyanarayanan, M. (1999). PowerScope: A Tool for Profiling the Energy Usage of Mobile Applications. *In WMCSA '99: Proceedings of the Second IEEE Workshop on Mobile Computer Systems and Applications* (p. 2).

Jain, R. (1991). *The Art of Computer Systems Performance Analysis: Techniques for Experimental Design, Measurement, Simulation, and Modeling* (p. 36). New York: John Wiley & Sons.

Jbenchmark, New Estimator Software Measures True Mobile Phone Performance. (2004). Retrieved from http://www.jbenchmark.com/ jbacepr.jsp.

Keijzers, J., den Ouden, E., & Lu, Y. (2008). Usability benchmark study of commercially available smart phones: cell phone type platform, PDA type platform and PC type platform. In *MobileHCI '08: Proceedings of the 10th international conference on Human computer interaction with mobile devices and services* (pp. 265-272).

Kravets, R., & Krishnan, P. (1998). Power Management Techniques for Mobile Communication. In *MobiCom '98: Proceedings of the 4th annual ACM/IEEE international conference on Mobile computing and networking* (pp. 157-168).

Lifecycles, C., Fundamentals, A., & Guide, A. D. (2009). Retrieved from http://dev.android.com/ guide/topics/ fundamentals.html.

Paternò, F., Russino, A., & Santoro, C. (2007). *Remote Evaluation of Mobile Applications* (pp. 155–169). Task Models and Diagrams for User Interface Design.

Satyanarayanan, M. (1996). Fundamental challenges in mobile computing. In *PODC '96: Proceedings of the fifteenth annual ACM symposium on principles of distributed computing* (pp. 1-7).

Satyanarayanan, M. (2001). Pervasive computing: Vision and challenges. *Personal Communications, IEEE, 8*(4), 10–70. doi:10.1109/98.943998

Shapiro, S., Wilk, M., & Chen, H. (1968). A Comparative Study of Various Tests for Normality. *American Statistical Association, 63*(324), 134–137. doi:10.2307/2285889

Smith, J. E. (1988). Characterizing computer performance with a single number. *Communications of the ACM, 31*(10), 1202–1206. doi:10.1145/63039.63043

Sysmark2007 preview. (2007). Retrieved from http://www.bapco.com/ techdocs/SYSmark2007 Preview_WhitePaper.pdf.

Thompson, D. V., Hamilton, R. W., & Rust, R. T. (2005). Feature Fatigue: When Product Capabilities Become Too Much of a Good Thing. *JMR, Journal of Marketing Research, 42*, 431–442. doi:10.1509/jmkr.2005.42.4.431

UI/Application Exerciser Monkey, Android Developer Guide. (2009). Retrieved from http://dev.android.com/ guide/developing /tools/monkey.html.

Walsh, W. E., Tesauro, G., Kephart, J. O., & Das, R. (2004). Utility Functions in Autonomic Systems. In *International Conference on Autonomic Computing*. (pp. 70-77).

Yuan, W., & Nahrstedt, K. (2003). Energy-Efficient Soft Real-Time CPU Scheduling for Mobile Multimedia Systems. In *SOSP '03: Proceedings of the Nineteenth ACM Symposium on Operating Systems Principles* (pp. 149-163).

ENDNOTES

1 www.android.com

2 We downloaded two freeware Sudoku games from the internet

Section 3

Innovations in Multimedia Analysis, Modeling, Processing and Transformation

Chapter 12
Fast Vector Quantization Encoding Algorithms for Image Compression

Ahmed Swilem
Minia University, Egypt

ABSTRACT

Vector quantization (VQ) is a well-known compression method. In the encoding phase, given a block represented as a vector, searching the closest codeword in the codebook is a time-consuming task. In this paper, two fast encoding algorithms for VQ are proposed. To reduce the search area and accelerate the search process, the first algorithm utilizes three significant features of a vector that are, the norm, and two projection angles to two projection axes. The second algorithm uses the first two features as in the first algorithm with the projection value of the vector to the second projection axe. The algorithms allow significant acceleration in the encoding process. Experimental results are presented on image block data. These results confirm the effectiveness of the proposed algorithms.

1. INTRODUCTION

Thanks to the incredibly rapid advancement of the Internet, online information communication and human connection, where the delivery and storage of digital images of all kinds have become the main stream, seem to have been planning a more and more important part of our life. However, for massive quantities of multimedia data to get traveled online from place to place at least acceptable speed or to be stored without taking up too much of the memory space, especially when the network bandwidth is limited with small hardware storage, there is nothing more important than a good digital image compression scheme. Image compression methods can be categorized as two types: lossless compression and lossy compression. In lossless compression scheme, the original image can be recreated exactly from the compressed data but lossy compression scheme must exist some distortion between the original image and the reconstructed image. So far, vector quantization (VQ) (Gray, 1984; Linde, Buzo, & Gray, 1980) has long been a well-celebrated

DOI: 10.4018/978-1-60960-563-6.ch012

lossy compression technique that guarantees the achievement of a satisfactory balance between image quality and compression ratio (Chen, Chang, & Hwang, 1998). At the same time, because of its simple and easy implementation, VQ has been very popular in a variety of research fields such as speech recognition and face detection (Garcia, & Tziritas, 1999). Even in real-time video-based events detection (Liao, Chen, Su, & Tyan, 2006) and the anomaly intrusion detection systems (Zheng, & Hu, 2006), VQ has been exploited recently to learn and collect some representative patterns and then to identify similar feature vectors or detect unusual activities. Next, the procedure of vector quantization will be interpreted from the aspect of image compression.

VQ can be defined as a mapping Q from a k-dimensional Euclidean space R^k to a finite set $C = \{c_1, ..., c_N\}$ of vectors in R^k called the codebook. Each representative vector c_i in the codebook is called a codeword. Generally, VQ can be divided into three procedures: codebook design, encoding and decoding. The codebook design procedure is executed before the other two procedures for VQ. The goal of the codebook design is to construct a codebook C from a set of training vectors using clustering algorithms like the generalized Lloyd algorithm (GLA) (Linde, Buzo, & Gray, 1980). This codebook is used in both the image encoding/decoding procedures. In the encoding procedure, for each training vector q, we find the index i of the codeword c_i is the closest codeword to the vector q which gives minimum distortion and satisfy $d^2(q, c_i) < d^2(q, c_j), \{j = 1, 2, ..., N; i \neq j\}$, where

$$d^2(q, c_i) = \sum_{j=1}^{k} (q_j - c_{ij})^2, \tag{1}$$

so the codeword c_i now represents the vector q. The decoding procedure is simply a table look-up procedure that uses the received index i to deduce the reproduction codeword c_i, and then uses c_i to represent the input vector q.

Using the squared Euclidean distance criteria in (1), each distortion computation requires k multiplications and $2k - 1$ additions. For an exhaustive full search (FS) algorithm, encoding each input vector requires N distortion calculations and $N - 1$ comparisons. Therefore, it is necessary to perform kN multiplications and $(2k - 1)N$ additions and $N - 1$ comparisons to encode each input vector. When and/or k become larger, the computation complexity problem occurs for full codebook search. Many algorithms were proposed later in this field to reduce the computational calculation needed to accomplish the VQ technique. Some of these algorithms use one or more constraint inequality in the spatial domain to reduce the computational calculation needed to accomplish the VQ technique (Bake, Bae, & Sung, 1999; Cao, & Li 2000; Cardinal, 1999; Guan, & Kamel, 1992; Huang, Bi, Stiles, & Harris, 1992; Imamura, Swilem, & Hashimoto, 2003; Imamura, Swilem, & Hashimoto, 2004; Johnson, Ladner, & Riskin, 1996; Johnson, Ladner, & Riskin, 2000; Lai, & Lue, 1996; Lee, & Chen, 1994; Orchard, 1991; Swilem, Imamura, & Hashimoto, 2002; Swilem, Imamura, & Hashimoto, 2004; Wu, & Lin, 2000). Other algorithms exploit the topological structure of vectors to accomplish the same target (Lee, & Chen, 1995; Pan, Lu, & Sun, 2000; Song, & Ra, 2002; Swilem, Imamura, & Hashimoto, 2005).

An algorithm for fast nearest neighbor search presented by Orchard (Orchard, 1991) precomputes and stores the distance between each pair of codewords. Given an input vector q, the current best codeword c_b, and a candidate codeword c_j, if $d(q, c_j) \leq d(q, c_b)$, then $d(c_j, c_b) \leq 2d(q, c_b)$. Graphically, this constrains the search area within a sphere centered on the current best codeword, with a radius of twice the smallest distortion calculated so far. In (Huang, Bi, Stiles, & Harris, 1992), an additional constraint on codewords by sorting their distances from the origin was introduced. The distance between the current best codeword and the input vector constrains the search to codewords within an area about the origin, which is represented by an annulus in two dimensions. In three dimensions, this is the area between two concentric spheres. This constraint is known as the annular constraint. In the same paper, a combination of the spherical and annular constraints with an efficient search method was proposed. Lee and Chen (Lee, & Chen, 1994) introduced a projection method, which uses the mean and the variance of the vector as two constraints to reject the codewords. We described a lossy design method in (Swilem, Imamura, & Hashimoto, 2004), which employs a hyperplane decision rule to separate the search areas in Lee and Chen method (Lee, & Chen, 1994) and is considered as the extension of it. Johnson et al. (Johnson, Ladner, & Riskin, 2000) generalized the techniques in (Orchard, 1991; Huang, Bi, Stiles, & Harris, 1992) to apply them to class of vector quantizers using Lagrangian distortion measure, in which a sum of the Euclidean distance and some constant assigned to each codeword is incorporated. Another technique added an additional endpoint for the annular constraint was proposed in the same paper. This constraint is known as the double annulus constraint.

This paper introduces two new algorithms to reduce the computational complexity of the encoding process. The algorithms achieve equivalent performance to the full search VQ. They use two perpendicular projection axes. In the first algorithm, three constraints are utilized to accelerate the search process; the annular constraint and two angular constraints. For every input vector or codeword, its norm, its projection angle to the first projection axe and its projection angle to the second projection axe are calculated. In the second algorithm, three constraints are used; the annular constraint and one angular constraint with a projection value constraint. For every vector or codeword, its projection value to the second projection axe is calculated with its norm and its projection angle to the first projection axe. These algorithms are considered as enhancement versions of the double annulus search algorithm (DAS) (Johnson, Ladner, & Riskin, 2000). They search a smaller number of codewords than the previously known methods.

This paper is organized as follows: In section 2, some of the nearest neighbor search algorithms are introduced and analyzed. In section 3, the proposed algorithms are discussed in detail. In section 4, the simulation results are given. Section 5 concludes the paper.

2. NEAREST NEIGHBOR SEARCH ALGORITHMS

2.1. Equal Average Equal Variance Nearest Neighbor Search (EENNS)

Lee and Chen (Lee, & Chen, 1994) introduced an efficient codeword search algorithm, which uses the mean and the variance of the vector as two tests to reject the codewords. In the mean test, a projection axe l_1 in an Euclidean space R^k is defined as a line on which the coordinates of any point have the same value. For a given input vector q with mean value m_q, and a current best codeword c_b with distance $d_{\min} = d(q, c_b)$, any codeword that is closer to q than c_b has to be located inside the hypersphere centered at q with radius $d_{\min}$. Two boundary points

$$L_{\max} = (m_{\max}, m_{\max}, \ldots, m_{\max})$$

and $L_{\min} = (m_{\min}, m_{\min}, \ldots, m_{\min})$

can be obtained by projecting the hypersphere on the projection axe l_1, where

$$m_{\max} = m_q + d_{\min} / \sqrt{k} \quad (2)$$

and

$$m_{\min} = m_q - d_{\min} / \sqrt{k}. \quad (3)$$

Thus, by (2) and (3) only the codewords that are bounded by the two hyperplanes S_1 and S_2 which are perpendicular to the projection axe l_1 and pass through $L_{\max}$ and $L_{\min}$, respectively, will be searched.

In the variance test, the squared root of variance of the vector q, v_q, is used as the distance $d(q, L_q)$ between q and its projection point L_q on the projection axe l_1. The closest codeword c_i with squared root of variance v_{c_i} will satisfy the following inequality:

$$(v_q - v_{c_i})^2 < d_{\min}^2. \quad (4)$$

The inequalities (2), (3), and (4) constrain the distortion calculation to the codeword whose is completely contained in the search region shown in Figure 1.

Figure 1. Geometrical interpretation of the EENNS for 2-dimensional case

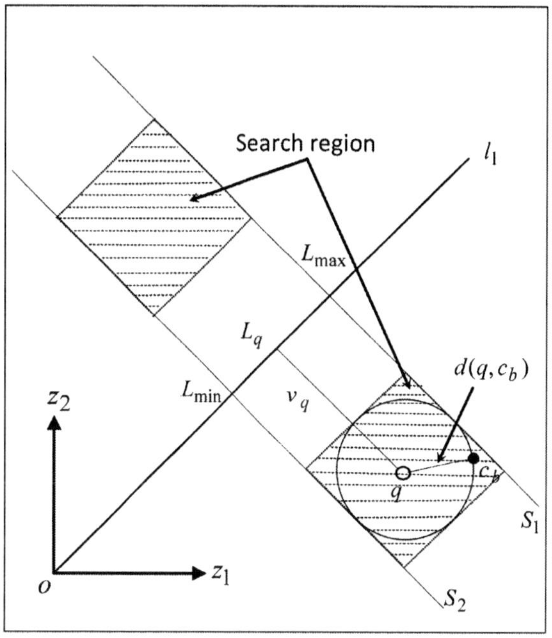

2.2. Double Annulus Search Algorithm (DAS)

The double annulus search algorithm is another nearest neighbor search algorithm designed specifically for VQ (Johnson, Ladner, & Riskin, 2000). It is based on a geometric observation. As shown in Figure 2, the first annulus is centered at the origin that is the first reference point. For a given input vector q of distance $|| q ||$ from the origin and a current best codeword c_b with distance $d(q, c_b)$, any closer codeword c_i to q than c_b will satisfy the following inequalities:

$$|| c_i || < || q || + d(q, c_b) \quad (5)$$

and

$$|| c_i || > || q || - d(q, c_b), \quad (6)$$

where $|| c_i ||$ is the Euclidean distance of c_i from the origin. Thus, for any codeword c_i satisfying (5) and (6), must be fully contained in the annulus defined by $|| q || + d(q, c_b)$ and $|| q || - d(q, c_b)$ (Huang, Bi, Stiles, & Harris, 1992).

The second annulus is centered at the farthest codeword from the origin, which is the second reference point, c_r. By using the distance to this codeword, the following inequalities can be defined:

$$d(c_i, c_r) > d(q, c_r) - d(q, c_b) \quad (7)$$

and

Figure 2. Geometrical interpretation of the DAS for 2-dimensional case

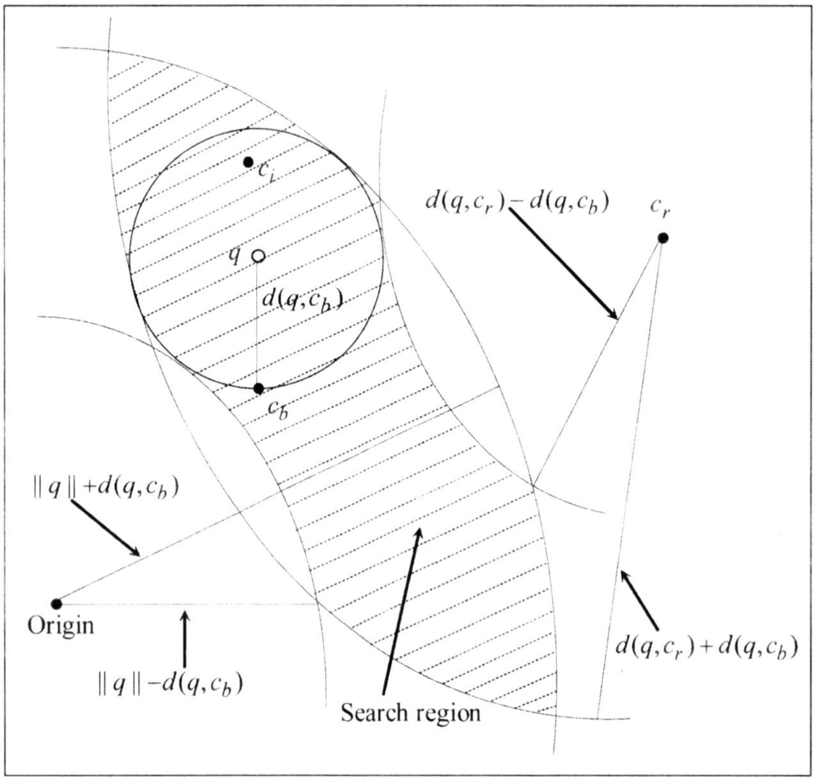

$$d(c_i, c_r) > d(q, c_r) - d(q, c_b). \quad (8)$$

The inequalities (5), (6), (7) and (8) constrain the distortion calculation to the codeword whose is completely contained in the search region shown in Figure 2.

The overlapped region between these two annulus, which is the DAS region, depends on two factors; the values of $d(q, c_b)$ and $d(q, c_r)$. If any of these two factors increases, the overlapped region will increase also. The problem of minimizing the value of $d(q, c_b)$ is easy to solve by choosing a good best codeword to q, while the problem of minimizing the value of $d(q, c_r)$ depends on the distribution of codewords. This problem will be solved in the following section by using two proposed algorithms.

3. THE PROPOSED ALGORITHMS

3.1 Annular-Double Angular Search (ADAS)

In previous work (Swilem, Imamura, & Hashimoto, 2004), we introduced a good solution for the problem of the DAS by using a new constraint called the angular constraint instead of the second annulus constraint. This method is called the annular with angular search (AAS). In this method, let l_1 be a projection axe in search space and it contains a unit vector $u_1 = (1, 1, \ldots, 1) / \sqrt{k}$ on it. For any vector p, we define the angle between p and the projection axe l_1 as:

$$\alpha_{1p} = \cos^{-1} \frac{u_1 p^T}{\| p \|}. \quad (9)$$

Figure 3. Geometrical interpretation of the AAS for 2-dimensional case

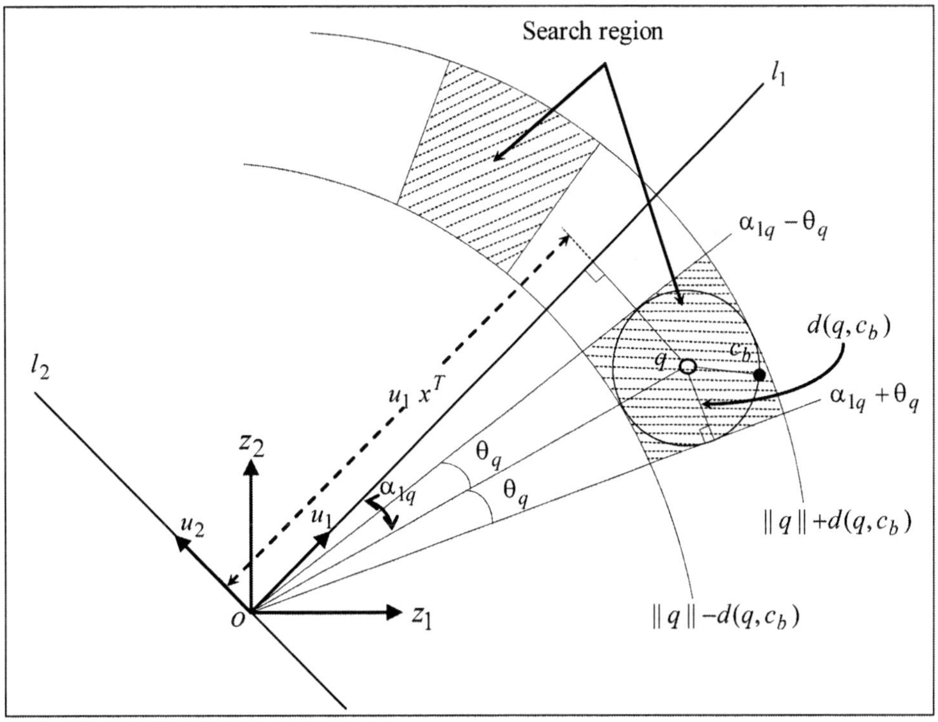

Because the values of all vector components are nonnegative, then the angle $\alpha_1 \in [0, \frac{\pi}{4}]$. The angle α_1 is called the projection angle to the projection axe l_1. We define another angle between the input vector q and the tangent from the origin to the hypersphere centered at q with radius $d(q, c_b)$ as:

$$\theta_q = \sin^{-1} \frac{d(q, c_b)}{||q||}. \qquad (10)$$

For a given input vector q with its projection angle α_{1q} to l_1 and the closest codeword c_i with its projection angle α_{1c_i}, the following inequalities should be satisfied:

$$\alpha_{1c_i} < \alpha_{1q} + \theta_q \qquad (11)$$

and

$$\alpha_{1c_i} > \alpha_{1q} - \theta_q. \qquad (12)$$

Thus, any codeword c_i satisfying (11) and (12) must be contained in the region bounded by the two angles $\alpha_{1q} + \theta_q$ and $\alpha_{1q} - \theta_q$. Using the first annular constraint in (5) and (6) with the angular constraint in (11) and (12), the search region will be reduced to the region shown in Figure 3.

For more reduction of the number of distortion calculations, we must be able to eliminate sections of the codebook from consideration. In order that, we add more projection angle constraint for the above angular constraint using another projection axe perpendicular on the reference line l_1. Let l_2 be a projection axe in the search space and it is perpendicular on l_1. It contains a unit vector

$u_2 = (1,1,\ldots,1,-1,-1,\ldots,-1) / \sqrt{k}$ (the first half components are 1's and the second half are −1's) on it. For any input vector q with its projection angle $\alpha_{2q} = \cos^{-1} \frac{u_2 q^T}{\|q\|}$ to l_2 and the closest codeword c_i with its projection angle α_{2c_i}, the following inequalities should be satisfied:

$$\alpha_{2c_i} < \alpha_{2q} + \theta_q \tag{13}$$

and

$$\alpha_{2c_i} > \alpha_{2q} - \theta_q . \tag{14}$$

Thus, any codeword c_i satisfying (13) and (14) must be contained in the region bounded by the two angles $\alpha_{2q} + \theta_q$ and $\alpha_{2q} - \theta_q$.

Using the constraints in equations (5), (6), (11), (12), (13) and (14), the number of distortion calculations will be reduced and hence, the execution time will be reduced also.

To speed up the search process, the codebook is sorted according to increasing order of the codeword norms and the search is performed up and down iteratively as shown in Figure 4.

3.2 Annular-Angular-Projection Value Search (AAPVS)

In the projection value constraint (Bake, Bae, & Sung, 1999), let the projection axe l_2 in search space is defined as in the last subsection. For any vector z, we define its projection value to l_2 as:

$$p_z = u_2 . z^T . \tag{15}$$

For a given input vector q and a current best codeword c_b with distance $d_{\min} = d(q, c_b)$, any closer codeword c_i to q than c_b will satisfy the following relationships:

$$p_{c_i} < p_q + d_{\min} / \sqrt{k} , \tag{16}$$

and

$$p_{c_i} > p_q - d_{\min} / \sqrt{k} , \tag{17}$$

where p_{c_i} is the projection value of the codeword c_i to l_2. Thus, by (16) and (17) only the codewords that are bounded by the two hyperplanes:

$$S_1^* : u_2 z^T = p_q + d_{\min} / \sqrt{k}$$

and $S_2^* : u_2 z^T = p_q - d_{\min} / \sqrt{k}$

will be searched.

The inequalities (5), (6), (11), (12), (16) and (17) constrain the distortion calculation to the codewords that are contained in the search region shown in Figure 5. To speed up the search process, the codebook is sorted according to increasing order of the codeword norms and the search is performed up and down iteratively as shown in Figure 4.

Figure 4. Searching path

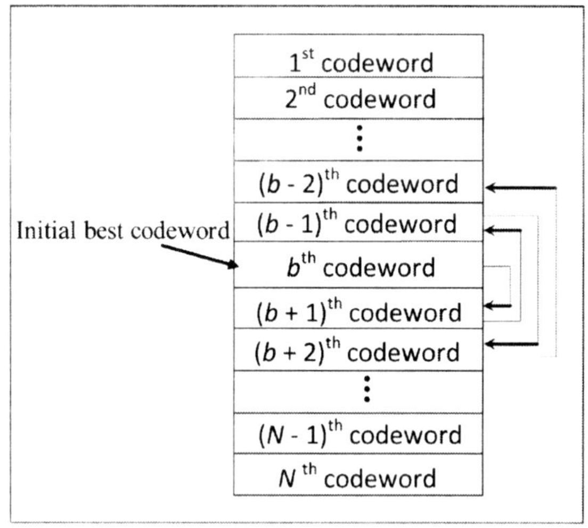

Figure 5. Geometrical interpretation of the AAPVS for 2-dimensional case

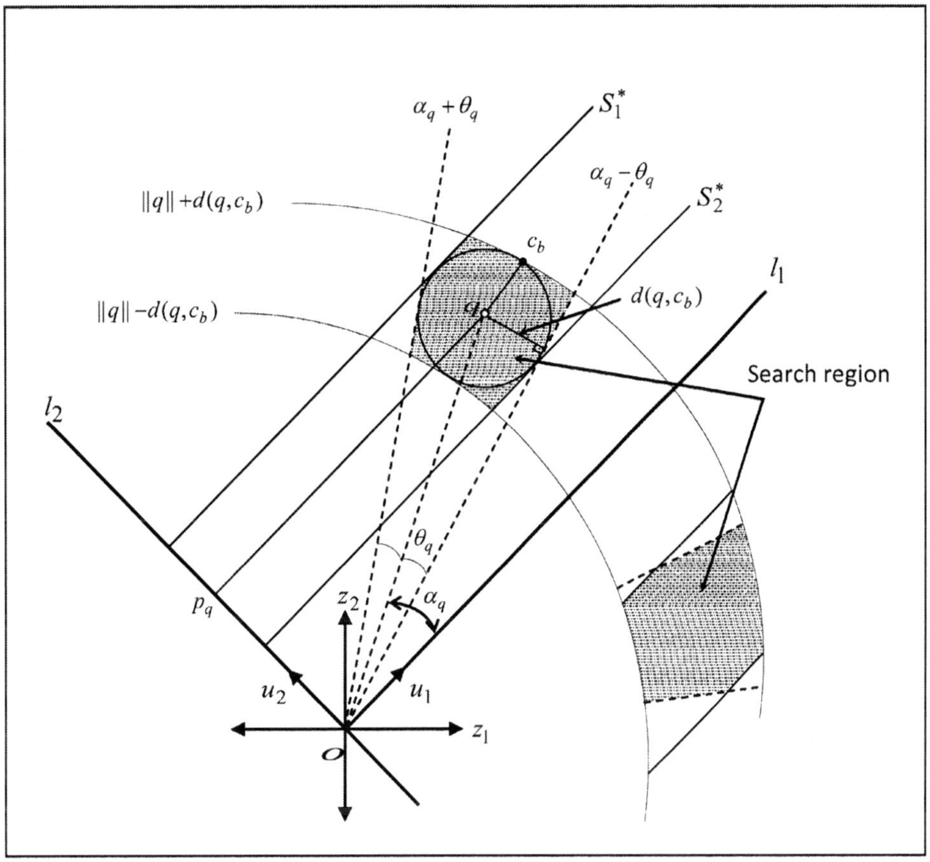

When $d(q,c_b) > \|q\|$, this means we can not calculate the angle θ. Hence, the angular constraint is not appropriate to find the closest codeword. In this case, we will search by using the annular constraint in the first proposed algorithm and by the annular and the projection value constraints in the second proposed algorithms.

4. EXPERIMENTAL RESULTS

Experiments were carried on vectors taken from the USC grayscale image set. We used four images, Lena, Baboon, Peppers and Jet airplane with size 512×512 and 256 gray levels as shown in Figure 6. Each image was divided into 4×4 blocks. Codebooks with different codebook sizes (N = 512, 1024, 2048) were designed using the full search algorithm for each image. The tested methods are full search (FS), the annular search (AS), double annulus search (DAS), equal-average equal-variance nearest neighbor search (EENNS), annular with angular search (AAS), annular-double angular search (ADAS), and annular-angular-projection value search (AAPVS).

Table 1 presents a comparison of the exact computing time (in seconds) for various codebook sizes. The timings were made on Pentium III (701 MHZ). Compared with the FS algorithm, the ADAS and the AAPVS algorithms reduce the computational time by 80.7-90.3%. Compared with AS, DAS, EENNS, and AAS algorithms, the proposed algorithms reduce the computational time by 25.5-43.2%, 22.2-34.5%, 11.7-27.8% and

Figure 6. The tested images

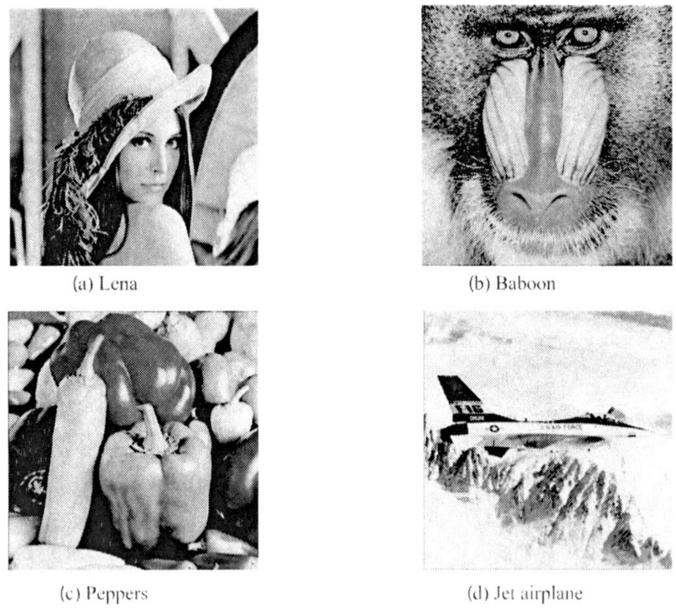

Table 1. Comparison of the exact computing time (in seconds) for various codebook sizes

Tested Image					
Jet airplane	**Peppers**	**Baboon**	**Lena**	**Method**	**Codebook size**
9.86	9.86	9.86	9.86	FS	512
1.74	1.51	3.04	1.51	AS	
1.50	1.37	2.72	1.44	DAS	
1.4	1.24	2.51	1.28	EENNS	
1.41	1.24	2.52	1.28	AAS	
1.12	1.06	1.90	1.10	ADAS	
1.09	1.03	1.83	1.09	AAPVS	
19.74	19.74	19.74	19.74	FS	1024
3.28	2.81	5.54	2.82	AS	
2.94	2.60	4.88	2.70	DAS	
2.69	2.36	4.44	2.38	EENNS	
2.69	2.36	4.48	2.38	AAS	
2.14	1.99	3.65	2.10	ADAS	
2.12	1.95	3.15	1.99	AAPVS	
39.44	39.44	39.44	39.44	FS	2048
6.29	5.49	10.75	5.60	AS	
5.72	5.17	9.32	5.34	DAS	
5.21	4.64	8.45	4.64	EENNS	
5.21	4.65	8.53	4.64	AAS	
4.06	3.89	6.19	3.91	ADAS	
4.01	3.82	6.1	3.84	AAPVS	

Table 2. Comparison of the average distortion computations for various codebook sizes

Tested Image				Method	Codebook size
Jet airplane	**Peppers**	**Baboon**	**Lena**		
9.86	9.86	9.86	9.86	FS	512
1.74	1.51	3.04	1.51	AS	
1.50	1.37	2.72	1.44	DAS	
1.4	1.24	2.51	1.28	EENNS	
1.41	1.24	2.52	1.28	AAS	
1.12	1.06	1.90	1.10	ADAS	
1.09	1.03	1.83	1.09	AAPVS	
19.74	19.74	19.74	19.74	FS	1024
3.28	2.81	5.54	2.82	AS	
2.94	2.60	4.88	2.70	DAS	
2.69	2.36	4.44	2.38	EENNS	
2.69	2.36	4.48	2.38	AAS	
2.14	1.99	3.65	2.10	ADAS	
2.12	1.95	3.15	1.99	AAPVS	
39.44	39.44	39.44	39.44	FS	2048
6.29	5.49	10.75	5.60	AS	
5.72	5.17	9.32	5.34	DAS	
5.21	4.64	8.45	4.64	EENNS	
5.21	4.65	8.53	4.64	AAS	
4.06	3.89	6.19	3.91	ADAS	
4.01	3.82	6.1	3.84	AAPVS	

11.7-28.4%, respectively. From this table, the proposed algorithms have the best performance in terms of computational time for all cases.

Table 2 lists the average distortion computations for various codebook sizes. Compared with the FS algorithm, ADAS and AAPVS algorithms could be reduce the distortion computation by 88.2-97.9%. Compared with AS, DAS, EENNS, and AAS algorithms, the proposed algorithms could be further reduce the distortion computation by 46.0-62.3%, 33.3-50.7%, 13.8-31.5% and 14.1-32.2%, respectively. From Table 2, we can see that the average distortion computations of the proposed algorithms are obviously much less than all other algorithms in all kind of cases. Compared with other available approaches, the proposed methods can reduce the computing time and distortion calculation significantly for VQ.

Table 3 depicts memory requirements for the tested methods of standard VQ encoding process with the vector dimension $k = 16$. Here, memory overhead is defined by comparing with the FS method where the required memory size is the codebook size N. In AS algorithm, we need extra memories to store the norm values of codewords. The DAS algorithm requires for all codeword storing norm values and the distances to the farthest codeword. The EENNS algorithm needs to store the mean and variance values of codewords. Also, the AAS needs to store the norm and the projection angles of codewords. The proposed algorithms ADAS and AAPVS require extra memories as in

Table 3. Comparison of total memory size required in encoding (N is the codebook size)

Tested method	Total memory size	Memory overhead
AS	$N + \frac{N}{16}$	6.25%
DAS	$N + \frac{2N}{16}$	12.5%
EENNS	$N + \frac{2N}{16}$	12.5%
AAS	$N + \frac{2N}{16}$	12.5%
ADAS	$N + \frac{3N}{16}$	18.75%
AAPVS	$N + \frac{3N}{16}$	18.75%

EENNS and AAS algorithms and another additional memory to store the second projection angle in ADAS and the projection value in AAPPS. Table 3 shows that the proposed algorithms have acceptable memory overhead for reducing of computational complexity.

5. CONCLUSION

This paper presents two fast codeword search algorithms based on using two perpendicular projection axes. The first algorithm uses three constraints for reducing the search area; the annular constraint and two angular constraints. It employs the norms and two projection angles to two projection axes of the vectors. In the second one, three constraints also are used; the annular constraint, one angular constraint and projection value constraint. The proposed algorithms (ADAS and AAPVS) can dramatically reduce the complexity of image vector quantization. They achieve equivalent performance to full search VQ. They have more efficiency than AS, DAS, EENNS and AAS algorithms. Experimental results demonstrate that the proposed algorithms are superior to the other existing methods.

REFERENCES

Bake, S. J., Bae, M. J., & Sung, K. M. (1999). A fast vector quantization encoding algorithm using multiple projection axes. *Signal Processing*, *75*, 89–92. doi:10.1016/S0165-1684(99)00035-3

Cao, H. Q., & Li, W. (2000). A fast search algorithm for vector quantization using a directed graph. *IEEE Transactions on Circuits and Systems for Video Technology*, *10*, 585–593. doi:10.1109/76.845003

Cardinal, J. (1999, September). Fast search for entropy-constrained VQ. Preceding of (ICIAP) the 10th International Conference on Image Analysis and Processing, (pp. 1038-1042).

Chen, T. S., Chang, C. C., & Hwang, M. S. (1998). A virtual image cryptosystem based on vector quantization. *IEEE Transactions on Image Processing*, *7*(10), 1485–1488. doi:10.1109/83.718488

Garcia, C., & Tziritas, G. (1999). Face detection using quantized skin color regions merging and wavelet packet analysis. *IEEE Transactions on Multimedia*, *1*(3), 264–277. doi:10.1109/6046.784465

Gersho, A., & Gray, R. M. (1991). *Vector quantization and signal compression*. Boston: Kluwer.

Gray, R. M. (1984). Vector quantization. *IEEE Acoustics, Speech, and Signal Processing*, *1*(2), 4–29.

Guan, L., & Kamel, M. (1992). Equal-average hyperplane partitioning method for vector quantization of image data. *Pattern Recognition Letters*, *13*(10), 693–699. doi:10.1016/0167-8655(92)90098-K

Huang, C. M., Bi, Q., Stiles, G. S., & Harris, R. W. (1992). Fast full search equivalent encoding algorithms for image compression using vector quantization. *IEEE Transactions on Image Processing*, *1*(3), 413–416. doi:10.1109/83.148613

Imamura, K., Swilem, A., & Hashimoto, H. (2003, May). Fast codeword search algorithm for ECVQ using hyperplane decision rule. Proceeding of (ISCAS) International Symposium on Circuits and Systems, vol. 2, (pp. 476-479).

Imamura, K., Swilem, A., & Hashimoto, H. (2004, October). Fast VQ encoding algorithms using angular constraint. Proceeding of (ICIP) International Conference on Image Processing, (pp. 3161-3164).

Johnson, M. H., Ladner, R., & Riskin, E. A. (1996, September). Fast nearest neighbor search for ECVQ and other modified distortion measures. Proceeding of (ICIP) International Conference on Image Processing, Vol. 3, (pp. 423-426).

Johnson, M. H., Ladner, R., & Riskin, E. A. (2000). Fast nearest neighbor search of entropy-constrained vector quantization. *IEEE Transactions on Image Processing*, *9*, 1435–1437. doi:10.1109/83.855438

Lai, J. Z. C., & Lue, C. C. (1996). Fast search algorithms for VQ codebook generation. *Journal of Visual Communication and Image Representation*, *7*(2), 163–168. doi:10.1006/jvci.1996.0016

Lee, C. H., & Chen, L. H. (1994). Fast closest codeword search algorithm for vector quantization. *IEE Proceedings. Vision Image and Signal Processing*, *141*(3), 143–148. doi:10.1049/ip-vis:19941140

Lee, C.-H., & Chen, L.-H. (1995). A fast search algorithm for vector quantization using mean pyramids of codewords. *IEEE Transactions on Communications*, *43*, 1697–1702. doi:10.1109/26.380218

Liao, H. Y. M., Chen, D. Y., Su, C. W., & Tyan, H. R. (2006). Real-time event detection and its applications to surveillance systems. Proceedings IEEE International Symposium on Circuits and Systems, (pp. 509-512).

Linde, Y., Buzo, A., & Gray, R. M. (1980). An algorithm for vector quantizer design. *IEEE Transactions on Communications*, *COM-28*(1), 84–95. doi:10.1109/TCOM.1980.1094577

Orchard, M. D. (1991). A fast nearest-neighbor search algorithm. Proceeding of (ICASSP) International Conference on Acoustics, Speech, and Signal Processing, vol. 4, (pp. 2297-2300).

Pan, J. S., Lu, Z. M., & Sun, S. H. (2000). Fast codeword search algorithm for image coding based on mean-variance pyramids of codewords. *Electronics Letters*, *36*(3), 210–211. doi:10.1049/el:20000237

Song, B. C., & Ra, J. B. (2002). A fast search algorithm for vector quantization using L2-norm pyramid of codewords. *IEEE Transactions on Image Processing*, *11*(1), 10–15. doi:10.1109/83.977878

Swilem, A., Imamura, K., & Hashimoto, H. (2002). A fast search algorithm for vector quantization using hyperplane decision rule. *Journal of the Institute of Image Information and Television Engineers*, *56*(9), 1513–1517.

Swilem, A., Imamura, K., & Hashimoto, H. (2004). A fast codebook design algorithm for ECVQ based on angular constraint and hyperplane decision rule. IEICE Transactions on Fundamentals of Electronics, Communications and Computer Sciences. *E (Norwalk, Conn.)*, *87-A*(3), 732–739.

Swilem, A., Imamura, K., & Hashimoto, H. (2005). A high-speed closest codeword search algorithm using the pyramid structure of codewords. (IIEEJ) *Journal of the Institute of Image Electronics Engineers of Japan*, *34*(5), 653–662.

Wu, K. S., & Lin, J. C. (2000). Fast VQ encoding by an efficient kick-out condition. *IEEE Transactions on Circuits and Systems for Video Technology*, *10*(1), 59–62. doi:10.1109/76.825859

Zheng, J., & Hu, M. (2006). An anomaly intrusion detection system based on vector quantization. IEICE Transactions on Information and Systems. *E (Norwalk, Conn.)*, *89-D*(1), 201–210.

This work was previously published in International Journal of Mobile Computing and Multimedia Communications (IJMCMC) 1(1), edited by Ismail Khalil & Edgar Weippl, pp. 16-28, copyright 2009 by IGI Publishing (an imprint of IGI Global).

Chapter 13
Mobile Video Streaming Over Heterogeneous Networks

Ghaida A. Al-Suhail
University of Basra, Iraq

Martin Fleury
University of Essex, UK

Salah M. Saleh Al-Majeed
University of Essex, UK

ABSTRACT

All-IP networks are under development with multimedia services in mind. Video multicast is an efficient way to deliver one video simultaneously to many users over such heterogeneous wired-to-wireless networks, such as in wireless IP applications where a mobile terminal communicates with an IP server through a wired IP network in tandem with a wireless network. Unicast video streaming is also an attractive way to deliver time-shifted TV to mobile devices. This Chapter presents a simple cross-layer model that leads to the optimal throughput to multiple users for multicasting video over a heterogeneous network. An adaptive forward-error-correction scheme is applied at the byte-level as well as at the packet-level to reduce channel errors. The results show that a server can significantly adapt to the bandwidth and FEC codes to maximize the video quality of service. For unicast streaming, the Chapter presents a single negative acknowledgment scheme in which a video stream is transmitted over a heterogeneous network from a streaming server to a mobile device in a WiMAX network. The broadband streaming system is compared to several candidate solutions based on originally wired network congestion controllers. Multi-connection streaming is also investigated.

INTRODUCTION

Video multicast is an efficient way to deliver one video simultaneously to many users over heterogeneous wired-to-wireless networks, such as in wireless IP applications where a mobile terminal communicates with an IP server through a wired IP network in tandem with a wireless network as in Figure 1. Such a network is commonly called an all-IP network (Lin & Pang, 2005) in

DOI: 10.4018/978-1-60960-563-6.ch013

Figure 1. Video multicast system over heterogeneous wired-to-wireless network

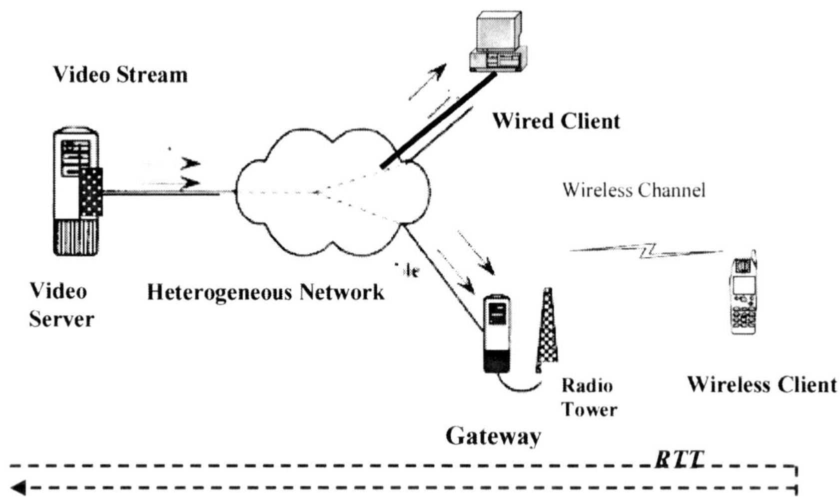

which Internet Protocol (IP) framing is standardized. Compared to unicast, multicast improves bandwidth efficiency by sharing video packets delivered through network. However, it suffers some particular problems arising from the use of wireless network applications. For example, a multicasting wireless network is often characterized by having a physical channel that is highly error-prone and time-varying. In addition, users in such a network can often have diverse channel conditions (Liu et al., 2007). Unicast video streaming, also over a heterogeneous network as in Figure 1, is generally introduced to provide a value-added addition to multicast video streaming. At the protocol level for heterogeneous networks, various extensions of traditional congestion control using TCP-Friendly Rate Control (TFRC) (Handley et al., 2003) have occurred. Some extensions (Fu et al., 2006; Görkemli et al., 2008), employ a cross-layer approach as in the multicast case study in this Chapter, while others use multi-connections (Chen & Zakhor, 2006; Al-Majeed & Fleury, 2010). Another approach is to use corruption-aware TCP (Balan et al., 2001; Tickoo et al., 2005; Cui et al., 2007), though there is a risk to throughput on the wired channel if TCP transport is used when congestion is high, because of TCP's reliability mechanism. Another scheme for unicast streaming is multi-connection TCP as in the Stream Control Transmission Protocol (STCP) (Stewart, 2007). In the unicast case study, a specific study of IEEE 802.16e (mobile WiMAX) (IEEE, 2005; Andrews et al., 2007) occurs, though the approach is general. The Chapter considers single negative acknowledgment broadband video streaming (Al-Majeed & Fleury, 2010a), which is a simplified method of approaching the problems of heterogeneous networks.

This Chapter introduces two case studies, the first of which considers multicast distribution of video and the second of which examines unicast distribution. The first study is wireless technology agnostic, while the second concentrates on wireless broadband, specifically IEEE 802.16e (mobile WiMAX). The first case introduces a framework of TCP-Adaptive Forward Error Correction (FEC) scheme to improve the link reliability via achieving a maximum TCP throughput for each client of MPEG-4 video multicast over a hybrid network. The cross-layer design considerably depends on the end-to-end bandwidth between a client and the server; a client (receiver) may progressively improve the video quality by adapting Reed Solomon (RS) FEC codes at the packet-level as well as byte-level at the data-link (radio-link) layer. The cross-layer model can eventually estimate the wireless channel state at the physical layer for each client and return feedback to the server to adapt the desired FEC codes. For sake of simplicity, time-invariant wireless channel is assumed corrupted by high error-bits in terms of channel Signal-to-Noise Ratio (SNR). As a result, the model can predict the quality of MPEG-4 video by adapting a playable frame rate (temporal scaling) under various packet errors and channel errors conditions.

The second of the case studies investigates the situation when mobile users are connected by WiMAX to a wired network infrastructure. For example, this could allow one mobile device user to stream over WiMAX through the Internet to another mobile user also connected to a WiMAX base station. As previously mentioned this Chapter proposes broadband video streaming as an effective ARQ (Automatic Repeat reQuest)-based way to improve unicast streaming. In Degrande et al., 2008, ways to improve IPTV (IP framed network delivery of TV) quality were discussed with the assumption that intelligent content management would bring popular video content nearer to the end viewer. If the round-trip-time is reduced as a consequence then a negative acknowledgment (NACK), through ARQ, become attractive as a way of mitigating channel errors. NACK error mitigation is particularly attractive for video-on-demand services as the complete file can be cached in a retransmission buffer at the 'video serving office' that lies at the intersection of the metro and access network.

Before looking in detail at the case studies, we now turn to background information that puts the studies into context.

VIDEO STREAMING OVER HETEROGENEOUS NETWORKS

Quality of Service

Many heterogeneous networks cannot provide a guaranteed of quality of service (QoS) for video streaming traffic. To this end, it is essential to rely on QoS metrics of a connection (flow or session) in terms of data throughput, packet error/loss rate, and delay performance. In practice, for QoS guarantees in high-rate multimedia applications, many major challenges of video traffic are faced on heterogeneous wired and wireless Internet links (Lee et al., 2002; Pei et al., 2004; Liu et al., 2004; Zhang et al., 2006; Chiasserini & Meo, 2002). Some of these challenges are to do with the high packet loss rate due to buffer overflow arising from traffic congestion over wired networks, and others are mainly faced by the characteristic of wireless links, which mostly suffer from low bandwidth and high bit error rates due to noise, interference, unpredictable user mobility (Doppler effects) and multi-path fading. The 'bottleneck' common to both military and civilian networks is the wireless link, not only because wireless resources (bandwidth and power) are more scarce and expensive than their wired counterparts, but also because the overall system performance degrades markedly due to time- and frequency-dispersive fading effects introduced by the wireless air interface. In fact, these link errors may result in packet (segment) losses, and a TCP sender interprets such losses as a signal of network congestion and consequently decreases the transmission rate (Chen & Zakhor, 2006). These reductions in transmission rates are unnecessary and lead to resource inefficiency. Unlike wired networks, even if large bandwidth is allocated to a certain connection, the loss and delay requirements may not be satisfied when the wireless channel experiences deep fades or high noise levels.

Congestion Control

In video streaming across a heterogeneous IP network unreliable UDP transport serves to reduce delay at the expense of considerable packet loss, while application-layer TCP emulation (Widmer et al., 2003), such as TFRC (Handley et al., 2003), acts as a form of cooperative congestion control (assuming most other traffic is carried through TCP transport). However, TCP emulation by the application is *not* the same as TCP. TCP itself if unmodified is unsuitable for delay-variation intolerant video streaming, because it introduces unbounded delay in support of a reliable service. Instead, TFRC emulation mimics the average behavior of TCP, but is not 'reliable' and does not result in the 'saw-tooth'-like rate fluctuations that arise from TCP's aggressive congestion control algorithms, which can cause disconcerting quality fluctuations at an end-user's display if the compression ratio is varied according to the available bandwidth (through bitrate transcoding of stored video or altering the quantization parameter of live video).

The main point of TFRC is to maintain compatibility with TCP within the conventional wired Internet but to also avoid the deficiencies of TCP on a wireless network. TCP's congestion control is a co-operative process which seeks to avoid the acquisition of too much bandwidth by any one transport protocol. The concept TCP-friendliness (Floyd & Fall, 1999) on the conventional Internet was introduced to avoid the risk of congestion collapse arising from the use of uncooperative protocols. In (Chen & Zakhor, 2006), the issue of whether multiple TFRC connections were fair to TCP was examined in detail. It was found that a single TCP connection reduced its throughput in proportion to the increase in roundtrip caused by the extra TFRC connections. However, the same behavior would occur if extra TCP connections had been introduced. As an alternative approach to the multiple TFRC connections, there have been many attempts to improve the behavior of TCP by intervening at the interface between the wired and wireless Internet. For example, in the Snoop approach (Balakrishnan et al., 1997) a module resides on a base station. The Snoop module checks all traffic for TCP flows and carries out local retransmissions when a packet is lost on the wireless link. It is necessary for the Snoop module to suppress ACKs while retransmission is going on. Though the Snoop approach can be extended to TFRC it implies the introduction of additional delay to a real-time service by the introduction of retransmissions. It also increases the complexity of the implementation, especially if uplink rather than downlink streaming takes place. Another approach, for example (Cen et al., 2003), is to gather end-to-end statistics in order to distinguish congestion loss from wireless channel loss. Unfortunately, these techniques do not appear to be sufficiently accurate. A promising alternative is to modify TFRC to respond to explicit loss notification (Balakrishnan & Katz, 1998) of channel losses as opposed to congestion losses.

Research in Tappayuthpijam et al., 2009 applied single connection TFRC over the emerging Long Term Evolution (LTE) 4G cellular wireless system. TFRC was combined with the Scalable Video Coding (SVC) extension to the H.264/AVC (Advanced Video Coding) codec, in such a way that the layering was adapted to the bitrate. The results are encouraging in terms of reduced packet losses, reduction of streaming interruptions and end-to-end delay, and buffering levels compared to not using TFRC. However, at the time of writing, these results for LTE rely on a stand-alone emulator, which does not include the effect of transport across the core network. Another issue is to what extent this approach is dictated by the need to avoid loss of core packets within the base layer of SVC, after which it becomes difficult to reconstruct the stream.

It is also possible to modify TFRC to detect packet losses occurring due to wireless channel conditions. In Fu et al., 2006, a reassembly failure at the Radio Link Control (RLC) layer signals such losses to a 3G SS. Feedback packets contain an estimate of the wireless channel packet loss rate. However, this approach (Fu et al., 2006) assumes an absence of congestion at the base station. It is also reliant on the safe arrival of the feedback packets. The Data Congestion Control Protocol (DCCP) (Kohler et al., 2006) includes a TFRC-like component. Görkemli et al., 2008 used DCCP in a scenario in which there are wireless access links to the core network at the sender and receiver sites. Therefore, this work investigatedu plink as well as downlink streaming to a mobile devices. Cross-leyer information was employed to modify TFRC's estimate of packet loss rates. In this case, the PHY layer ARQ information.

Error Control

In video multicasting due both to the varied nature of channel conditions, i.e., bandwidth and error rate, among the clients and the fact such conditions mainly varies over time, the video server has to continuously adapt the error-recover mechanisms and bit rate in order to optimize the overall temporal video quality in frame per second at clients. Specifically, to recover for packet loss, feedback recovery or forward-error correction code (FEC) may be used (Chan et al., 2006; Bajie, 2006; Al-Suhail, 2008). In general, feedback recovery does not work very well over long distances with real-time guarantees. When multicast groups grow large, simple reliable multicast protocols (Chan et al, 2006; Rubenstein et al., 1998) suffer from a condition known as *feedback implosion*: an overload of network resources due to the attempts of many receivers trying to send repair requests (referred to as NAK) for a single packet. A number of approaches exist to avoid this implosion effect such as randomized timers, local recovery whereby receivers can also send repair packets, and hierarchical recovery. However, while such approaches are effective, providing reliability without implosion, they can result in significant and unpredictable delays making them unsuitable for real-time applications which eventually must conform to stringent real-time constraints. Many studies have been devoted to hybrid schemes in combining FEC and automatic repeat request (ARQ) to reliably deliver data with an emphasis on reducing delay and meeting real-time constraints without using the solutions of randomized delays, local recover, or hierarchical recovery (e.g., Rubenstein et al., 1998; Marco et al., 2006; Liu et al., 2007).

The first case study in this Chapter uses FEC (Chan et al, 2006; Liu et al, 2004). The scheme illustrated in Figure 2 uses a channel code, i.e. an (n,k) linear code, that is designed for simultaneous error detection and error correction. It consists of arranging the data and redundancy symbols (bits) in such a way that even when not all the symbols (bits) are received, the original data may still be recovered. When the received block is detected with errors, the receiver first attempts to locate and correct the errors. If the presence of an error pattern is corrected that exceeds correction capabilities, the receiver rejects the received block and requests a retransmission. For the scheme, a Reed-Solomon (RS) code was adopted at byte and packet levels to provide a robust error cross-layer model for multicasting over a TCP transport layer. By adapting concatenated FECs according to the network conditions video quality at client can be efficiently maintained (Liu et al, 2002). The model, in Figure 2, is effective in terms of FEC parity and allocated bandwidth required to each user (client) over multicast-capable IP heterogeneous network.

Multi-Connection Video Streaming

In multi-connection unicast TFRC video streaming, a *single* video source is multiplexed onto several connections across the wireless link in order to increase the throughput, thereby improving wireless channel utilization. By multiplexing a video stream across multiple connections it is hoped that the impact of packet loss on one or

Figure 2. Concatenated FEC codes for video multicast system over heterogeneous network

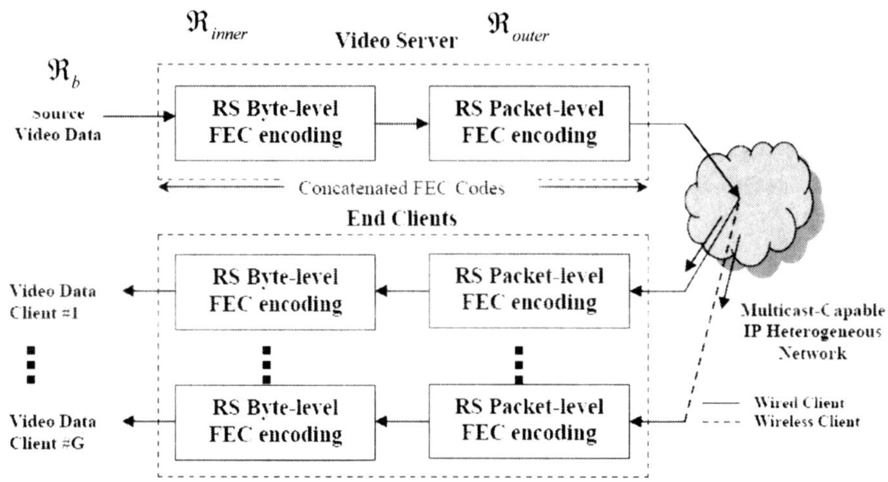

more of these connections will be mitigated by the data-rate across the remaining connections. TFRC's main role when congestion occurs across the network path is to reduce the video streaming data rate across the wired portion of the concatenated network. It does this in response to packet drops at intermediate routers, which signal the presence of contending traffic. Unfortunately, TFRC can misinterpret as congestion packet losses due to wireless interference and noise, leading to a reduction in wireless channel utilization.

In pioneering work on multi-connection TFRC, in MULTTFRC (Chen & Zakhor, 2005) improved video quality comes about by increasing the quantity of video data that can be sent over the multiple connections. Of course, increased video data implies a lower compression ratio and, hence, higher quality delivered video, provided the rise in packet losses across the wireless channel does not degrade the quality. If burst errors occur then during the time that they occur all connections are affected, leading to a rise in packet losses, which was countered in Chen & Zakhor, 2005 by means of application layer FEC. Unfortunately, if the number of connections varies, as it does in Chen & Zakhor, 2005, then sending rate oscillations can occur. If the compression rate was varied at the source (either by changing the quantization parameter at the codec if live video or through a bit-rate transcoder) then oscillations in bitrate run the risk of disconcerting changes in displayed video quality.

The research in (Chen & Zakhor, 2005) originally proposed MULTTFRC as a form of downlink control. Misinterpretation of channel loss as congestion is the source of under-utilization if a single TFRC connection were to be used. Any single TFRC connection responds to packet loss by reducing its output rate by increasing the inter-packet gap and reducing its throughput. In MULTFRC, other connections not affected by the packet loss can balance out the drop in throughput over the wireless part of the path. MULTTFRC represents a lightweight way to retain TFRC for the Internet path but avoid more complex means of suppressing channel loss feedback to TFRC over the wireless link. In the most definitive account of MULTTFRC so far, the work in Chen & Zakhor, 2006, there is no account of how a single video stream is multiplexed onto the multiple connections using dynamic scheduling or what the resulting video quality is in quantitative terms. Only generic packet loss and delay statistics are reported, even though the type of error pattern is

Figure 3. Layer structure over an heterogeneous wired-to-wireless network

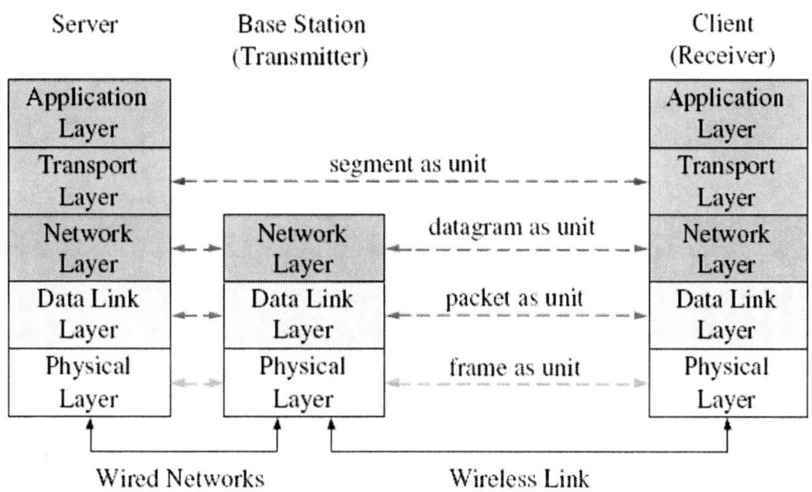

known to change the video quality (PSNR) by several dBs. Other work on MULTTFRC and its variants such as (Yogesh et al., 2006) appears largely confined to analysis of a generic link without other traffic.

MULTICAST VIDEO STREAMING CASE STUDY

Wireless Channel Model

The layer structure of the system under consideration and the processing units at each layer in the first cast study are shown in Figure 3. Hence, we consider a wireless link model across the physical (hardware-radio), data link, and transport layers for each wireless client, which enables the desired QoS metrics to be derived analytically in terms of BER, packet loss and throughput for each wireless client.

Let G be the multicast group size and the feedbacks for client g ($1 \leq g \leq G$) are in terms of the estimated end-to-end available bit rate B_g, and packet drop rate $P_{l,g}$ for wired line clients (due to permanently missing sequence numbers of the packets), and for wireless clients, the BER of wireless hop $p_{e,g}$(estimated after accounting for limited Automatic Repeat reQuest (ARQ) recovery in the wireless hop or by using a Markov process) (Chiasserini and Meo, 2002). To obtain $p_{e,g}$, frequent and random bit errors of a simple noisy wireless channel are considered without taking any fast fading effect into account. We define the physical layer packet loss rate as a function of the bit error probability $p_{e,g}$ (or γ_b which is SNR per bit) for a given modulation mode and packet length in L as in Yoo et al, 2004,

$$P_{e,g}(\gamma_b, L) \leq 1 - (1 - p_{e,g}(\gamma_b))^L \qquad (1)$$

where L denotes a packet length (in bits), and the inequality in (1) represents the fact that one can recover from bit errors in a packet, due to the coding scheme. Furthermore, we can define at the radio data link layer the maximum throughput (goodput) of a channel coding as the number of payload bits per second received correctly for a simple modulation scheme such as Binary-Phase Shift Keying (BPSK) (Yoo et al, 2004),

$$G_{Phy.g} = \frac{L - C}{L} \Re_b \left[1 - P_{e.g}(\gamma_b, L)\right] \quad (2)$$

Assume C may not only involve error-correction bits, but any *extra bits* which are related to a header of ARQ packet scheme (if ARQ scheme effect is taken into account) (Al-Suhail, 2008). The term $[1\text{-}P_{e,g}(\gamma, L)]$ denotes the packet success rate (PSR) (i.e., the probability of receiving a packet correctly), R_b is the source bit rate (in bps) excluding FEC code, and γ_b is the channel SNR per bit (in dB) given by Yoo et al., 2004 as

$$\gamma_b = E_b / N_o = P / (N_o \Re_b) \quad (3)$$

where E_b, N_o, and P represent the bit energy, the one-sided noise power spectral density, and the received power respectively. In non-fading or slowly fading channels where the fade duration is longer than the packet period, the system throughput can be optimized. In this case, the packet error in burst-error condition cannot easily be modeled by a single equation. The reason is that the distribution of error bits is not uniform. To simplify the estimation of BER performance, a BPSK scheme over an Additive Gaussian White Noise (AWGN) channel can be applied. Since $p_{e,g}$ in AWGN channel decays exponentially as γ_b increases, the probability of bit error can be given by (Yoo et al., 2004):

$$p_{e.g}(\gamma_b) = Q\left(\sqrt{2\,\gamma_b}\right), \quad (4)$$

$Q(.)$ is a Gaussian cumulative distribution function.

Adaptive TCP Throughput Formula

Despite the complex behavior of TCP due to its various mechanisms such as slow start, congestion, timeout, etc., it has been shown in Chen & Zakhor, 2006 that the throughput of a TCP-Friendly connection is a simple expression in the absence of timeouts. Hence, the steady state TCP goodput (i.e. maximum throughput) of a long-lived connection can simply be obtained by scaling the throughput by a factor of $(1\text{-}\varepsilon_g)$

$$G_{TCP.g} \approx \min\left(W_{\max}, \frac{k \cdot S}{\sqrt{\varepsilon_g}}\right) \frac{(1 - \varepsilon_g)}{RTT}, \quad (5)$$

where $W_{\max}$ is the maximum congestion window size of the TCP sender, ε_g is the residual end-to-end packet loss rate for wireless client, S is the packet size (MSS), k is a constant that is usually set to either 1.22 or 1.31, depending on whether the receiver uses delayed acknowledgments, and RTT is the round trip time experienced by the connection per packet sent (Liu et al., 2002, Barman et al., 2004). Since (5) does not account for timeouts, it usually overestimates the connection throughput as loss rate increases. It is reported in (Padhye et al., 2000) that (5) is not accurate for loss rates higher than 5%. Upon equation (5), Figure 4 therefore explains the end-to-end analytical TCP model of wired-to-wireless connection which is required to achieve the optimal performance in transporting video for each wireless client end (Chiasserini & Meo, 2002).

Optimal FEC Codes

According to Figures 2– 4, we present a possible end-to-end solution which employs a TCP-based adaptive concatenated FEC encoding to provide an error-resilient video service for each client over a heterogeneous IP network. We use the TCP goodput formula in (5) to analyze the tradeoff between the gain of the TCP goodput and the reduction of effective channel bandwidth through the application of FEC codes. We provide a cross-layer algorithm to obtain the optimum FEC codes that

Figure 4. An end-to-end analytical TCP model of wired-wireless connection for each wireless client

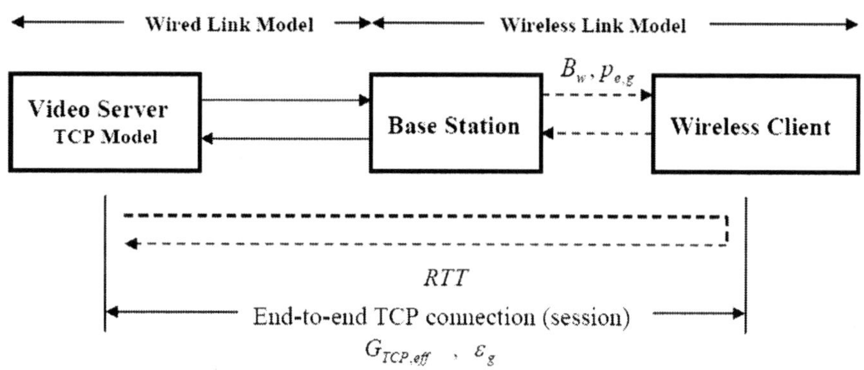

maximizes TCP goodput and consequently the resultant video play-out frame rate at end clients.

Application layer concatenated byte-level (*inner*) and packet-level (*outer*) FECs protect the video layers with Reed Solomon (RS) codes performed at the video server and error correction *only* performed at the end clients (Pei et al., 2004; Liu et al., 2004). It is possible to introduce a transcoding gateway between the wired and wireless components of the network path. The transcoding gateway can perform operations that adapt the packets for the differing error conditions across the wireless link. These operations may include fragmentation and re-packetization, as well as introducing additional FEC. This does not occur in this system because it removes end-to-end control without noticeably improving performance ((Lee et al., 2002).

Link-layer agents are assumed in the server, the base station and the mobile host to improve the TCP throughput. That is each wireless client effectively receives the video sequence with the bit rate allocated including FEC encoding (done at video server) equal to the minimum end-to-end bit rate (as output at the gateway), whereas the gateway (at the base station) *does not* recover any dropped packets through packet-level FEC or even pad the video packets with byte-level FEC parity. Note that with this system, byte-level FEC does not effectively help wired clients (where packet drops do occur) in improving their error resilience capability. Thus, the *only* concern at the video server is how much error control (FEC codes) should be applied to serve both wireless and wired clients so that their overall temporal quality at frame-level is optimized.

In this work, we consider an (n, k) RS codes to adjust the level of redundancy at the byte and packet level. Thus, $(n - k)$ parity symbols are added to n information symbols to form a codeword of size n. The size of code length n is *fixed* and the information symbols per codeword k is *varied* to adjust the redundancy level of the code. Here a symbol is the basic information unit used in a RS code, and is composed of a certain number of bytes (bits) or packets. Now, to generate the byte-level FEC based in an RS code, the encoder processes symbols, whereby each symbol consists of m bits (m=8 in general). Given a packet size n_b (in bytes), k_b (in bytes), k_b (≥ 1) bytes of source data are packed with $n_b - k_b$ parity bytes, where $k_b = n_b, n_b$-2... This is the so-called RS(n_b , k_b) code, which is able to correct up to t_b symbol errors in a packet, where $t_b = \lceil (n_b - k_b) / 2 \rceil$. The packet size n_b is limited by 2^m-1 symbols; therefore, for m=8, $n_b \leq 255$.

With every k_b of these byte-encoded video packets, a packet level FEC is then applied to generate $n_p - k_p$ parity packets to form a block of n_p packets. Here, the ith byte of each of the k_p

video packets ($1 \le i \le n_b$) is taken out to generate parity bytes. The generated parity bytes are then redistributed as the *i*th byte of each $n_p - k_p$ parity packets corrected up to $t_p = n_p - k_p$ packet losses.

On the other hand, the server computes the optimal allocation between the video data rate, the packet-level FEC rate (number of packet parity bits per second, i.e., the outer rate, $R_{outer}=k_p/n_p$, and the byte-level FEC rate (number of byte parity bits per second, i.e., the inner rate, $R_{inner}=k_b/n_b$, for given feedbacks from the end clients. Here, the server has first to decide the packet-level and byte-level FEC rates with its transmission rate R_b, including all the redundant bits is equal to the least end-to-end bit rate in the multicast group, i.e., $R_b=\arg\min_g(B_g)$. Consequently, the resultant video source rate G_C excluding all the FEC is given as (Pei et al., 2004),

$$G_{C,g} = \Re_b \times \Re_{outer} \times \Re_{inner} \tag{6}$$

Let $\Gamma_{C,g}$ be the goodput of the *gth* client. We have studied the following byte-level and packet-level FEC allocation problem: Given n_b and n_p, find the optimal k_p *and* k_b in order to maximize the end-to-end goodput of all clients,

$$\Gamma = \sum_{g=1}^{G} \Gamma_{C,g} \le B_w \tag{7}$$

such that the end-to-end packet loss rate after error correction is no more than a certain threshold value, $\varepsilon_{Threshold}$ (say, 1%-3%) for loss rate over all clients. Further, $\Gamma \le B_w$ and $\Gamma_{C,g} \le B_w$, where B_w is a limited wireless channel capacity. We consider that all clients in the system have the same priority or importance.

Let us consider a particular client *g* and obtain its goodput given $P_{l,g}$ and $P_{e,g}$. In the wireless hop, the symbol error rate is computed from (1) for *m*=8 bits and L= n_b =255 bytes. Since the RS (n_b, k_b) code corrects up to t_b symbol errors, the probability that a random packet cannot recovered by byte-level FEC is given by (Lee et al, 2002),

$$\alpha_g = \sum_{j=t_b+1}^{n_b} \binom{n_b}{j} P_{e,g}^{\ j} (1 - P_{e,g})^{n_b - j} \tag{8}$$

Notice that for wired clients, $\alpha_g=0$ as $p_{e,g}= 0$ by definition.

In contrary, a packet is correctly received by the client, only if it is not dropped from overflow or blocking in wired networks with probability (1- $P_{l,g}$) and is correctly received through the wireless channel with probability (1- α_g). Hence, we express the packet (segment) loss rate, as in (Liu et al., 2004):

$$p_g = 1 - (1 - P_{l,g})(1 - \alpha_g) \tag{9}$$

Note that the dropped packets may be recovered by the packet-level FEC. Then, the probability that a random packet is permanently "lost", i.e., the end-to-end packet loss rate *after* error correction is given by,

$$\varepsilon_g = \sum_{k=t_p+1}^{n_p} \frac{k}{n_p} \binom{n_p}{k} p_g^{\ k} (1 - p_g)^{n_p - k} \tag{10}$$

To reduce the complexity of optimization, we use the same two-step procedures mentioned in (Lee et al., 2002) as follows:

- **Step (1)**: Packet-level FEC optimization is used to find the value of k_p* so that the residual loss rate over wired network is no more than $\varepsilon_{Threshold}$ ignoring wireless links errors by setting $\alpha_g = 0$ for all clients. Let $P_l = \max_g P_{l,g}$ be the maximum packet drop rate for all the clients. If $P_l \le \varepsilon_{Threshold}$, STOP and proceed to the next step, otherwise the packet drop rate is low so that k_p*= n_p. For all the clients with $P_{l,g} > \varepsilon_{Threshold}$, we need to search for the largest $k_p \leq n_p$ such that $\varepsilon_g \leq \varepsilon_{Threshold}$.

- **Step (2)**: Byte-level FEC optimization follows to find the value of k_b*, for given k_p* and re-introducing α_g, that is the largest $k_b \leq n_b$, such ε_g in (9) for all the wireless clients is no more than $\varepsilon_{Threshold}$. Hence, the effective link bandwidth (goodput) of the client g can be obtained after manipulating formula (2) to match the model requirements. Then,

$$\Gamma^*_{C,g} = G^*_{C,g}\left(1-\varepsilon_g\right) = \Re_b \times \Re^*_{outer} \times \Re^*_{inner}\left(1-\varepsilon_g\right) \quad (11)$$

Optimal TCP-Adaptive FEC over Wireless

According to the model in Figure 4, given the estimate of ε_g in (10) and a certain value of *RTT*, we can compute the available TCP goodput ($G_{TCP,g}$) using (5). Then, to achieve the optimal TCP-adaptive FEC for wireless clients, the real TCP goodput ($G_{TCP,eff}$), i.e., the optimal allocated bandwidth required for the TCP protocol, can be accounted for by the minimum of the achievable TCP goodput ($G_{TCP,g}$) and the effective link bandwidth ($\Gamma_{C,g}$), as in (Liu et al, 2002),

$$G^*_{TCP,eff}(k^*_p, k^*_b, \varepsilon_g) = \min\left(\Gamma_{C,g}(k^*_p, k^*_b, \varepsilon_g), G_{TCP,g}(k^*_p, k^*_b, \varepsilon_g)\right) \quad (12)$$

As a result, the optimal RS codes (n_p ,k_p*) and (n_b ,k_b*) are the codes that maximizes TCP throughput after error correction at end clients (i.e., TCP Goodput), and are computed as follows (Liu et al, 2004):

$$k^*_p, k^*_b, \varepsilon^*_g = \arg\max \; G_{TCP,eff}(k_p, k_b, \varepsilon_g) \quad (13)$$

$$k_p \in n_p, k_b \in n_b, \varepsilon_g \in p_g$$

In point of fact, the TCP throughput is maximized when the achievable TCP throughput equals the effective channel bandwidth. Ideally each k_p and k_b is just the solution to the equation $\Gamma_{C,g}=G_{TCP,g}$. Thus the protocol, which is based on the client feedback via the base station (gateway), uses a lookup table at the video server generated a priori to find the code that yields the largest goodput as a function of the channel SNR estimate.

Temporal Video Model

Using a TCP-adaptive FEC scenario, the predicted optimal playable frame rate (PFR) based on maximum effective TCP throughput in (12) can be evaluated as (Wu et al., 2005; Al-Suhail, 2008),

$$R^*_{g,eff} = G.W_I \cdot [1 + \chi_P + N_{BP} \cdot W_B(\chi_P + W_I \cdot W_P^{N_P})] \quad (14)$$

where,

$$\chi_P = \frac{W_P - W_P^{N_P+1}}{1-W_P}, W_i = \left(1-\varepsilon^*_g\right)^{S_i}, \; and$$
$$G^*_{GoP} = \frac{G^*_{TCP,eff} / L}{S_I + N_P S_P + N_B S_B}. \quad (15)$$

W_i stands for the successful transmission probability of the *i-th* frame type (I, P, and B) in a GoP pattern taking into account the end-to-end packet loss rate after error correction given in (10). S_i denotes packet size of the *i-th* frame type. In our analysis, packet length must be fixed at 255 bytes in the case of byte-level FEC. $G^*_{TCP,eff}$ is defined as the effective network throughput received at the client in (bps) or in other words, the optimal allocated bandwidth required for TCP protocol under the constraint of (12), and G^*_{GoP} corresponds to the optimal number of GoPs per second. S_I,S_P, and S_B are the frames' sizes of the I, P, and B frames in GoP pattern (in packets), respectively.

Table 1. Clients' profile used for proposed cross-layer model

Client #	$P_{l,g}$ (%)	$p_{e,g}$(10-4), Error Type
C#1 wireless	2.7698	1.0134, random error AWGN
C#2 wireless	2.4790	0.8594, random error AWGN
C#3 wireless	2.0572	0.9993, random error AWGN
C#4 wireless	1.8248	1.3363, random error AWGN
C#5 wireless	1.7179	0.5460, random error AWGN
C#6 wired	1.1049	0.0
C#7 wired	1.3341	0.0
C#8 wired	2.1079	0.0
C#9 wired	2.4529	0.0
C#10 wired	2.7578	0.0

Table 2. Wireless network and GoP pattern parameters used in simulation

Parameter	Wireless Network Design Parameters
RTT	168 [ms]
k	1.22
B_w	1Mbps (1xRTT CDMA)
L_{max}	255 byte*
Modulation	BPSK (upload/download)
γ_b(AWGN)	5...10 [dB] Channel SNR/bit
	GoP Pattern Design parameters
F_o	30 [fps] reference frame rate at video server
GoP(2,3)	*I-BBB-P-BBB-P-BBB*
(S_I, S_P, S_B)	Typical values (25,8,1), (10,3,1), (7,2,1) [packet]

* We assume that a TCP packet size (MSS) is equal to the maximum length of packet (block) at the data-link layer (Barman et al, 2004).

Evaluation

We assume the clients' profile of loss rates used for non-transcoding gateway as in Table 1 (Lee et al, 2002). As a reference, we fix a typical set of parameters as: threshold error $\varepsilon_{Threshold}$=1%, n_b=255 byte, n_p=40 packet, and R_b=100 kbps. We consider also a baseline system of G=10 clients, with half of them being wireless and the remaining being wired according to a uniform distribution with mean $P_{l,g}$ = 2% and $p_{e,g}$=10-4 (i.e., average SNR, γ_b=8.4 (dB) for a simple random error of AWGN channel) , respectively.

Table 2 contains the wireless network and GoP parameters used in simulation. Optimal FEC allocation has been conducted given the parameters using the two-step procedures. For example, the resultant predicted PFR of (14) is 9.32 [fps] in the case of client **C#3** using the optimization procedure defined in Steps 1 and 2, respectively. The corresponding optimal values of the design parameters are achieved as follows: optimal TCP goodput ($G_{TCP,eff}$) equals 93 kbps, for the obtained values of k_p^* =38, k_b^* =249 (251), and ε_g^* = 3.7×10^{-3} (8.57×10^{-4} at a minimum given RTT of 168 ms in 1xRTT CDMA network (Chen and Zakhor, 2006; Al-Suhail, 2008). Note that for $p_{e,g}$=10-4, the byte-level FEC occurs giving a packet error with probability of 0.1845 using (1), whereas L=$m.n_b$. With this packet-loss rate, only a few parity bytes (about 4-6) are enough to bring this error rate down to a low level given by $\varepsilon_{Threshold}$(Lee et al, 2002).

Effectively, this indicates the efficiency of byte-level FEC. In fact, once k_p^* is increased, the packet-level error correction efficiency decreases, and hence a lower k_b^* (i.e., stronger byte-level correction efficiency) is needed to adapt the overall bandwidth. Hence, the optimal allocated bandwidth for TCP in (12) is analytically affected by the two factors: the optimal end-to-end packet loss rate ε_g^* (i.e., optimal bandwidth required in (12) decreases if ε_g^* is increased), and the optimal parameters k_p^* and k_b^* (i.e., optimal bandwidth required in (12) increases if both parameters are increased). Accordingly, the effective TCP throughput can be evaluated through the optimal allocated bandwidth that is required to compensate the high values of residual packet loss rate when low values of FEC parity symbols (at byte-level as well as packet-level) are allocated for video bitstream. Figure 5 illustrates the

Figure 5. Optimal allocated bandwidth vs. the end-to-end resultant packet error rate. $P_{l,g}$ = 2% and $p_{e,g}$ = 10^{-4}, R_b=100 kbps, n_b=255, and n_p=40 for various FEC codes

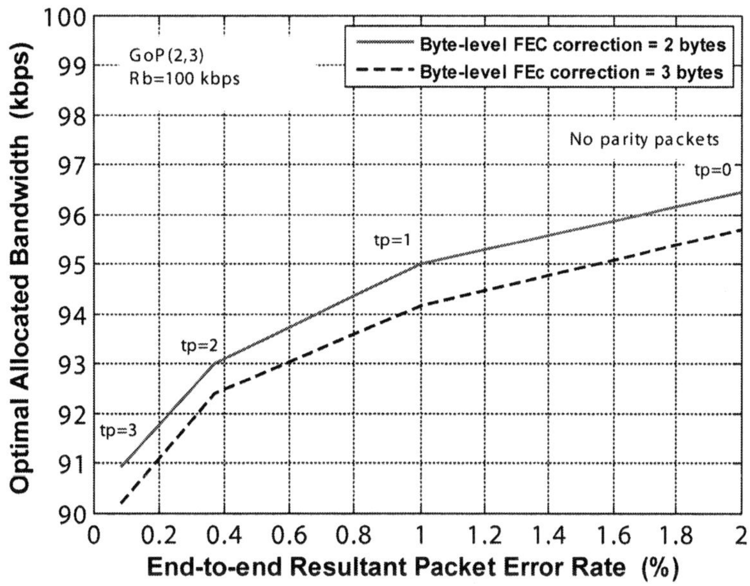

ε_g

optimal allocated bandwidth vs. the end-to-end residual packet loss rate for various values of FEC codes. Meanwhile, Figure 6 displays the corresponding system threshold required of a residual error rate vs. optimal allocated bandwidth for some significant values of packet-level FEC. It is clearly noticed that when $\varepsilon_{Threshold}$ increases then the resultant ε_g^* increases and consequently the optimal allocated bandwidth required in this case considerably increases according to our model in (12). It means that we should choose only a minimum required bit rate as an optimal allocated bandwidth for TCP protocol for each wireless client. In contrast, the throughout of each client will be limited by the upper bound of the video source data R_b=100 kbps.

On the other hand, it is found that when the optimal t_p^* is increased the end-to-end optimal bandwidth required is clearly reduced for a given byte-level FEC codes. The results obtained are compatible with corresponding ones in (Liu et al, 2004), but with different network conditions. In our model, Figure 7 reveals the optimal PFR vs. the system threshold of residual error rate under different packet size for each *i-th* frame type of GoP. Specifically, PFR in (14) depends on two design parameters: ε_g^* and the optimal allocated bandwidth of TCP protocol. The later parameter can basically depend indirectly on ε_g^* in case of $\Gamma_{C,g}$ is being greater than $G_{TCP,g}G_{TCP,g}$ in (12).

Since the optimal allocated bandwidth based TCP is limited by the upper bound of video source rate, R_b, it is noticed that for a value of k_p^* = 38, there is no significant change in PFR when t_b^* is increased from 2 to 3 bytes (Lee et al, 2002). In our model, given feedback of clients the video server needs to adapt the resultant PFR at the client by choosing either the certain threshold value $\varepsilon_{Threshold}$ or appropriate packet size for each frame type in GoP pattern, or both. Therefore, it is noticed clearly that the play-out frame rate outperforms as far as the frame's packet size is shortened at the video server in order to compen-

Figure 6. Optimal allocated bandwidth vs. the system's threshold value of residual error rate. $P_{L,g} = 2\%$ and $p_{e,g} = 10^{-4}$, $R_b = 100$ kbps, $n_b = 255$, and $n_p = 40$ for various FEC codes

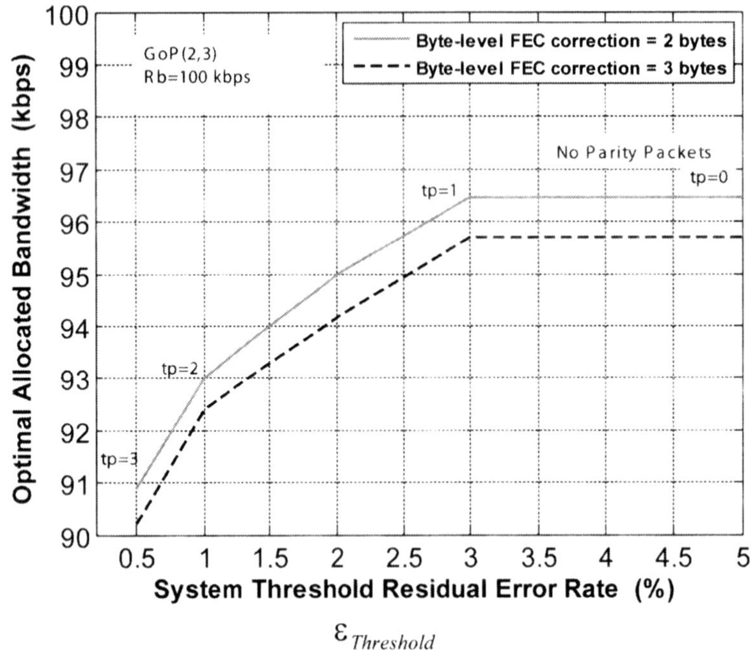

Figure 7. Optimal play-out frame rate vs. the system's threshold value of residual error rate. $P_{L,g} = 2\%$ and $p_{e,g} = 10^{-4}$, $R_b = 100$ kbps, $n_b = 255$, and $n_p = 40$ for various packet frame size of GoP.)

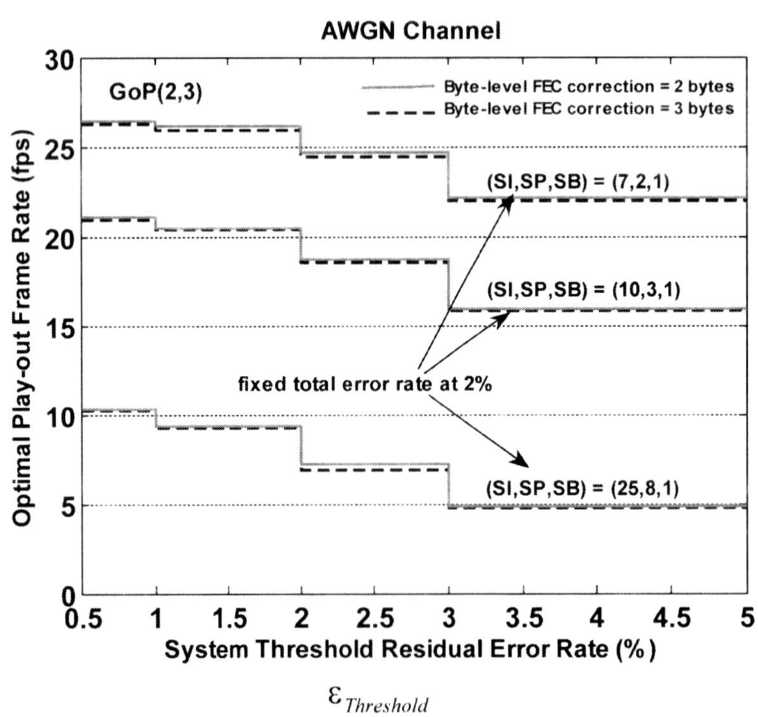

sate any packet drop in each frame of GoP. An example is given in Table 3. The results obtained are compared depending on the maximum throughput of BCH channel coding in (Al-Suhail, 2008). The effective PFR can significantly rise to achieve 20.52 [fps] and 26.16 [fps] at the wireless client **C#3** for different packet size settings (generated at video server) of I, P, and B frames, such as (10, 3, 1) and (7, 2, 1) respectively.

Thus, we can deduce that as the packet size of each *i*th frame type (I, P, and B) decreases the play-out frame rate considerably increases at the client ends in order to compensate the frame dropping process in our model compared to example of approach in (Al-Suhail, 2008). However, the system threshold value of residual error rate has a significant effect on video quality. As this certain value (at the video server) increases to be greater than 3%, the quality degradation will be constant due to the fixed achievable value of residual packet error of (10) at 2%. Moreover, extra TCP goodput results can also be obtained when the reference video source rate is considered to be 160 kbps or 200 kbps for UMTS networks.

In Table 3, it is also noticeable that our proposed approach clearly introduces reasonable performance compared to the cross-layer design in (Liu et al, 2004). The optimal values of the residual packet loss rate after correction (PLR) can achieve the values of 0.0857%-0.37% to provide optimal PFR; meanwhile the optimal TCP throughput improvement by using cross-layer design in (Liu et al, 2004) under different channel conditions and effects of design parameters illustrates that the resultant PLR values are no less than 0.3%-0.6%.

As a result, we can conclude that compared to the other related work (Chan et al., 2006; Lo et al., 2005; Lee et al., 2002; Liu et al, 2004; and Al-Suhail, 2008), one of the key features of this cross-layer model is that it uses a formula-based approach to analytically derive the optimal adaptive FEC that maximizes end-to-end TCP throughput and consequently multicasting the optimal video quality at the clients. The required modifications to implement TCP adaptive FEC include some link layer operations at the video server, base station and mobile host. Since the modified link layer operations are transparent to the TCP at the end clients, the end-to-end semantics of TCP is preserved.

MULTICAST VIDEO STREAMING CASE STUDY

The success of IPTV services such as the BBC's iPlayer in the UK suggests that delivery of video streams to the user will be important. The iPlayer

Table 3. A comparative example of MPEG-4 video transport under different network conditions

Approach	Channel State	Error Type*	FEC Parity, Correction Symbols	Residual PLR *after* error correction (%)	Optimal TCP Bandwidth (kbps)	S_I, S_P, S_B (packets)	*PFR* (fps)
Hybrid FEC and ARQ-Based TFRC (AL-Suhail, 2008)	C1 9.40 dB (wireless link)	1×10^{-4} Random Error AWGN	9 (t=1) BCH at data link layer	0.126	80.11	25,8,3	26.13
	C2 7.35 dB (wireless link)	5×10^{-4} Random Error AWGN	18 (t=2) BCH at data link layer	0.228	71.08	25,8,3	22.13
Proposed Cross-layer based byte-level and Packet-level FECs	C# 3.8dB-8.4dB (wired & wireless links)	$P_{e,g} = 1\times10^{-4}$, Random Error AWGN $P_{l,g}$ =2%, Congestion Error	2(tp =2 packets) 4-6 (tb =2-3 bytes) (RS coding)	0.37-0.0857	93.0 (on average)	25,8,1 10,3,1 7,2,1	9.32 20.52 26.16

* Error Type= Random Error over AWGN channel, GOP(2,3), RTT=168 ms.

Figure 8. (a) Downlink and uplink streaming scenarios, (b) schematic IPTV distribution

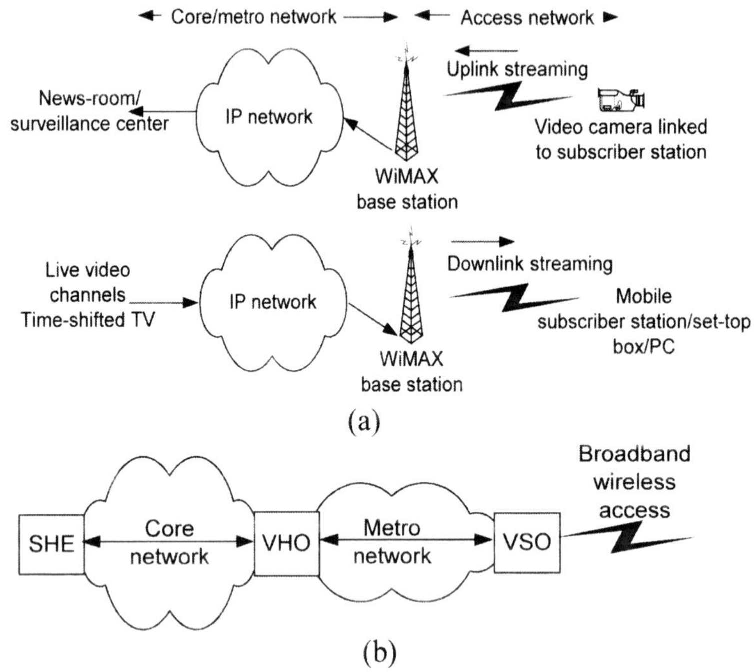

allows TV programmes to be streamed on demand, either live programmes or time-shifted TV. Though currently based on Adobe Flash Player technology, it could be that because of the limitations of TCP transport (unbounded delays and fluctuating bitrates) this will be superseded by other transport protocols, especially if mobile TV is supported on WiMAX (IEEE. 2005; Andrews et al., 2007).

In Issa et al., 2009, IP/UDP/RTP (Real Time Protocol) IPTV streaming was evaluated on a WiMAX testbed for downlink delivery of TV channels and uplink delivery of either TV news reports or video surveillance; refer to Figure 8a. However, that research (Issa et al., 2009) did not consider the impact of the intervening core wired network connecting the WiMAX base stations. In Degrande et al., 2008, ways to improve IPTV quality were discussed with the assumption that intelligent content management would bring popular video content nearer to the end viewer. The typical IPTV architecture considered in Degrande et al., 2008, Figure 8b, assumes a super head-end (SHE) distributor of content across a core network to regional video hub offices (VHOs). VHOs are connected to video serving offices (VSOs) over a regional metro network. It is a VSO that interacts with users over an access network. In this Section, the various streaming protocols are tested through simulation across the path over the metro network to the mobile user subscriber station (SS).

Figure 9 shows theheterogeous network simulated in which node C represents the source or sink of downlink or uplink streaming according to Figure 8. The WiMAX channel is between the basestation (BS) and SS shown. In the Figure, all links except a bottleneck link within the metro network are set to 100 Mbps to easily accommodate the traffic flows entering and leaving the network. The link delays are minimal (2 ms) to avoid confusing propagation delay with re-ordering delay in the results. A bottleneck link with capacity set to 5 Mbps is set up between the two routers. This arrangement is not meant to physically correspond to a network layout but to rep-

Figure 9. Video streaming scenario for IPTV

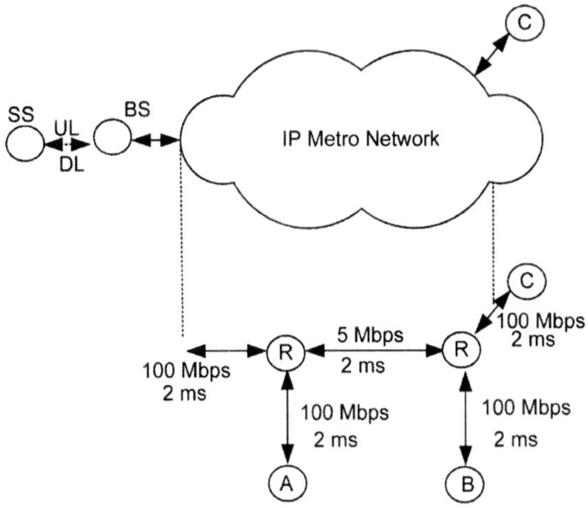

resent the type of bottleneck that commonly lies at the network edge. Node A sources to node B a Constant Bit-Rate (CBR) stream at 1.5 Mbps with packet size 1 kB and sinks a continuous TCP File Transfer Protocol (FTP) flow sourced at node B. Node B also sources an FTP flow to the BS and a CBR stream at 1.5 Mbps with packet size 1 kB.

VIDEO STREAMING PROTOCOLS

We now outline the characteristics of the streaming protocols that are compared. This is followed by details of the simulation model used in the evaluation.

After connection negotiation has taken place, DCCP (See the Introduction) (Kohler et al., 2006) adopts TFRC for streaming purposes. However, in the wireless domain, several attempts, e.g. (Fu et al., 2006; Görkemli et al., 2008), have taken place to improve TFRC's utilization of the wireless channel, which reduces sharply when packet loss occurs. The result can be that considerable interruption to the stream may occur if packet losses, causing TFRC to reduce its streaming rate without reducing video quality. In Fu et al., 2006 and Görkemli et al., 2008, cross-layer intervention occurs in one way or another to mask channel packet loss from TFRC. Alternatively in Tappayuthpijam et al., 2009, it is assumed that the data-link layer transparently retransmits packets until successful receipt occurs. The potential problem of this approach is that the application loses control of packet latency, which implies that the mobile device will require large buffers to compensate, with resulting lengthy start-up times. Another approach, as mentioned in the Introduction is to employ multi-connection TFRC (Chen & Zakhor, 2006). In the experiments reported in this Section, the number of connections was set to four, as this value has been found to be Al-Majeed & Fleury, 2010) the point when the value of the multi-connection approach is maximized. It is possible that other numbers of connections could improve the performance of multi-connection TFRC. However, finding the correct number of connections then becomes a further problem.

In contrast, after observing that UDP at least succeeds in good wireless channel utilization (Issa et al, 2009), without any protection from channel loss, the simple broadband video streaming scheme introduces a single negative acknowledgment (NACK) to UDP with a remarkable improvement across metrics of interest. Figure 10 is a representation of the processing involved, showing the NACK response of the receiver. At a mobile SS a record is kept of packet sequence numbers available through the RTP header and, if an out of sequence packet arrives, a NACK is transmitted to the base station in the next sub-frame. The base station prevents transmission from its input buffer until a single retransmission of the missing packet in the sequence has taken place. The base station continues its transmissions with the next packet in sequence. Further retransmissions do not take place, as waiting packets could be delayed and because the failure of one retransmission may indicate continuing poor channel conditions.

It is possible to elaborate this simple scheme by assigning priorities to the compressed video data and varying the number of retransmissions accordingly. Priority decisions could be based on:

Figure 10. Operation of broadband video streaming

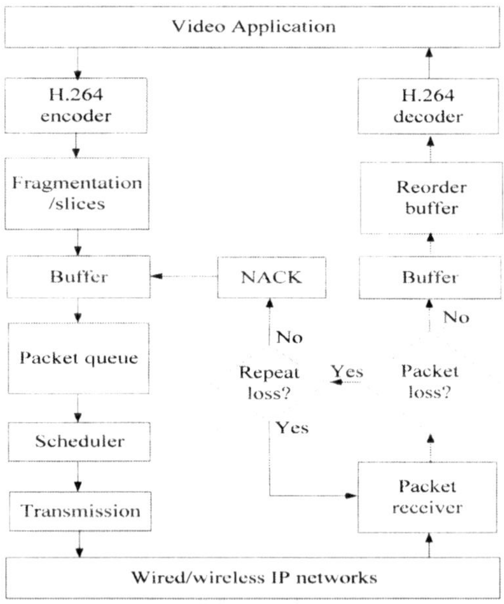

picture type, or display and coding deadlines (Razavi et al., 2008), or on some form of layering such as H.264/AVC (Advanced Video Coding) codec data-partitioning (Liu et al., 2006) or through H.264/SVC (Scalable Video Coding). The contribution of this Section is to directly illustrate the advantage of a broadband video streaming scheme by showing how use of a simple NACK delivers good quality video. However, we also explore the potential of priority retransmission based on picture type.

When seeking for an architecture that is suitable for video streaming over heterogeneous wireless networks, the video control protocol (VCP) (Suhoen et al., 2002) also included NACK capability. In fact, different classes of video traffic are parameterized in VCP, with Automatic Repeat reQuests (ARQs) reserved for a video stream that does not have a strict scheduling regime (unlike interactive video applications such as video conferencing) and for which buffering can be employed to maximize video quality. However unlike in broadband video streaming, VCP either sends an ARQ after every packet or groups several missing packets in an NACK, which obviously either increases overhead or delay.

WiMAX Model

The WiMAX system operating in point-to-multipoint mode was simulated by well-known ns-2 simulator (v. 2.29) augmented by a WiMAX module (Tsai et al., 2006). Mean data points are the arithmetic mean of twenty-five runs. These points were found with 95% confidence to be statistically independent of equivalent points. The simulator was allowed to reach steady-state over 20 s before commencing video streaming.

The PHY settings selected for WiMAX simulation are given in Table 4, with additional MAC settings defaulted from (Tsai et al., 2007). The antenna is modeled for comparison purposes as a half-wavelength dipole. In practice, a directional sector antenna might be used. Other antenna settings are defaulted from the standard. The modulation and PHY FEC rates are appropriate to a donlink/uplink sub-frame ratio of 3:1, which is standard and gives the BS more capacity to cope with a number of mobile stations. The frame length is significant, as a longer frame reduces delay at the MS by permitting more data to be removed from any queues at each polling time. The value of 20 ms is at the high end of the available durations in the Standard (IEEE, 2005) in order to reduce this source of queuing delay. The buffer sizes at the base station and mobile station were set to fifty packets, as it is unlikely that mobile stations will support large buffers. Similarly, router buffers were also set to fifty packets. In a WiMAX setting, a packet corresponds to a MAC Service Data Unit (MSDU) within a MAC Protocol Data Unit (MPDU).

A trace file was input to ns-2 and packet losses recorded in the output. The output serves to calculate the objective video quality (PSNR). Video quality comparisons were made under the EvalVid environment (Klaue et al., 2003). As a test, we used the *Paris* sequence H.264/AVC

Table 4. Simulated WiMAX settings

Parameter	Value
PHY	OFDMA
Frequency band	5 GHz
Duplexing mode	TDD
Frame length	20 ms
Max. packet length	1024 B
Raw data rate	10.67 Mbps
IFFT size	1024
Modulation	16-QAM 1/2
Guard band ratio	1/8
DL/UL ratio	3:1
Path loss model	Two-ray ground
Channel model	Gilbert-Elliott
MS transmit power	250 mW
BS transmit power	20 W
Approx. range to MS	0.7 km
Antenna type	Omni-directional
Antenna gains	0 dBD
MS antenna height	1.5 m
BS antenna height	32 m

OFDMA=Orthogonal Frequency Division Multiple Access, QAM = Quadrature Amplitude Modulation, TDD = Time Division Duplex

Variable Bit-Rate (VBR)-encoded at 30 frame/s with Common Intermediate Format (CIF) (352×288 pixel/frame) with quantization parameter (QP) set to 26 (from a range 0 to 51). The video quality (PSNR) for this sequencewithout packet loss is 38 dB. The slice size was fixed at the encoder at 900 B. In this way the risk of network segmentation of the packet was avoided, which could result in loss of synchronization at the decoder. *Paris* consists of two figures seated round a table in a TV studio setting, with high spatial-coding complexity and moderate motion. Quality-of-Experience tests show (Agboma & Liotta, 2006) that this type of content is favored by users of mobile devices as it does not stretch the capabilities of the screen display (as, for instance, sport sequences would do). The Intra-refresh rate was every 15 frames with an IPBB… I coding structure. 1065 frames were transmitted resulting in a video duration of 35.5 s. Simple previous frame replacement was set for error concealment at the decoder as a point of comparison with others' work.

A Gilbert-Elliott two-state, discrete-time, ergodic Markov chain (Haβlinger & Hohlfeld, 2008) modeled the wireless channel error characteristics at the ns-2 physical layer. A two-state model reproduces conditions experienced during fast fading but does not model slow fades, implying the model is valid for nearby mobile nodes. The probability of remaining in the good state was set to 0.95 and of remaining in the bad state was 0.94, with both states modeled by a Uniform distribution. The packet loss probability in the good state was fixed at 0.01 and the bad state default was 0.05.

Evaluation

As a form of identification, broadband video streaming is called BVS in this Section. In Figure 11, UDP streaming suffers unacceptable packet losses (above 10%) in the downlink streaming direction because the stream not only suffers some losses due to congestion as it enters the buffers of the two routers but further losses occur across the WiMAX link. For uplink streaming packet losses from congestion are reduced. This is because in uplink streaming more packets may be lost traversing the wireless link compared to downlink streaming across a congested core network. Once the wireless link is crossed, for uplink streaming the stream is less likely to suffer loss from congestion. This is because its packet arrival rate has already been reduced by losses arising from wireless channel conditions and consequently self-congestion in the intervening router buffers is reduced. BVS exhibits a similar asymmetric packet loss pattern between downlink and uplink, as it is essentially an improved version of UDP. Notice that the BVS totals in Figure 11 are the losses after retransmissions and do not directly show packet losses across the transmission paths. For DCCP and multi-connection TFRC, downlink streaming, the majority of packet losses occur across the wireless link, as these protocols are able to respond to congestion across the core net-

Figure 11. Percentage overall packet loss according to streaming direction

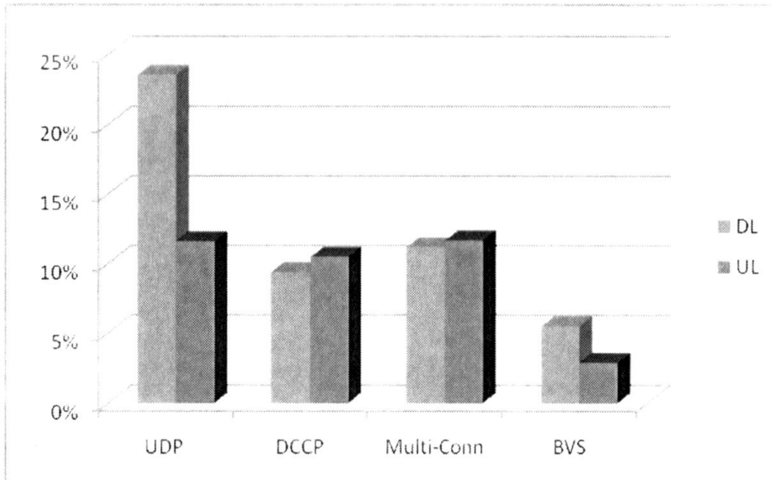

work to some extent but cannot prevent wireless channel losses. However, the number of packets available to be dropped at the wireless stage is reduced because of earlier losses from congestion. Figure 12 shows the breakdown. The number of packets dropped is greater in uplink streaming for these two protocols. The number of packets dropped is greater in uplink streaming for these two protocols, as all packets are dropped over the wireless link, which is encountered first.

From Table 5, the percentages of packet losses for UDP transport fro downlink streaming are much higher than the other methods. Though DCCP and multi-connection TFRC are able to reduce the packet loss levels, in this IPTV distribution network, the levels are too high as they are around 10%. This implies that only the introduction of application-layer forward error control or some form of error resilience could improve the situation. The net result of these packet losses, Table 5, is that UDP transport results in poor video quality. Only uplink streaming video quality passes above 25 dB when quality is 'fair' (according to an approximate mapping between the ITU's mean opinion score rankings and PSNR). However, BVS uplink streaming results in 'good' quality video (just). The mean end-to-end delay of DCCP and multi-connection TFRC is lower again than UDP and BVS. This is because both DCCP and multi-connection TFRC reduce their sending rate, resulting in less queuing time. From Table 5, UDP and BVS's sending period is approximately the same and close to the duration of the *Paris* sequence. However for DCCP, packet losses on the wireless link again cause excessive delay, as DCCP introduces large inter-packet gaps. Multi-connection TFRC is able to increase wireless utilization but this can be at a cost of greater packet losses across the connections. BVS still almost matches the sending period of the video sequence, by virtue of reduced end-to-end delay, despite sending more packets through retransmissions than UDP. The levels of inter-arrival-time packet jitter confirm that DCCP decreases congestion by increasing the inter-packet gap to too high a duration. Multi-connection TFRC can reduce the jitter but not enough compared to UDP and BVS. Similarly, multi-connection TFRC with four connections increases throughput but greater net throughput is achievable with BVS.

An interesting feature of our analysis, Figure 13, was that in downlink streaming proportionally more of the larger intra-coded I-frame pack-

Figure 12. Proportion of wired/wireless network packet losses for downlink streaming

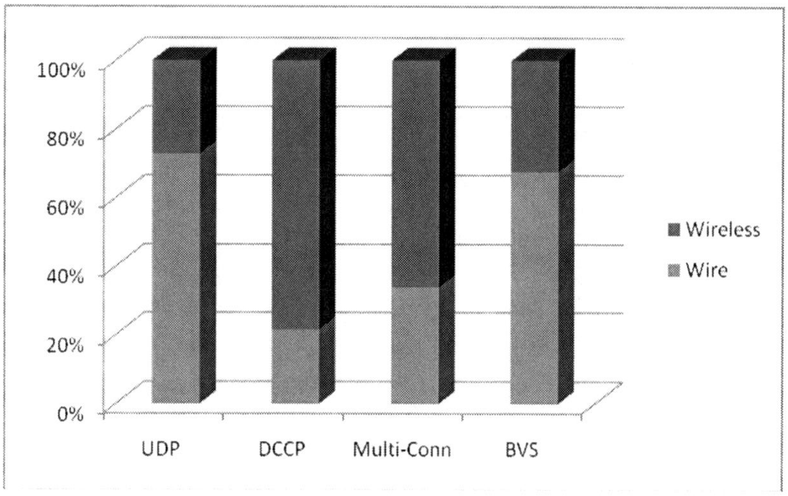

Table 5. Mean performance metrics when streaming Paris over an IPTV delivery network

	UDP	DCCP	Multi-Conn	BVS
	Packets lost			
DL	23.4	9.37	11.18	5.49
UL	11.6	10.46	11.67	2.87
	PSNR (dB)			
DL	18.01	24.55	24.18	27.62
UL	24.81	25.46	25.02	31.18
	End-to-end delay (s)			
DL	0.029	0.018	0.029	0.042
UL	0.049	0.016	0.020	0.062
	Sending period (s)			
DL	35.63	139.18	91.18	36.32
UL	35.62	134.00	69.81	35.77
	Jitter (s)			
DL	0.0097	0.0349	0.0097	0.0079
UL	0.0084	0.0314	0.0071	0.0076
	Throughput (kbps)			
DL	627	189	271	773
UL	751	197	360	809

ets are lost in UDP and BVS streaming. Notice from Figure 10 that for UDP and BVS, essentially without congestion control, more losses occur in the wired portion of the IPTV delivery network than occur in the wireless part. Larger I-frame packets on average contribute to 26 packets after segmentization at the encoder, all arriving together at a router buffer. In contrast, predictively-coded P-frames on average are broken into three packets at the encoder. The total number of

Figure 13. Breakdown by frame type of packet losses when downlink streaming

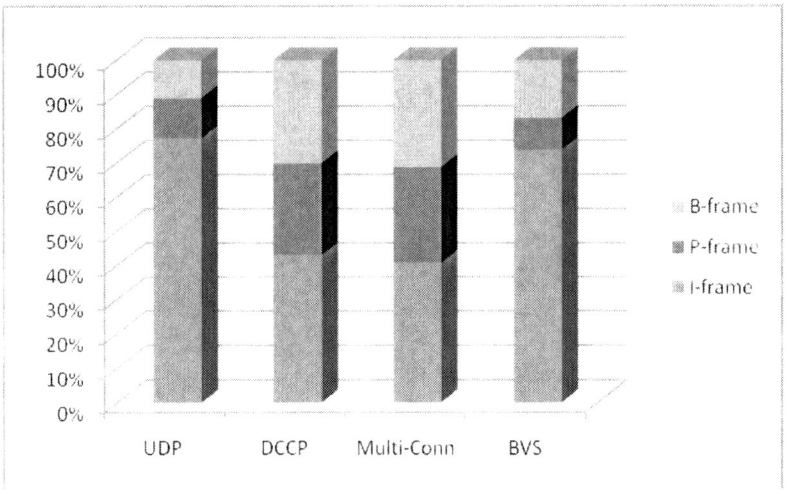

Figure 14. Breakdown by frame type of packet losses when uplink streaming

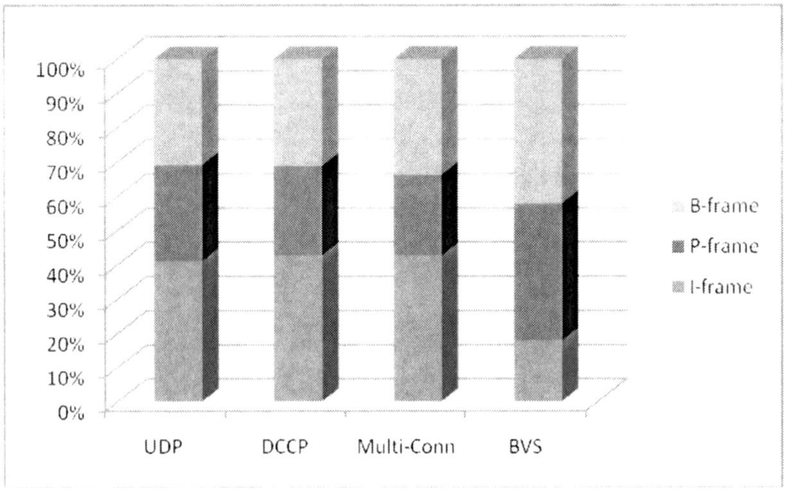

P-frame packets in a 15 frame group-of-pictures (GOP) is 12, but it is the arrival pattern that is significant. Bi-predictively-coded B-frame contribute two packets in the mean, leading to bursts of four packets (and 20 packets per GOP). The pattern of wireless packet channel losses is not so selective of I-frame packets, as the breakdown in Figure 14 by frame type for uplink streaming illustrates. Notice again that in Figure 13, the BVS packet losses include retransmissions and, therefore, do not directly reflect the packet loss pattern.

From Figures 13 and 14, it is apparent that video quality for BVS streaming can be further improved by avoiding bursts of I-frame packets. In fact, if packet loss levels increased compared to the comparatively low levels of Table 5, then this pattern of packet losses would be a problem for BVS downlink streaming. At a cost in delay, this can be achieved by packet reordering between the frame type packets, as occurred in Jammeh et al., 2004 as a form of video smoothing.

CONCLUSION

In this Chapter, we presented a cross-layer model to achieve the optimal TCP throughput for video multicasting over heterogeneous networks using adaptive forward-error correction (FEC). The model integrates the TCP throughput at the transport layer with link layer error control over wireless links. A model based TCP goodput formula is combined with adaptive FEC at byte and packet levels to select the optimal code that maximizes TCP goodput and consequently multicast the optimal video quality to end mobile devices. The results show a good video quality that can be achieved when a maximum TCP throughput is reached at appropriate system settings for the threshold residual error rate and frames size of group pattern (GoP) of MPEG-4 video.

Turing to unicast streamingto mobile devices, TFRC in its DCCP guise is a version of TCP's congestion control adapted for delay intolerant applications such as video streaming. It is adopted for wireless networks because: a) it is an industry standard form of congestion control and may be suitable for wireless transmission if packet losses can be hidden from the controller at the MAC layer; and b) it can work as an end-to-end congestion controller for heterogeneous networks consisting of wired and wireless path elements, all or several of which elements are subject to congestion. For the latter type of network, multiple-connection TFRC may be preferable as it avoids the need for packet loss hiding. It should be mentioned that several broadband network technologies have built-in error control, though in WiMAX this is optional. When applied a form of hybrid ARQ takes place, as the wireless channel is sampled to determine the level of ARQ required. The problem with this is twofold: it is not an end-to-end solution and the reliability of channel measurements is not strongly established.

For IPTV with intelligent placement of content close to the access network then TFRC/DCCP has some problems. For example, there may be long pauses in transmission when handoff occurs unless intervention occurs such as a fast retransmission scheme. Poor wireless channel utilization is partially solved by multiple-connection TFRC/DCCP in better channel conditions but video quality is reduced in comparison to broadband video streaming and buffer management is required at the mobile station.

Therefore, in this Chapter we have demonstrated broadband video streaming which is a simple broadband wireless scheme based on negative acknowledgments. Broadband video streaming achieves effective wireless channel utilization without the scale of packet losses that badly affect fragile compressed video streams.

ACKNOWLEDGMENT

Part of this work was originally supported by the Austraining International under Endeavour Award 687-2008, Australia. The first author also acknowledges the work in this area of Liansheng Tan of the Department of Computer Science at the Central China Normal University, Wuhan, China and Rodney Kennedy of the Research School of Information Science and Engineering at the Australian National University, Canberra, ACT, Australia.

REFERENCES

Agboma, F., and Liotta, A. (2007). Addressing user expectations in mobile content delivery. *Mobile Information Systems, 3*(3/4), 153:164.

Al-Majeed, S.S., & Fleury, M. (2010). Options for WiMAX uplink media streaming. *International Journal of Mobile Computing and Multimedia Communications, 2*(2), 49:66

Al-Majeed, S. S., & Fleury, M. (2010a). A simple wireless broadband video streaming scheme for IPTV. *IFIP Wireless and Mobile Networking Conference*.

Al-Suhail A.G., (2008). An efficient error-robust wireless video transmission using link-layer FEC and low-delay ARQ schemes. *Journal of Mobile Multimedia, 3 & 4*(3), 275:292.

Andrews, J. G., Ghosh, A., & Muhamed, R. (2007). *Fundamentals of WiMAX*. Upper Saddle River, NJ: Prentice Hall.Bajie I. V., (2006). Efficient error control for wireless video multicast. In *8th IEEE workshop on Multimedia Signal Processing,* (pp. 306-309).

Balakrishnan, H., & Katz, R. (1998). Explicit loss notification and wireless Web performance. In *IEEE GLOBCOM Internet Mini-Conference.*

Balakrishnan, H., Padmanabhan, V., Sehan, S., & Katz, R. (1997). A comparison of mechanisms for improving TCP performance over wireless links. *IEEE/ACM Transactions on Networking, 5*(6), 756–769. doi:10.1109/90.650137

Balan, R. K., Lee, B. P., Kumar, K. R., & Jacob, L. (2001). TCP header checksum option to improve performance over lossy links. In *IEEE INFOCOM* (pp. 309–318). TCP HACK.

Barman, D., Matta, I., Altman, E., & El Azouzi, R. (2004). TCP optimization through FEC, ARQ, transmission power tradeoff. In. *Proceedings of Wired-Wireless Internet Communication, LNC, 2957*, 87–98. doi:10.1007/978-3-540-24643-5_8

Cen, S., Cosman, P., & Voelker, G. (2003). *End-to-end differentiation of congestion and wireless losses, 11*(5), 703-717.

Chan, S.-H., Zheng, X., Zhang, Q., Zhu, W., & Zhang, Y. (2006). Video loss recovery with FEC and stream replication. *IEEE Transactions on Multimedia, 8*(2), 370–381. doi:10.1109/TMM.2005.864340

Chen, M., & Zakhor, A. (2005). Rate control for streaming video over wireless. *IEEE Wireless Communications, 12*(4), 32–44. doi:10.1109/MWC.2005.1497856

Chen, M., & Zakhor, A. (2006). Multiple TFRC connection based rate control for wireless networks. *IEEE Transactions on Multimedia, 8*(5), 1045–1062. doi:10.1109/TMM.2006.879837

Chiasserini, C.-F., & Meo, M. (2002). A reconfigurable protocol setting to improve TCP over wireless. *IEEE Transactions on Vehicular Technology Multimedia, 51*(6), 1608–1620. doi:10.1109/TVT.2002.804863

Cui, L., Koh, S. J., Cui, X., & Kim, Y. J. (2007). Adaptive increase and decrease algorithm for wireless TCP. In *International Conference on Natural Computation*, (pp. 392-398).

Degrande, N., Laevens, K., & De Vleeschauwer, D. (2008). Increasing the user perceived quality for IPTV services. *IEEE Communications Magazine, 46*(2), 94–100. doi:10.1109/MCOM.2008.4473090

Floyd, S., & Fall, A. (1999). Promoting the use of end-to-end congestion control in the Internet. *IEEE/ACM Transactions on Networking, 7*(4), 43–56. doi:10.1109/90.793002

Fu, Y., Hu, R., Tian, G., & Wang, T. (2006). TCP-Friendly Rate Control for streaming service over 3G network. In *International Conference on Wireless Communications., Networking and Mobile Computing*, 4 pages.

Görkemli, B., Sunay, M. O., & Tekalp, A. M. (2008). Video streaming over wireless DCCP. In *IEEE International Conference. on Image Processing*, (pp. 2028-2031).

Handley, M., Pahdye, J., Floyd, S., & Widmer, J. (2003). *TCP-Friendly Rate Control (TFRC): protocol specification.* RFC 3448.

Haßlinger, G., & Hohlfeld, O. (2008). The Gilbert-Elliott model for packet loss in real time services on the Internet. In *14th GI/ITG Conference on Measurement, Modelling, and Evaluation of Computer and Communications Systems.*, (pp. 269–283).

IEEE, 802.16e-2005. (2005). IEEE Standard for Local and Metropolitan Area Networks. Part 16: Air Interface for Fixed and Mobile Broadband Wireless Access Systems.

Issa, O., Li, W., & Liu, H. (2009). Performance evaluation of TV over broadband wireless access networks. *IEEE Transactions on Broadcasting, 56*(2), 201–210. doi:10.1109/TBC.2010.2046979

Jammeh, E., Fleury, M., & Ghanbari, M. (2004). Smoothing transcoded MPEG-1 video streams for Internet transmission. *IEE Proceedings. Vision Image and Signal Processing, 151*(4), 298–306. doi:10.1049/ip-vis:20040747

Klaue, J., Rathke, B., & Wolisz, A. (2003). EvalVid - A framework for video transmission and quality evaluation. In *International Conference on Modeling Techniques and Tools for Computer Performance*, (pp. 255–272).

Kohler, E., Handley, M. & Floyd, S. (2006). Datagram Congestion Control Protocol. *Internet Engineering Task Force*, RFC 4340.

Lee, T., Chan, S., Zhang, Q., Zhu, W., & Zhang, Y. (2002). Allocation of layer bandwidths and FECs for video multicast over wired and wireless networks. *IEEE Transactions on Circuits and Systems for Video Technology, 12*(2), 1059–1070. doi:10.1109/TCSVT.2002.806816

Lin, Y.-B., & Pang, A.-C. (2005). *Wireless and Mobile: All-IP networks*. Chichester, UK: Wiley and Sons.

Liu, H., Zhang, W., & Yang, X. (2006). Cross-layer conditional retransmission for layered video streaming over cellular networks. *Computer Communications, 29*(11), 2066–2073. doi:10.1016/j.comcom.2006.01.002

Liu, Q., Zhou, S., & Giannaki, G. B. (2004). TCP performance in wireless access with adaptive modulation and coding. *IEEE International Conference on Communications*, pp. 3989-3993, 2004.

Liu, Z., Wu, Z., Liu, H., & Stein, A. (2007). A layered hybrid-ARQ scheme for scalable video multicast over wireless networks. In *Asilomar Conference on Signals, Systems and Computers*, (pp. 914-919).

Lo, A., Heijenk, G., & Niemegeers, I. (2005). Performance evaluation of MPEG-4 video streaming over UMTS network using an integrated tool environment. *International Symposium on Performance Analysis of Wireless Networks and Communications Systems.*

Marco, P. D., Rinaldi, C., Santucci, F. M., Johansson, K. H., & Moller, N. (2006). Performance analysis and optimization of TCP over adaptive wireless links. In *17th Annual IEEE International Symposium on Personal, Indoor and Mobile Radio Communications.*

Padhye, J., Firoiu, V., Towsley, D. F., & Kurose, J. F. (2000). Modeling TCP Reno performance: a simple model and its empirical validation. *IEEE/ACM Transactions on Networking, 8*(2), 133–145. doi:10.1109/90.842137

Pei Y. & Modestino J.W. (2004). Interactive video coding and transmission over heterogeneous wired-to-wireless IP networks using an edge proxy. *EURASIP Journal on Applied Signals*, 2004 volume [online], 253-264.

Razavi, R., Fleury, M., & Ghanbari, M. (2008). *Unequal protection of video streaming through adaptive modulation with a tri-zone buffer over Bluetooth enhanced data rates*. EURASIP Journal on Wireless Communications and Networking. [online]

Rubenstein, D., Kurose, J., & Towsley, D. (1998). *Real-time reliable multicast using proactive forward error correction. Technical Report 98-19*. Amherst, MA: Department of Computer Science, University of Massachusetts.

Stewart, R. (Ed.). (Sept. 2007). Stream Control Transmission Protocol. *Internet Engineering Task Force*, RFC 4960.

Suhoen, J., et al. (2002). Video transfer control protocol for a wireless video demonstrator. In *International Conference on Information Technology: Coding and Computing*, (pp. 462–467).

Tappayuthpijam, K., Liebl, G., Stockhammer, T., & Steinbach, T. (2009). Adaptive video streaming over a mobile network with TCP-Friendly Rate Control. In. *International Conference on Wireless Communications and Mobile Computing*.

Tickoo, O., Subramanian, V., Kalyanaraman, S., & Ramakrishnan, K. K. (2001). LT-TCP: End-to-end framework to improve TCP performance over networks with lossy channels. In *13th International Workshop on Quality of Service*.

Tsai, F. C.-D., et al. (2006). The design and implementation of WiMAX module for ns-2 simulator. In *Workshop on NS2: The IP Network Simulator*, article no. 5.

Widmer, J., Denda, R., & Mauve, M. (2001). A survey on TCP-friendly congestion control. *IEEE Network*, *15*(3), 28–37. doi:10.1109/65.923938

Wu, H., Claypool, M., & Kinicki, R. (2005). Adjusting Forward Error Correction with quality scaling for streaming MPEG. In *International Workshop on Network and Operating Support for Digital Audio and Video*, (pp. 111-116).

Yogesh, Y., Bose, S., & Kannan, A. (2006). Effective TCP Friendly Rate Control scheme for video streaming over wireless networks. *Asian Journal of Information Technology*, *5*(4), 442–447.

Yoo, T., Lavery, R. J., Goldsmith, A., & Goodman, D. J. (2004). *Throughput optimization using adaptive techniques. Technical Report*. US: Stanford University.

Zhang, X., Tang, J., Chan, H., Ci, S., & Guizani, M. (2006). Cross-layer-based modeling for quality of service guarantees in mobile wireless networks. *IEEE Communications Magazine*, *44*(1), 100–106. doi:10.1109/MCOM.2006.1580939

Chapter 14
Automatic Talker Identification Using Optimal Spectral Resolution:
Application in Noisy Environment and Telephony

Siham Ouamour
USTHB University, Algeria

Halim Sayoud
USTHB University, Algeria

Mhania Guerti
ENS.Polytechnique, Algeria

ABSTRACT

This chapter deals with the problem of speaker characterization, for which the principal interest is the improvement of the techniques of talker identification. For this purpose, the authors investigate the effect of spectral resolution in the speaker identification performance. This investigation employs an approach based on the second order statistical measures using the Mel Frequency Spectral Coefficients (MFSC) and looks for the best spectral resolution (optimal number of MFSC). In fact, researchers do prefer using low spectral resolutions for many justifiable reasons, but we do not know what is the best resolution to adopt, especially in talker identification and we do not know what are the performances got with high spectral resolutions either.

To find that optimal resolution, in microphonic and telephonic bandwidth, the authors have experimented several dimensions for the MFSC coefficients and several types of additive noises, at several SNR ratios.

Results show the importance of the high spectral resolution in noisy environment and telephonic bandwidth, while the current research works have always favoured the low resolution of 24 coefficients in

DOI: 10.4018/978-1-60960-563-6.ch014

such tasks. For example, the authors notice an improvement of about 11% on the identification score, since they increase the resolution from 24 to 48 MFSC, in the telephonic bandwidth.

INTRODUCTION

The vocal expression is a particular characterization for the speaker (Reynolds, 2008): thus it is possible, in normal conditions, to recognise his corresponding talker during a direct or telephonic conversation (Feng, 2004). Speaker characterization is a generic term indicating the possibility to discriminate between several persons thanks to their voices, and in this research domain we want to recognise, not what it was said, but the identity of the speaker which is talking by his vocal characteristics.

In this context, several works are done using different spectral resolutions for the good characterisation of the vocal expression of the speaker. But often researchers favour low spectral resolutions, lower or equal to 24 spectral coefficients, for several justifiable reasons.

However, after a long reflection on the problem and several works we conducted on this field (Sayoud, 2003; Ouamour, Sayoud & Guerti, 2009), a question was raising: *What is the optimal spectral resolution in talker identification?*

To answer this question and to find the optimal dimension of the spectral characteristics (Kinnunen, 2003), we tested several spectral resolutions between 12 and 60 Mel Frequency Spectral Coefficients (Kinnunen, 2003) during the speech signal analysis (Kinnunen, 2003; Lee, Yoon & Kang, 2006; Razik, Fohr, Mella, & Parlangeau-Vallès, 2004). Furthermore, the experiments are done in a quiet and noisy environment to see all the possible cases, by using a statistical classifier called Second Order Statistical Measures (Magrin-Chagnolleau, Bonastre, & Bimbot, 1995). Moreover, for the noisy case, we tested three types of noises with five different Signal-on-Noise Ratios (SNRs) between 0 and 24 dB (Haque, Togneri & Zaknich, 2006; Hu & Loizou, 2007; Kim & Stern, 2006).

Thus, this chapter is organised as follows:

In the 2nd section, we describe the speech database used in the experiments. In the 3rd section, we try to find the advantages/ disadvantages of the low and high spectral resolution. The 4th section presents the statistical classifier used in talker identification. The 5th and 6th sections (respectively) describe the results of talker identification evaluated in quiet and noisy environment (respectively), and finally, we end our chapter with a general conclusion and some useful references.

SPEECH DATABASE

The speech database is extracted from TIMIT (Fisher, Zue, Bernstein & Pallet, 1986) and FTIMIT: TIMIT in which we only preserve the telephonic bandwith (Liu & Fu, 2007) corresponding to the 300-3400 Hz bandwidth (Magrin-Chagnolleau, Wilke, & Bimbot, 1996). There are 37 speakers: 22 males and 15 females. The approximate duration of an utterance is 9 s for the training and 7 s for the test. The recordings are done with a high quality microphone, at 16 bits and with a sampling frequency of 16 kHz.

A second investigation is made in noisy environment (Sayoud, 2003) and with three types of noise (Haque, Togneri & Zaknich, 2006; Hu & Loizou, 2007; Kim & Stern, 2006).

- the Gaussian White Noise: GWN (Paninski, 2006),
- the car noise (Jabloun & Enis Cetin, 1999),
- the babble noise (Elhilali & Shamma, 2008).

Figure 1. 3D-Mel spectrum of a speech signal with 12 coefficients

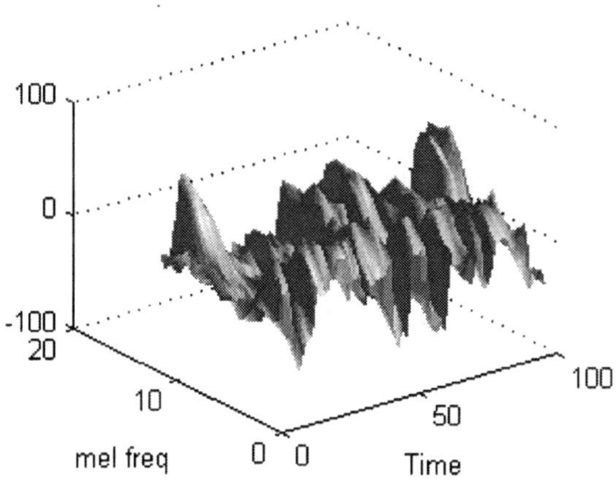

Figure 2. 3D-Mel spectrum of a speech signal with 24 coefficients

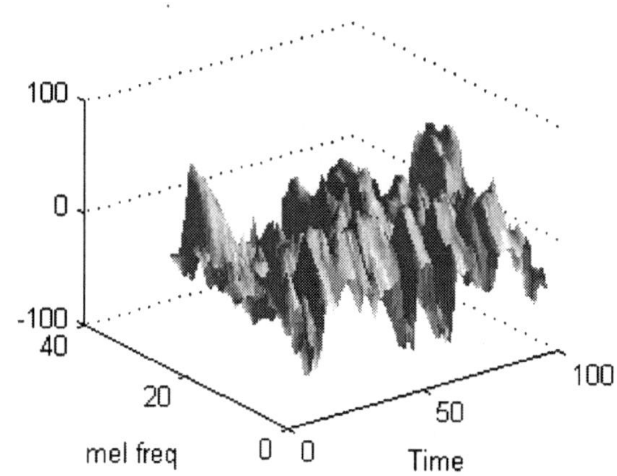

These noises are added during the training and test (Sayoud, 2003) at the following rates:

- 0 dB,
- 6 dB,
- 12 dB,
- 18 dB,
- 24 dB (without noise).

Each database is processed 5 times in order to extract the speaker's features, because we use 5 different spectral resolutions: from 12 to 60 filters (MFSC) inside the filter bank (Lee, Yoon & Kang, 2006):

- 12 filters,
- 24 filters,
- 36 filters,
- 48 filters,
- 60 filters.

Furthermore, we must recall that the computation of the MFSC coefficients is done for each segment of 32 ms, by using an energy normalisation.

Compromise between Low and High Spectral Resolution

An important question may be asked during the choice of the optimal dimension of the speaker acoustic features.

The question is:

Does it exist a relationship between the spectral resolution and the modelization reliability of the speaker characteristics?

For this purpose, we first try to investigate several 3D spectrums (Ouamour, Sayoud & Guerti, 2009) for the same speech signal, obtained with different spectral resolutions, in order to find a visual link between these resolutions and the accuracy of the speaker characterization (see Figures 1 to 5).

- **Observation 1**: In Figure 1 (*and Figure 6*), obtained with 12 MFSC, we can see that the shape of the formantic envelope is sharply displayed.
- **Observation 2**: In Figure 2 (*and Figure 6*), obtained with 24 MFSC, the formantic en-

Figure 3. 3D-Mel spectrum of a speech signal with 36 coefficients

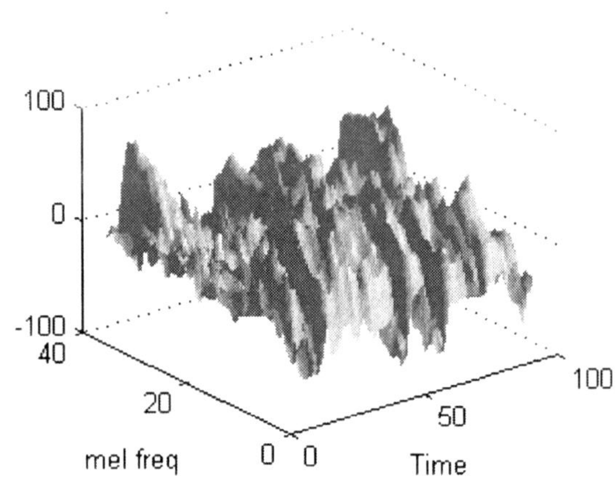

Figure 4. 3D-Mel spectrum of a speech signal with 48 coefficients

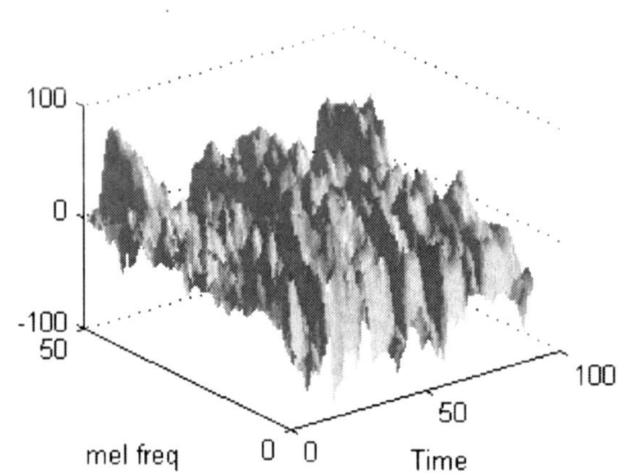

velope is very well modelized, but no harmonic is visible yet.

- **Observation 3**: In Figure 3 (*and Figure 6*), obtained with 36 MFSC, we can observe the formantic envelope well enough and we can also observe some harmonics in the same time.
- **Observation 4**: In Figure 4 (*and Figure 6*), obtained with 48 MFSC, the principal shape of the formantic envelope is lost, but the harmonics and the fine spectral details are displayed much better, as we can see in the 3D illustration.
- **Observation 5**: In Figure 5 (*and Figure 6*), obtained with 60 MFSC, the spectral envelope tends to disappear, but the harmonics and the fine spectral details are displayed in a very sharp contrast, almost similar to a typical FFT spectrum.

According to the previous figures and observations, we clearly notice that there is a strong dependence between the MFSC dimension and both the formantic envelope and the fine spectral details. Thanks to this dependence, we can adjust the MFSC dimension in order to get a balanced spectral characterization for the speaker.

In the same way, Figure 6 below, which represents the Mel-spectrum of a unique speech segment, provides a sharp representation of this phenomenon: we can easily observe the good formantic envelope in the low resolutions of 12 or 24 MFSC curves. But on the other hand, we really find more spectral details in the spectrums of the curves with 60 MFSC.

The curves are spaced and shifted vertically in order to make the representation sharper.

SUMMARY

Since the information contained in the formantic envelope is important, we have a great interest in reducing the number of the Mel filters (Kinnunen, 2003), but as described in our previous investigations (Sayoud, 2003), the information contained in the fine spectral details can bring an appreciable improvement in talker identification (especially for the corrupted speech) (Sayoud & Selmane, 1998), it will be then beneficial to introduce the effect of that information in the MFSC parameterisation, if we want to obtain a higher identification quality. This means that the filter-bank size should be high.

Figure 5. 3D-Mel spectrum of a speech signal with 60 coefficients

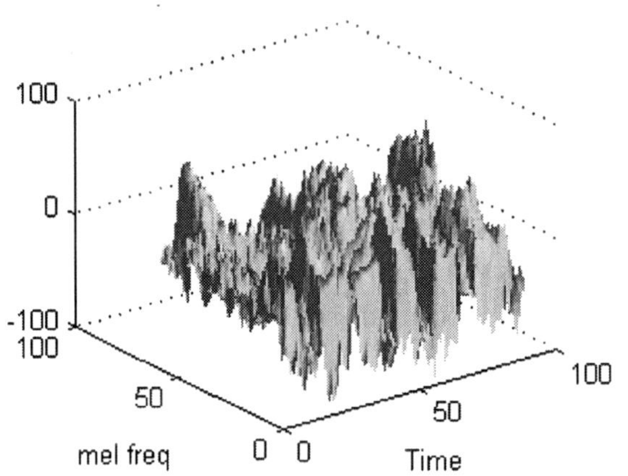

Figure 6. Representation of the Mel-spectrum of a unique speech segment, with several resolutions

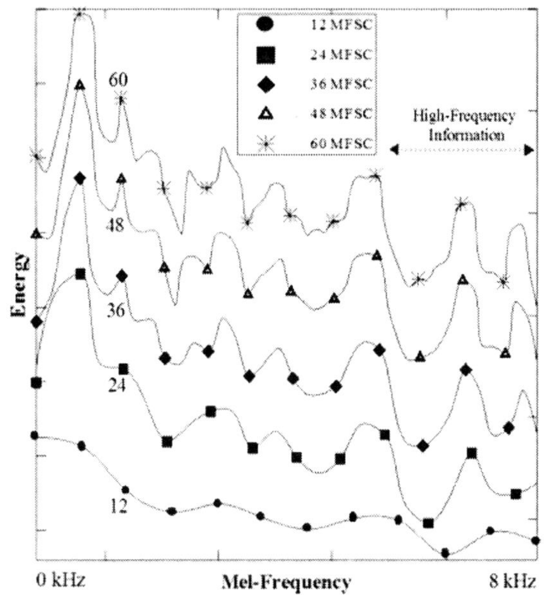

However, we must recall that when we use very high dimensions we shall deal with two problems: the complexity of calculation and the bad modelization of the covariance matrix (Kinnunen, 2003).

Therefore, a good speaker characterization needs a balanced compromise between the low and high spectral resolution.

Therefore, a second question is raising now:

What is the optimal dimension for the spectral coefficients?

This problem prompts us to test and experiment different spectral resolutions: from the classic model with 12 spectral channels to the complex model with 60 spectral channels.

Results of this second investigation (experimental test of talker identification) are discussed in the fifth and sixth sections.

STATISTICAL MEASURE FOR TALKER IDENTIFICATION

For the test of speaker identification (Feng, 2004), we chose a statistical measure based on mono-Gaussian models (Sayoud, 2003). The talker identification technique (Sayoud, 2003), using such measures of similarity, is usually called Second Order Statistical Measures (SOSM) and is used in order to recognise the speaker in each speech signal.

We recall bellow the most important properties of the approach.

Let $\{x_t\}_{1 \leq t \leq M}$ be a sequence of M vectors resulting from the *P*-dimensional acoustic analysis of a speech signal uttered by speaker $\boldsymbol{x}$. These vectors are summarised by the mean vector $\bar{x}$ and the covariance matrix X:

$$\bar{x} = \frac{1}{M}\sum_{t=1}^{M} x_t \tag{1}$$

and

$$X = \frac{1}{M}\sum_{t=1}^{M}(x_t - \bar{x})(x_t - \bar{x})^T \tag{2}$$

Similarly, for a speech signal uttered by speaker *y*, a sequence of N vectors $\{y_t\}_{1\le t\le M}$ can be extracted.

By assuming that all acoustic vectors extracted from the speech signal uttered by speaker *x* are distributed like a Gaussian function, the likelihood of a single vector y_t uttered by speaker *y* is

$$G(y_t / \mathbf{x}) = \frac{1}{(2\pi)^{p/2}(\det X)^{1/2}} e^{(1/2)(y_t-\bar{x})^T X^{-1}(y_t-\bar{x})} \tag{3}$$

If we assume that all vectors y_t are independent observations, the average log-likelihood of $\{y_t\}_{1\le t\le M}$ can be written as

$$\overline{Lx}(y_1^N) = \frac{1}{N}\log G(y_1...y_N|\mathbf{x}) = \frac{1}{N}\sum_{t=1}^{N}\log G(y_t|\mathbf{x}) \tag{4}$$

We also define the minus-log-likelihood $\mu(\mathbf{x}, y_t)$ which is equivalent to similarity measure between vector y_t (uttered by *y*) and the model of speaker *x*, so that

$$\underset{x}{Arg\max}\ G(y_t / \mathbf{x}) = \underset{x}{Arg\min}\ \mu(\mathbf{x}, y_t) \tag{5}$$

We have then:

$$\mu(\mathbf{x}, y_t) = -\log\ G(y_t / \mathbf{x}) \tag{6}$$

The similarity measure between test utterance $\{y_t\}_{1\le t\le M}$ of speaker *y* and the model of speaker *x* is then

$$\mu(\mathbf{x}, \mathbf{y}) = \mu(\mathbf{x}, y_1^N) = \tfrac{1}{N}\sum_{t=1}^{N}\mu(\mathbf{x}, y_t) \tag{7}$$

$$= -\overline{Lx}(y_1^N) \tag{8}$$

After simplifications, we obtain

$$\mu(\mathbf{x}, \mathbf{y}) = \frac{1}{P}\left[-\log(\frac{\det(Y)}{\det(X)}) + tr(YX^{-1}) + (\bar{y}-\bar{x})^T X^{-1}(\bar{y}-\bar{x})\right] - 1 \tag{9}$$

This measure is equivalent to the standard Gaussian likelihood measure (asymmetric μ_G) defined in (Bimbot, Magrin-Chagnolleau & Mathan, 1995; Sayoud, 2003).

A variant of this measure called μ_{Gc} is deduced from the previous one by assuming that

$\bar{y} = \bar{x}$ (i.e. the inter-speaker variability of the mean vector is negligible).

Thus, the new formula becomes:

$$\mu_{GC}(\mathbf{x}, \mathbf{y}) = \frac{1}{P}\left[-\log(\frac{\det(Y)}{\det(X)}) + tr(YX^{-1})\right] - 1 \tag{10}$$

A symmetric measure can be constructed by combining $\mu_{GC}(\mathbf{x}, \mathbf{y})$ with its dual term $\mu_{GC}(\mathbf{y}, \mathbf{x})$, leading to $\mu_{GC0.5}(\mathbf{x}, \mathbf{y})$ (see formula 11).

$$\mu_{GC0.5}(\mathbf{x}, \mathbf{y}) = \frac{\mu_{GC}(\mathbf{x}, \mathbf{y}) + \mu_{GC}(\mathbf{y}, \mathbf{x})}{2} \tag{11}$$

The two measures ($\mu_{GC0.5}$ and $\mu_{G0.5}$) can be used in speaker identification, thanks to the nearest neighbour technique.

FIRST EXPERIMENT

Talker Identification in Quiet Environment

This investigation consists in the identification (Lee, Yoon & Kang, 2006) of 37 speakers of

TIMIT (Fisher, Zue, Bernstein & Pallet, 1986) by the SOSM method (Sayoud, 2003) in a quiet environment: without noise. Two cases are investigated: the identification in the [0-8 kHz] bandwidth and the identification in the telephonic bandwidth (Jankowski, Kalyanswamy, Basson & Spitz, 1990; Vary & Martin, 2007). The MFSC size varies from 12 to 60 coefficients.

Results of this Investigation

The errors of talker identification are shown in Figure 7. We can note two important points:

- **For the 0-8000 Hz bandwidth,** the error of talker identification is 5.4% with 12 MFSC and it is 0% for all the other cases (24, 36, 48 et and 60 MFSC), which involves a good identification from 24 channels.
- **For the telephonic bandwidth** (Liu & Fu, 2007), the error of talker identification is 56.7% with 12 MFSC, it is 13.5% with 24 MFSC, it is 8.1% with 36 MFSC, it is 2.7% with 48 MFSC and it is 8.1% with 60 MFSC. Therefore, the size of 48 channels seems to be the best one on the telephonic bandwidth.

Discussion about the First Experiment

Since the information contained in the formantic envelope is important in speaker identification, it is in our best interest to reduce the spectral resolution. However, since the information contained in the fine spectral details is also useful for the discrimination between two different speakers (Sayoud, Ouamour & Guerti, 2007; Mühlera, Ziesea & Rostalskib, 2009), it is interesting to introduce their effect by increasing the spectral resolution, which requires to find an optimal compromise for a good talker identification. This experiment shows the importance of the

Figure 7. Error of talker identification in the case of quiet environment

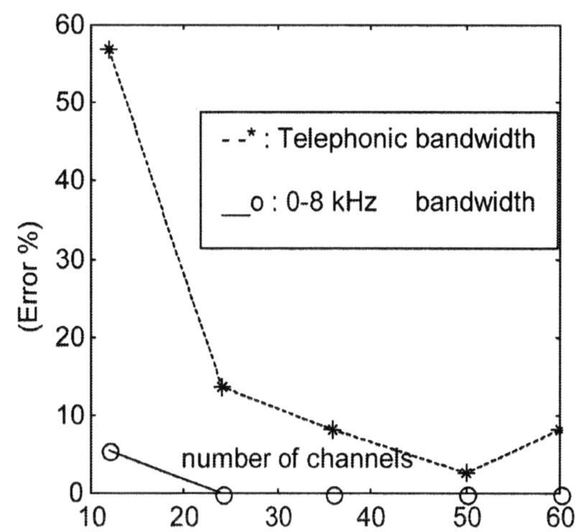

high spectral resolution by finding the optimal size of the filter bank (Lee, Yoon & Kang, 2006) to use in speaker recognition. So, according to Figure 7, the optimal size of the filter bank, in quiet environment, is commonly between 24 and 60 for the 0-8 kHz bandwidth and it is 48 for the telephonic bandwidth.

Note that this special result concerns the quiet environment (*without noise*) only. In the next section we will see the effect of noises on the optimal spectral resolution.

SECOND EXPERIMENT

Talker Identification in Noisy Environment

The second investigation consists in identifying all the previous speakers in noisy environment. The used noises (see below) have variable weightings ranging from 0 dB to 18 dB (Sayoud, 2003).

Thus, two cases are experimented:

- the identification in the [0-8 kHz] bandwidth,
- the identification in the telephonic bandwidth [300-3400 Hz].

The MFSC size varies from 12 to 60 coefficients.

Types of Noise

Noise can have different causes (Haque, Togneri & Zaknich, 2006; Hu & Loizou, 2007; Kim & Stern, 2006). In our experiment, three types of noise are used:

- Gaussian white noise: centred, stationary and denoted GWN (Sayoud, 2003);
- Babble noise: obtained with a summation of 7 signals of dialogue recorded from different channels of television and with a high SNR, through an input / output line;
- Car noise: which consists in a mixture of three car noises, recorded by a high quality microphone in a medium traffic road.

The corruption rate is controlled by the selected SNR (Hu & Loizou, 2007; Sayoud, 2003), from 0 dB to 18 dB. We note that the original signal (without noise) will be represented by a SNR of 24 dB in order to simplify the graphic display of the results.

In this way, the energy mean of the speech signal is estimated firstly and, then, the normalised noise is mixed with the speech signal at a specific ratio corresponding to the selected SNR (Sayoud, 2003). This procedure is made during both training and test.

OBSERVATION OF THE RESULTS

Results of this second investigation (*noisy environment*) are shown in Figures 8 to 13 with interpolated curves between the five fixed points (*corresponding to the five different SNRs*).

- In the audible bandwidth 0-8 kHz, the best scores are got with 60 coefficients, which means that the high spectral resolution provides more protection against noises (see Figures 8, 9 and 10).
- In the telephonic bandwidth 300-3400 Hz, the best scores are get once with 48 coefficients and once with 60 coefficients. We can see an oscillation of the curve of 48 channels around the 60 channels curve; which implies an intermediate optimal score (see Figures 11, 12 and 13).
- In 0-8 kHz, the most disruptive noise is the babble noise followed by the GWN noise and finally the car noise, which seems not very disruptive with respect to the previous types (see Figures 8, 9 and 10).
- In the telephonic bandwidth, the most disruptive noise is also the babble noise followed by the two other noises (see Figures 11, 12 and 13).
- In the telephonic bandwidth, we notice the paradoxical reduction of the recognition score when the SNR is growing from 18 to 24 dB in the case of 12 channels curve.
- In the 0-8 kHz bandwidth, a very strong noise at 0 dB causes a devaluation of the identification score for over 20%, except for the case of the car noise where the score remains high (97.3%) even at 0 dB and with 60 channels (see Figures 8, 9 and 10).
- In the telephonic bandwidth, the GWN and the car noise at 0 dB provoke a score devaluation of over 20%. Concerning the babble noise, the score devaluation is more than 40%, which involves a failure of the identification system in this case (see Figures 11, 12 and 13).

Figure 8. Recognition scores in noisy environment (corrupted by the GWN), in the 0-8 kHz bandwidth

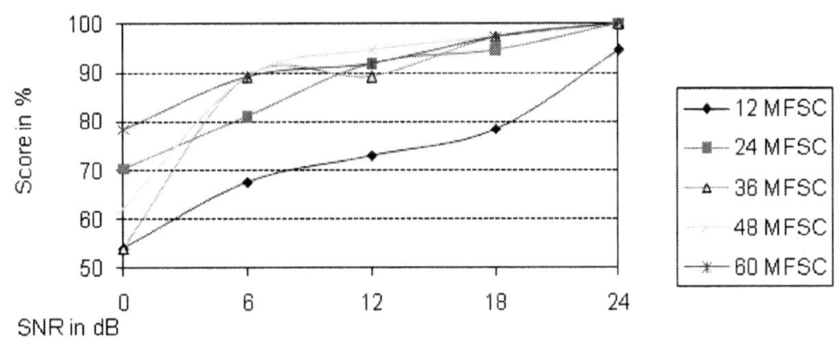

Figure 9. Recognition scores in noisy environment (corrupted by the babble noise), in the 0-8 kHz bandwidth

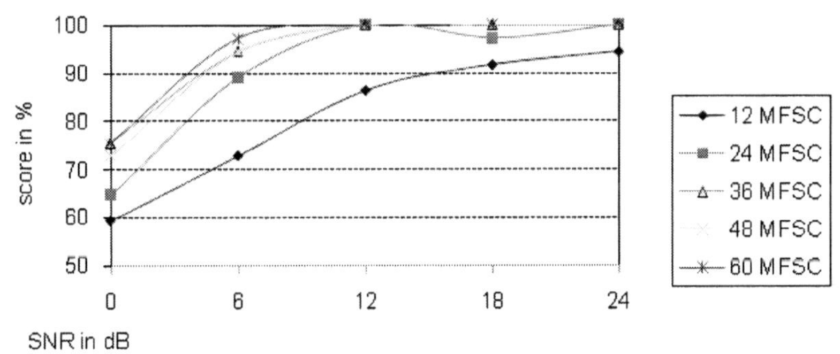

Figure 10. Recognition scores in noisy environment (corrupted by the car noise), in the 0-8 kHz bandwidth

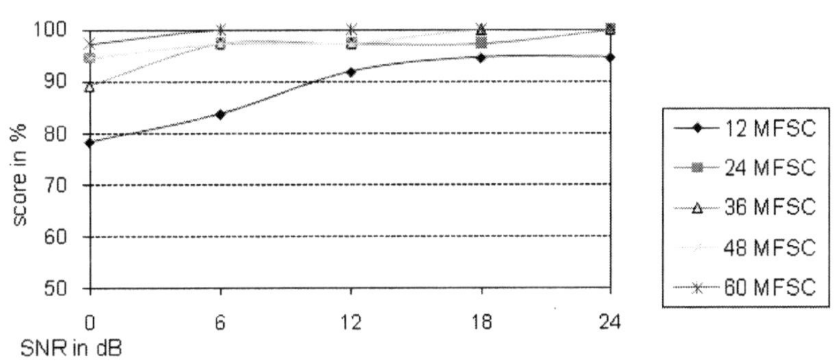

Figure 11. Recognition scores in noisy environment (corrupted by the GWN), in the telephonic bandwidth

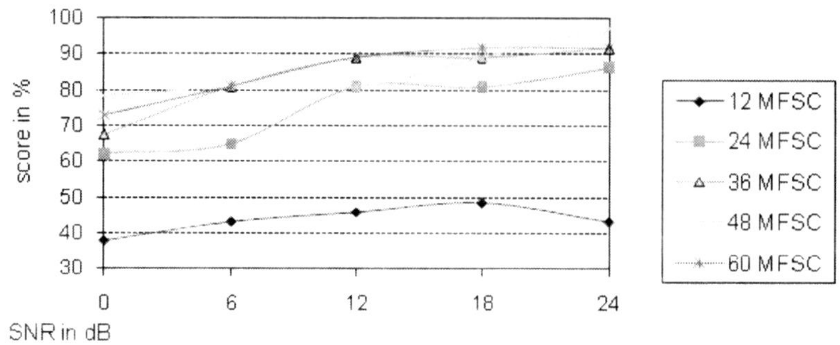

Figure 12. Recognition scores in noisy environment (corrupted by the babble noise), in the telephonic bandwidth

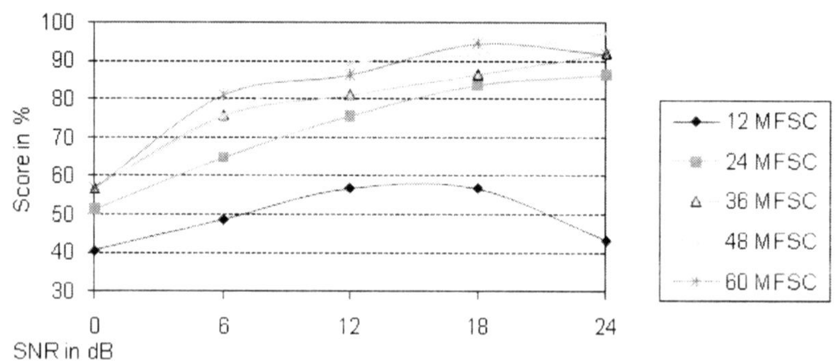

Figure 13. Recognition scores in noisy environment (corrupted by the car noise), in the telephonic bandwidth

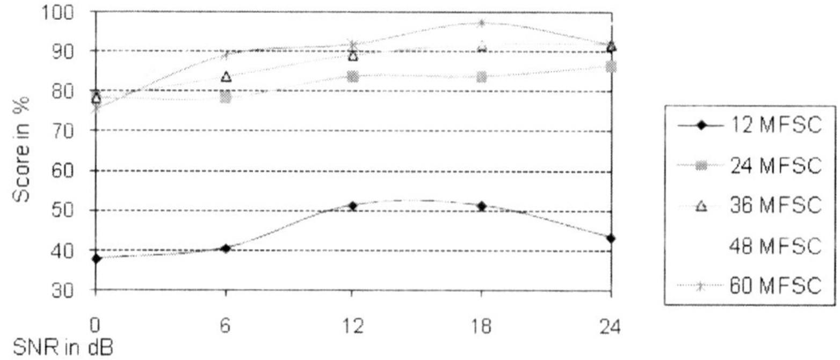

Discussion about the Second Experiment

The first conclusion we can deduce is that the high spectral resolution provides a lot of information characterizing the speaker and which improve the speaker identification performance, especially in noisy environment. Thus the 60 channels resolution should be an excellent resolution for systems of speaker recognition which are located in noisy environment. However, in the telephonic bandwidth, the optimal resolution should be between 48 and 60 channels.

The second conclusion we can deduce is that the car noise is not disruptive in speaker recognition (when using the statistical classifier), unlike we could think, because of the auditory disturbance provoked by that noise. On the other hand, the babble noise, which is well filtered by our brain, seems to be extremely disruptive in speaker recognition, even more disruptive then the GWN. The most probable cause of the failure encountered with the babble noise, is that this one and the speech signal have the same type of characteristics (speech), which implies a strong degradation of the own speaker characterization.

CONCLUSION

Usually, researchers prefer using models with low spectral resolution for the speech signal analysis, with dimensions ranging from only 12 to 24 spectral (or cepstral) coefficients. This low spectral resolution has two advantages: the simplification of the computing operations and the good representation of the formantic envelope in the spectrum.

When we use high dimensions, we shall deal with two problems: the complexity of calculation and the bad modelization of the covariance matrix. However, since the information in the fine spectral details can enhance the discrimination between the speakers, it will be interesting to introduce them by increasing the spectral resolution.

This issue, needing to find an optimal compromise in speaker characterization, prompted us to do some experiments in quiet and noisy environment and with different spectral resolutions. During this investigation, we have shown the importance of the high spectral resolution in speaker characterization and we have also found that the optimal spectral resolution depends on several parameters: the spectral bandwidth, the SNR and the type of noise corrupting the speech signal (if any).

The two experiments, done in both quiet and noisy environment (corrupted by GWN, babble and car noise), show that the high spectral resolution provides very important information for the speaker and helps recognise him more accurately, particularly in noisy environment.

In the case of noisy environment, the mel-spectral resolution of 60 coefficients / 8 kHz seems to be very interesting in speaker recognition. So in the audible bandwidth of 0-8 kHz, the best resolution is 60 coefficients / 8 kHz.

For the telephonic speech, where the speech is filtered by the [300-3400 Hz] filter (Liu & Fu, 2007), the optimal would be 48 coefficients / 8 kHz in non-noisy environment, and it would be a resolution between 48 and 60 coefficients, when the speech is corrupted. We think that this little reduction is caused by the limitation of the telephonic filtering which rejects the entire part of the spectrum over 3400 Hz and which also involves a significant loss of the high-frequency information.

Results obtained with car noise show that this one is not disruptive in speaker recognition, unlike we could think due to the severe auditory disturbance provoked by this noise on the human brain. Therefore, systems of speaker verification can easily be implemented near to motorways, for example: we can imagine self-service bank terminals using speaker verification techniques in petrol stations, bus stations etc. For the babble noise, which is so well filtered by our brain, it appears to be extremely disruptive in speaker recognition. The most probable cause of this failure is that this noise and the speech signal have the same type of features, which involves a strong degradation in speaker characterization. We deduce that, before any procedure of speaker verification, we must be sure that the surrounding does not contain any type of babble noise (*like popular markets, post offices, cafes, etc.*).

Finally, we think that our new result can easily contribute to improve the reliability of existing speaker verification systems, since this improvement does not require any modification in the architecture of those systems, except the adjustment of the input features size.

ACKNOWLEDGMENT

We would like to thank the editors who accepted the publication of this book chapter. We also wish to thank all the researchers who contributed during the synthesis of this chapter.

REFERENCES

Bimbot, F., Magrin-Chagnolleau, I., & Mathan, L. (1995). Second-Order Statistical Measures for text-independent Broadcaster Identification. *Journal of Speech Communication, 17*(1-2), 177–192. doi:10.1016/0167-6393(95)00013-E

Elhilali, M., & Shamma, S. (2008). Information-bearing components of speech intelligibility under babble-noise and bandlimiting distortions. In *Proceedings of Acoustics, Speech and Signal Processing (ICASSP).* March 31 -April 4, pp 4205 - 4208.

Feng, L. (2004). *Speaker Recognition.* IMM thesis, Technical University of Denmark, Denmark.

Fisher, W., Zue, V., Bernstein, J. & Pallet, D. (1986). An acoustic-phonetic database. *Journal JASA, suppl. A, 81(S92).*

Haque, S., Togneri, R., & Zaknich, A. (2006). *Zero- Crossings with Adaptation for Automatic Speech Recognition.* Paper presented at the Eleventh Australasian International Conference on Speech Science and Technology 2006, University of Auckland, Auckland, New Zealand.

Hu, Y., & Loizou, P. C. (2007). A Comparative Intelligibility Study of Speech Enhancement Algorithms. In *proceedings of IEEE International Conference on Acoustics, Speech and Signal Processing: Vol. 4,* April 15-20.

Jabloun, F., & Enis Cetin, A. (1999). *The Teager energy based feature parameters for robust speech recognition in car noise.* In proceedings of IEEE International Conference on Acoustics, Speech and Signal Processing, 15-19 Mar 1999, (Vol. 1, pp 273 – 276).

Jankowski, C., Kalyanswamy, A., Basson, S., & Spitz, J. (1990). NTIMIT: a phonetically balanced, continuous speech, telephone bandwidth speech database. In proceedigs of IEEE International Conference on Acoustics, Speech, and Signal Processing,3-6 Apr. (pp 109 – 112).

Kim, W., & Stern, R. M. (2006). *Band-Independent Mask Estimation for Missing-Feature Reconstruction in the Presence of Unknown Background Noise.* In proceedigs of IEEE International Conference on Acoustics, Speech, and Signal Processing, Toulouse, France.

Kinnunen, T. (2003). *Spectral Features for Automatic Text-Independent Speaker Recognition.* Licentiate's thesis, University of Joensuu, Department of Computer Science, Joensuu, Finland.

Lee, B., Yoon, S., & Kang, H. (2006). *On the Use of Voting Methods for Speaker Identification Based on Various Resoulution Filterbanks.* In Proceedings of ICASSP'06, Toulouse, France.

Liu, C., & Fu, Q. (2007). *Effects of Bandwidth Extension on Telephone Speech Recognition by Cochlear Implant Users.* Paper presented at the conference on Implantable Auditory Prostheses, Granlibakken, Conference Center Lake Tahoe, California.

Magrin-Chagnolleau, I., Bonastre, J. F., & Bimbot, F. (1995). *Effect of Utterance Duration and phonetic Content on Speaker identification Using Second-Order Statistical Methods.* In Proceedings of the ESCA EUROSPEECH'95, Madrid, 1, 337-340.

Magrin-Chagnolleau, I., Wilke, J., & Bimbot, F. (1996). *A Further Investigation on AR-Vector Models for Text-Independent Speaker Identification.* In Proceedings of the ICASSP'96 conference, Atlanta, GA.

Mühlera, R., Ziesea, M. & Rostalskib, D. (2009). Development of a Speaker Discrimination Test for Cochlear Implant Users Based on the Oldenburg Logatome Corpus. *Journal of oto-rhino-laryngology (ORL), 71*(1), pp 14-20.

Ouamour, S., Sayoud, H., & Guerti, M. (2009, April-June). Optimal Spectral Resolution in Speaker Authentication, Application in noisy environment and Telephony. [IJMCMC]. *International Journal of Mobile Computing and Multimedia Communications, 1*(2), 36–47. doi:10.4018/jmcmc.2009040103

Paninski, L. (2006). *The Spike-Triggered Average of the Integrate-and-Fire Cell Driven by Gaussian White Noise.* Neural computation, 2006 - MIT Press, pp 2592-2616.

Razik, J., Fohr, D., Mella, O., & Parlangeau-Vallès, N. (2004). *Segmentation Parole/Musique pour la Transcription Automatique.* Paper presented at the JEP'04, Journées d'Etude sur la Parole, Fez, Morocco.

Reynolds, D. (2008). *Speaker and language recognition: a guided safari odyssey workshop.* Paper presented at the Odyssey Speaker and Language Characterization Interest Group, Stellenbosch.

Sayoud, H. (2003). *Automatic speaker recognition using neural approaches.* PhD thesis, USTHB University, Algiers, Algeria.

Sayoud, H., Ouamour, S., & Guerti, M. (2007). *Authentification Discriminative du Locuteur Basée sur une Fusion Statistique – Connexionniste.* Paper presented at the 4rth International Conference on Sciences of Electronic, Technologies of Information and Telecommunications, Tunisia.

Sayoud, H., & Selmane, M. K. (1998). *Méthodes comparatives en reconnaisance du locuteur.* Paper presented at the JTEA'98 conference, Nabeul, Tunis, Tunisia.

Vary, P., & Martin, R. (2007). Digital Speech Transmission: Enhancement, Coding, and Error Concealment. *The Journal of the Acoustical Society of America, 121*(1), 10–11. doi:10.1121/1.2400671

Chapter 15

A Fast Image Encoding Algorithm Based on the Pyramid Structure of Codewords

Ahmed A. Radwan
Minia University, Egypt

Ahmed Swilem
Minia University, Egypt

Mamdouh M. Gomaa
Minia University, Egypt

ABSTRACT

This article presents a very simple and efficient algorithm for codeword search in the vector quantization encoding. This algorithm uses 2-pixel merging norm pyramid structure to speed up the closest codeword search process. The authors first derive a condition to eliminate unnecessary matching operations from the search procedure. Then, based on this elimination condition, a fast search algorithm is suggested. Simulation results show that, the proposed search algorithm reduces the encoding complexity while maintaining the same encoding quality as that of the full search algorithm. It is also found that the proposed algorithm outperforms the existing search algorithms.

INTRODUCTION

A standard vector quantization (VQ) is an efficient data compression technique that has been widely applied to image and speech coding (Gray, 1984; Linde, Buzo, & Gray, 1980). A vector quantizer of rate r bits/sample and dimension k is a mapping from a k-dimensional vector space R^k into some finite subset of it, $Y = \{y_i;\ i = 1, 2, \ldots, N\}$ where, $N = 2^{kr}$. The subset Y is called a codebook and its elements are called codewords or reproducing vectors. In the conventional full search method to encode an input vector $x = (x_1, x_2, \ldots, x_k)$, we have to find its distance from each of N codewords, and then compare these distances to find the best-matched codeword.

A complete description of a vector quantization process includes three phases, namely training,

DOI: 10.4018/978-1-60960-563-6.ch015

encoding and decoding. The original signal is first segmented into individual vectors. The training phase is a codebook design procedure which tries to find the best set of representatives from a large set of input vectors using clustering algorithms like the generalized lioyd algorithm (GLA) (Gray, 1984; Linde, Buzo, & Gray, 1980). The GLA consists of alternatively applying the nearest neighbor condition and centroid condition to generate a sequence of codebooks with decreasing average distortion. The algorithm operates as follows: It begins with an initial codebook of size N. It partitions the training set into N sets of vectors using the nearest neighbor rule which associates each training vector with one of the N codeword's in the codebook. This partitioning defines the vector encoder mapping $Q(.)$. Then, a new codebook is created with the new codeword's being the centroids of each of the N partition regions of input vectors. This new codebook defines the vector decoder mapping $\hat{Q}$ (.). The algorithm repeats these steps and calculates the average distortion for the new codebook over the entire training set at the end of each iteration. When the average distortion fails to decrease by a certain threshold amount, it is assumed that a local minimum in distortion has been reached and the algorithm terminates.

The encoding phase finds the best matched codeword for a test vector and uses the index of the codeword to represent it. A full codebook search could be used in an encoder of vector quantization to find the codeword which is the nearest neighbor to test vector. The decoding phase is simply a table look-up procedure which uses the received index to deduce the reproduction codeword. The particularly simple table look-up decoding procedure makes VQ an attractive method of data compression in practice. Both of the training and encoding phases are computation-intensive procedures. This limits the applicability of VQ.

To find the best-matched codevector, we need a matching criterion and a codebook search algorithm. The most popular matching criterion is the Euclidean distance. When the squared Euclidean distance is used as the distortion measure, the distance between x and $y_i = (y_{i1}, y_{i2}, \ldots, y_{ik})$ can be expressed as

$$d^2(x, y_i) = \| x - y_i \|^2 = \sum_{j=1}^{k} (x_j - y_{ij})^2; i = 1, 2, \ldots, N, \quad (1)$$

where x is the current image block, y_i is the ith codeword, j represents the jth element of a vector, $k(= n \times n)$ is the vector dimension and N is the codebook size. Then a best-matched codeword with minimum distortion, which is called winner afterwards, can be determined straightforwardly by

$$d^2(x, y_w) = \min[d^2(x, y_i)]; i = 1, 2, \ldots, N \quad (2)$$

where y_w means winner and subscript "w" is the index of the winner. Once winner has been found, VQ only transmits the index of winner instead of the winner itself so as to reduce the amount of image data. Because the exact same codebook is also available at the receiving end, an image can be decoded by using the received winner index via inverse look-up table method easily and ultimately the winner itself is pasted to the corresponding position of the original image block to reconstruct it. Therefore, VQ features a very heavy encoding process due to a lot of distance computations and an extremely simple decoding process. VQ is asymmetric and the time-consuming encoding is a critical time bottleneck for its practical applications. Based on VQ property, it can be predicted that image compression in VQ method is very applicable to the broadcasting-type communication systems since they need a lot of low-cost decoding implementations at the receiving end.

From (1), each distance calculation requires k multiplications and $2k-1$ additions (and subtractions). It is necessary to perform kN multiplications $(2k-1)N$ additions, and $N-1$ comparisons to encode each input vector. The complexity of the VQ can alternatively be expressed in terms of N multiplications, $(2k-1)N$ additions, and $N-1$ comparisons per sample. The need for a larger codebook size and higher dimension for high performance in VQ encoding system results in increased computation load during the codeword search.

Despite the popularity of VQ coding, the complexity of the nearest neighbor search in high-dimensional space can become prohibitively expensive in some applications. A straightforward approach for the nearest neighbor codeword search is to search in the codebook by comparing the target vector with each codeword exhaustively using some distortion measures, and choosing the one which is nearest to the target vector. The computational cost of finding the nearest neighbor codeword in the codebook design and encoding imposes practical limits on the codebook size and the vector dimension. The excessive cost of this full search approach makes it impractical to use large codebooks for the application requiring high-dimensional vectors.

For example, if the mean-square error (MSE) is used for distortion measure of vector with dimension k, the exhaustively searching in a codebook with size N requires computation of order kN for each target vector. In most applications, the search has to be done for a large number of target vectors.

To release the computation burden, a lot of fast codeword search algorithms have been presented to speed up the searching process while maintaining the reconstructed performance same as that of the full search. These algorithms can be grouped into four categories: spatial inequality based (Bake, Jeon, & Sung, 1997; Bake, Bae, & Sung, 1999; Cao, & Li 2000; Cardinal, 1999; Guan, & Kamel, 1992; Huang, Bi, Stiles, & Harris, 1992; Imamura, Swilem, & Hashimoto, 2003; Imamura, Swilem, & Hashimoto, 2004; Johnson, Ladner, & Riskin, 1996; Johnson, Ladner, & Riskin, 2000; Lee, & Chen, 1994; Orchard, 1991; Swilem, Imamura, & Hashimoto, 2002; Swilem, Imamura, & Hashimoto, 2004; Wu, & Lin, 2000), pyramid structure based (Lee, & Chen, 1995; Pan, Lu, & Sun, 2000; Pan, kotani & Ohmi, 2004; Song, & Ra, 2002; Swilem, Imamura, & Hashimoto, 2005), sub-vector based (Pan, Lu, & Sun, 2003; Pan, kotani & Ohmi, 2006; Shan, Fang, Weile & Tian, 2008), and transform domain based (Hwang, Jeng, & Chen, 1997; Lu, Pan, & Sun, 2000; Jiang, Lu, & Wang, 2003; Chu, Lu, & Pan, 2007).

The spatial inequality based algorithms eliminate unlikely codewords by utilizing the inequalities based on the characteristic values of the input vectors and codewords. The equal-average nearest neighbor search (ENNS) algorithm (Guan, & Kamel, 1992) uses the mean as a constraint value to reject impossible codewords. The equal-average equal-variance nearest neighbor search (EENNS) (Lee, & Chen, 1994) uses the mean and the variance as two individual inequalities to reduce the search area and reject the codewords that are not contained in this area. The improved algorithm termed (IEENNS) (Baek, Jeon, & Sung, 1997) uses the mean and the variance in one inequality to reduce the search area. Wu and Lin (Wu, & Lin, 2000) presented a new kick out condition based on the norms of codewords. A lossy design method was described in (Swilem, Imamura, & Hashimoto, 2002) using a hyperplane partitioning rule. Lu and Sun (Lu, & Sun, 2003) have presented the equal-average equal-variance equal-norm nearest neighbor search (EEENNS) algorithm, which uses three significant features to reject many impossible codewords. A fast codebook design algorithm for entropy constrained vector quantization was introduced in (Swilem, Imamura, & Hashimoto, 2004) using a new constraint called the angular constraint.

With some memory overhead, the fast nearest-neighbor search (FNNS) algorithm and the projection method saves great deal of computation time. FNNS algorithm uses the triangle inequality and can reject a great many unlikely codeword's.

However, this algorithm requires an additional memory of size $N(N-1)/2$ to store the distances of all pairs of codeword's, where N is a codebook size. When N is large, the memory requirement can be a serious problem. The projection method such as equal-average nearest neighbor search (ENNS) algorithm uses the mean of an input vector to cancel the unlikely codeword. This method shows a great deal of computation time savings over conventional full-search algorithm with only N additional memory. The improved algorithm using the variance as well as the mean of an input vector, it called the equal-average equal variance nearest-neighbor search (EENNS) algorithm shows more computation time savings $2N$ with additional memory.

All of these previous works adopted a multi-resolution distortion check method, which refers to that a series of rejection checks are conducted from top level towards bottom level one by one until a rejection decision is made or the bottom level is achieved. They are very search-efficient. However, all of these previous works ignored a multi-resolution computation method, which implies that the distortion at a higher resolution level can be computed by completely reusing the obtained distortion at a previous lower resolution level and then just adding some improvements to enhance the resolution instead of computing it from the very beginning once again. As a result, there is no waste to the distortion obtained at a lower resolution level occurs.

The pyramid structure based algorithms exploit the topological structure of codeword to avoid unnecessary codeword matching procedure. Lee and Chen (Lee, & Chen, 1995) proposed a fast search algorithm based on the mean pyramid search for codebook design using the squared Euclidean distance as distortion measure. This algorithm uses the mean pyramids of codewords to reject many unmatched codewords, thus drastically speeds up the search process in encoding and codebook design. Pan el al. (Pan, Lu, & Sun, 2000) improved the encoding search process by adopting the variance pyramid in addition to the mean pyramid using the squared Euclidean distance as the distortion measure. The method uses a virtual distance between the input vector and the tested codeword at any level of the pyramid structure. This distance consists of the squared mean distance and the variance distance. Song and Ra (Song, & Ra, 2002) provided another technique using L_2-norm pyramid is heavy from many multiplication operations which increase the complexity. In (Swilem, 2005), a high-speed closest search algorithm for VQ using the projection pyramid of the vectors was established. In this algorithm, a multilevel inequality for a simple and much lighter modified distortion measure was derived based on the pyramids of codewords. By employing this inequality the procedures of codeword search for encoding or codebook design are speeded up. Another algorithm used the sum and norm pyramid data structure to reject many codeword (Shinfeng, Shih, & Chi, 2003).

The sub-vector based algorithms divide the vector to subvectors. Pan et al. (Pan, Lu, & Sun, 2003) proposed an efficient algorithm to reject unlikely codewords using sums and variances of an input vector and its two subvectors. Pan's method is better method of using subvector. To improve the Pan's method (Pan, Lu, & Sun, 2003), the algorithm (Pan, kotani & Ohmi, 2006) uses the sum, the variance, and the subvector sum to construct rejecting inequalities. Recently, another algorithm (Shan, Fang, Weile & Tian, 2008) uses the sum and the partial norms of a vector and the subvector to reject more unnecessary codewords.

In a transform domain, there are also some algorithms presented. For example wavelet transform based partial distortion search (WTPDS) algorithm (Hwang, Jeng, & Chen, 1997) which uses wavelet transform and hadamard transform based partial distortion search (HTPDS) (Lu, Pan, & Sun, 2000). Also, Jiang et al. (Jiang, Lu, & Wang, 2003) introduced another algorithm which uses hadamard transform based on norm-ordered search (NOS). An improved technique based on hadamard transform was introduced in (Chu, Lu, & Pan, 2007).

In this article, we present a high-speed closest codeword search algorithm for VQ. The proposed algorithm is developed by using the idea of the 2-pixel merging pyramid structure. A new multilevel inequality was derived for a new lighter modified distortion based on the pyramid structure of the codewords. By employing this inequality, the procedures of encoding can be speeded up.

This article is organized as follows. In the following section, some of previous work pyramid data structure are introduced and analyzed. Section 3 introduces the proposed encoding algorithm in detail. Intensive simulation results are given in Section 4. Finally, the conclusion is drawn in Section 5.

PREVIOUS WORK

Pyramid Data Structure

The image pyramid data structure was originally developed for image coding (Burt, & Adelson, 1983). In this data structure, an image is represented hierarchically, with each level corresponding to a reduced- resolution approximation. Given an image U_L of size $2^L \times 2^L$, its pyramid can be defined as a sequence of matrices $\{U_0, U_1, \ldots, U_{L-1}, U_L\}$. A pyramid data structure can be formed by successively performing appropriate operations over pixels on the higher levels. Note that U_0 has only one pixel. And the bottom level is the original image. There are many different types of image pyramids. The simplest pyramid data structure is L_1-norm pyramid, i.e., 4-pixel merging mean pyramid and 2-pixel merging sum pyramid.

Pixel Merging Mean Pyramid (4PMMP)

A 4-pixel merging mean pyramid data structure can be formed by successively operating over 2 $\times$ 2 neighboring pixels on the higher levels. That is the value of a pixel $X_{m-1}(i, j)$ on level $m - 1$ can be obtained from the values of the corresponding 2 $\times$ 2 neighboring pixels $X_m(2i - 1, 2j - 1)$, $X_m(2i - 1, 2j)$, $X_m(2i, 2j - 1)$, and $X_m(2i, 2j)$, on level m. In other words, $X_{m-1}(i, j)$ can be obtained by $X_{m-1}(i, j) = f(X_m(2i-1, 2j-1), X_m(2i-1, 2j), X_m(2i, 2j-1), X_m(2i, 2j))$, where f is an operating averaging function. From Equation 1 the following lemma can be defined:

Lemma (1)

$$d^2(x, y_i) \geq k(m_x - m_{yi})^2 .$$

Assume the vector dimension $k = 2^n \times 2^n$, then for each vector x and each codeword y_i, we can express them as two $2^n \times 2^n$ vectors. Two mean pyramid $\{X_0, X_1, \ldots, X_{r-1}, X_r, \ldots, X_n\}$ and $\{Y_{i0}, Y_{i1}, \ldots, Y_{ir-1}, Y_{ir}, \ldots, Y_{in}\}$ for x and y_i respectively, can then be constructed. Let $d_r^2(x, y_i)$ represents the squared Euclidean distance between X_r and Y_{ir}, i.e.,

$$d_r^2(x, y_i) = \sum_{j=1}^{2^r} \sum_{h=1}^{2^r} [X_r(j, h) - Y_{ir}(j, h)]^2 ,$$

where $X_r(j, h)$ and $Y_{ir}(j, h)$ represent the values of the $(j, h)^{th}$ pixels on X_r and Y_{ir}, respectively. Thus, on the top level, $d_0^2(x, y_i) = (m_x - m_{yi})^2$. From the above definitions and lemma. 1, the following theorem can be defined:

$$d^2_{m,u_1}(x, y_i) \geq \ldots \geq 4^{u_1 - v} d^2_{m,v}(x, y_i) \geq \ldots 4^{u_1 - 1} d^2_{m,1}(x, y_i) \geq 4^{u_1} d^2_{m,0}(x, y_i), \tag{3}$$

where $u_1 = \log_4(n \times n)$. Suppose $M_{x_{v,m}}$ is the mth pixel (the mean after round off) at the vth level in mean pyramid for x and $M_{y_{i,v,m}}$ implies the same thing to y_i for $m \in [0 \sim 4v]$, then Euclidean distance at the vth level for $v \in [0 \sim u_1]$ between the two mean pyramids is

$$d_{s,v}^2(x, y_i) = \sum_{m=1}^{4^v} (M_{x_{v,m}} - M_{y_{i,v,m}})^2,$$

where the real squared Euclidean distance $d^2(x, y_i) = d_{m,u_1}^2(x, y_i)$ holds (based on the definition in (1)) and $u_1 = 2$ for a 4×4 block. At any v^{th} level for $v \in [0 \sim u_1]$, if $4^{u_1 - v} d_{m,v}^2(x, y_i) > d_{\min}^2$ holds, then the current real squared Euclidean distance $d_{m,u_1}^2(x, y_i)$ would be definitely larger than $d_{\min}^2$. Thus, the search can be terminated at this v^{th} level and y_i can be rejected safely.

Using Equation. 1, for a training vector x, the mean pyramid of x is first established and the current best codeword y_b with minimum mean difference from x is found. The current minimum distortion $d_{\min}^2 = d^2(x, y_b)$ is evaluated. For any other codeword y_i the algorithm starts from the top level of the mean pyramid. It first calculates $d_0^2(x, y_i)$ and checks if $4^n d_0^2(x, y_i) \geq d_{\min}^2$. If the answer is yes, this codeword will not be the closest one and can be rejected. Otherwise, $d_1^2(x, y_i)$ on the second level is calculated and checked. If $4^{n-1} d_1^2(x, y_i) \geq d_{\min}^2$, from similar reason as above, the codeword y_i can be rejected. If it is not rejected, the third level is tested. This process is repeated until y_i is rejected or the bottom level is reached. If the bottom level is reached, then the distortion $d^2(x, y_i)$ is calculated and checked. If $d^2(x, y_i) < d_{\min}^2$, the current minimum distortion $d_{\min}^2$ is replaced by $d^2(x, y_i)$ and the closest codeword to x is set to be y_i.

Originally, this mean pyramid data structure is used for a progressive image transmission. In this $(u_1 + 1)$ level mean pyramid, each pixel value is computed by averaging the neighboring 2×2 pixels at the corresponding lower level and then taking the round off.

Generally speaking, it needs $1 + 4^1 + 4^2 + \ldots + 4^{u1-1}$ more memory units. For a 4 × 4 block, (1 +4) = 5 memory units are necessary in addition. Two reasons of introducing this type of meantype pyramid can be predicted for image transmission applications. First, because only the pixels with integer values in [0, 255] can be displayed for 8-bit, 256-level image coding, the averaging and roundoff are necessary. Second, in order to display the reconstructed image rightly, the aspect ratio, which implies the averaged pixels in horizontal and vertical direction, has to be the same like 2 × 2(4- PM way) other than 2 × 1 or 1 × 2. It can be concluded that this 4-PM mean pyramid is an appropriate data structure for progressive image transmission. The similar 4-PM mean-type pyramid data structure can also be directly applied to fast encoding for VQ as proposed in (Lee, & Chen, 1995; Lin, Chung, Chang, 2001; Song, & Ra, 2002). Under the condition of using arithmetical mean, an accompanying mean pyramid for each codeword is generated off-line and in (Lin, Chung, & Chang, 2001) the real means of all codewords are saved and sorted at top level as well. When an image block is to be encoded by VQ, its mean pyramid is generated online first. Then winner search is started from the position at which the real mean difference between the image block and all codewords is minimum and goes up and down in interleaving way. During the search, in order to check the difference between the input image block and a codeword, the distance is computed level by level from the top towards the bottom in a pyramid way. The distance corresponding to the bottom level between two pyramids is called real distance afterwards. The min real distance "so far" is kept as $d_{\min}^2$.

Pixel Merging Sum Pyramid (2PMSP)

A 2-pixel merging sum pyramid data structure can be formed by successively performing appropriate operations over 2-pixels on the higher levels. Therefore, the value of a pixel $U_{L-1}(m, n)$ in level $L - 1$ is obtained from the values of the corresponding 2 neighboring pixels $\{U_L(2m-1, 2n-1), U_L(2m-1, 2n)\}$ or $\{U_L(2m-1, 2n-1), U_L(2m, 2n-1)\}$ or $\{U_L(2m-1, 2n), U_L(2m, 2n)\}$ in level L. During a winner search process, suppose the "so far" minimum squared Euclidean distance is $d^2_{\min}$. Based on statistical features of a vector, a 2-pixel merging sum pyramid has been proposed in (Pan, Kotani, & Ohmi, 2004), to realize a multi-resolution description for a vector. A new hierarchical rejection rule can be obtained based on the 2PMSP data structure as

$$d^2_{s,u_2}(x, y_i) \geq \ldots \geq 2^{-(u_2-v)} d^2_{s,v}(x, y_i) \geq \ldots \geq 2^{-(u_2-1)} d^2_{s,1}(x, y_i) \geq 2^{-u_2} d^2_{s,0}(x, y_i). \quad (4)$$

where $u_2 = \log_2(n \times n)$, and squared Euclidean distance at the vth level for $v \in [0 \sim u_1]$ is

$$d^2_{s,v}(x, y_i) = \sum_{j=1}^{2^v} (S_{x_{v,j}} - S_{y_{i,v,j}})^2,$$

where $S_{x_{v,m}}$ is the jth pixel (sum) at the vth level in a sum pyramid for x and $S_{y_{i,v,j}}$ implies the same thing to y_i for $j \in [1 \sim 2^v]$. The real squared Euclidean distance $d^2(x, y_i) = d^2_{s,u_2}(x, y_i)$ holds. For a 4×4 block, $u_2 = 4$. At any v^{th} level for $v \in [0 \sim u_2]$, if $2^{-(u_2-v)} d^2_{s,v}(x, y_i) > d^2_{\min}$ holds, then the current real squared Euclidean distance $d^2_{s,u_2}(x, y_i)$ will be definitely larger than $d^2_{\min}$. Therefore the search can be terminated at this v^{th} level and y_i can be rejected safely.

Comparing Equation.4 with Equation.3, it is obvious that Equation.4 has doubled the number of tests for a possible rejection. The tests of Equation.4 at the even levels (i.e. v = 0, 2, 4, …, u2) have the same rejection power as Equation.3 but the tests of Equation.4 at the odd levels are newly inserted.

If Equation.4 is directly used, when a test at the vth level fails, the obtained $d^2_{s,v}(x, y_i)$ is also discarded completely and $d^2_{s,v+1}(x, y_i)$ will be computed once again from the very beginning for the following test at $(v + 1)^{th}$ level. It is certainly a waste to discard $d^2_{s,v}(x, y_i)$. How to reuse the obtained $d^2_{s,v}(x, y_i)$ to compute $d^2_{s,v+1}(x, y_i)$ recursively is the key of method 2-pixel merging sum pyramid.

In fact, a 2-PM sum pyramid needs $(2^{u2-1} + \ldots + 2^2 + 2^1 + 1) = 2^{u2} - 1 = n \times (n-1)$ extra memories. For a 4 × 4 block ($u2 = 4$), either an image block or a codeword), (8 + 4 +2 + 1) = 15 extra memories are needed. In contrast, a 4-PM mean pyramid only needs

$$(4^{u1-1} + \ldots + 4^2 + 4^1 + 1) = (4^{u1} - 1) / 3 = n \times (n-1) / 3$$

extra memories. Based on the discussions above, it can be concluded that a 4-PM mean pyramid is more memory efficient than a 2-PM sum pyramid. Because the target of this article is the VQ encoding speed, thus extra memory requirement will not be specially treated in this method afterwards.

Obviously, the search process based on Equation 4 only used the multi-resolution distortion check method but completely ignored the multi-resolution distortion computation method. It is a waste to computation. This method aims at introducing the multi-resolution distortion computation method into Equation 4 in order to fully exploit the potential power of the two aspects included in the multi-resolution concept.

Three major modifications are made. Because sum can be computed accurately using integer operation, mismatched winners due to round off error of taking the mean can be overcome firstly by using sum instead of the mean. Secondly, since sum is computed by 2-pixel-merging other than 4-pixel-merging way, more levels can be constructed in a sum pyramid so that the rejection to current codeword can be realized easier. Finally, codebook is off-line rearranged by sorted real sums directly, and the lower and upper bound for promising codeword class are updated dynamically to narrow search scope further once a better-matched codeword has been found during winner search process.

PROPOSED ALGORITHM

In this section, we propose a fast closest codeword search algorithm by deriving a robust inequality condition based on the norm pyramid of codewords see Figure 1. As mentioned in Section 1, to find the closest codeword to an input vector, all the codewords in a codebook will be compared with x. This process is described as $y_{b_m} = Q(x)$, where Q is a mapping operator to find y_{b_m} satisfying (2). For a distortion measure of squared Euclidean distance, (2) can be rewritten as follows:

$$d^2(x, y_{bm}) = \min_{i=1,2,\ldots,N} d^2(x, y_i) = \min_{i=1,2,\ldots,N} \left(\| x \|^2 + \| y_i \|^2 - 2\sum_{j=1}^{k} x_j y_{ij} \right) \quad (5)$$

Hence, we can easily see that the problem of finding the codeword that satisfies (5) can be replaced with a simpler one that satisfies

$$d'^2(x, y_{bm}) = \min_{i=1,2,\ldots,N} d'^2(x, y_i), \quad (6)$$

where

$$d'^2(x, y_i) = d^2(x, y_i) - \| x \|^2 = \| y_i \|^2 - 2\sum_{j=1}^{k} x_j y_{ij}. \quad (7)$$

This is because $\|x\|^2$ is a common term in the distortion measure of (5). It should be noted that evaluating (7) is much faster than evaluating (1) since $\| y_i \|; 1 \le i \le N$ can be predetermined and stored in the preprocessing procedure. Let us assume that parts of codewords have been inspected, and, let the "so far" minimum distance be $d^2_{\min}$. From (7), the following property can be derived by using the Cauchy–Schwarz inequality:

Figure 1. 2-pixel merging norm pyramid

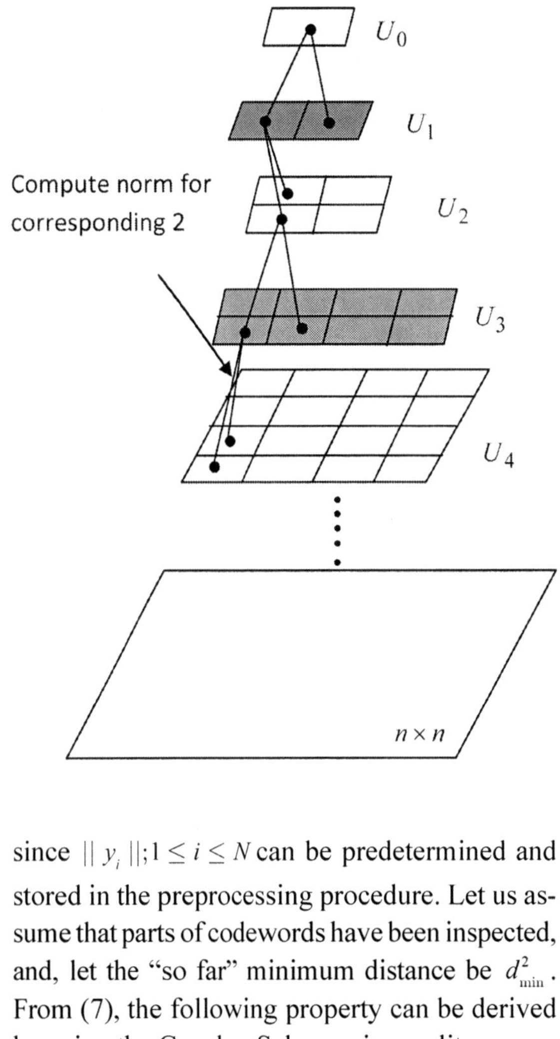

$$d'^2(x, y_i) = \| y_i \|^2 - 2\sum_{j=1}^{k} x_j y_{ij} \ge \| y_i \|^2 - 2\| y_i \| \cdot \| x \| = \| y_i \| \cdot (\| y_i \| - 2\| x \|). \quad (8)$$

Therefore, if a codeword y_i satisfies:

$$\| y_i \| \cdot (\| y_i \| - 2\| x \|) \ge d^2_{\min}, \quad (9)$$

then $d'^2(x, y_i) \ge d^2_{\min}$ is always guaranteed. Hence, it can be eliminated since it cannot be closer to x than the "so far" nearest codeword.

For a query vector x, the computation of $\| y_i \| . (\| y_i \| - 2 \| x \|)$ is quite simple, because $\{\| y_i \|; 1 \le i \le N\}$ can be computed in advance in the preprocessing procedure. We can extend the inequality property in (8) to a general case, where tighter decision boundaries for eliminating search operations are obtained in a multilevel manner. Let us assume that the vector dimension k is 2^L. Then, for an input vector x and a codeword y_i, their corresponding vectors in level L are denoted as x_L and y_{iL}, respectively. Here, we note that $x_0 = \|x\|$. From the Cauchy–Schwarz inequality, we can derive the following inequality:

$$\sum_{j=1}^{k} x_j y_{ij} = \sum_{j=1}^{2^L} x_{jL} y_{ijL} \le \sum_{j=1}^{2^{L-1}} x_{jL-1} y_{ijL-1} \le \dots \le x_0 y_{i0} = \| x \| . \| y_i \| . \quad (10)$$

where x_{jL} denotes the jth element of x_L and it is the norm of the corresponding 2 neighboring pixels in level $L + 1$. From (10), we can generalize (8) as follows:

$$d'^2(x, y_i) \equiv d_L'^2(x, y_i) \ge d_{L-1}'^2(x, y_i) \ge \dots \ge d_0'^2(x, y_i), \quad (11)$$

where

$$d_L'^2(x, y_i) = \| y_i \|^2 - 2\sum_{j=1}^{2^L} x_{jL} y_{ijL}. \quad (12)$$

Based on (11), we propose a fast search algorithm. Prior to the encoding process, codewords in a codebook are stored in the ascending order of their norm values. The searching proceeds in the alternating order starting from y_{ib_m} as depicted in Figure 2. To determine whether the next candidate $y_{ib_{m+1}}$ is closer to x than the current best match y_{ib_m}, we examine $y_{ib_{m+1}}$, starting from the top level of its 2-pixel merging norm pyramid. First, $d_0'^2(x, y_{ib_{m+1}})$ is computed, and is compared

Figure 2. Searching path

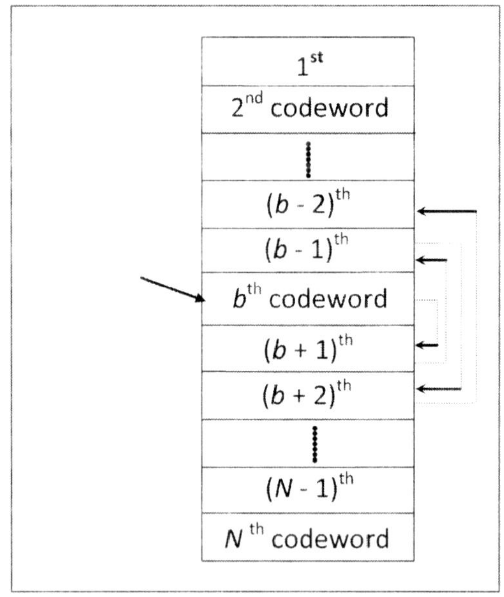

with $d_{\min}^2$. If it is larger than or equal to $d_{\min}^2$, from (11). Thus, codeword $y_{ib_{m+1}}$ cannot be the closest one and is rejected. Otherwise, $d_1'^2(x, y_{ib_{m+1}})$ is calculated and compared with $d_{\min}^2$. If $d_1'^2(x, y_{ib_{m+1}}) \ge d_{\min}^2$, $y_{ib_{m+1}}$ is rejected due to a similar reason as above. Otherwise, the next level is considered $d_2'^2(x, y_{ib_{m+1}})$ is calculated and compared with $d_{\min}^2$. If $d_2'^2(x, y_{ib_{m+1}}) \ge d_{\min}^2$, $y_{ib_{m+1}}$ is rejected. Otherwise $d_3'^2(x, y_{ib_{m+1}})$ is calculated and compared with $d_{\min}^2$. If $d_3'^2(x, y_{ib_{m+1}}) \ge d_{\min}^2$, $y_{ib_{m+1}}$ is rejected. This process is repeated until $y_{ib_{m+1}}$ is rejected or the bottom level is reached. If the bottom level is reached, $d_L'^2(x, y_{ib_{m+1}})$ is computed and compared with $d_{\min}^2$. If $d_L'^2(x, y_{ib_{m+1}}) \ge d_{\min}^2$, $y_{ib_{m+1}}$ is rejected. Otherwise, $d_{\min}^2$ is replaced with $d_L'^2(x, y_{ib_{m+1}})$ and the current best match codeword to x is set to $y_{ib_{m+1}}$. This procedure is repeated for all candidate codewords

Table 1. Comparison of the exact computing time (in seconds) for various codebook sizes

Codebook size	Method	Tested Images	
		Lena	Baboon
512	FS	2.17	2.12
	4PMMP	0.77	0.54
	2PMSP	0.87	0.63
	2PMNP	0.58	0.39
1024	FS	4.22	4.23
	4PMMP	1.27	0.93
	2PMSP	1.58	1.19
	2PMNP	0.99	0.69
2048	FS	8.45	8.44
	4PMMP	2.28	1.7
	2PMSP	2.90	2.29
	2PMNP	1.81	1.31

Table 2. Comparison of the average distortion computations for various codebook sizes

Codebook size	Method	Tested Image	
		Lena	Baboon
512	FS	1	1
	4PMMP	0.055	0.014
	2PMSP	0.016	0.009
	2PMNP	0.017	0.010
1024	FS	1	1
	4PMMP	0.043	0.011
	2PMSP	0.011	0.007
	2PMNP	0.012	0.007
2048	FS	1	1
	4PMMP	0.033	0.009
	2PMSP	0.007	0.005
	2PMNP	0.008	0.005

y_i in the codebook, starting from $y_{ib_{m+1}}$ according to the search order shown in Figure 2.

EXPERIMENTAL RESULTS

Experiments were carried on vectors taken from the USC grayscale image set. We used two images, Lena and Baboon with size 512×512 and 256 gray levels. Each image was divided into 4×4 blocks. The well-known LBG algorithm (Gray, 1984; Linde, Buzo, & Gray, 1980) is used to design codebooks with different codebook sizes (N = 512, 1024, 2048) for each image. The tested methods are full search (FS), 4-pixel merging mean pyramid (4PMMP), 2-pixel merging sum pyramid (2PMSP) and the proposed method the 2-pixel merging norm pyramid (2PMNP).

Table 1 presents a comparison of the exact computing time (in seconds) for various codebook sizes. The timings were made on Pentium IV (3 GHZ). Compared with the FS algorithm, the proposed algorithm reduces the computational time by 73.2%-84.4%. Compared with 2-sum merge algorithm, the proposed algorithm reduces the computational time by 33.3-42.79%. From this table, the proposed algorithm has the best performance in terms of computational time for all cases than the FS and 2PMSP algorithms.

Table 2 lists the average distortion computations for various codebook sizes. Compared with the FS algorithm, the proposed algorithm can reduce the distortion computation by 98.2-99.4%. Compared with 2PMSP algorithm, the proposed algorithm has almost the same number of distortion calculation.

From Table 2, we can see that the average distortion computation of the proposed algorithm is obviously much less than FS in all kind of cases. Compared with other available approaches, the proposed method can reduce the computing time and distortion calculation significantly for VQ.

Table 3 summarizes computation complexities of various fast algorithms in terms of the average number of operations per pixel. We can see that for those complex arithmetical operations, the proposed algorithm requires the least number of

Table 3. Comparison of computation complexity pixel

Tested image	Codebook Size	Method	The Average Number Of Operation				
			×	±	CMP	SQRT	Total
Lena	512	2PMSP	3.006	3.784	2.141	0	8.931
		2PMNP	3.107	3.207	2.141	0.029	8.484
	1024	2PMSP	2.800	3.384	2.124	0	8.038
		2PMNP	2.851	2.909	2.123	0.014	7.897
	2048	2PMSP	2.641	3.087	2.107	0	7.835
		2PMNP	2.670	2.703	2.107	0.007	7.487
Baboon	512	2PMSP	4.370	5.955	2.426	0	12.751
		2PMNP	4.452	4.676	2.438	0.029	11.595
	1024	2PMSP	3.889	5.094	2.361	0	11.344
		2PMNP	4.010	4.086	2.374	0.014	10.484
	2048	2PMSP	3.528	4.458	2.311	0	10.297
		2PMNP	3.611	3.653	2.322	0.007	9.593

operations. This means that using the elimination inequality in (11) reduces the number of operations.

CONCLUSION

This article presents a fast search algorithm for VQ encoding. First, by using the Cauchy–Schwarz inequality, a robust elimination inequality is derived on the basis of 2-pixel merging norm pyramid data structure. Then, using this inequality property, the algorithm can remove many unreliable candidates by examining their norm pyramids without calculating full distances, and hence dramatically speeds up the search process in VQ encoding. Simulation results show that the encoding time of the proposed algorithm is reduced significantly while encoding quality remains the same with respect to the exhaustive search algorithm.

REFERENCES

Bake, S. J., Bae, M. J., & Sung, K. M. (1999). A fast vector quantization encoding algorithm using multiple projection axes. *Signal Processing, 75*, 89–92. doi:10.1016/S0165-1684(99)00035-3

Bake, S. J., Jeon, B. K., & Sung, K. M. (1997). A fast encoding algorithm for vector quantization. *IEEE Signal Processing Letters, 4*(2), 325–327. doi:10.1109/97.650035

Burt, P. J., & Adelson, E. (1983). The Laplacian pyramid as a compact image code. *IEEE Transactions on Communications, COM-31*, 532–540. doi:10.1109/TCOM.1983.1095851

Cao, H. Q., & Li, W. (2000). A fast search algorithm for vector quantization using a directed graph. *IEEE Transactions on Circuits and Systems for Video Technology, 10*, 585–593. doi:10.1109/76.845003

Cardinal, J. (1999, September). Fast search for entropy-constrained VQ. In *Proceeding of (ICIAP) the 10th International Conference on Image Analysis and Processing* (pp. 1038-1042).

Chu, S. C., Lu, Z. M., & Pan, J. S. (2007). Hadamard transform based fast codeword search algorithm for high-dimensional VQ encoding. *Elsevier Science Inc., 177*(3), 734-746

Gray, R. M. (1984). Vector quantization. *IEEE Acoustics, Speech, and Signal Processing, 1*(2), 4–29.

Guan, L., & Kamel, M. (1992). Equal-average hyperplane partitioning method for vector quantization of image data. *Pattern Recognition Letters, 13*(10), 693–699. doi:10.1016/0167-8655(92)90098-K

Huang, C. M., Bi, Q., Stiles, G. S., & Harris, R. W. (1992). Fast full search equivalent encoding algorithms for image compression using vector quantization. *IEEE Transactions on Image Processing, 1*(3), 413–416. PubMeddoi:10.1109/83.148613

Hwang, W. J., Jeng, S. S., & Chen, B. Y. (1997). Fast codeword search algorithm using wavelet transform and partial distortion search techniques. *Electronics Letters, 33*(5), 365–366. doi:10.1049/el:19970249

Imamura, K., Swilem, A., & Hashimoto, H. (2003, May). Fast codeword search algorithm for ECVQ using hyperplane decision rule. In *Proceeding of (ISCAS) International Symposium on Circuits and Systems, 2*, 476-479.

Imamura, K., Swilem, A., & Hashimoto, H. (2004, October). Fast VQ encoding algorithms using angular constraint. In *Proceeding of (ICIP) International Conference on Image Processing* (pp. 3161-3164).

Johnson, M. H., Ladner, R., & Riskin, E. A. (1996, September). Fast nearest neighbor search for ECVQ and other modified distortion measures. In *Proceeding of (ICIP) International Conference on Image Processing, 3*, 423-426.

Johnson, M. H., Ladner, R., & Riskin, E. A. (2000). Fast nearest neighbor search of entropy-constrained vector quantization. *IEEE Transactions on Image Processing, 9*, 1435–1437. PubMeddoi:10.1109/83.855438

Lee, C. H., & Chen, L. H. (1994). Fast closest codeword search algorithm for vector quantization. *IEEE Proceedings Vision, Image &. Signal Processing, 141*(3), 143–148.

Lee, C.-H., & Chen, L.-H. (1995). A fast search algorithm for vector quantization using mean pyramids of codewords. *IEEE Transactions on Communications, 43*, 1697–1702. doi:10.1109/26.380218

Lin, S. J., Chung, K. L., & Chang, L. C. (2001). An improved search algorithm for vector quantization using mean pyramid structure. *Pattern Recognition Letters, 22*(3/4), 373–379. doi:10.1016/S0167-8655(00)00136-7

Linde, Y., Buzo, A., & Gray, R. M. (1980). An algorithm for vector quantizer design. *IEEE Transactions on Communications, COM-28*(1), 84–95. doi:10.1109/TCOM.1980.1094577

Lu, Z. M., & Sun, S. H. (2003). Equal-average equal-variance equal-norm nearest neighbor search algorithm for vector quantization. *IEICE Transactions on Information and systems. E (Norwalk, Conn.), 86D*(3), 660–663.

Lu, Z. M., Pan, J. S., & Sun, S. H. (2000). Efficient codeword search algorithm based on hadamard transform. *Electronics Letters, 36*(16), 1364–1365. doi:10.1049/el:20000972

Orchard, M. D. (1991). A fast nearest-neighbor search algorithm. In *Proceeding of (ICASSP) International Conference on Acoustics, Speech, and Signal Processing, 4*, 2297-2300.

Pan, J. S., Lu, Z. M., & Sun, S. H. (2000). Fast codeword search algorithm for image coding based on mean-variance pyramids of codewords. *Electronics Letters*, *36*(3), 210–211. doi:10.1049/el:20000237

Pan, J. S., Lu, Z. M., & Sun, S. H. (2003). An efficient encoding algorithm for vector quantization based on sub-vector technique. *IEEE Transactions on Image Processing*, *12*(3), 265–270. PubMed-doi:10.1109/TIP.2003.810587

Pan, Z., Kotani, K., & Ohmi, T. (2004). An improved fast encoding algorithm for vector quantization using 2-pixel-merging sum pyramid data structure. *Pattern Recognition Letters*, *25*, 459–468. doi:10.1016/j.patrec.2003.12.009

Pan, Z., Kotani, K., & Ohmi, T. (2006). Subvector-based fast encoding method for vector quantization without using two partial variances. *Optical Review*, *13*(6), 410–416. doi:10.1007/s10043-006-0410-1

Shan, X. C., Fang, W. L., Wei, L. Z., & Tian, Q. Z. (2008). Fast search algorithm for vector quantization based on subvector technique. *IEICE Transactions on Information & Systems. E (Norwalk, Conn.)*, *91-D*(7), 2035–2040.

Shinfeng, D. L., Shih, C. S., & Chi, Y. H. (2003). A Fast Codebook Search Algorithm Based on Sum-Norm Pyramid of Codewords. In *Proceeding of (CVGIP) 16th IPPR Conference on Computer Vision, Graphics and Image Processing*.

Song, B. C., & Ra, J. B. (2002). A fast search algorithm for vector quantization using L2-norm pyramid of codewords. *IEEE Transactions on Image Processing*, *11*(1), 10–15. PubMed-doi:10.1109/83.977878

Swilem, A., Imamura, K., & Hashimoto, H. (2002). A fast search algorithm for vector quantization using hyperplane decision rule. *Journal of the Institute of Image Information and Television Engineers*, *56*(9), 1513–1517.

Swilem, A., Imamura, K., & Hashimoto, H. (2004). A fast codebook design algorithm for ECVQ based on angular constraint and hyperplane decision rule. *IEICE Transactions on Fundamentals of Electronics, Communications and Computer Sciences. E (Norwalk, Conn.)*, *87-A*(3), 732–739.

Swilem, A., Imamura, K., & Hashimoto, H. (2005). A high-speed closest codeword search algorithm using the pyramid structure of codewords. *(IIEEJ). Journal of the Institute of Image Electronics Engineers of Japan*, *34*(5), 653–662.

Wu, K. S., & Lin, J. C. (2000). Fast VQ encoding by an efficient kick-out condition. *IEEE Transactions on Circuits and Systems for Video Technology*, *10*(1), 59–62. doi:10.1109/76.825859

This work was previously published in International Journal of Mobile Computing and Multimedia Communications (IJMCMC) 1(4), edited by Ismail Khalil & Edgar Weippl, pp. 1-13, copyright 2009 by IGI Publishing (an imprint of IGI Global)

Chapter 16
Analysis and Modeling of H.264 Unconstrained VBR Video Traffic

Harilaos Koumaras
Business College of Athens (BCA), Greece

Charalampos Skianis
University of Aegean, Greece

Anastasios Kourtis
Institute of Informatics and Telecommunications NCSR, Greece

ABSTRACT

In future communication networks, video is expected to represent a large portion of the total traffic, given that especially variable bit rate (VBR) coded video streams, are becoming increasingly popular. Consequently, traffic modeling and characterization of such video services is essential for the efficient traffic control and resource management. Besides, providing an insight of video coding mechanisms, traffic models can be used as a tool for the allocation of network resources, the design of efficient networks for streaming services and the reassurance of specific QoS characteristics to the end users. The new H.264/AVC standard, proposed by the ITU-T Video Coding Expert Group (VCEG) and ISO/IEC Moving Pictures Expert Group (MPEG), is expected to dominate in upcoming multimedia services, due to the fact that it outperforms in many fields the previous encoded standards. This article presents both a frame and a layer (i.e. I, P and B frames) level analysis of H.264 encoded sources. Analysis of the data suggests that the video traffic can be considered as a stationary stochastic process with an autocorrelation function of exponentially fast decay and a marginal frame size distribution of approximately Gamma form. Finally, based on the statistical analysis, an efficient model of H.264 video traffic is proposed.

INTRODUCTION

Multimedia applications and services have already possessed a major portion of the today traffic over computer and mobile communication networks. Among the various types of multimedia, video services (transmission of moving images and sound) are proven dominant for present and future broadband networks.

DOI: 10.4018/978-1-60960-563-6.ch016

Raw video data has very high bandwidth and storage requirements making its transmission and storage impractical and economically unaffordable. For this reason, a lot of research has been performed on developing techniques that exploit both temporal and spatial redundancy in video sequences, in order to succeed efficient data compression.

From the advent of video coding, two main encoding schemes were proposed and are still in use: The Constant Bit Rate (CBR) and the Variable Bit Rate (VBR) modes. The choice of VBR mode for video services over communication networks prevails against CBR mode due to a number of advantages such as

- Better video quality for the same average bit rate without the need to adjust the quantization parameters during the encoding as in CBR
- Shorter delay since the buffer size in the encoder side can be reduced without encountering an equivalent delay in the network
- Increased call-carrying capacity due to the fact that the bandwidth per call for VBR video may be lower than for equivalent quality of CBR source.
- Although a CBR transmission mode makes the network management easier, mainly due to the predictable traffic patterns, on the other hand it prevents a possible traffic gain via statistical multiplexing, which means that does not efficiently exploits the available capacity of the transmission channel as VBR does.

Efficient network utilization and constant picture quality can be achieved by VBR mode. However, when transmission and statistical multiplexing of VBR-coded video traffic is considered over a shared medium, like the Internet, the improvement in network utilization cannot be determined only by the compression ratio. VBR coding results in large fluctuations in bit rate and high correlation among the bit rates in successive time intervals due to the video content and the abrupt scene changes (Yegenoglu, 1993). This complex nature of VBR-coded video traffic creates a challenge in the efficient design of communications/transmission networks and the associated traffic control. Therefore an accurate traffic study is necessary for the prediction of the network performance. A method of doing this is to perform real experiments using existing networks and actual sources. However, testing real networks is quite impractical, while performing tests with real video clips, although it is possible, the deduced results may be very video-content specialized and therefore not general and scalable. Thus, major surge of interest in the topic of VBR video traffic modeling has appeared, because it provides information on how VBR mode affects network performance and besides is a useful tool for traffic engineering of communication networks in order to optimize admission control, to perform short-term traffic forecast and optimize buffer lengths.

Generally speaking, an analysis of video traffic is required in order to develop an efficient video traffic model. Such models can be evaluated (Al-heraish, 2004) by the fact that they must satisfy some criteria, namely: They must match certain characteristics of a real video sequence, such as probability density function, mean, variance, peak, autocorrelation etc. Moreover, the deduced generated video traffic must be similar to real video data, in order to be able to be used for predicting a desired performance metric (i.e. delay, buffer, size etc.). Furthermore, the proposed models should be simple and able to generate video traffic with low computational power.

Early studies in unconstraint VBR models examined various characteristics of VBR video traffic, such as differences in successive frame sizes and cluster lengths (Chin, 1989) or scene duration distributions (Verbiest, 1988). Also recently introduced efficient modeling tools and techniques of VBR MPEG-1/H.261 coded video at frame and GOP level (Skianis, 2003), (Doulamis, 2000).

Results from these and other works indicate that the frame sizes exhibits a bell-shape (e.g. (Heyman, 1992), (Maglaris, 1988), (Nomura, 1989)). Furthermore, in certain cases correlations in the video bit rate are found to decay exponentially (Heyman, 1992), (Maglaris, 1988), (Cohen, 1993), (Haskel, 1972), (Lucantoni, 1994) while other studies (Nomura, 1989), (Ramamurthy, 1990), (Rodriguez-Dagnino, 1991) observe a more complex phenomenon, in which the correlation decay is rapid for the initial lags and then continues at a lower rate.

The most popular and widely used encoding algorithms are the ones developed by the Moving Picture Experts Group (MPEG) and the Video Coding Expert Group (VCEG) of the ITU. Recently these two organizations jointly developed a new codec, the H.264 or MPEG-4 Part 10 Advanced Video Coding (AVC) codec (Wiegand, 2003). Featuring updated capabilities, the new codec can achieve 40-50% compression efficiency gain over today's optimized MPEG-2 codecs. Due to the advances of H.264 in comparison to earlier standards, e.g. H.263 (Girod, 1997) (Steinbach, 1997), it is expected that it will prevail in future networks and mobile application systems, making traffic modeling and characterization of H.264 video streams a useful tool for network managers and designers.

Following this trend, this work presents a detailed frame and layer (i.e. I, P and B) level analysis of H.264 video traffic and proposes an adequate traffic model.

The rest of the article is organized as follows: Section 2 outlines the new characteristics and enhancements of the H.264 standard, Section 3 presents the statistical analysis of the H.264 video stream. Section 4 discusses on the video traffic modeling, presenting related work and a novel H.264 model. Finally, Section 5 concludes the article.

THE H.264/AVC STANDARD: ESSENTIAL ISSUES AND CURRENT STATUS

In 1998 the ITU-T VCEG issued a call for proposals (H.26L project), with main scope to double the coding efficiency in comparison to the already existing coding standards. In 2001, VCEG and ISO/IEC MPEG formed a Joint Video Team (JVT) in order to finalize the standard and submit for formal approval as H.264/AVC (Wiegand, 2003).

The new video coding standard known as H.264/MPEG-4 Advanced Video Coding (AVC), now in its fourth version, has demonstrated significant achievements in terms of coding efficiency, robustness to a variety of network channels and conditions, and breadth of applications (Sullivan, 2004). Some essential indicative enhancements are:

- Variable block size support for motion compensation with luma block sizes down to 4x4, in conjunction with 4x4 level transformations.
- Quarter-sample motion vector accuracy.
- Extended reference frame selection for P frames, among various previously decoded frames.
- De-blocking filter within the motion-compensated prediction loop.
- New context-based adapted entropy coding methods: CAVLC and CABAC.

The main target of the aforementioned enhancements is the perceived quality improvement and the high-compression efficiency. With the expected wide breadth of applications, from videoconferencing and entertainment to streaming video and digital cinema, where the new coding standard is expected to be implemented, three basic feature sets (called profiles) were established to address these application domains:

- Baseline profile (BP): Designed to minimize complexity and provide high robustness and flexibility for use over a broad range of network environments and conditions.
- Main Profile (MP): Designed with an emphasis on compression coding efficiency capability.
- Extended Profile (XP): Designed to combine the robustness of the Baseline profile with a higher degree of coding efficiency and greater network robustness.

At present, the Baseline profile seems that it provides a good solution for its target application area. The JVT is working on incorporating a Scalable Video Coding (SVC) amendment into the design of the existed H.264 standard. In terms of coding structure, a scalable bit stream will be composed of a base layer and one or more enhancement layer bit streams. The base layer will be conforming to one of the profiles of the prior H.264/MPEG-4 AVC design. Additional key issues are fidelity range extensions (Su, 2005), (Sullivan, 2005), which addressed the issue of more demanding applications of H.264 in resolution, bits/sample and chroma sampling, and improvement on H.264 encoder performance (Lai, 2005).

STATISTICAL ANALYSIS OF THE H.264 ENCODED DATA

For the statistical analysis of H.264/AVC encoded data, the reference encoder JM is used, considering encodings without rate control and fixed quantization parameters for all test sequences. In H.264, the three common different frame modes are adopted, namely: Intra-frame (I), Predictive (P) and Bidirectional predictive (B), widely referred as I, P and B. In particular, the I frames are also called Intra frames, while B and P are known as Inter frames. The combination of successive types of frames forms a Group Of Pictures (GOP), whose length is mainly described by the distance of two successive I frames. In the described work, the frame rate is set constant at 25fps, coding GOP structure is set as IPBPBPBPB… and Intra-period adopts values between 3 and 12. Finally, a video segment from the film "Spider-man II" is used as reference signal. This segment consists of 18357 frames of YUV 4:2:0 format in 528x384 resolution.

Frame Level Analysis

Focusing on the Frame Level analysis, Figure 1 illustrates the size of 1100 frames of an H.264 test signal (encoded with quantization scale 20 for all the frames and GOP length 12), where it can be noticed that the large frame sizes (periodical peaks in the figure) correspond to I frames, while the smaller ones are B frames and the intermediate frame sizes are P frames. Moreover, the periodicity that seems to appear in the peaks of I frames, corresponds to the distance of two successive I frames, which reveals the length of the used GOP. It is also noted that the frame size follows the spatial and temporal activity of the test signal, where more complex frames require more bits for their description, while static and simple frames are described by fewer bits.

Also another interesting observation is that inter-frames (especially P frames) present more intense fluctuation in comparison with the Intra frames. This stems from the fact that according to the content dynamics of the video signal, some Macro-Blocks (MBs) of the inter-frames may be intra-coded, which results in lower compression ratio and therefore higher frame sizes. Figure 2 depicts the total number of Intra MBs for the P frames of the total 1100 frames of Figure 1. It can be observed that the shape of the Intra MBs vs. Inter-frames graph (Figure 2) plays a major role in the form of the frame size graph (Figure 1). In other words, inter-frames appear to influence largely the actual video traffic.

A principal issue regarding the modeling of unconstraint Variable Bit Rate traffic is whether or not the encoded traffic can be considered as

Figure 1. The frame level analysis over a time-window of 1100 frames

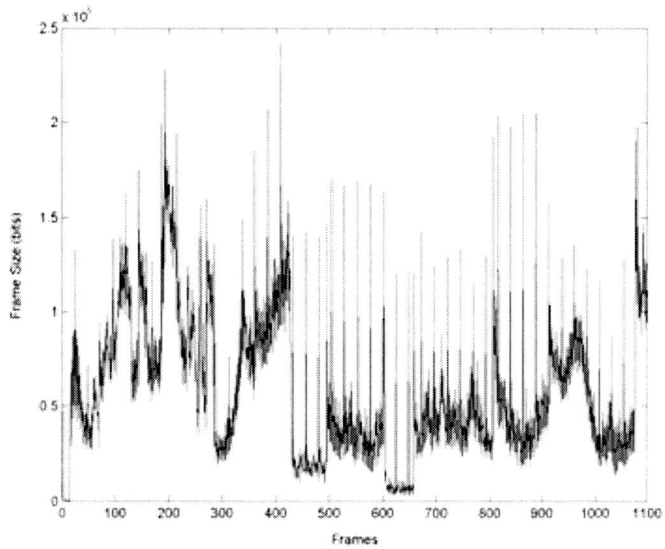

Figure 2. The Intra MBs for the inter-frames (i.e. P) over a time of 1100 frames

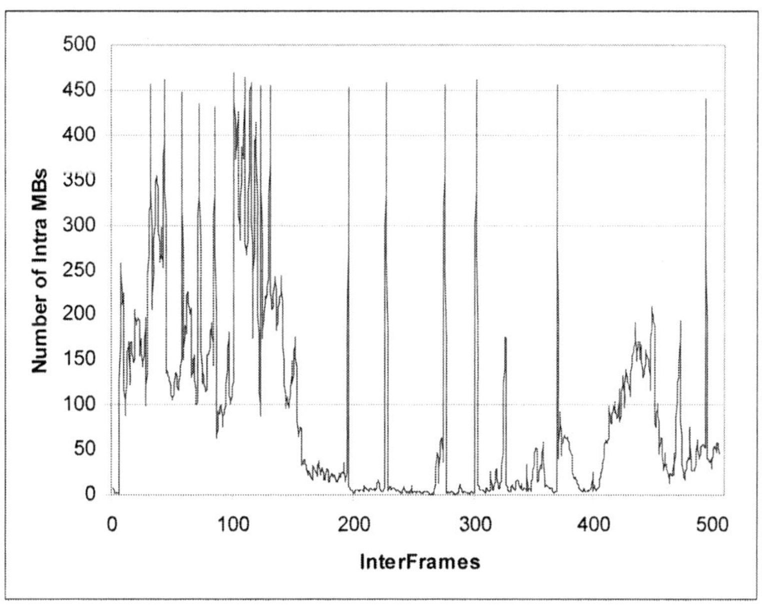

stationary process. In this respect, an encoding frame sequence from "Spider-man 2" was split in a moderate number of windows (actually four) and the empirical density function for the frame size was calculated from the samples of each window.

These windows densities, which are depicted in figures 3(a), (b)), where found very similar, a property directly suggesting that the sequence is

Figure 3. Frame size histograms in different time windows (a), (b) and autocorrelations of such histograms (c), (d)

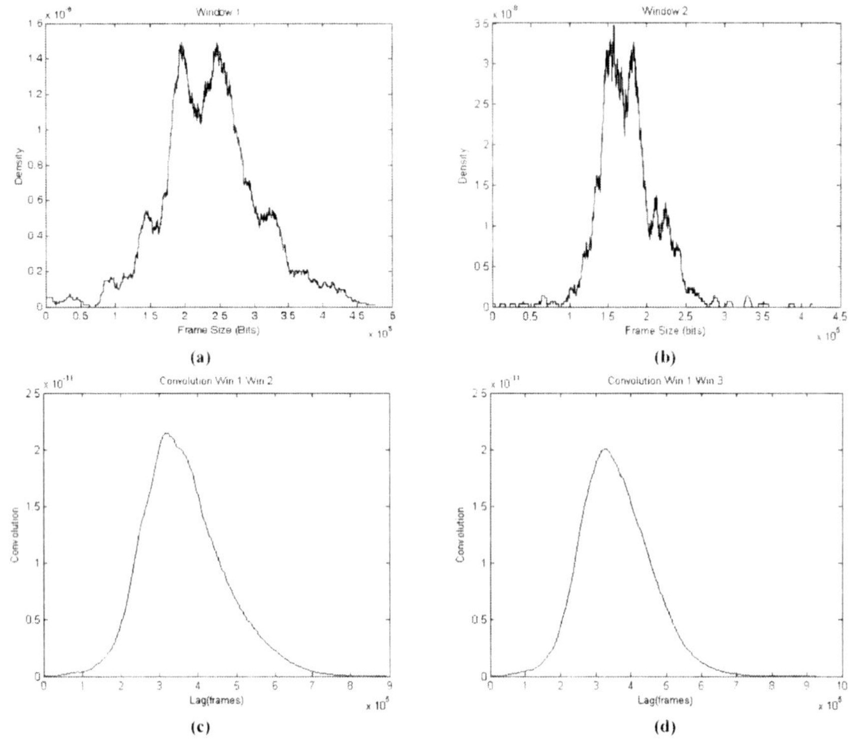

stationary (Skianis, 2003), (Haag, 2002). In order to expand further the second-order stationary (Skianis, 2003), (Haag, 2002), the autocorrelations of these empirical densities were constructed for pairs of time windows, showing almost identical shape across window combinations (figures 3(c), (d)). Therefore, the aforementioned result about stationary is further reinforced.

Figure 4 illustrates the autocorrelation function for the 1100 frames. It can be observed that the autocorrelation graph consist of periodic spikes that are superimposed on a decaying curve. The highest peaks correspond to the autocorrelation of the Intra frames of the video sequence, which are followed by 11 lower spikes before the next "Intra" peak. The lower spikes between two successive "Intra" peaks correspond to P frames, which are typically smaller than the I-frames. Finally, the wells between I and P peaks, correspond to the B frames of the test sequence, which are the smaller frames of all.

Based on the already discussed results, it can be deduced that the behavior of the H.264 encoded signal can be described as a superimposition of three different distributions, which result from three different frames modes (i.e. I/B/P). Therefore, elaborating each frame type separately is more efficient and produces more detailed description of the H.264 video traffic. The next section presents an I/B/P layer analysis of the encoded signal.

I/B/P Level Analysis

For the I/P/B level analysis again the same video segment from the film "Spider-man II" is used as reference signal. In order to study the nature of the video stream, intra-frame period and quantization

Figure 4. The autocorrelation of the 1100 frame sizes

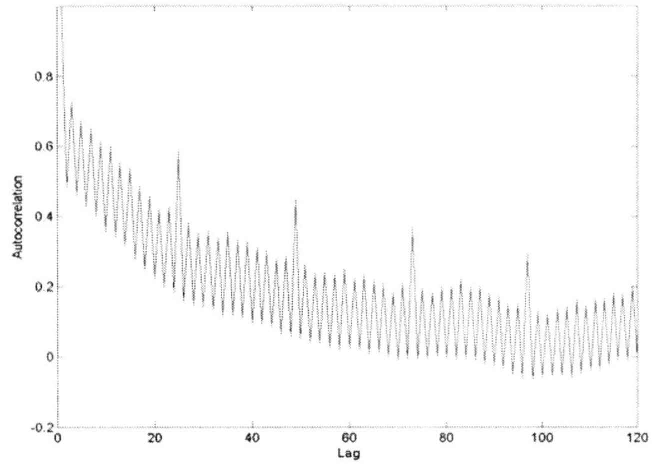

Table 1. Frame statistics overview of the encoded signal in Kbits

Quantization Settings / Frame Types	I Frames (in Kbits)				B Frames (in Kbits)				P Frames (in Kbits)			
	Mean	σ	min	max	Mean	σ	min	max	Mean	σ	min	max
(10,10,10)-12	354.47	87.29	35.7	610.1	227.65	57.67	6.3	576.5	271.34	65.67	24.2	588.7
(20,20,20)-12	148.16	58.25	2.62	325.8	43.02	33.17	0.26	291.5	67.01	41.98	0.62	294.8
(30,30,30)-12	53.91	25.68	1.70	146.7	7.86	8.71	0.24	109.1	16.33	13.93	0.24	116.9
(20,20,20)-3	147.21	57.32	2.62	325.8	42.86	33.25	0.28	290.6	67.22	42.15	0.51	294.8
(20,20,20)-6	148.46	58.03	2.62	325.8	43.10	33.30	0.26	291.5	67.06	42.08	0.50	294.8

parameters are altered during the experiments. During each encoding process, video traces are captured, containing data on the type and the size of each encoded frame. As a result, frame statistics based on specific quantization scale and encoding settings are derived and depicted in Table 1 in the form of mean values and variances of I/P/B frame sizes. The notation (x,y,z)-I is used for the quantization scales of I,B,P frames and the selected intra-frame period.

From Table 1, it can be derived that higher encoding parameters, which cause coarser encoding quality, result in lower mean frame sizes and variations in comparison with lower quantization parameters, which produce better encoding quality. On the contrary, the alternation of Intra-frame period does not affect frame sizes, which remain practically constant. Table 1 depicts the mean value, the standard deviation and the min and max values of the same experimental sets expressed in Kbits.

Finally, Table 2 contains the variation coefficients for the various quantization schemes, which represent a metric of the variation and the shape of the deduced frame size distribution. The variation coefficient is defined as $S_x / \bar{x}$, where S_x is the standard deviation and $\bar{x}$ the mean value. Their deduced values are consistent with the aforementioned observations.

Table 2. Variation coefficients of the various frame types

Quantization Settings / Frame Types	I Frames Variation Coefficient	B Frames Variation Coefficient	P Frames Variation Coefficient
(10,10,10)-12	0.2463	0.2533	0.2420
(20,20,20)-12	0.3932	0.7710	0.6265
(30,30,30)-12	0.4763	1.1081	0.8530
(20,20,20)-3	0.3894	0.7758	0.6271
(20,20,20)-6	0.3909	0.7726	0.6275

Figure 5. Representative frame size histograms and Gamma models

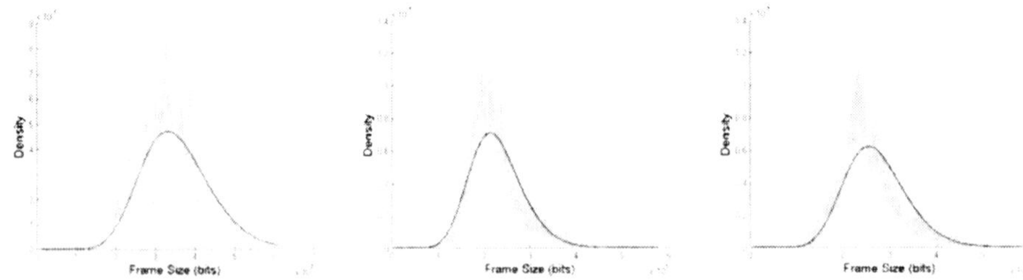

(a) (10-10-10)-12 (I/B/P)

In order to study the statistical behavior of the encoding stream, the Probability Density Functions (PDFs) for each frame type of the encoded signal at various quantization scales are drawn. Figure 5 depicts the representative case of (10-10-10)-12.

Observing that the derived graphs follow the expected bell-like shape, then the well adopted (Skianis, 2003) method of moments is used in order to fit a gamma density (the equivalent of the negative binomial in the continuous domain) to the data output. The Gamma density is given by the following formula

$$f_k = \frac{\left(x/\mu \right)^{p-1}}{\mu\Gamma(p)} e^{-x/\mu}, \ \mu,\ p > 0,\ x \geq 0,$$

$$\Gamma(p) = \int_0^{\infty} t^{p-1} e^{-t} dt \qquad (1)$$

Where $\mu>0$ is the scale parameter, and $p>0$ the shape parameter respectively.

The Gamma distribution has mean $p\mu$ and variance $p\mu^2$ and by equating to the mean and sample variance, denoted as m and v respectively, it can be deduced that $\mu=v/m$ and $p=m^2/v$. Table 3 contains the Gamma distribution parameters for each quantization scale and Figure 5 illustrates the frame histograms in conjunction with the corresponding Gamma models.

As a next step, the autocorrelation function is derived for each frame type. Three representative graphs for the case of quantization scale 10-10-10 and Intra-frame period equal to 12, appear in figure 6, suggesting that the autocorrelation exhibits a reduced decay rate beyond the initial lags.

It is observed that in the case of H.264 streams, the autocorrelation functions follow the same decaying shape, as in previously studied encoding formats, i.e. H.261 (Skianis, 2003) and MPEG-1(Doulamis, 2000). In principle, this phenomenon can be successfully captured by a weighted sum of two geometric terms

Table 3. Gamma model statistics overview of the encoded signals

Quantization Settings / Frame Types	I Frames		B Frames		P Frames	
	p	μ	p	μ	p	μ
(10,10,10)-12	16.487	21499	15.584	14608	17.071	15895
(20,20,20)-12	6.468	22905	1.682	25572	2.549	26287
(30,30,30)-12	4.406	12235	0.813	7857	1.376	11874
(20,20,20)-3	6.597	22316	1.661	25808	2.542	26446
(20,20,20)-6	6.545	22683	1.676	25720	2.540	26404

Figure 6. Representative Autocorrelation Graphs for each scaling case of I/B/P frames

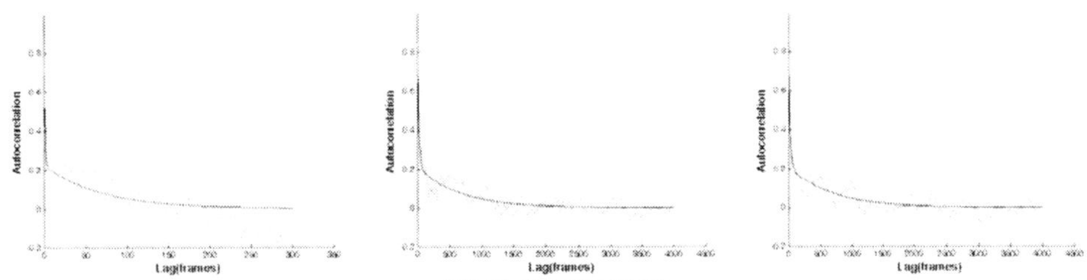

(10-10-10)-12 (I/B/P)

$$\rho_k = w\lambda_1^k + (1-w)\lambda_2^k \text{ with } |\lambda_2|<|\lambda_1|<1 \quad (2)$$

where $\lambda_{1,2}$ are the decay rates, which can be complex in general. Moreover, equation 2 corresponds to the general form of the autocorrelation function for AR(2) models (Cox, 1965), (Heyman, 1992) which have been observed to match well some aspects of video conferencing traffic.

In order to further investigate the appropriateness of equation 2, the corresponding relevant parameters were estimated (in the domain of real-valued decay rates) through a least-squares fit to the autocorrelation samples for the first 300 lags for the case of I frames and the first 4000 lags for the case of B and P frames, respectively. Numerical results appear in Table 4, while the graphs of the fitted models are compared to the sample autocorrelations in figure 6, demonstrating a satisfactory fit between the experimental and the modeling functions. Furthermore, in all cases the matched model represented in 2 captures the long-term trends of the autocorrelation decay.

Since the aforementioned model of autocorrelation functions approximates successfully the long term decay rate, which is the most important for queuing design, there is no point for further pursuing the issue towards more complex models.

Having performed I/P/B level analysis for various encoding schemes of H.264 video traffic and showing satisfactory matching between the experimental data and the corresponding quantitative mathematical data, a video traffic modeling of H.264 unconstraint video traffic can be performed.

VIDEO TRAFFIC MODELING

Related Work

During the last years, video traffic modeling has been an active research area, where several efficient models for VBR traffic have been proposed in the literature. Broadly speaking, these models

Table 4. The parameters of the autocorrelation model

Autocorrelation Type / Quantization Scale	I	B	P
10-10-10-12	w = 0.2248 λ_1=0.9855 λ_2=0.3884	w = 0.2132 λ_1=0.9985 λ_2=0.9526	w = 0.2014 λ_1=0.9985 λ_2=0.9467
20-20-20-12	w = 0.2472 λ_1=0.9828 λ_2=0.7216	w = 0.1859 λ_1=0.9987 λ_2=0.9658	w = 0.1712 λ_1=0.9987 λ_2=0.9594
30-30-30-12	w = 0.2011 λ_1=0.9819 λ_2=0.7417	w = 0.2650 λ_1=0.9981 λ_2=0.9457	w = 0.1923 λ_1=0.9984 λ_2=0.9465
20-20-20-3	w = 0.1931 λ_1=0.9973 λ_2=0.9093	w = 0.1686 λ_1=0.9980 λ_2=0.9683	w = 0.1590 λ_1=0.9982 λ_2=0.9482
20-20-20-6	w = 0.2124 λ_1=0.9942 λ_2=0.8377	w = 0.1746 λ_1=0.9988 λ_2=0.9672	w = 0.1648 λ_1=0.9986 λ_2=0.9566

can be classified into the following categories (Alheraish, 2004):

i. Auto-Regressive Models

The Discrete Auto Regressive(1)–DAR(1) model is the most primitive model that was proposed for modeling of video conference traffic. In (Maglaris, 1988), such a model is tested for 10-second long sequences based on the AR process:

$$x(n) = a_1 x(n-1) + e(n) \tag{3}$$

where x(n) is the bit rate of the coded video during the nth frame, e(n) is a Gaussian process with variance σ^2 – usually described as residuals- and α_1 is the autocorrelation coefficient at lag-1, when the sequence is considered as stationary. Moreover, this model can be considered as a continuous-state, discrete-time Markov process, which is a special case of the general form

$$x(n) = \sum_{i=1}^{p} a_i x(n-i) + e(n) \tag{4}$$

where p is the order of the AR process. In this statement, the residuals *e(n)* are assumed uncorrelated and normally distributed, while Xu *et al* proposed a Gamma distribution residual process instead of a normal distribution in the AR process, analyzing a first-order Gamma AR (GAR) model of the form:

$$\phi_e(s) = a^3 + 3a^3(1-a)\frac{\beta}{\beta+2} + 3a(i-a)^2\left(\frac{\beta}{\beta+s}\right)^2 + (1-a)^3\left(\frac{\beta}{\beta+s}\right)^3 \tag{5}$$

where $\phi_e(s)$ is the Laplace transform of the residual process, α is the AR coefficient and β is the scale parameter. Although the GAR model outperformed in some cases the DAR(1), the fact that the residual generation is too complicated if the order of the AR is increased discourages its use and practical implementation.

Moreover, the Gamma Beta AR (GBAR) process has been proposed in (McKenzie, 1986), where it is assumed that the AR coefficients A_n are Beta distributed and the residuals B_n are Gamma distributed. More specifically, it is proposed that

$$X_n = A_n X_{n-1} + B_n \quad (6)$$

Although the GBAR model is shown to be more efficient than GAR model, it is not suitable for studying admission control algorithms particularly in ATM networks.

Finally, a general AR model has been described by (Zhang, 1999), where firstly the model decomposes the given Gamma process into a weighted sum of a number of $x^2(1)$ sequences. Afterwards, each sequence element is obtained by squaring a Gaussian process, which is efficiently generated by using an AR model from the given covariance matrix.

ii. Markov Models

Furthermore, in the literature many Markov chains based models have been presented, which use states for representing bit-rate range. These models demonstrate that the current state of a Markov process depends only on its previous state, and not on any additional previous states. This can be briefly described by the Markovian property: a stochastic process X_k with state space S={1,2,3,…} is Markovian if for every n and all states i_1, i_2,… satisfies the Markov property

$$P[X_n \mid X_{n-1} = i_{n-1}, X_{n-2} = i_{n-2}, \ldots, X_1 = i_2] = P[X_n = i_n \mid X_{n-1} = i_{n-1}] \quad (7)$$

A MC based model has been proposed in (Sen, 1989), which can also capture multiple video activity levels and scene changes. The transition matrix $\boldsymbol{P}$ of this Markov chain is given by

$$\boldsymbol{P} = \rho \boldsymbol{I} + (1-\rho)\boldsymbol{Q} \quad (8)$$

Where ρ is the autocorrelation coefficient, $\boldsymbol{I}$ is the identity matrix and each row of $\boldsymbol{Q}$ consists of the negative binomial probabilities. The frame sizes remain constant during a scene but vary from one scene to another according to negative-binomial probabilities.

iii. Long Range Dependent Models

Other works, in an attempt to track more closely the complexities observed in the long range autocorrelation decay, employ more elaborate means such as the autoregressive moving-average process (ARMA), which adds a moving average giving

$$x(n) = a_0 + \sum_{i=1}^{p} a_i x(n-i) + \sum_{j=0}^{p} b_j e(n) \quad (9)$$

where p is the order of the AR part and q is the order of moving average part. Such a model was described in (Grunenfelder, 1991).

iv. Hybrid Models that Combine both Markov and Regression Models

Also, more complicated models that combine both Gaussian AR(1) models with parameters modulated according to the transitions of a MC have been proposed in the literature. (Ramamurthy, 1990), (Shim, 1994), (Yegenoglu, 1993).

In comparison with the aforementioned models and methods, this article will introduce a novel model that exploits three discrete DAR(1) processes for the generation of each I, B and P stream in conjunction with a Markov based decision mechanism for bit rate range and scene change simulation. Therefore, due to its novelty and hybrid nature, the proposed model is called by the authors as Markov Modified Model.

THE PROPOSED H.264 MARKOV MODIFIED MODEL

We discussed in the previous sections the statistical properties of I, B and P frame types. This detailed traffic study and mathematical quantitative approach of H.264 VBR video is necessary for understanding the properties of its characteristics, which will be used for generating synthetic H.264 traffic by an appropriate video model of H.264 Unconstraint VBR traffic. Thus, in order the proposed model to be efficient, it must fulfill the following criteria (Alheraish, 2004):

- It must adopt specific statistical characteristics of the real video traffic, like PDF and ACF.
- The characteristics of the synthetic video model must be similar to the ones of the real video, so that it can be used instead of real video traffic for predicting desired performance metrics. This means that the proposed model must simulate successfully the long range dependencies (LRD) and short range dependencies (SRD) of the real video traffic.
- It must be as simple as possible and able to generate synthetic video traffic with low computational complexity.
- It must generate efficient synthetic video traffic for a wide range of video sources ranging from static video content up to very active.

According to the aforementioned criteria and based on the statistical analysis and quantitative mathematical approach of section 3, it is possible to generate sequences of stationary discrete random variables with Markov properties using the discrete autoregressive process of order 1, DAR(1).

The Markov properties are reflected by the fact that the distribution of F_n, of the size of the nth frame, only depends on F_{n-1}. Such process is specified by the stationary marginal distribution of F_n and several other parameters which, independently of the marginal distribution, determine the correlation structure of the sequence.

According to the DAR(1) model, the first order autoregressive form is given by

$$F_n = V_n F_{n-1} + (1 - V_n) Y_n \; for \; n=1,2,... \qquad (10)$$

where $\{V_n\}$ are independent and identically distributed (iid) binary random variables with $P(V_n = 1) = 1 - P(V_n = 0) = \alpha \; with \; 0 \leq \alpha < 1$ and $\{Y_n\}$ are iid random variables with a marginal distribution π.

In other words the model defines the current observation to be a mixture of two independent random variables: It is either the last observation with probability α, or another independent sample from the same distribution. It is a very simple and general model since π is the distribution of any random variable and the correlation structure is independent of π. The autocorrelation function (ACF) of $\{F_n\}$ as defined by (3) is given by $\rho_F(k) = \alpha^k, k = 0, 1, ...$ and $\{F_n\}$ is a Markov chain with transition probability matrix given by $aI + (1 - a)Q$, where I is the identity matrix and Q is a matrix of whose rows are the distribution π.

This approach is used independently for the study of the I, B and P frames trying to capture the fact that a video sequence consists of various scenes with different spatial and temporal activity levels and thus a candidate video traffic model should capture this *inter* and *intra* scene state. More specifically, during a scene, the sizes of the same frame type remain typically constant, while on the contrary follow different sizes over scenes changes.

In this respect, two discrete processes for frame generation are considered:

- The first process simulates the intra-scene state, which means that the frame size, referring to frames of the same type, retains the characteristics of the previous frame.
- The second process, which models the inter-scene state, the frame size is generated using a AR(1) process of the form $x(n) = a_l x(n-1) + e(n)$, based on the size of the previous frame size, where α_1 is the au-

Table 5. Correlation coefficient of intra/inter frames

Quantization Scale	I-P Correlation Coefficient	I-B Correlation Coefficient	P-B Correlation Coefficient
(10-10-10)-12	0.6412	0.1750	0.1721
(20-20-20)-12	0.6250	0.0653	0.0948
(30-30-30)-12	0.5180	-0.0381	0.0916

tocorrelation parameter at lag-1 and e(n) a residual following the normal distribution.

Especially for the case of I frames, which are strongly responsible for the *inter*-GOP correlation of the video stream, which is well characterized by the ACF of the I-frames, the first process is considered as $I(n)=\alpha_1 I(n-1)$, where α_1 is the autocorrelation coefficient at lag-1, in order to better capture this phenomenon.

Although *inter*-GOP correlation, described by the ACF of the I frames, is an important measure, another aspect of video traffic is the correlation between I/P/B frames within the same GOP (the *intra*-GOP correlation). If uncorrelated *I*, *P* and *B* component generation is considered, then the estimated aggregate sequence will not represent successfully the burstness of the real data, resulting in poor estimation of the network resources and poor simulation of the *intra*-GOP dependence. So, a candidate model must adapt the correlation properties between the various frame types within every GOP.

For this reason, in our case the correlation coefficients are calculated between the first neighboring *I-P, I-B* and *P-B* frames of each GOP structure. Table 5 shows the corresponding results for various encoding schemes, showing that there is high correlation between the neighboring first *I-P* frames of each GOP and lower between the *P-B*, and even lower between *I-B*.

Figure 7 depicts the proposed algorithm for the generation of synthetic H.264 traffic. The proposed model, except from the use of the afore-

Figure 7. Block diagram of the proposed model

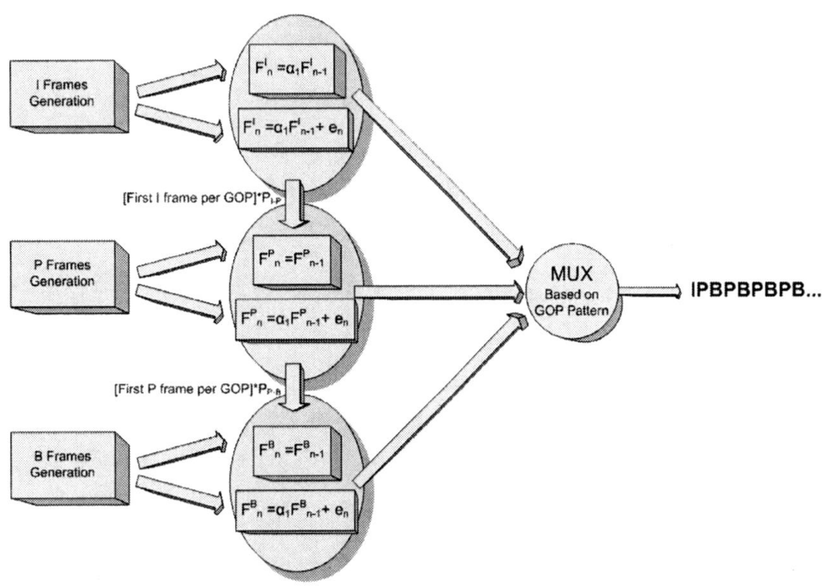

mentioned DAR(1) models, it also exploits the *I-P* correlation coefficient and the *P-B* correlation coefficient for the generation of P and B streams respectively. More specifically, based on the GOP length and structure that the generated traffic has, the proposed algorithm, during a GOP initialization, propagates the first I and P frame size, multiplied by the corresponding correlation coefficient of Table 5, as feed to the respective P and B DAR processes. By this way, it manages to captures the *intra*-GOP dependences of the actual traffic, which is an important issue for the modeling of real traffic video characteristics.

In the proposed model, the MUX component of the described block diagram is responsible for the appropriate multiplexing of I, B and P streams according to a specific GOP pattern.

For the case of (20-20-20)-12, various synthetic traffics of 3 min duration each were generated by the proposed model and compared against various actual samples. Finally the corresponding Q-Q plots are depicted in Figure 8, reporting the good behavior of the proposed model.

Finally, Figure 9 shows the ACF of the actual I-frame sizes and that of synthetic traffic. As observed, the proposed model captures satisfactorily the *inter*-coded correlation of the actual H.264 traffic both at the cases of LRD and SRD.

Therefore, it has been shown that the latest H.264 video encoding standard can be satisfactorily modeled by adapting already known video traffic techniques on the characteristics of the new standard. This is expected because the synthesis of the statistical distribution of H.264 traffic remains similar to previous standards. More specifically, H.264 encoded traffic remains a superimposition of three different distributions that come from three different frames modes (i.e. I/B/P), exactly like earlier standards. Therefore, elaborating on this triple model separately makes possible the use of already proposed models and techniques, like the DAR(1) model and Markov

Figure 8. Q-Q plots of real and generated H.264 traffic

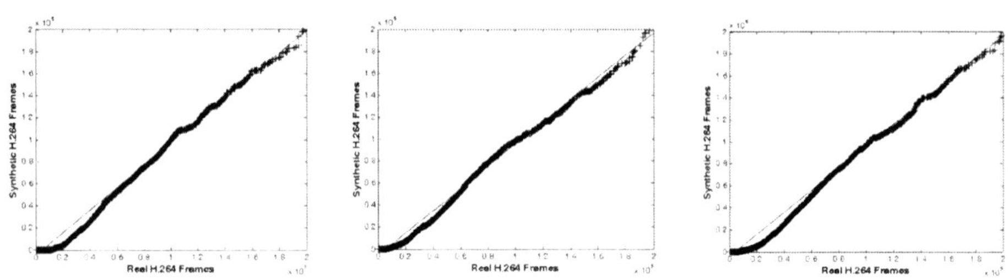

Figure 9. ACF of actual I-frame sizes and that of synthetic for the case of LRD and SRD

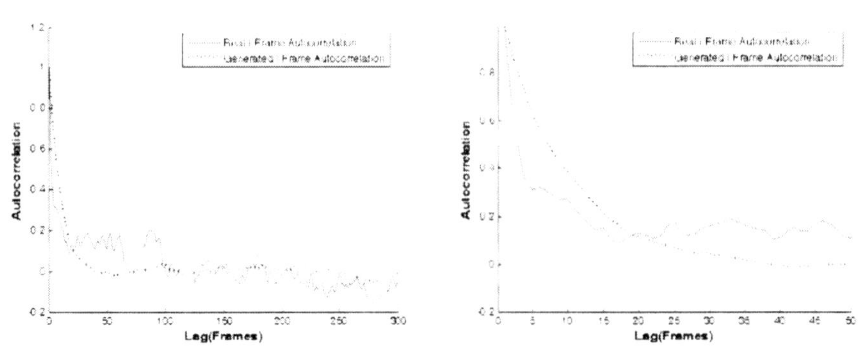

chains. All the new features of the H.264 standard aim at increasing compression efficiency in conjunction with improvement of the deduced perceived quality level for a specific encoding bit rate. Thus, H.264 by featuring variable block size support down to 4x4, quarter-sample motion vector accuracy, extended reference frame selection for P frames, de-blocking filter within the motion-compensated prediction loop and new context-based adapted entropy coding methods: CAVLC and CABAC, it succeeds better compression efficiency in comparison to the previous standards, without altering the distribution characteristics of the generated stream and the encoding scheme, which remains similar to the previous MPEG standards.

CONCLUSION

This article reports on an experimental study of H.264 encoded video streams where additional statistical analysis established general results about the video traffic. The experiments covered cases with different quantization scales and GOP lengths, showing that the derived data can be expressed as superimposition of three discrete frame contributions. In this respect, the density functions of the I/B/P frame sizes were derived and it was shown that they can be successfully represented by Gamma distributions. Moreover, I/B/P autocorrelation functions were drawn, showing that they exhibit an exponentially decaying shape. Finally, a novel GOP adaptive model for generating and simulating H.264 VBR traffic is presented and evaluated by comparing real and generated video traffic.

ACKNOWLEDGMENT

Part of this work was undertaken in the context of the ICT project ADAMANTIUM FP7-ICT/214751 project, which is partially funded by the Commission of the European Union.

REFERENCES

Alheraish, A. (2004). Autoregressive video conference models. *International Journal of Network Management*, *14*, 329–337. doi:10.1002/nem.528

Chin, H. S., Goodge, J. W., Griffiths, R., & Parish, D. J. (1989). Statistics of video signals for viewphone-type pictures. *IEEE Journal on Selected Areas in Communications*, *7*(5), 826–832. doi:10.1109/49.32346

Cohen, D. M., & Heyman, D. P. (1993). Performance modeling of video teleconferencing in ATM networks. *IEEE Transactions on Circuits and Systems for Video Technology*, *3*(6), 408–422. doi:10.1109/76.260197

Cox, D. R., & Miller, H. D. (1965). *The Theory of Stochastic Processes*. London, UK: Chapman & Hall.

Doulamis, N. D., Doulamis, A. D., Konstantoulakis, G. E., & Stassinopoulos, G. I. (2000). Efficient Modeling of VBR MPEG-1 Coded Video Sources. *IEEE Transactions on Circuits and Systems for Video Technology*, *10*(1), 93–112. doi:10.1109/76.825864

Girod, B., Steinbach, E., & Faerber, N. (1997). Performance of the H.263 Video Compression Standard. *Journal of VLSI Signal Processing Systems for Signal, Image and Video Technology. Special issue on recent development in video: Algorithms* [Springer Editions.]. *Implementation and Applications*, *17*, 101–111.

Grunenfelder, R., Cosmas, J. P., Manthorpe, S., & Odinma-Okafor, A. (1991). Characterization of Video Codecs as Autoregressive Moving Average processes and Related Queuing System Performance. *IEEE Journal on Selected Areas in Communications*, *9*(3), 284–293. doi:10.1109/49.76626

Haag, M. (2002). *Stationary and Nonstationary Random Processes, Version 2.1*. Online available at http://cnx.rice.edu/content/m10684/latest/.

Haskell, B. G. (1972). Buffer and channel sharing by several interframe picturephone coders. *The Bell System Technical Journal, 51*(1), 261–289.

Heyman, D. P., Tabatabai, A., & Lakshman, T. V. (1992). Statistical analysis and simulation study of video teleconference traffic in ATM networks. *IEEE Transactions on Circuits and Systems for Video Technology, 2*(1), 49–59. doi:10.1109/76.134371

Lai, Y. L., Tseng, Y. Y., Lin, C. W., Zhou, Z., & Sun, M. T. (2005). H.264 Encoder Speed-UP via Joint Algorithm/Code-Level Optimization. *Proceedings of VCIP 2005*, Beijing, China.

Lucantoni, D.M., Neuts, M.F., & Reibman, A.R. (1994). Methods for performance evaluation of VBR video traffic models. *IEEE/ACM Transactions on Networking, 2*(2), 176–180.

Maglaris, B., Anastassiou, D., Sen, P., Karlsson, G., & Robbins, J. D. (1988). Performance models of statistical multiplexing in packet video communications. *IEEE Transactions on Communications, 36*(7), 834–843. doi:10.1109/26.2812

McKenzie, E. (1986). Autoregressive moving average processes with negative binomial and geometri mariginal distributions. *Advances in Applied Probability, 18*, 679–705. doi:10.2307/1427183

Nomura, M., Fujii, T., & Ohta, N. (1989). Basic characteristics of variable rate video coding in ATM environment. *IEEE Journal on Selected Areas in Communications, 7*(5), 752–760. doi:10.1109/49.32338

Ramamurthy, G., & Sengupta, B. (1990). Modeling and analysis of a variable bit rate video multiplexer. *Proceedings of the 7th Internat. Teletraffic Congress Seminar*, Morristown, NJ.

Rodriguez-Dagnino, R. M., Khansari, M. R. K., & Leon-Garcia, A. (1991). Prediction of bit rate sequences of encoded video signals. *IEEE Journal on Selected Areas in Communications, 9*(3), 305–314. doi:10.1109/49.76628

Sen, P., Maglaris, B., Rikli, N. E., & Anastasiou, D. (1989). Models for packet switching of variable bit rate video in ATM networks. *IEEE Journal on Selected Areas in Communications, 7*(5), 865–869. doi:10.1109/49.32350

Shim, C., Ryoo, I., Lee, J., & Lee, S. (1994). Modeling and call admission control algorithm of variable bit rate video in ATM Networks. *IEEE Journal on Selected Areas in Communications, 12*(2), 332–344. doi:10.1109/49.272884

Skianis, C., Kontovasilis, K., Drigas, A., & Moatsos, M. (2003). Measurement and Statistical Analysis of Asymmetric Multipoint Videoconference Traffic in IP Networks. *Kluwer. Telecommunication Systems, 23*, 95–122. doi:10.1023/A:1023620903222

Steinbach, E., Faerber, N., & Girod, B. (1997). Standard Compatible Extension of H.263 for robust video transmission in Mobile environments. *IEEE Transactions on Circuits and Systems for Video Technology, 7*(6), 872–881. doi:10.1109/76.644067

Su, Y., & Sun, M. T. (2005). Encoder Optimization for H.264/AVC Fidelity Range Extensions. *Proceedings of VCIP 2005*, Beijing, China.

Sullivan, G. (2005). The H.264/MPEG-4 AVC vdeo coding standard and its deployment status. *Proceedings of VCIP 2005*, Beijing, China.

Sullivan, G. J., Topiwala, P., & Luthra, A. (2004). The H.264/AVC Advanced Video Coding Standard: Overview and Introduction to the Fidelity Range Extensions. *SPIE Annual Conference on Applications of Digital Image Processing XXVII, Special Session on Advances in the New Emerging Standard H.264/AVC* (pp. 454-474).

Verbiest, W., Pinnoo, L., & Voeten, B. (1988). The impact of the ATM concept on video coding. *IEEE Journal on Selected Areas in Communications*, *6*(9), 1623–1632. doi:10.1109/49.12890

Wiegand, T., Sullivan, G., Bjontegaard, G., & Luthra, A. (2003). Overview of the H.264/AVC Video Coding Standard. *IEEE Transactions on Circuits and Systems for Video Technology*, Special Issue in H.264.

Xu, S., & Hung, Z. (1996). A gamma autoregressive video model on ATM networks. *IEEE Transactions on Circuits and Systems for Video Technology*, *8*(4), 138–142.

Yegenoglu, F., Jabbari, B., & Zhang, Y.-Q. (1993). Motion-classified autoregressive modeling of variable bit rate video. *IEEE Transactions on Circuits and Systems for Video Technology*, *3*(1), 42–53. doi:10.1109/76.180689

Zhang, Q. T. (1999). A general AR-based technique for the generation of arbitrary gamma VBR video traffic in ATM networks. *IEEE Transactions on Circuits and Systems for Video Technology*, *9*(10), 1130–1137. doi:10.1109/76.795063

This work was previously published in International Journal of Mobile Computing and Multimedia Communications (IJMCMC) 1(4), edited by Ismail Khalil & Edgar Weippl, pp. 14-31, copyright 2009 by IGI Publishing (an imprint of IGI Global).

Chapter 17
Speaker Discrimination on Broadcast News and Telephonic Calls Based on New Fusion Techniques

Halim Sayoud
USTHB University, Algeria

Siham Ouamour
USTHB University, Algeria

ABSTRACT

This chapter describes a new Speaker Discrimination System (SDS), which is a part of an overall project called Audio Documents Indexing based on a Speaker Discrimination System (ADISDS). Speaker discrimination consists in checking whether two speech segments come from the same speaker or not. This research domain presents an important field in biometry, since the voice remains an important feature used at distance (via telephone). However, although some discriminative classifiers do exist nowadays, their performances are not enough sufficient for short speech segments. This issue led us to propose an efficient fusion between such classifiers in order to enhance the discriminative performance. This fusion is obtained, by using three different techniques: a serial fusion, parallel fusion and serial-parallel fusion. Also, two classifiers have been chosen for the evaluation: a mono-gaussian statistical classifier and a Multi Layer Perceptron (MLP). Several experiments of speaker discrimination are conducted on different databases: Hub4 Broadcast-News and telephonic calls. Results show that the fusion has efficiently improved the scores obtained by each approach alone. So, for instance, the authors got an Equal Error Rate (EER) of about 7% on a subset of Hub4 Broadcast-News database, with short segments of 4 seconds, and an EER of about 4% on telephonic speech, with medium segments of 10 seconds.

DOI: 10.4018/978-1-60960-563-6.ch017

INTRODUCTION

While fingerprints (Youssif, Chowdhury, Ray & Nafaa, 2007) and retinal scans (Daugman, 2007) are more reliable means of authentication, speech can be seen as a non-evasive biometric that can be collected with or without the person's knowledge or even transmitted over long distances via telephone. Furthermore, a person's voice cannot be stolen, forgotten, guessed, given to another person or lost. Then, voice based speaker discrimination represents a secure and efficient way in biometry.

Speaker discrimination (Koreman, Wu, & Morris, 2007) consists in checking whether two different pronunciations (speech segments) belong to the same speaker or not. One means used to compare these utterances is to extract the vocal characteristics from each speech signal, in order to detect the degree of similarity between them. Speaker discrimination has several applications such as: speaker verification, audio signal segmentation and speaker based clustering.

In this domain, several classifiers do exist but most of them do not accept very short speech segments as required in some applications (e.g. audio stream segmentation). That is why; we have proposed some techniques of fusion between those classifiers. The principal goal of this investigation is to develop a fusion-based speaker discrimination system for the task of speaker changes detection in multi-speaker audio streams (i.e. SDS system). A second goal will concern the application of this discriminative system in the task of audio document indexing (i.e. ADISDS system). In fact, audio documents indexing represents respectively the process of speech segmentation, which divides the audio flow into homogeneous segments (each segment contains only one speaker) and the process of clustering, which gathers all the speech segments belonging to the same speaker together. Thus, the SDS system will be used firstly for detecting the speaker changes in the audio stream and secondly for gathering all the segments of a same speaker, in order to obtain the overall intervention of each speaker, at the end of the process.

In this research work, we are interested in investigating two classifiers:

- The Statistical classifier based on mono-gaussian model and employing a symmetric measure of similarity (third section);
- The Multi-Layer Perceptron (MLP) using a new characteristic called "RSC" (fourth section), which is developed in order to reduce the neural network input size, minimize the training database and optimize its convergence;

Although it does exist many other classifiers giving high performances in this domain, a lot of them require long speech segments during either the training or the testing step, which makes them not suitable for discrimination applications using short segments, like in speech segmentation (Meignier, 2002).

The particular choice of the statistical mono-gaussian classifier is due to its easy implementation, low cost in computation, and good discrimination on short speech segments: even with only 2 seconds (Sayoud, Ouamour & Boudraa, 2003) unlike the Gaussian Mixture Models (GMM) for example, which require long speech duration at least for one segment (the model). However, the MLP has been chosen for its high discriminative performance, as it is the case for most of the Neural Networks (NNs). That is why we have decided to choose both of these two classifiers.

In order to enhance the discrimination accuracy, we try to combine those two classifiers by a fusion technique. Theoretically, the fusion should get the advantages of the statistical classifier and the advantages of the neural classifier, corresponding respectively to a high resolution (working on short speech segment) and a high discriminative capacity, which is well adapted in speech indexing (Meignier, 2002; Sayoud, Ouamour & Boudraa, 2003).

In this way, we have developed three methods of fusion, which we called: serial fusion, parallel fusion and serial-parallel fusion. These fusion techniques are evaluated on the following databases: sub-sets of Hub4 Broadcast-News database (Woodland, 1997) and sub-sets of telephonic speech database. Results show that the fusion gives the best discrimination scores comparatively to the simple classifiers alone, especially the parallel fusion and the serial-parallel one.

The different sections of this chapter are presented as follows: In the second section, we define, in details, the speaker discrimination. The third and fourth sections describe the two simple classifiers (statistical and neural classifiers). A new relativity notion is introduced in the fourth section, for the task of speaker characterization, and which is used for the optimization of the NN. The description of the different fusion architectures is provided in the fifth section and the experimental results are detailed in the sixth section. Finally, an overall conclusion is given at the end of this chapter.

SPEAKER DISCRIMINATION

Speaker Recognition, Applications and Some Problems

Speaker recognition is the ability to recognize who is speaking, by using the vocal characteristics of his or her speech signal (Bonastre, 2005). Speaker recognition is divided into several specialties: speaker identification, speaker verification, speaker indexing and speaker discrimination.

- Speaker identification is the ability to identify the identity of a speaker among others;
- Speaker verification is the process of accepting or rejecting the identity claim of a speaker;
- Speaker indexing consists in segmenting and labeling a multi speaker audio document into homogenous segments containing only one speaker each;

Speaker discrimination is the ability to recognize whether two utterances come from the same or different speakers. This field is an important component of segmenting an audio stream into meaningful subunits, because the location of the speaker changes is crucial for dialogue understanding. Speaker discrimination is also related to speaker verification, but this last process is based on prior knowledge about a limited number of speaker identities, whereas in speaker discrimination, only knowledge about the speech signal is provided.

Speaker recognition has several practical applications (Bonastre, 2005) as:

Banking transactions by telephone, database access services, voice mail, biometric secure access, forensic applications, etc....

The problems encountered in speaker recognition are usually due to the intra-speaker variability of the speech, effect of noise and reduction of the spectral bandwidth as for the telephonic speech: [300-3400Hz]. These problems led us to choose two types of speech databases during our experiments, namely Hub4 Broadcast-News for the corrupted speech and telephonic calls for the reduced spectral bandwidth, in order to evaluate our approaches.

Overview of Some Techniques Related to Speaker Discrimination and Parameterization

Several techniques were tested so far, for the task of speaker discrimination, or more precisely for the task of speaker verification:

In the past, different recognition models were proposed: Dynamic Time Warping (DTW) (Furui, 1981-b), Hidden Markov Models (HMM) (Tishby, 1991), Vector Quantization (VQ) (Furui, 1991) and Gaussian Mixture Models (GMM) (Miyajima, Hattori, Tokuda, Masuko, Kobayashi & Kitamura, 2001; Morris, Wu & Koreman, 2005). Let us briefly summarize and criticize some of them:

DTW technique is effective in text-dependent speaker recognition, but not suitable for text-independent (Furui, 1981-b);

- HMM models are widely used in speech recognition, and are also commonly used in text-dependent speaker verification. Furui and Matsui (Matsui & Furui, 1993) proposed some methods to design speaker dependent phoneme models for speaker verification. Those methods are based on either phoneme adaptation or speaker adaptation of an initial set of phoneme HMMs; in 1995, Flavigna (Flavigna, 1995) proposed three different speaker verification methods. All of them are based on Hidden Markov Models (HMM); the first one is of type text independent, the two others are of type text prompted, and he proved, on a subset of 138 speakers extracted from APASCI database, that the text prompted methods provide widely better performance than text independent ones. This result shows once again that the HMM approach is not well adapted for text-independent applications.
- Concerning the VQ technique, it has been shown that the method is effective for speaker recognition, although the performance of VQ is not as good as that of the recent classifiers like GMM (Reynolds 1995).
- GMM (Reynolds, 1995) provides a probabilistic model of the underlying sounds of a person's voice. It is computationally more efficient than HMM and has been widely used in text-independent speaker recognition (Chen, Hsieh & Lai, 2004). In fact during the last several years, Gaussian Mixture Models (GMMs) became very popular in Speaker Recognition systems and have proven to perform very well for clean wideband speech. However, in noisy environment or for noisy band-limited telephone speech the performance degrades considerably. Another difficulty is that GMMs require at least several minutes of training speech, which is not suitable for real world applications. On the other hand, despite certain disadvantages they present mainly at the training stage, the NN based classifiers, show better performance for telephone speech and need less training data than the GMM- based ones (Viv, Ortega- Garcia, Alonso & Moro, 2001; Ganchev, Tsopanoglou, Fakotakis & Kokkinakis, 2002).
- Recently, Support Vector Machines (SVM) have become very popular in speaker verification, however the size of training data presents a real disadvantage, because SVM training algorithms scale poorly with the size of the training documents, and the best training algorithms still have a running time proportional to the cube of the size of the training documents (Strathmeyer, 2005).

In this research work, we propose to solve the discrimination problem by using the fusion between a MLP neural network and a mono-gaussian statistical classifiers. The NN classifier has been chosen for its good discriminative performance and because it needs few training data (when using a special reduced characteristic). In addition, the mono-gaussian statistical classifier has been chosen because it allows a high resolution (works with very short speech segments). This proposal is suitable to our main application: *discrimination system for audio stream segmentation.*

Hence, the different techniques proposed in this chapter are summarized as follows:

- Multi Layer Perceptron, using a new relative characteristic (RSC) as input;
- Mono-gaussian statistical measure;
- Serial fusion technique, resulting from the two previous classifiers, by using the scores of the statistical classifier as an additional input at the MLP, in order to enhance the NN training;
- Parallel fusion technique, which combines the statistical classifier scores and the neural classifier scores, by summing them with some weighting confidence coefficients. The values of the confidence coefficients depend on the importance of each technique in the fusion process;
- Serial-Parallel technique, which considers the serial fusion as a third classifier, and sums its scores with the other classifier scores (statistical and neural) by using a weighted sum and a parallel architecture.

Concerning the parameterization, we use 37 Mel Frequency Spectral Coefficients (MFSC) obtained from the calculation of the log-energies in the Mel spectral scale (Lee & TSOI, 1995; Sayoud, Ouamour & Boudraa, 2003) for each speech segment of 32 ms (ensuring the stationarity of the signal). This dimension of 37 MFSC has been chosen after a thorough investigation on the optimal spectral resolution (Sayoud & Ouamour, 2000).

Statistical Classifier

The mono-gaussian statistical method is based on some measures of similarity, which are called Second Order Statistical Measures (SOSM). These measures are used in order to recognize the speaker at each segment of the speech signal.

We recall below the most important properties of this approach (Gish, 1990; Bimbot, Magrin-Chagnolleau & Mathan, 1995; Bonastre & Besacier, 1997).

Let $\{x_t\}_{1 \le t \le M}$ be a sequence of M vectors resulting from the P-dimensional acoustic analysis of a speech signal uttered by speaker $\boldsymbol{x}$. These vectors are summarized by the mean vector x_m and the covariance matrix X:

$$x_m = \frac{1}{M} \sum_{t=1}^{M} x_t \tag{1}$$

and

$$X = \frac{1}{M} \sum_{t=1}^{M} (x_t - x_m)(x_t - x_m)^T \tag{2}$$

Similarly, for a speech signal uttered by speaker $\boldsymbol{y}$, a sequence of N vectors $\{y_t\}_{1 \le t \le N}$ can be extracted.

By assuming that all acoustic vectors extracted from the speech signal uttered by speaker $\boldsymbol{x}$ are distributed like a gaussian function, the likelihood of a single vector y_t uttered by speaker $\boldsymbol{y}$ is:

$$G(y_t / \boldsymbol{x}) = \frac{1}{(2\pi)^{P/2} (\det X)^{1/2}} e^{-(1/2)(y_t - x_m)^T X^{-1} (y_t - x_m)} \tag{3}$$

If we assume that all vectors y_t are independent observations, the average log-likelihood of $\{y_t\}_{1 \le t \le N}$ can be written as

$$G_m \boldsymbol{x}(y_1^N) = \frac{1}{N} \log G(y_1 \ldots y_N | \boldsymbol{x}) = \frac{1}{N} \sum_{t-1}^{N} \log G(y_t | \boldsymbol{x}) \tag{4}$$

By replacing $y_t - x_m$ by $y_t - y_m + y_m - x_m$ and using the property

$$\frac{1}{N} \sum_{t-1}^{N} \left[(y_t - y_m)^T X^{-1} (y_t - y_m) \right] = tr\left(YX^{-1}\right) \tag{5}$$

we get:

$$\frac{2}{P}G_{m^{\mathbf{x}}}(y_1^N)+\log 2\pi+\frac{1}{P}\log\left(\det Y\right)+1= \frac{1}{P}\left[\log(\frac{\det(Y)}{\det(X)})-tr(YX^{-1})-(y_m-x_m)^T X^{-1}(y_m-x_m)\right]+1 \quad (6)$$

Then, the Gaussian likelihood measure μ_G is defined by:

$$\mu_G(\mathbf{x},\mathbf{y})= \frac{1}{P}\left[-\log(\frac{\det(Y)}{\det(X)})+tr(YX^{-1})+(y_m-x_m)^T X^{-1}(y_m-x_m)\right]-1 \quad (7)$$

since we have:

$$\underset{\mathbf{x}}{Arg\max}\, G_{m^{\mathbf{x}}}(y_1^N)=\underset{\mathbf{x}}{Arg\min}\,\mu_G(\mathbf{x},\mathbf{y}) \quad (8)$$

A variant of this measure (Bimbot, Magrin-Chagnolleau & Mathan, 1995) called μ_{Gc} is deduced from the Formula (7) by subtracting the term depending on mean vectors:

$$\mu_{Gc}(\mathbf{x},\mathbf{y})=\psi(\mathbf{x},\mathbf{y})= \mu_G(\mathbf{x},\mathbf{y})-\frac{1}{P}(y_m-x_m)^T X^{-1}(y_m-x_m)= \frac{1}{P}\left[-\log(\frac{\det(Y)}{\det(X)})+tr(YX^{-1})\right]-1 \quad (9)$$

this measure is non-symmetric, which means:

$$\mu_{Gc}(\mathbf{x},\mathbf{y})\neq\mu_{Gc}(\mathbf{y},\mathbf{x}) \quad (10)$$

One possibility for symmetrizing this measure is to construct the average between this measure and its dual term:

$$\mu_{Gc0.5}(\mathbf{x},\mathbf{y})=\frac{1}{2}\mu_{Gc}(\mathbf{x},\mathbf{y})+\frac{1}{2}\mu_{Gc}(\mathbf{y},\mathbf{x})=\mu_{Gc0.5}(\mathbf{y},\mathbf{x}) \quad (11)$$

For reason of simplification, we will denote this measure by μ_{Gc}.

NEURAL CLASSIFIER

Knowing the high discriminative capacities of neural networks (Bennani, 1992; Sayoud, 2003) and their good performance in telephonic speech, we opted for the use of a MLP neural network with two hidden layers and with only one output (representing the discrimination decision).

The training set is composed of speech segments of 3 or 4 seconds, extracted from Hub4 Broadcast-News database in the first experiment, and it is composed of speech segments of 10 seconds, extracted from a real telephonic database, in the second experiment. The training is performed by the back-propagation algorithm. The NN output (NN_{output}) will give then an indication on the correlation between the two speech segments presented at its input.

If $NN_{output}=0$, then the speech segments belong to the same speaker,

If $NN_{output}=1$, then the speech segments belong to different speakers.

This neural network (Sayoud, 2003) is used to discriminate the different speakers thanks to their speech features (Figure 1). For this goal, we used a reduced input vector (see the fourth section) that is extracted from the MFSC coefficients.

Notion of RSC

Most of existing systems, in speaker recognition, used to model the speaker by his statistical components, like: the mean vector or the covariance matrix, which are extracted from the acoustic features as: the MFSC, Linear Frequency Cepstral Coefficients (LFCC), Mel Frequency Cepstral Coefficients (MFCC), Auto Regressive (AR) coefficients and their derivatives. Thus, in order to model a speaker, we need only his own speech

Figure 1. Principle of the RSC-NN based discrimination system

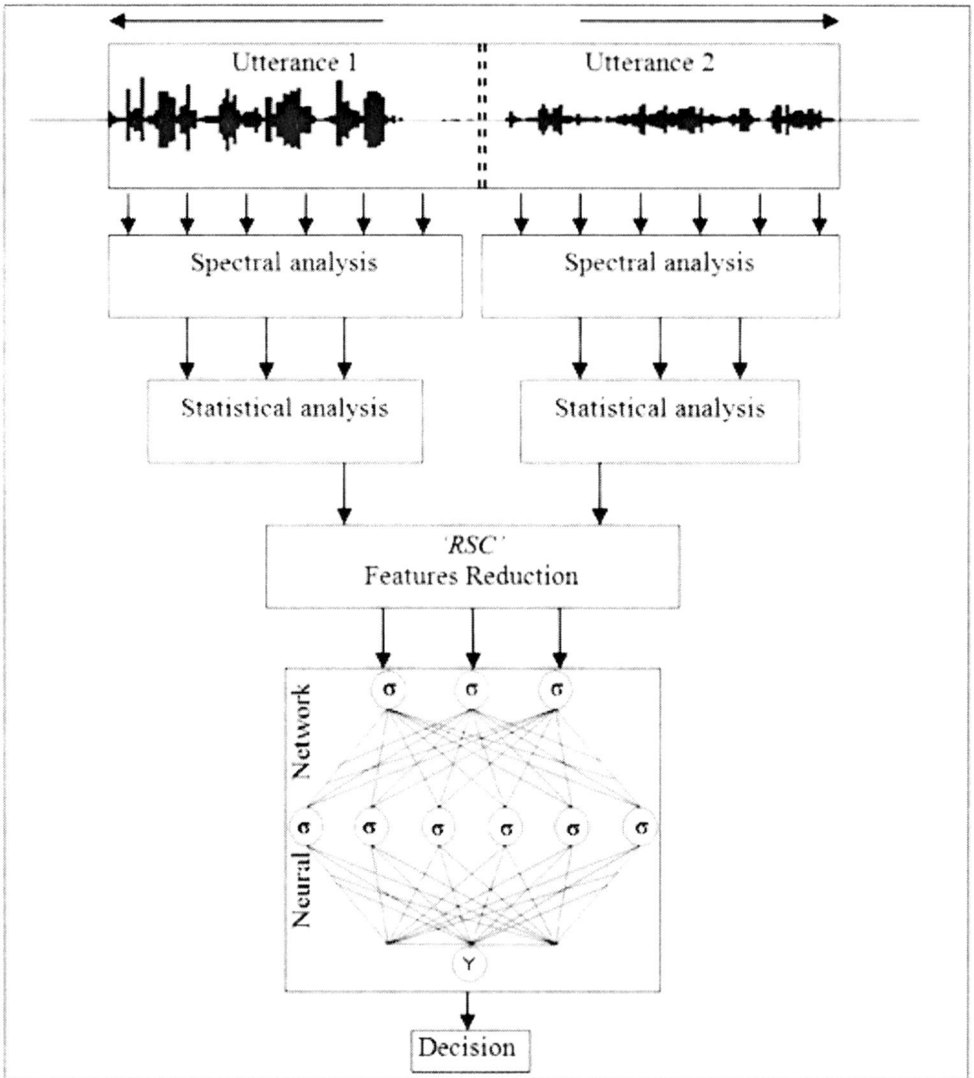

signal. However, in speaker discrimination, we are always in presence of two segments (usually short) to compare. These segments may belong to a same speaker or to two different speakers. For this reason, we thought to model each speaker relatively to another one. Thus, we have defined a new characteristic used at the MLP input, which allows us to improve the performance of the NN and which we called: Relative Speaker Characteristic (RSC).

Furthermore, by using this new characteristic, we reduce the size of the NN input, speed up the learning step and then provide a well-adapted feature. A description of this new characteristic is given here below.

Given two speech segments, to compare, belonging to speakers x and y. Let us assume that X and Y are their respective covariance matrices and x_m and y_m their mean vectors. The similarity measure between them is given by formula 9 (see section 3).

$$\psi(\mathbf{x},\mathbf{y}) = \frac{1}{P}\left[-\log(\frac{\det(Y)}{\det(X)}) + tr(YX^{-1})\right] - 1 \quad (12)$$

But, since

$$\frac{\det(Y)}{\det(X)} = \det(Y\ /\ X) \quad (13)$$

Then

$$\psi(\mathbf{x},\mathbf{y}) = \frac{1}{P}\left[-\log(\det(Y/X) + tr(Y/X))\right] - 1 \quad (14)$$

and if we introduce the relative R ratio by:

$$\mathrm{R}(x, y) = \frac{\mathrm{X}}{\mathrm{Y}} = \mathrm{X}{*}\mathrm{Y}^{-1}$$

Then

$$\psi(\mathbf{x},\mathbf{y}) = \frac{1}{P}\left[-\log(\det(\Re(y,x)) + tr(\Re(y,x)))\right] - 1 \quad (15)$$

So, $\psi(\mathbf{x},\mathbf{y})$ is a function of the relative ratio R(y, x).

We have called the ratio R: Relative Speaker Characteristic (RSC). The acronym DRSC will denote the diagonal of the RSC matrix:

$$\mathrm{DRSC}(x, y) = \mathrm{diag}(\mathrm{R}(x, y)) \quad (16)$$

The new relative characteristic DRSC (*Diagonal of the RSC*), computed in this way, contains a lot of information normally able to discriminate between two different utterances. Experiments done on speech signals extracted from Hub4 Broadcast-News showed the good performance of the MLP neural network using this characteristic, comparatively to other results obtained on the same corpus and using other characteristics namely: the mean vector, the diagonal of the covariance and the first two eigen vectors of the covariance (see Table 1).

Table 1. Equal Error Rates (EERs) obtained for different characteristics on a subset of Hub4 Broadcast-News. The EER definition is given in the sixth section.

Characteristic	EER %
First 2 eigen vectors of the covariance	33.67
Mean vector	25.19
Diagonal of the covariance	13.9
DRSC (*Diagonal of the RSC*)	7.2

Neural Network Inputs

Neural networks must have at their inputs a number of receptive cells equal to the dimension of the training example vector (Master, 1992; Sayoud, 2003). Thus, in case of using a features vector with P coefficients (Sayoud, 2003; Lee & Tsoi, 1995) representing one speech segment (of one speaker), the number of the NN input cells should be 2P, because we have two different speech segments to compare (two speakers to compare).

For instance, if we use 37 acoustic features (in our case: 37 MFSC) for each speaker, we should need 37x37 (= 1369) components in the covariance matrix to model each speech segment, the number of NN inputs becomes too great: 2x1369 (= 2738) for each input example. Whereas if we use the DRSC we only need a vector of 2x37 (= 74) elements at the input of the NN, which represents 2.7% (74 / 2738) of the total size of the previous memory, without decreasing the discrimination accuracy of the NN. We can see in Table 1 the performances got with different reduced input features.

FUSION

In order to enhance the discrimination performance, we thought to use several classifiers, which are combined in order to get a lower discrimination error: this combination is called Fusion.

The fusion in the broad sense can be performed at different hierarchical levels or processing stages. A very commonly encountered taxonomy of data fusion is given by the following three-stage hierarchy (Dasarathy, 1994; Verlinde, 1999; Jain, Ross & Prabhakar, 2004):

1. *Feature level* where the feature sets of different modalities are combined. Fusion at this level provides the highest flexibility but classification problems may arise due to the large dimension of the combined (concatenated) feature vectors.
2. *Score (matching) level* is the most common level where the fusion takes place. The scores of the classifiers are usually normalized and then they are combined in a consistent manner.
3. *Decision level* where the outputs of the classifiers establish the decision via techniques such as majority voting. Fusion at the decision level is considered to be rigid for information integration (Stylianou, 2005).

Combining the partial decisions from the different classifiers (Dasarathy, 1994; Verlinde, 1999) in a score or decision fusion level, the fusion model could be done using one of the three following basic architectures:

- *Serial fusion architecture,* which consists of a set of different classifiers whose outputs are combined in series or tandem;
- *Parallel fusion architecture,* which consists of a set of different classifiers that are interrogated in parallel. The outputs derived from these classifiers are combined in parallel by the fusion module.
- *Serial-parallel fusion architecture,* which consists of a combination of the two previous types.

In the present chapter, we are interested in some types of fusion resulting from the two individual classifiers: the statistical and neural ones.

Serial Fusion

Since it has been proved that NNs has an excellent discriminative property (Bennani & Gallinari, 1995), we thought to mix the statistical measure with the neural inputs in order to improve the NN performance. Thus, we have added to the NN a new input, in which we injected the discrimination result given by the statistical measure (μ_{Gc}) for each couple of speech segments. This new measure is injected to the NN input together with the DRSC characteristic, and we launch the training of the MLP neural network with this new mixed input vector (μ_{Gc} score and DRSC features), as shown in Figure 2.

The serial fusion is performed as follows: First, the features are extracted from the two speech segments to compare, and then the statistical measure μ_{Gc} and the DRSC vector are computed and injected to the input of the MLP classifier together. The training is then enhanced by the information brought by the statistical approach (μ_{Gc} score).

Parallel Fusion

The parallel fusion is done at the score level by using a weighted sum of the scores (Kittler, 2005) obtained by each individual classifier alone. In fact, each classifier is applied to the couple of speech signals to compare, in order to have an individual

Figure 2. Serial fusion between the statistical and neural classifiers

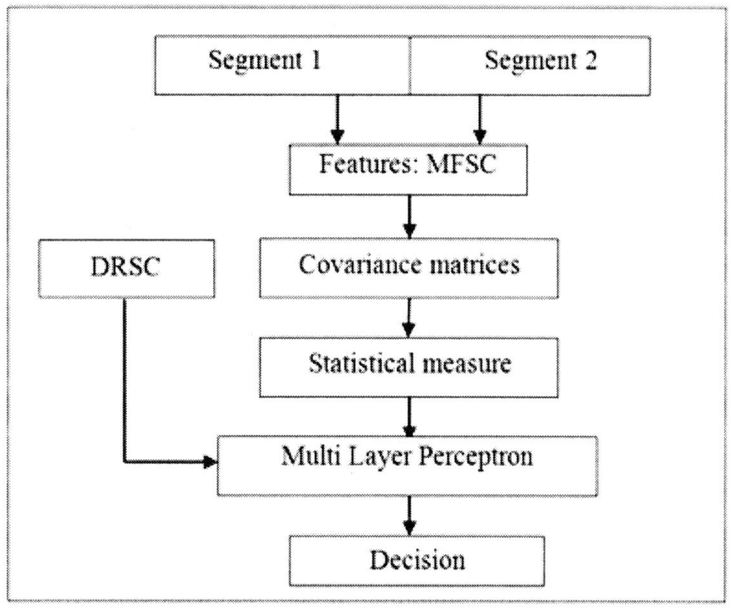

discrimination decision (or score). After that, the different scores are combined to give a final decision. This combination consists in using a special weighted sum, as described below:

If the individual scores are denoted by S_j, then the fusion score Sf is given by:

$$Sf = \sum_{j=1}^{N} C_j S_j \tag{17}$$

where C_j is the weighing coefficient for the classifier "j" and N denotes the number of classifiers. With

$$\sum_{j=1}^{N} C_j = 1 \text{ and } C_j \in]0, 1[\tag{18}$$

In practice, C_j represents the relevance of the classifier j, and the more this classifier is accurate, the more its weighing coefficient is high.

In Figure 3, we describe the principle of the weighted sum used to combine the statistical classifier with the neural one. The outputs of the statistical and neural classifiers are represented respectively by score 1 and score 2. These last ones are multiplied respectively by the weights C_1 and C_2, and then the sum of the weighted scores provides the final result of the parallel fusion.

Serial-Parallel Fusion

In the serial-parallel fusion, we consider the serial fusion (see the fifth section) as a third classifier which provides another partial decision (serial score). Hence, with another parallel fusion, we combine all the different scores: statistical score, neural score and serial-fusion one, in order to obtain the combined parallel decision as shown in Figure 4.

The principle of the proposed serial-parallel fusion can be summarized as follows (see Figure 4).

Firstly, each classifier is applied to get a discrimination decision between the two speech segments to compare; we get then two decisions represented by score 1 for the statistical classifier and score 2 for the neural classifier. Secondly,

Figure 3. Parallel fusion (at the score level) of the two classifiers

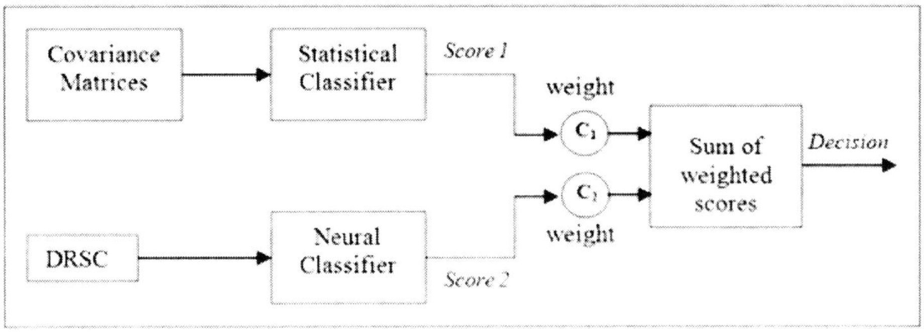

Figure 4. Serial-Parallel fusion

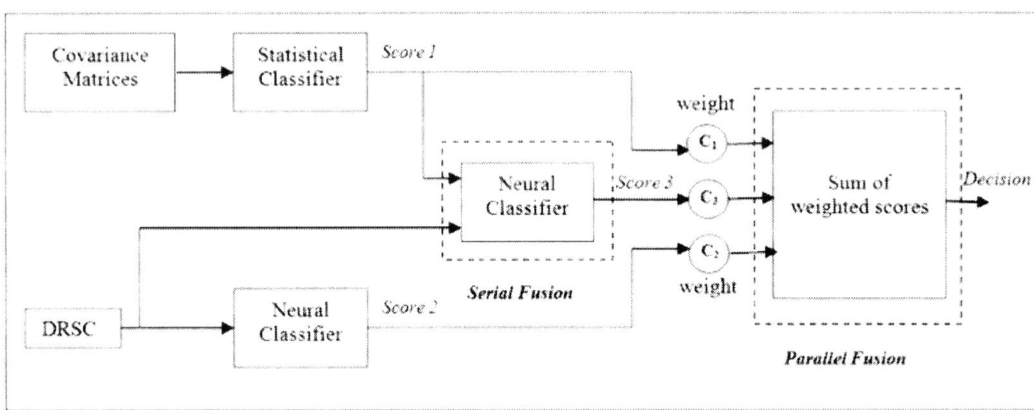

the statistical measure is injected with the DRSC vector to the MLP input in order to obtain the serial fusion decision (section 5.1) represented by score 3. Thirdly, the different scores: score 1, score 2 and score 3 are multiplied by their corresponding weights represented respectively by the coefficients C_1, C_2 and C_3. And finally, the weighted scores are summed to provide a final discrimination decision.

EXPERIMENTS

Databases

The first part of the experiment concerns a comparative evaluation of speaker discrimination with the two individual classifiers. The second one concerns a comparative evaluation of speaker discrimination with the different fusion techniques, seen in section 5.

The experiments of speaker discrimination are conducted on four databases, as described below:

1. Two sub-sets (DB1 and DB2) of "Hub4 Broadcast-News 96" database, containing some recordings from the "CNN early edition" and composed of clean speech, music, telephonic calls, noises, etc. The sampling frequency is 16 kHz. The speech signals are extracted and arranged into segments of about 4 seconds each.
2. Two other sub-sets (TB1 and TB2) containing some real telephonic recordings with a

sampling frequency of 8 kHz. The duration of each speech segment is about 10 seconds.

In all the databases, the testing examples are different from the training ones. In addition, the experiments are done with two speech types: The first type is extracted from Hub4 Broadcast-News, which has a large bandwidth of [0-8000Hz] with sometimes some sequences of advertisement, music or noise. The second type is collected from real telephonic conversations, which has a reduced bandwidth of [300-3400Hz]. Usually (not always) speaker recognition is more difficult in telephony, due to the limited bandwidth. However, the presence of noise and music in Hub4 Broadcast-News make the discrimination task rather difficult in this case.

The databases are organized into speaker combinations, namely: pairs of two speech segments to discriminate. The sizes of the different databases are indicated below:

- DB1 contains 14 different speakers (most of them journalists, speaking about the news) organized into 259 speaker combinations for the training and 195 speaker combinations for the test.
- DB2 contains 14 different speakers (most of them journalists, speaking about the news) organized into 518 combinations for the training and 390 combinations for the test.
- TB1 contains 24 different speakers: 12 males and 12 females (speaking by telephone about different topics), organized into 670 speaker combinations for the training and 334 speaker combinations for the test.
- TB2 contains 24 different speakers: 12 males and 12 females (speaking by telephone about different topics), organized into 1340 combinations for the training and 668 combinations for the test.

RESULTS AND COMPARISON

Herein we describe the different experiments of speaker discrimination (quoted in the previous sub-section) and show their results.

In order to evaluate the different classifiers during the comparative experiments, we use some common error rates for the performance evaluation. Their definitions are given here below:

- **False Alarms (FA):** represents the errors in case the discrimination system decides that the two speech signals (to compare) do not belong to the same speaker, whereas they really come from the same one.
- **Missed Detections (MD):** represents the errors in case the system cannot detect the difference between two speech signals belonging to two different speakers.
- **Equal Error Rate (EER):** represents the error of speaker discrimination when the FA ratio is equal to the MD ratio. Then the EER is equal to both FA and MD.

Thus, in each experiment we compute the corresponding error rates, as described previously, and represent them graphically to get a clear information about the discrimination precision of every method.

The overall results are represented in Table 2, Figure 5 and Figure 6. The figures represent the different Receiver Operating Characteristic (ROC) curves (namely: FA versus MD) for the two individual classifiers: statistical classifier (μ_{Gc} measure) and MLP, and the three different fusion methods: serial, parallel and serial-parallel fusion.

Table 2 shows the different Equal Error Rates in % obtained on the different databases.

Concerning the simple classifiers, we can observe the following points:

- The comparison between the 2 simple classifiers: statistical and Neural, shows that the EERs obtained by the MLP on the dif-

Table 2. Equal Error Rates obtained on the different databases: DB1, DB2, TB1 and TB2.

	EER % in Hub4 Broadcast-News with segments of 4 seconds		EER % in real Telephonic talks with segments of 10 seconds	
Classifier	**DB1**	**DB2**	**TB1**	**TB2**
Mono-gaussian classifier	11.75	11.75	5.74	5.74
MLP-DRSC	7.20	9.25	5.02	3.83
Serial Fusion	7.70	9.95	4.65	*3.28*
Parallel Fusion	*6.77*	*7.88*	*4.29*	3.65
Serial-Parallel Fusion	*6.77*	*7.88*	*4.29*	*3.28*

Figure 5. ROC curves of the two simple classifiers and the three fusion techniques on DB2

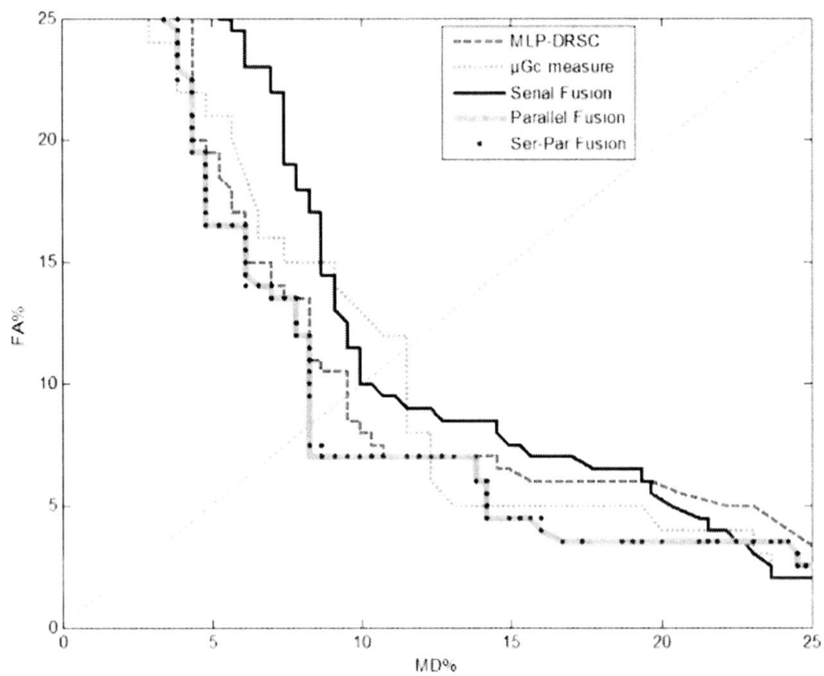

ferent databases: 7.2% on DB1, 9.25% on DB2, 5.02% on TB1 and 3.83% on TB2, are better than their corresponding errors given by the statistical classifier: 11.75% on DB1, 11.75% on DB2, 5.74% on TB1 and 5.74% on TB2. See Table 2.

- The NN appears better than the statistical method in the medium area of the ROC curve, whereas the statistical classifier appears better at the borders (Figure 5). This means that the MLP is more accurate than the statistical classifier for applications requiring close values of FA and MD. However, in high secure applications (where the MD is close to 0%) the statistical classifier will perform better. Therefore, a fusion between them may get the advantages of the two classifiers without decreasing the precision.

Concerning the different techniques of fusion, we can notice the following points:

Figure 6. ROC curves of the two simple classifiers and the three fusion techniques on TB2

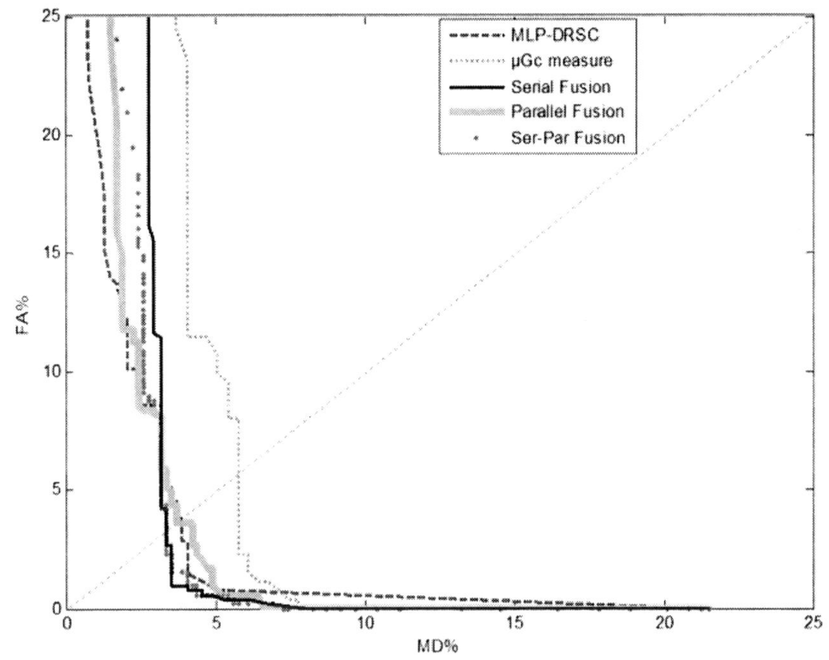

The serial fusion does not improve the results comparatively to the simple classifiers on DB1 and DB2 (7.70% on DB1 and 9.95% on DB2), but shows a good improvement on the telephonic databases: TB1 and TB2 (4.65% on TB1 and 3.28% on TB2). On the other hand, the parallel fusion, done at the score level, enhances the performances on all the databases: 6.77% on DB1, 7.88 on DB2, 4.29% on TB1 and 3.65% on TB2 (see Table 2).

The best results are given by the serial-parallel fusion as defined in this chapter (fifth section). These results are equivalent to those given by the parallel fusion on DB1, DB2 and TB1, and are equivalent to those obtained by the serial fusion on TB2. This means that the serial-parallel fusion gets the best scores given by both the serial and parallel fusion (see Figure 5, Figure 6 and Table 2).

The comparison between the different fusion architectures shows that the parallel fusion between the two simple classifiers is better than the serial fusion on the three databases: DB1, DB2 and TB1, whereas on TB2 the serial fusion is better. On the telephonic database, the serial fusion appears interesting, but in Broadcast- News speech (with short segments) the parallel fusion is better. In the whole, the serial-parallel fusion ensures the best EERs on all databases tested in this work (see Figure 5, Figure 6 and Table 2) and remains the best fusion technique of this investigation.

Concerning the new relative speaker characterization, introduced in this chapter for modeling the speaker relatively to another one, and which is used as input of the neural classifier, we have shown that it optimizes its training, reduces the size of the training database, and enhances the discriminative precision of the classifier. See Table 1 for a comparison with other speaker features tested in the same conditions. Furthermore, this new characteristic gives to the speaker a flexible model (i.e. the model of the speaker changes when his dual speaker changes).

CONCLUSION

In this research work, we have proposed a new Speaker Discrimination System (SDS), which is a part of an overall project called "Audio Documents Indexing based on a Speaker Discrimination System". As explained in the second section, speaker discrimination consists in checking whether two speech segments come from the same speaker or not. For this purpose, we have employed two discriminative classifiers:

- A Statistical classifier based on mono-gaussian model and employing a symmetric measure of similarity (section 3);
- A Multi-Layer Perceptron using a new characteristic called "DRSC", which is developed in order to reduce the neural network input size, minimize the training database and optimize its convergence.

In order to enhance the discrimination performance, we have proposed several fusion architectures between those classifiers: serial fusion, parallel fusion and serial-parallel fusion, as described in the fifth section.

To evaluate our system, several experiments are done. The first part of the experiments concerns a comparative evaluation of speaker discrimination with the two simple classifiers. The second one concerns a comparative evaluation of speaker discrimination with the different fusion techniques. These experiments are conducted on four databases, mainly: two sub-sets of "Hub4 Broadcast-News 96" database, and two other sub-sets containing real telephonic speech.

Results show several important points: On one hand, the DRSC characteristic (*Diagonal of the RSC*) causes a good enhancement in the MLP performance, comparatively to other features tested in the same conditions. Furthermore, this new characteristic gives to the speaker a flexible model, since this model changes every time his corresponding dual speaker changes. On the other hand, results show that the MLP is more accurate than the statistical classifier, and that the fusion gets a better performance than the individual classifiers.

Concerning the different techniques of fusion, we can notice that the parallel fusion is better than the serial fusion on the three databases: DB1, DB2 and TB1, whereas on TB2 the serial fusion is better. On the telephonic database, the serial fusion appears interesting, but in Broadcast-News speech (with short segments) the parallel fusion is better.

In the whole, the serial-parallel fusion ensures the best performances on the four databases and remains the best fusion technique of this investigation.

DISCUSSION

In fact, this chapter describes the speaker discrimination system used in the global project of speech indexing (speaker diarization). This research domain presents an important field in biometry, since the voice remains an important biometric key used at distance (via telephone). However, although some discriminative classifiers do exist and give good results, their performances are not enough sufficient for short speech segments (ie. less than 10 seconds). This issue led us to look for an efficient fusion between different classifiers in order to enhance the discriminative accuracy. The fusion is applied, by using three different techniques: a serial fusion, parallel fusion and serial-parallel fusion.

Two classifiers have been chosen for the evaluation: a classical statistical classifier and an artificial neural network. Experiments of speaker discrimination are done on different databases: Hub4 Broadcast-News, which contains news, advertisement and debates, and telephonic calls corresponding to discussions between couples of speakers via the telephone line.

We have noticed that the fusion permitted us to improve the performance of speaker discrimination, appreciably. The parallel fusion appears to be better than the serial one, but in the overall the parallel-serial fusion gives the best results at all.

Theoretically this result was expected, since we know that the fusion takes advantage of the combined classifiers, but we wanted to check this rule on the ground and see what are the real performances in real conditions (such as broadcast news or telephonic calls), where usually the discrimination accuracy is not appreciable.

ACKNOWLEDGMENT

The authors wish to thank the team of the LIA Laboratory, especially Pr. Jean Francois Bonastre, and the team of the ENST, especially Pr. Gerard Chollet, for many helpful discussions and assistances throughout this work. They would like to thank the editors who accepted to read and evaluate our book chapter.

REFERENCES

Bennani, Y. (1992). *Approches connexionnistes pour la reconnaissance du locuteur: modélisation et identification*. Phd thesis, Paris XI University, France.

Bennani, Y., & Gallinari, P. (1995). Neural Networks for discrimination and modelization of speakers. *Journal of Speech Communication*, *17*(1-2), 159–175..doi:10.1016/0167-6393(95)00014-F

Bimbot, F., Magrin-Chagnolleau, I., & Mathan, L. (1995). Second-Order Statistical Measures for text-independent Broadcaster Identification. *Journal of Speech Communication*, *17*(1-2), 177–192.. doi:10.1016/0167-6393(95)00013-E

Bonastre, F., & Besacier, L. (1997). *Traitement Indépendant de Sous-bandes Fréquentielles par des méthodes Statistiques du Second Ordre pour la Reconnaissance du Locuteur*. In Proceedings of the 4th « Congrès Français d'Acoustique », Marseille April 14-18, 357-360.

Bonastre, J. F. (2005). *L'authentification biométrique vocal*. Bordeaux, France: Tutorial PARISTIC Bordeaux.

Chen, W. C., Hsieh, C. T., & Lai, E. (2004). Multiband Approach to Robust Text-Independent Speaker Identification. *Journal of Computational Linguistics and Chinese Language Processing*, *9*(2), 63–76.

Dasarathy, B. V. (1994). *Decision Fusion. In proceedings of IEEE Computer Society Press*. CA: Los Alamitos.

Daugman, J. (2007). New Methods in Iris Recognition. *Journal of the IEEE Transactions on Systems. Man. Cybernetics*, *3*(5), 1167–1175. doi:10.1109/TSMCB.2007.903540

Falavigna, D. (1995). *Comparison of Different HMM Based Methods For Speaker Verification*. In proceedings of Eurospeech'95- Fourth European Conference on Speech Communication and Technology, Madrid, *Spain,* September 18-21.

Furui, S. (1981). -a. Cepstral analysis technique for automatic speaker verification. *Journal of IEEE Transactions on Acoustics, Speech, and Signal Processing*, *29*(2), 254–272..doi:10.1109/TASSP.1981.1163530

Furui, S. (1981). -b. Comparison of speaker recognition methods using statistical features and dynamic features. *Journal of IEEE Transactions on Acoustics, Speech, and Signal Processing*, *29*(3), 342–350..doi:10.1109/TASSP.1981.1163605

Furui, S. (1991). Vector-quantization-based speech recognition and speaker recognition techniques. In *Proceedings of Conference Record of the Twenty-Fifth Asilomar Conference on Signals, Systems and Computers*, November 4-6, 954-958.

Ganchev, T., Tsopanoglou, A., Fakotakis, N., & Kokkinakis, G. (2002). Probabilistic Neural Networks Combined with GMMs for Speaker Recognition Over Telephone Channels. *14th International Conference on Digital Signal Processing (DSP2002), 2*, Santorini, Greece, July 1- 3, 1081-1084.

Gish, H. (1990). Robust discrimination in automatic speaker identification. In *Proceedings of IEEE International Conference on Acoustics Speech and Signal Processing, New Mexico,* 289-292.

Jain, A. K., Ross, A., & Prabhakar, S. (2004). An Introduction to Biometric Recognition. *Journal of IEEE Transactions on Circuits and Systems for Video Technology*, *14*(1), 4–20..doi:10.1109/TCSVT.2003.818349

Kittler, J. (2005). *Multiple classifier systems in decision-level fusion of multimodal biometric experts*. 1st BioSecure residential workshop, Paris, France, August 1- 26.

Koreman, J., Wu, D., & Morris, A. C. (2007). Enhancing Speaker Discrimination at the Feature Level. *Journal of Lecture Notes In Artificial Intelligence: Springer Berlin / Heidelberg Volume 4343,* 260-277.

Lee, H. S., & Tsoi, A. C. (1995). Application of multi-layer perceptron in estimating speech / noise characteristics for speech recognition in noisy environment. *Journal of Speech Communication*, *17*(1-2), 59–76..doi:10.1016/0167-6393(95)00018-J

Master, J. (1992). *Pratical Neural Network Recipes in C*. New York: Academic press.

Matsui, T., & Furui, S. (1993). *Concatenated Phoneme Models for Text-Variable Speaker Recognition.* In Proceedings of ICASSP, Vol. 2, Minneapolis, 391-394.

Meignier, S. (2002). *Indexation en locuteurs de documents sonores: Segmentation d'un document et Appariement d'une collection.* PhD thesis, LIA Avignon, France.

Miyajima, C., Hattori, Y., Tokuda, K., Masuko, T., Kobayashi, T., & Kitamura, T. (2001). Text-independent speaker identification using Gaussian mixture models based on multi-space probability distribution. *Journal of IEICE Transactions on Information and Systems. E (Norwalk, Conn.)*, *84-D*(7), 847–855.

Morris, A., Wu, D., & Koreman, J. (2005). GMM based clustering and speaker separability in the Timit speech database. *Journal of IEICE Transactions Fundamentals/ Commun./ Electron./ Inf. & Syst., Vol. E85- A/ B/ C/ D, No. 1.*

Reynolds, D. A. (1995). Speaker identification and verification using Gaussian mixture speaker models. *Journal of Speech Communication*, *17*(1-2), 91–108..doi:10.1016/0167-6393(95)00009-D

Reynolds, D. A., & Rose, R. C. (1995). Robust test-independent speaker identification using Gaussian mixture speaker models. *Journal of IEEE Transactions on Speech and Audio Processing*, *3*(1), 72–83..doi:10.1109/89.365379

Sayoud, H. (2003). *Automatic speaker recognition using neural approaches.* PhD thesis, USTHB University, Algiers, Algeria.

Sayoud, H., & Ouamour, S. (2000). *Reconnaissance Automatique du Locuteur en Milieu Bruité.* In Proceedings of JEP'00, 345-348, Aussois, France.

Sayoud, H., Ouamour, S., & Boudraa, M. (2003). 'ASTRA' An Automatic Speaker Tracking System based on SOSM measures and an Interlaced Indexation. *Journal of Acta Acustica*, *89*(4), 702–710.

Strathmeyer, J. (2005). *Automatic Text Categorization. Internal Project Report, Pittsburgh FLAC_Projects Computer Science Department*. USA: Carnegie Mellon University.

Stylianou, Y., Pantazis, Y., Calderero, F., Larroy, P., Severin, F., Schimke, S., Bonal, R., Matta, F., & Valsamakis, A. (2005). *GMM- Based Multimodal Biometric Verification.* Final Project Report 1, Enterface'05, Mons, Belgium, JULY 18th- August 12th.

Tishby, N. Z. (1991). On the application of mixture AR hidden Markov models to text independent speaker recognition. *Journal of IEEE Transactions on Signal Processing*, *39*(3), 563–570.. doi:10.1109/78.80876

Verlinde, P. (1999). *Contribution à la vérification multimodale d'identité en utilisant la fusion de décisions*. PhD thesis, Ecole Nationale Supérieure des Télécommunications, Paris, France, September 17th.

Vivaracho, C. E. Ortega- Garcia, J., Alonso, L., & Moro, Q. I. (2001). A Comparative Study of MLP- based Artificial Neural Networks in Text- Independent Speaker Verification against GMM- based Systems. In *Proceedings of Eurospeech 2001- Scandinavia, volume 3*, 1753-1756, Aalborg-Denmark, September.

Woodland, P. C., Gales, M. J. F., Pye, D., & Young, S. J. (1997). *The Development of the 1996 HTK broadcast news transcription system*. In: DARPA Speech Recognition Workshop, 97-99.

Youssif, A. A. A., Chowdhury, M. U., Ray, S., & Nafaa, H. Y. (2007). Fingerprint Recognition System Using Hybrid Matching Techniques. In *Proceedings of International Computer and Information Science (ICIS 2007). 6th IEEE/ACIS International Conference,* July 11-13, 234-240.

Section 4
Innovations in Mobile Multimedia Applications and Services

Chapter 18
FCVW:
Experiments in Groupware

Ivan Tomek
Acadia University, Canada

Elhadi Shakshuki
Acadia University, Canada

ABSTRACT

Pervasiveness of Internet and increasing geographical dispersal of work teams result in continuously growing importance of groupware - software that supports collaboration within groups. Numerous applications have been developed to address collaboration needs and many are widely used but don't fully satisfy work team requirements and the Internet potential. This paper surveys several groupware products and describes FCVW (Federated Collaborative Virtual Workspace), an experimental project designed to explore certain groupware aspects that are not sufficiently addressed by existing products.

INTRODUCTION

One of the predominant uses of computer networks since their inception has been support for collaboration and social interaction. Initially, this took the form of remote access to documents, but soon e-mail became one of the most prevalent network applications. Then, applications addressing various aspects of the use of computers in support of social and work-related communication (socialware and groupware) began to appear. From the start it was clear that needs of social and work groups are closely related and that ideas used to support one area have their use in software supporting the other area as well.

A prime example of the overlap of social and recreational applications with groupware is the category of programs devised initially for fantasy games known as MUDs (Multi User Dungeons after the fantasy game of Dungeons and Dragons) (Curtis, P. & Nichols D. A., 1993). The spread of MUDs among recreational users and the recognition of their potential for work-related use lead to a new form of MUDs known as MOOs (MUD

DOI: 10.4018/978-1-60960-563-6.ch018

Object-Oriented) that took advantage of object-oriented technology. The flexibility of its basic concepts provided a new dimension– runtime user programmability - that is one of the essential features of the project described in this paper. Groupware and socialware research and production did not, of course, stop there and many other products based on a variety of different principles have been developed since the inception of MOOs.

Despite intensive research and development, many questions relating to groupware remain unanswered and intensive research is being conducted in the field. In our work, we selected an existing open-source non-commercial groupware product called CVW (Collaborative Virtual Workspace) (CVW Overview, 2007) as a starting point and used it to explore some of the most interesting groupware issues by adding functionality not present in the original. An overview of this exploration is the main subject of this paper. First, however, we present a classification of groupware to provide structure to the large number of existing groupware applications. We also give a brief overview of some of the most important or typical ones. This is followed by a summary of CVW, the basis of our own work, and a topic-by-topic presentation of the extensions that we added to CVW. A concluding section summarizes our work, reviews the lessons that we learned, and outlines our current work and plans.

A BRIEF SURVEY OF GROUPWARE

Applications designed for support of collaboration and social interaction initially focused on tools supporting individual work activities, such as shared access to documents, e-mail, scheduling of meetings, and synchronous and asynchronous communication. Eventually, it became clear that better results can be achieved by integrating tools supporting individual tasks into multifunctional environments. However, work on tools supporting parts of collaboration has not stopped. This is because research often needs to address specific major issues of collaboration, such as mutual awareness of team members, and because new concepts and technologies supporting collaboration, such as wikis (Leuf, B. & Cunningham W., 2001) and blogs (Maybury, M., 2004), emerge independently of groupware development and complement existing collaboration tools.

Groupware research and development intensified when technology made groupware more attractive by providing higher network communication bandwidth, new hardware and software platforms, faster CPUs, larger computer memories, better displays, and larger storage capacities at lower cost. The result of this effort is a long list of products that have been developed and used by researchers, work teams, and loosely connected individuals. Because the number of products is so large, we will begin our survey by offering a categorization of major approaches.

Categorizing Integrated Collaborative Environments

Surveying collaborative environments can be confusing because there are so many of them. To provide a perspective, different approaches are often characterized by the major conceptual principles that they use - their underlying metaphor. Although there is no generally accepted classification, the following is a reasonable scheme even though it does not clearly separate particular groupware products because some environments combine several of these metaphors.

- **Space-Based Environments:** These environments are organized around the principle of a persistent virtual space, an emulation of the real-world physical arrangement of workplaces into interconnected virtual buildings, floors and rooms. In these environments, virtual rooms contain virtual objects (mainly, but not exclusively, documents) and tools, and are inhabited by people represented by avatars (represented graphically or textually) who can navigate from one place to another, move objects around, communicate, and perform other activities associated with collaboration and social interaction. This approach has its origin in MUDs developed in late 1970s and MOOs developed in the 1980s and used mainly for social interaction, work, and in education. CVW (Collaborative Virtual Workspace) is a typical environment of this kind.

- **Document-Based Environments:** Designers of these environments start from the premise that work revolves around documents, such as reports, spreadsheets, diagrams, communication logs, etc. Foldera (2007) is a typical product representative of this metaphor. Its users organize their workspace around projects encompassed by persistent 'activity folders', and all work related to a project originates in them and is directed towards them.
- **Activity - Based Environments:** Environments of this kind are based on the view that the essence of teamwork is a meeting where team members exchange opinions, formulate plans, and make decisions. Users of environments based on this principle virtually meet in transient spaces where they can share access to documents and applications, communicate, and use various tools. Once finished, the meeting usually disappears without leaving a trace. Marratech (2007) is an environment of this kind.

After this general classification, the following section gives brief outlines of selected environments to illustrate how our conceptual classification maps into real products.

More on Selected Environments

To illustrate research and commercial work environments and provide background for the rest of this paper, this section briefly describes selected groupware products. Some are quite recent, many are in use, and others are included mainly for their historical significance.

BSCW (Basic Support for Cooperative Work) (BSCW, 2007) is a product widely used in commercial, research, and educational applications. It uses the concept of a shared workspace with documents shared by a group organized in a hierarchy of folders and made available through a Web browser. Available features include logging and notification of events and processes, document flow folders, search and polling facilities, threaded discussion groups, and activity logging for usage analysis. BSCW is under continuous active development and runs on several platforms including mobile devices. Its relative ease of modification lead, among other things, to the creation of user interfaces in multiple natural languages.

COAST and TUKAN. COAST (Cooperation Application System Technology) (Schuckmann, C., et al., 1996) is not a collaboration product but a framework for building collaborative environments. Its stated goal is to make development of 'synchronous groupware' as easy as development of single-user applications. This is achieved by offering shared base functionality and providing interfaces for application-specific modules that instantiate the framework for a specific groupware use. Facilities provided by the framework include storage of objects such as documents, network connection, transaction management, support for communication, and user interface widgets for groupware use. TUKAN (Schummer, T. & Schummer, J., 1999) is a collaborative environment for software development teams developed from COAST. In addition to COAST facilities, it provides an Interactive Development Environment (IDE) for single users and teams and supports real-time awareness (Figure 1) to prevent con-

Figure 1. TUKAN's cooperation-aware code browser

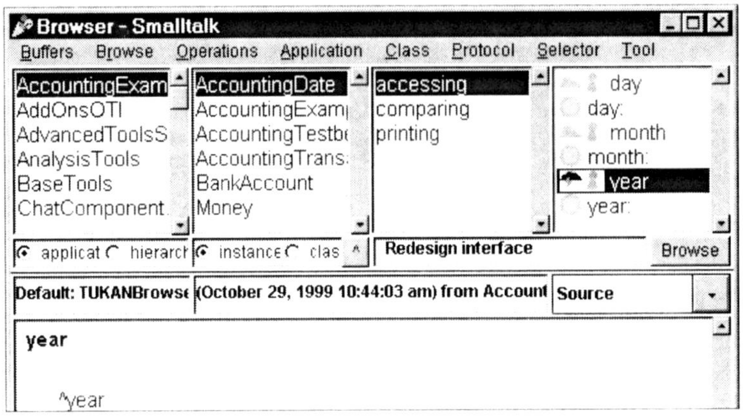

flicting updates of software and suggest possible cooperation among team members.

CURE (Collaborative Universal Remote Education) (Haake, J. M., et al., 2004; CURE 2007) is an open source product for support of distributed groups, mainly in distance education. It uses the room metaphor, uses a Web server, and clients are Web browsers. CURE users can create interconnected rooms and groups of users with permission to access them and add or edit their contents. CURE rooms can be interconnected to create course environments and users may subscribe to notifications of room events. Communication is via text chat, e-mail, and threaded discussion.

Foldera (2007) is a Web browser-based product that uses a central server to store all user data. The basis of its design is the observation that the most natural units of work are projects and that productivity tools such as e-mail and spreadsheets produce much information that requires laborious sorting and filing. Foldera thus centers on a project viewed as a folder containing documents, and assigns an Activity Folder to each project. Groups organize their work around these folders.

Habanero (2007) is a session oriented groupware project. It provides presence and event awareness and allows users to participate in multiple sessions at the same time. Habanero offers a set of tools that cover three dimensions: production, coordination, and communication (Figure 2).

Isabel (De Miguel T., et al., 1994). Isabel is a "session" or "meeting" oriented environment with three conceptual parts: Telepresence via audio and video conferencing, shared workspace, and interaction control. Each mode has a different shared workspace in which users are able to collaborate, using tools such as slides, pointer/pencil, and whiteboard.

Jazz (Cheng, Li-Te, et al., 2003; Jazz, 2007) integrates collaboration tools into IBM's ENVY environment to support small teams of programmers (Envy, 2007) during code development. Its main features include the "Jazz Band", chat, screen sharing, and "Concert Awareness". Jazz team members can use text-based chat and screen sharing for debugging and other activities. Jazz also provides a discussion board for asynchronous communication (Figure 3).

IBM Lotus Notes (Lotus, 2007). Lotus Notes were first released in 1989. Current Lotus Notes 7.0 provide instant messaging, e-mail, calendar, scheduling, discussion forums, a reference database, personal management tools, and other facilities, such as presence awareness and web conferencing. The product also provides a web-based space called "QuickPlace" for team's collaboration.

Figure 2. Habanero – part of user interface

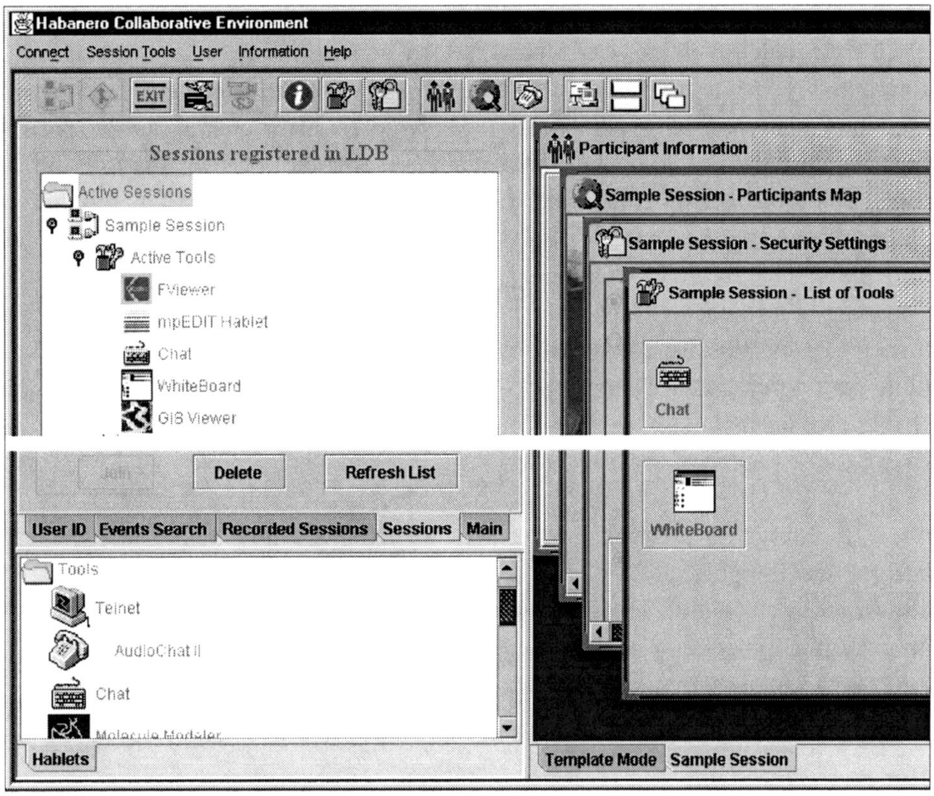

Figure 3. Jazz Band: Envy IDE augmented with team member and other icons

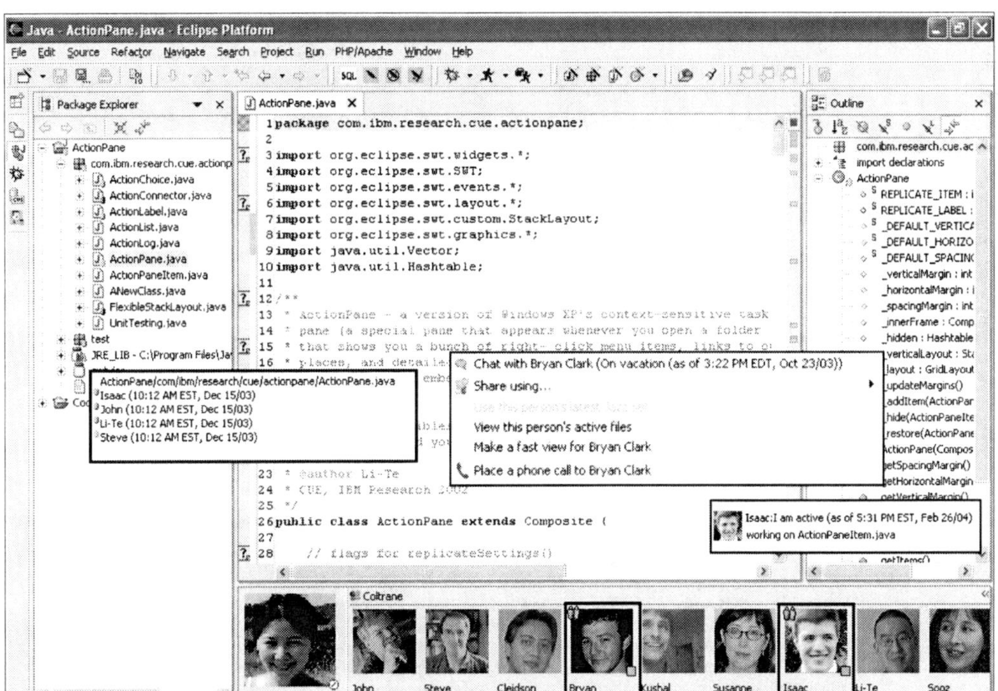

Marratech (2007). Marratech is a commercial product using the meeting metaphor via a 'virtual office' accessible through a Web browser. Users typically continuously run a Marratech client and use it to contact others to arrange personal or team meetings in a virtual office equipped with a variety of tools. The tools include an interactive whiteboard allowing annotation over the displayed background, video- and audio- conferencing, and instant messaging. Marratech users can also share applications, letting them to assume control of applications running on another participant's machine. Meeting media, including voice, video, and whiteboard, allow meetings to be recorded and played back. All functionality is available via a single window (Figure 4). The product works on multiple platforms and is highly scalable because it uses a distributed architecture where clients perform much of the processing and offload the server. As most collaborative applications, Marratech is a black-box product that does not allow customization, adaptation, or extension by its users.

Mushroom (Kindberg, T., 1996) was a research project aimed at creating a workspace and platform for collaboration over the Internet. The product is a workspace oriented, persistent collaboration platform. Figure 5 shows a "root place" for user "moonshake", containing documents and Mushlets - applications that use Mushroom platform. There are three types of spaces: place, work area (a place nested within another place), and Rubbish area containing deleted documents and Mushlets. Users can create their own Mushlets through the Mushlet API and built-in Mushlets in Mushroom Platform include the following:

- **MushMail:** a mail application for sending text message.
- **TypeTalk:** a chat application.

Figure 4. Marratech virtual office

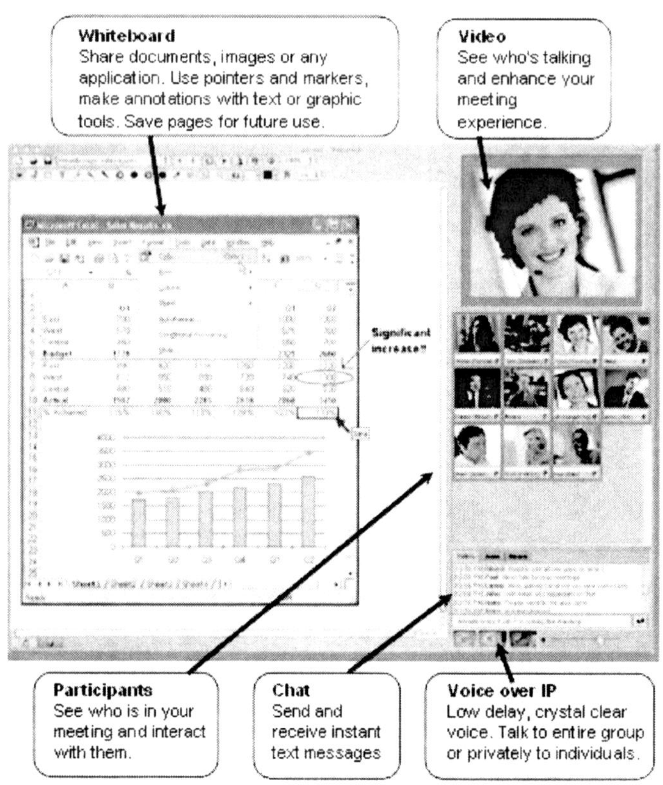

- **VideoPic:** for sharing images.
- **Whiteboard.**
- **Log:** an event-logging tool

The window in Figure 5 shows the following areas:

- A user panel showing who is currently in "This Place". The background color of each user image indicates the level of activity of that user. A greyed-out user image indicates an inactive user.
- A view of available Mushlets: Some Mushlets show a small icon indicating status, such as who the current user is and whether there is any update to current state.

Documents in the place: Documents in Mushroom include simple documents and synchronized documents with a "local version" (on client) and a "global version" (on server).

Orbit (Mansfield, T., et al. 1997) is an experimental environment that was developed to test the viability of the Locales Framework, an Activity Theory-based conceptual foundation whose goal is to bridge the distance between analysis, requirements formulation, and design of cooperative worlds. The Framework was developed from the experience with wOrlds (Fitzpatrick G., 2003), and is in this sense its successor. It uses the Locales metaphor - a generalization of the room metaphor in which the spatial view is replaced with a personalized view of an individual worker's perspective of work as a collection of locales containing 'furnishings'. The main interest

Figure 5. Basic browser Mushroom layout

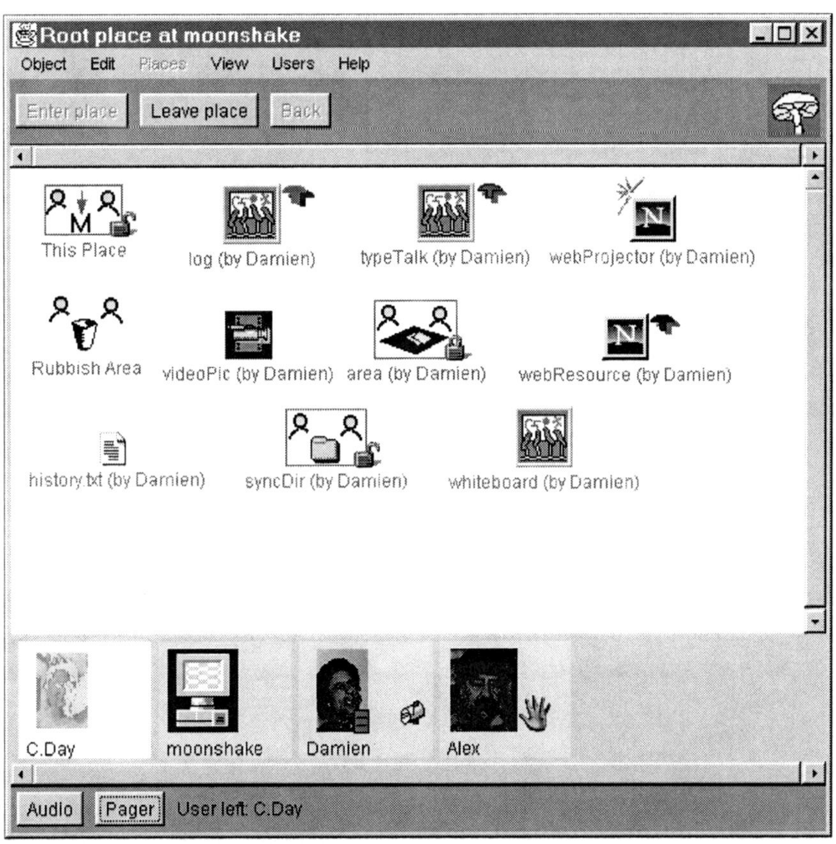

of Orbit and wOrlds is that they used ethnographic and other results rather than the *ad hoc* approach used to develop much other groupware.

TeamRooms (Roseman, M. & Greenberg, S., 1996), commercialized under the name TeamWave Workplace, is one of the early collaborative environments. The environment is place-based and consists of a collection of virtual rooms. TeamRooms (Figure 6) offers real-time and asynchronous communication and each room provides a shared whiteboard and books, reports and other documents can be imported into a room. TeamRooms also provide awareness by displaying a list of users with user idle time. Each room also offers a radar view so that a user can locate all tools in a room even when the room is larger than the display size.

NetMeeting / LiveMeeting (Microsoft Office LiveMeeting, 2007) is a Microsoft conferencing tool that supports point-to-point audio/video conferencing. The main NetMeeting features include video/audio conferencing, text-based chat, internet directory for finding other users on the internet, file transfer between users during a conference, program sharing, remote desktop sharing, and Whiteboard (Figure 7).

Many other environments have been developed and several are now production use but there is not enough space here to even outline their principles. We will, however, describe one more environment in the next section because we used it

Figure 6. TeamRooms meeting in a virtual room

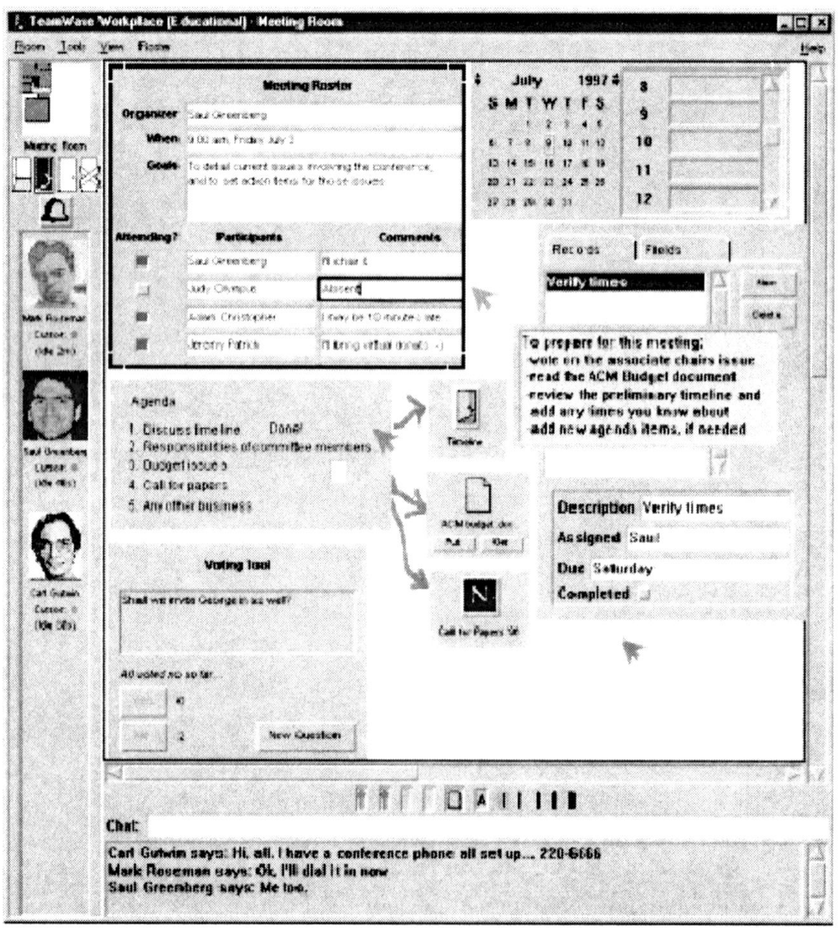

Figure 7. NetMeeting whiteboard

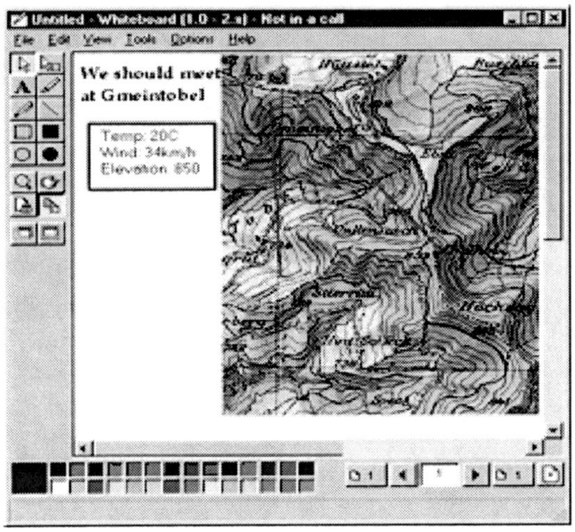

as the basis of the work that forms the core of this paper.

CVW: THE SEED OF FCVW

CVW (Collaborative Virtual Workspace) (Maybury, M., 2004; CVW Overview 2008) is an open-source product of MITRE Inc. Its user interfaces (Figure 8) provide access to a persistent virtual space emulating a physical workplace. In this space people, documents, and applications reside in rooms on floors inside a building and can move around, carrying and sharing documents. Rooms can be used to meet and communicate, and as repositories of objects, documents, and tools. Access is controlled by access rights and the persistent virtual space remains unchanged even when users log out.

CVW document types can be divided into 'native' (specific to CVW), and 'foreign' (created by applications, such as word processors, and imported). Native documents are stored as a part of the CVW universe, along with users, rooms, and tools. Foreign documents are stored on a document server, accessed by reference, and manipulated by the corresponding client-resident application. Besides documents, rooms can contain objects defined via the built-in LambdaMOO programming language (LambdaMOO] interpreted at runtime by the server. A 'find' facility allows search over objects, rooms, and people. Users can also examine who is currently logged in, how long they have been active, and where they are. Topology and navigation of the virtual space are supported by graphical maps. CVW interfaces also include individual and group pop ups, audio phone, and video and audio conferencing, and users can define user groups to simplify communication and control access to documents and virtual rooms.

A feature that distinguishes CVW from most other groupware is its runtime programmability, accomplished as follows: User actions such as chat send commands to the server, which translates them into 'verbs' defined in LambdaMOO. An interpreter executes the verbs and translates them into actions that usually result in messages sent back to the clients.

LambdaMOO is a prototype-based language, a variation on object-oriented languages defined specifically for MOO applications. Because the language is reflective and the core is interpreted, all modifications take immediate effect, allowing authorized users to modify CVW operation at runtime without shutting down its operation. The essence of CVW can be summarized as follows:

- CVW is a client-server application.
- The metaphor is that of a virtual space.
- The server combines a LambdaMOO-based universe server and a general-purpose document server.
- The application offers all functionality typically required in a work team context.
- Besides its built-in features, CVW offers runtime programmability via a special-purpose programming language.
- The rich user interface is implemented as a standalone application.

Figure 8. CVW user interfaces

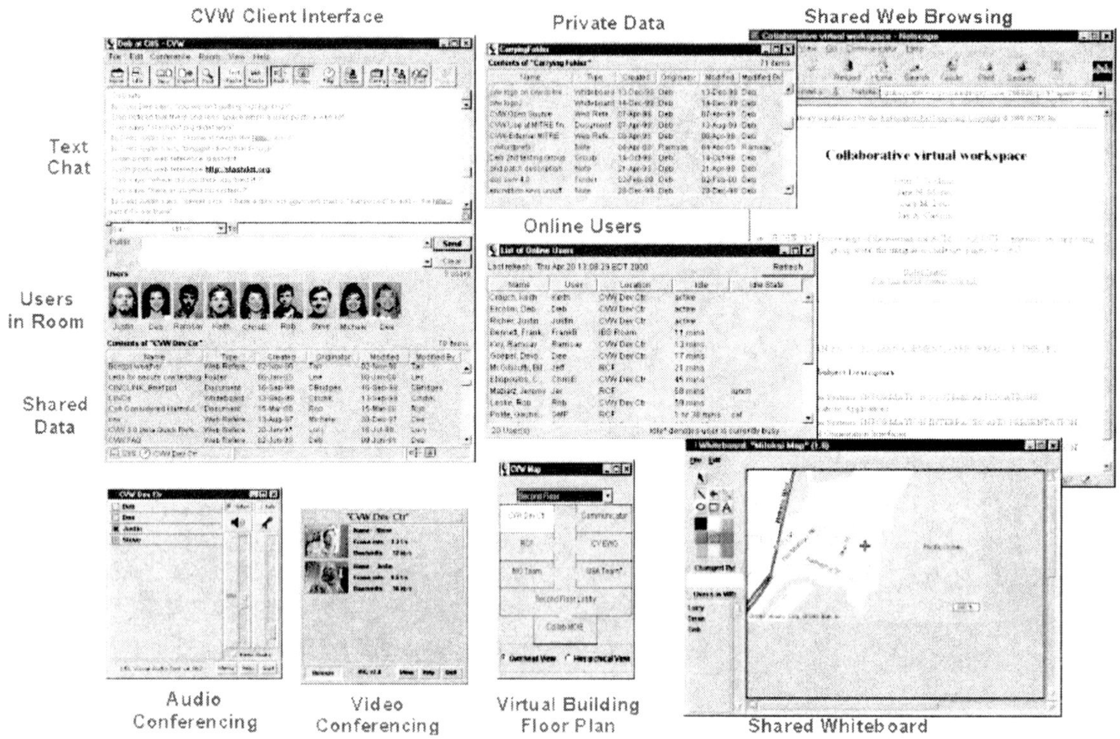

CVW offers an almost ideal combination of features but has shortcomings that include the following:

- Limited user scalability.
- Restrictive spatial metaphor - a single virtual building with a fixed layout.
- No IDE (Interactive Development Environment) for LambdaMOO programming.
- No support for code versioning.
- No sandbox for safe code development.
- No support for the development of software agents.
- Limited set of client platforms.
- Very poor on-line help.

A significant problem with CVW is documentation. The server is undocumented and contains many functions that were necessary at the time of its development but are superfluous today. The client is relatively new and written in Java but also undocumented and very poorly written. These factors make CVW very difficult to modify, extend, and improve, and contributed to some of our design decisions and the present direction of our research. As a result of both the advantages and shortcomings of CVW, we decided to develop a CVW extension called FCVW (Federated CVW), essentially a set of experiments with additional groupware features. A description of FCVW (Tomek, I., et al., 2006) features and their implementation is presented in the following sections.

FCVW: A FEDERATED COLLABORATIVE VIRTUAL WORKSPACE

FCVW is an experimental groupware application that implements a set of CVW extensions. Because

Figure 9. FCVW architecture

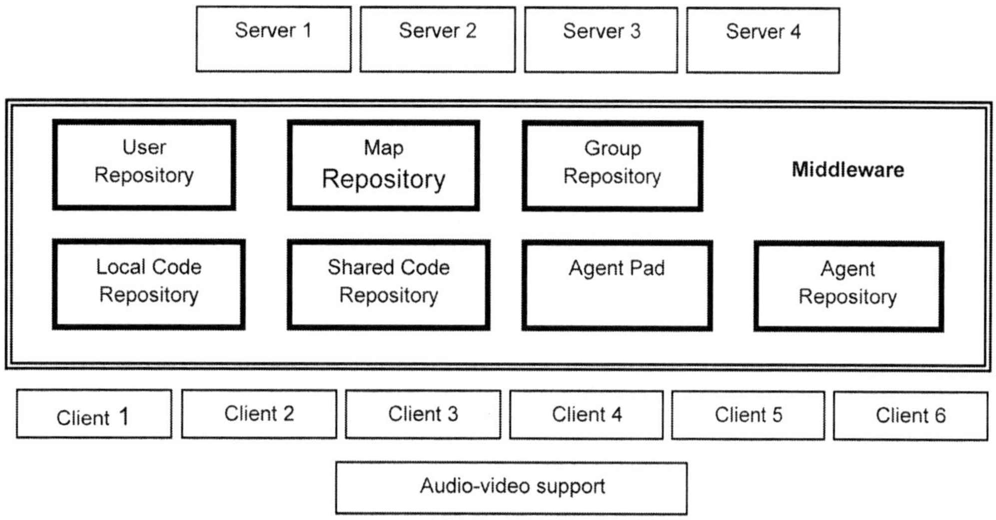

of the aforementioned difficulty of modification or extension of existing CVW code, FCVW has been implemented by middleware and client modifications. The features that constitute FCVW's extension of CVW are described next.

Support for Federated Servers

The view of CVW as a single virtual building containing multiple floors and rooms is very limiting and unintuitive. As an example, in an educational setting a view of the virtual environment as a campus consisting of multiple buildings dedicated to individual faculties, administration, library, and residences would be more natural and effective. And from the scalability perspective, limitation to a single server imposes limits on the number of users that can actively use CVW at the same time. One can, of course, install any number of independent CVW servers implementing individual campus buildings, but documents and all other information would remain restricted to the servers on which they are installed, and the corresponding virtual buildings would be disconnected, requiring users to log into each separately. Communication with users logged to other servers would be impossible as would transfer of objects from one building to another. The only viable solution is to create a campus using federated servers and make them transparently interconnected as a single virtual space. Instead of logging into a single virtual building, users should log into a virtual campus containing an arbitrary number of buildings, while the underlying infrastructure of many servers should remain invisible. Users should be able to move from one building to another with equal ease as within a building, transport documents from one building to another, search across all servers, and so on. Besides removing the conceptual constraint of a single-building metaphor, this arrangement should greatly improve scalability. Continuing with the virtual campus example, logged in users will, most of the time, be distributed among servers in the same way in which they are distributed among buildings on a physical campus. Server load should thus be naturally distributed.

Figure 9 shows the architecture of the FCVW solution. The middleware consists of multiple repositories, each containing a specific type of global information. Each database is complemented by software that handles network communication and performs the necessary searches and updates.

The following comments explain some of the components of the picture:

- Individual servers are unchanged original CVW servers. They communicate only with clients and are unaware of middleware and other servers.
- Clients are modified CVW clients. Their communication with CVW servers is unchanged but they also communicate with middleware. They also offer redesigned user interfaces and new functions that will be explained later.
    - Repositories contain information and functionality required to support federated server operation:
    - The User Repository contains information about all FCVW users. It is essentially a copy of CVW user information (login name, user name, etc.) augmented by information about the user's present virtual location (building/CVW server). In addition to supporting navigation among buildings, it provides information needed for operations such as searches and popup messages.
    - The Map Repository is a server directory and contains IP addresses and basic information about their virtual buildings. This makes it possible to find servers, perform searches, allow users to move from one virtual building to another, etc.
    - A Group Repository provides a central location for group information that must be shared. As an example, a new group is created on a CVW server, propagated to middleware, and then to all connected servers.
    - Functions of the remaining repositories will become clear when we describe additional FCVW features. Note that one repository that one would expect is missing - an Object Repository containing an inventory of all objects across all servers. This repository would require a system of unique Object IDs (OIDs) across FCVW and this, in turn, would require modification of individual servers. As we already mentioned, we did not modify CVW servers and this precluded redesign of the existing OID scheme.

Mobile Clients

One of the main developments in Internet usage is the invasion of mobile devices. Although CVW contains a limited early implementation of a rich mobile client (CVW architecture, 2006), the client is quite specialized and uses early mobile technology. Because of this, and the fact that modern collaboration is unimaginable without mobile devices, we built two mobile clients of our own (Shakshuki, E., et al., 2006; Madala, K., et al., 2006; Tomek, I., et al., 2007).

Rich Mobile Client

A rich mobile client is a stand-alone application installed on a mobile device and using its operating system's Application Programming Interfaces (APIs). The device's CPU speed and memory size directly affect its performance, maintenance, and functionality and we had to take these limitations into consideration and provide only a subset of the full desktop client. Given the exploratory nature of the project we have, in fact, restricted the client's functionality even further and implemented the following set of CVW functions and windows:

- Login/logout window.
- Main page with access to all other functionality.
- Settings (server address and other parameters).

Figure 10. Selected rich client windows: main, chat, people, and map

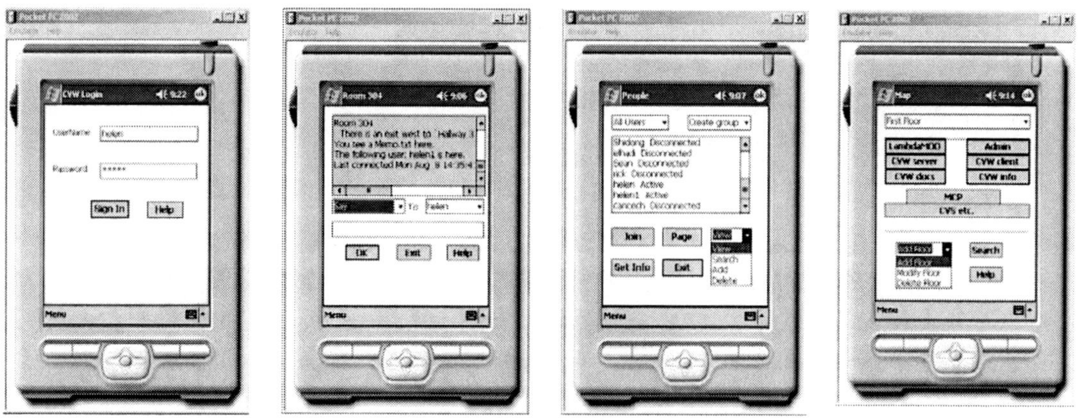

- Private and public chat.
- MCP command line for access to LambdaMOO commands.
- User window providing access to information on currently logged in users and all users with accounts.
- Document window with access to contents of carrying folder and current room.
- Map window with graphic layout providing access to floors, rooms and navigation.

Window design (Figure 10) was guided by the original CVW desktop client, but not strictly followed because of device limitations and because we considered some of the original windows poorly designed.

The rich client was implemented on a Hewlett-Packard Jornada 568 Pocket PC with its built-in custom version of Windows CE and uses Microsoft's eMbedded Visual C++. The architecture (Figure 11) revolves around the Data Center module in that when a new module is added, it must communicate with the Data Centre. This centralized design decouples functions, simplifies extension, and helps scaling. Via the NetCentre, the Data Center handles all messages from the server and distributes them to other modules. Messages from the server use the unchanged client-server interface and the server is not aware that the client is not a desktop.

Figure 11. CVW rich client architecture

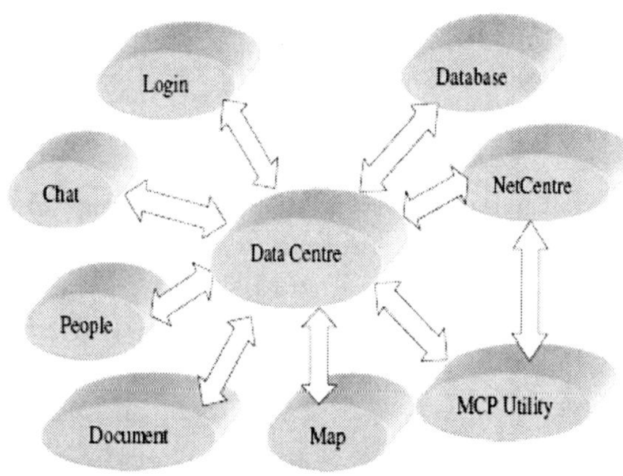

Thin Mobile Client

Our thin client runs on the BlackBerry, one of the most popular mobile devices and one on which thin clients are the most accepted technology for non-native applications. It can, however, run on any device with sufficient hardware resources and a compatible browser. Development options were again restricted by the hardware and software limitations of the device and browser.

The selected subset of CVW functionality was mapped into a set of windows (Figure 12)

Figure 12. Selected thin client windows: room, chat, users

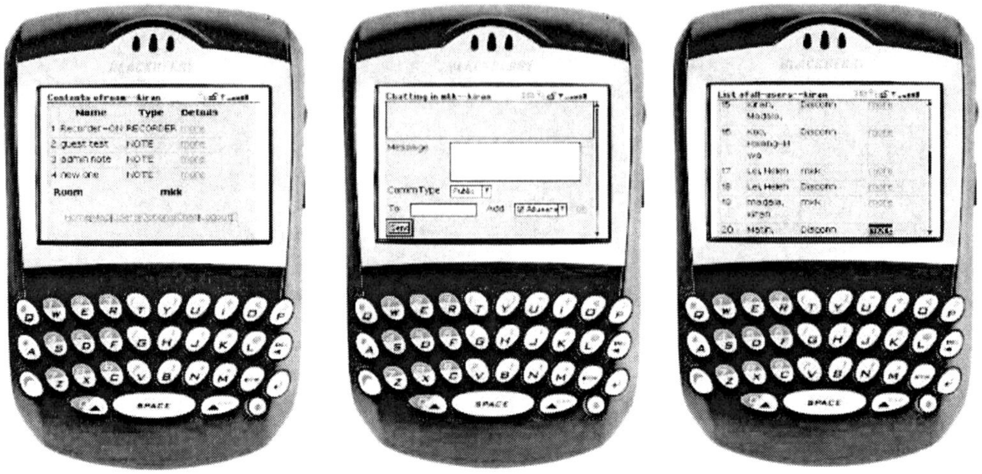

and a state diagram designed to make the use of the client as simple as possible and to optimize the experience of the user who has to rely on the relatively awkward input facilities (Figure 13).

The windows were then hard-coded using a markup language and their look and feel tested via a BlackBerry simulator (BlackBerry, 2007). The full architecture was then implemented (Figure 14).

The central piece of our architecture is the Application Server that acts as a client from the point of view of the server. It processes messages from the server and dynamically calculates XHTML-MP code, sending it to the client. From the client perspective, the Application Server generates messages to the server from client requests. The Application Server is based on the JSP Model-2 architecture (JSP, 2007). It serves dynamic content by combining servlet (Servlet, 2007) and JSP technologies. JSP generates display content and the servlet processes requests from the client, creating objects required by JSP. It also decides which JSP page to forward to the client. The Application Server uses an Apache Tomcat Web server (Apache Tomcat, 2007).

MOOIDE: Interactive Development Environment for FCVW

We consider end-user extendibility to be essential for collaborative environments. Unfortunately, most groupware applications are not end-user programmable and provide only a fixed set of functionalities. The fact that CVW is end-user programmable was one of the main reasons why we selected it as the basis of our work. Unfortunately, CVW lacks the following key features to make this facility truly useful: a suitable Integrated Development Environment (IDE), support for code versioning, and protection from programming errors in new code. We will now look at these shortcomings more closely.

- **Integrated Development Environment:** Most modern code development uses IDEs that provide facilities such as code browsing, editing, debugging, and other features. Unfortunately, the only way to program CVW is through a one-line input field in the main window and this makes end-user programming nearly impossible.
- **Code Versioning:** Application code evolves in iterations requiring occasional roll back to earlier versions but this is not

Figure 13. Thin client states, windows, and transitions

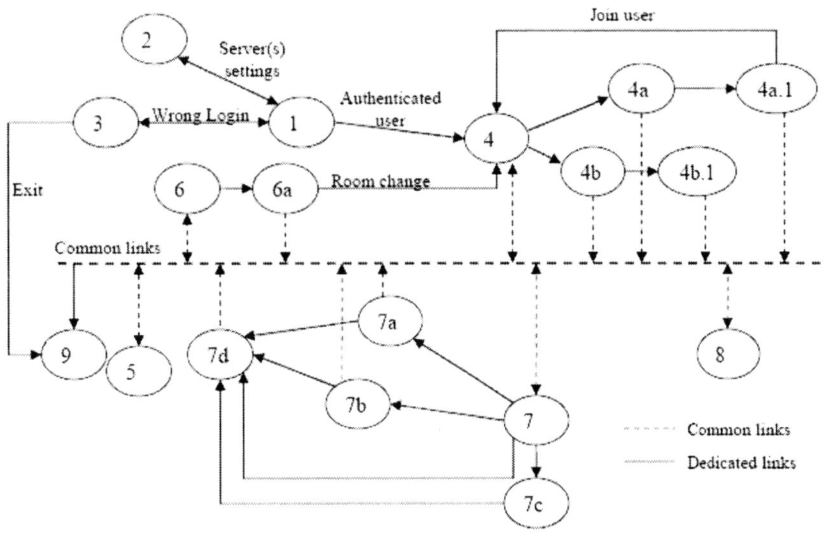

1 → Login 2→MOO/Document server settings 3→ Wrong login 4→ Home room page 4a →Users in room list. 4a.1 → Details of users 4b →Documents in room list 4b.1 → Details of documents 5→Chat	6 →Map 6a→List of rooms in a floor 7 →Options 7a→Search object 7b→Create document 7c→Change password 7d→ Options operation result. 8→Help 9→Logout

Figure 14. CVW thin client architecture

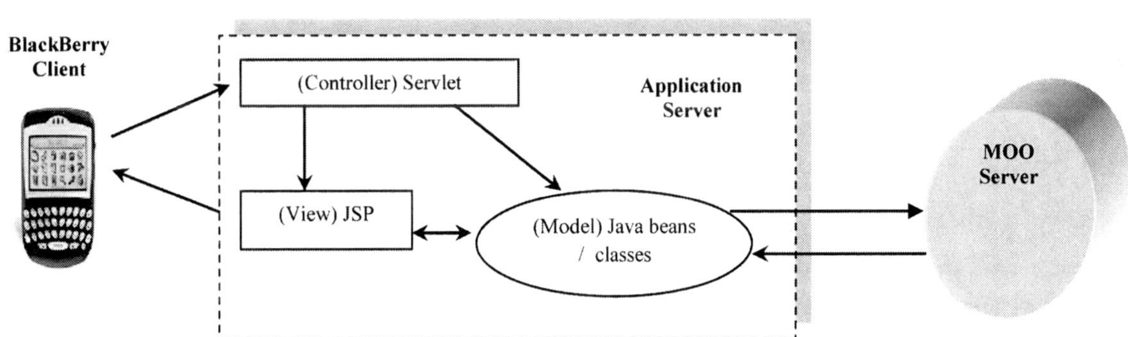

possible in CVW because there is always only one code version that resides in the code database on the server.

- **Safe Code Development:** Because CVW code development takes place at run-time and has immediate effect, programming mistakes can have disastrous consequences. CVW does not support safe development.

The federated nature of FCVW creates another problem - code is developed, compiled, and stored on the server connected to the client but needs to be propagated to all servers. Our MOOIDE (MOO

Figure 15. MOOIDE user interface

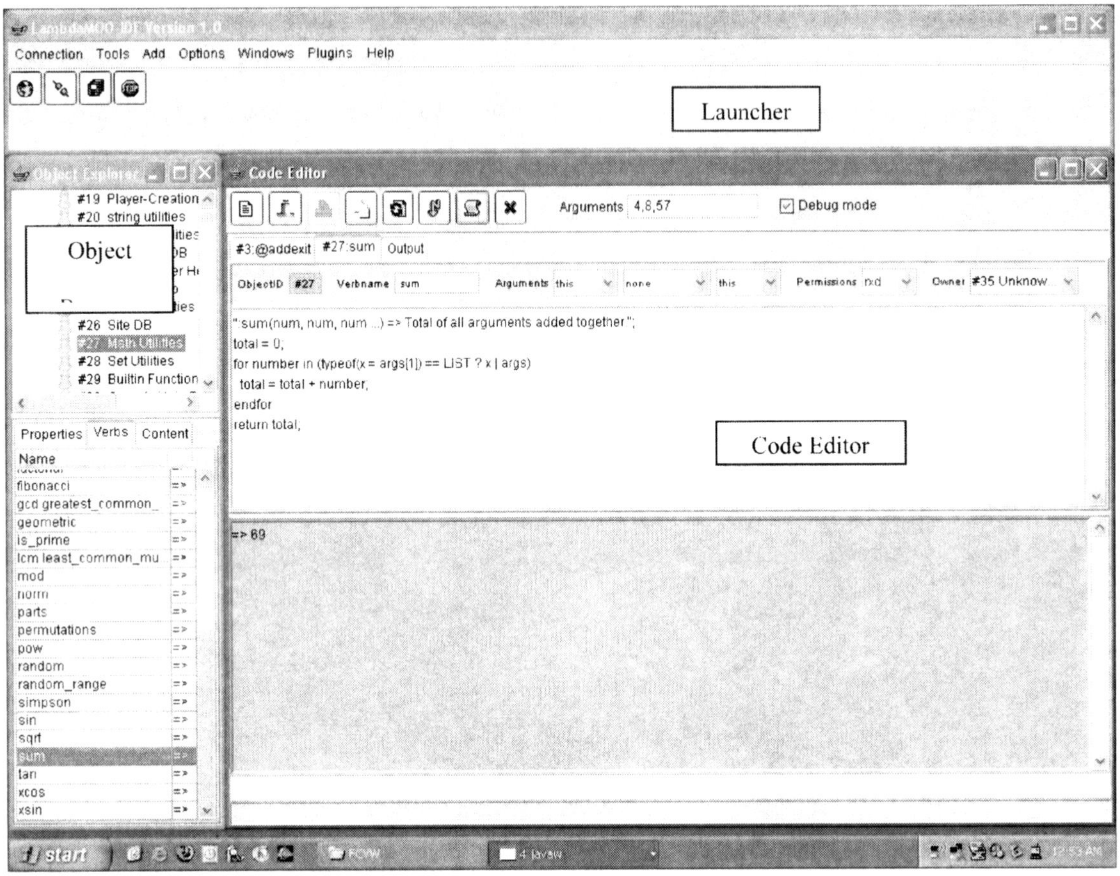

Integrated Development Environment) and its code versioning system address all these issues.

MOOIDE Functionality

MOOIDE provides browsing, creation, and editing of code residing on the current server via a three-part IDE (Figure 15). The Launcher allows the user to program, invoke online help, toggle between active windows, and offers other features. The Object Browser shows all objects in the code database. When the user selects an object, it displays its properties and their values, and the object's 'verbs' (messages). The Code Editor displays the definition of a selected verb and provides space for code definition and evaluation.

Support for code versioning is based on a middleware component called Local Code Repository (LCR). When a user creates a new version of an object or a verb and the server compiles it, the user can add the code to the server's LCR database. Each server has its own LCR and users can examine or compare existing versions using MOOIDE and its 'Diff' tool, and make current any of the existing versions.

To support the federated nature of FCVW servers, the Public Code Repository (PCR) stores all shared code. The combination of multiple LCRs and a single shared PCR makes it possible to develop code safely on one server and broadcast it to other servers when it is finalized. New code generates a notification to other servers and when their administrators receive the notification, they can accept the new code and recompile it on their machines. The present implementation does not yet address all issues related to code propagation.

As an example, it is possible that a server contains privately developed code that has not yet been broadcast and that clash with the new version. The new code may thus have a different effect on this server and may not even compile on it. Problems such as these are logistic and policy issues and have not been addressed because our work focuses on functional issues and their technical solution.

MOOIDE Implementation

Figure 16 shows MOOIDE architecture. The System Module contains client-side definitions of system classes and classes representing universe components, such as LambdaMOO objects, properties, and verbs. The Communication Module uses technologies such as Java pipelines and channels to maintain effective communication with servers and middleware. The GUI Module defines the user interface, and the Code Management Module is responsible for version control and code sharing among CVW servers. The Help Module provides programming help using help text obtained from the CVW server.

Figure 16. MOOIDE architecture

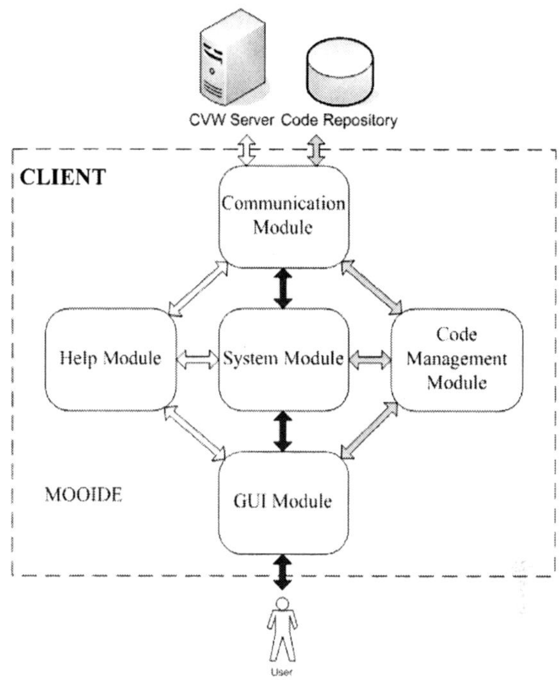

FCVW and Software Agents

Software agents can naturally and effectively complement other groupware components. They can make it more productive and enjoyable to use by performing functions such as delivering newsletters to subscribers, helping to collect 'garbage' such as outdated and unused documents, assisting in finding expert help among community members, providing personalized document filtering in the background, and fulfilling many other tasks. Software agents are a perfect conceptual fit for environments such as FCVW because of agents' person-like connotation and the real-world-like nature of FCVW. For all these reasons, software agents in FCVW were on our agenda from the start of the project. The essence of our adopted strategy was as follows: Software agents can be classified along several dimensions and agents in the same category share functional features. It is thus wasteful to create each agent from scratch and the appropriate approach is to develop a software agent framework that can help to avoid any unnecessary duplication of effort. We thus decided to develop an agent framework and test it by developing several representative agents.

In the rest of this section we survey several approaches and techniques for the development of frameworks and agents, describe how we developed our framework, and give an example of an agent created with it. We conclude by contrasting the presented approach with an alternative strategy that will be explored in our forthcoming research.

Agent Framework Development and Deployment

A framework is a reusable design implemented as a set of abstract classes and methods with well-defined interfaces (Fayad, M., et al., 1999). A framework consists of two basic parts - the stable and fully implemented Kernel, and a collection of Hot-spots that provide the framework's points of extension. For each particular application, the development of a framework consists of selecting a framework methodology and applying it to the model of the domain at hand. The development of an agent framework thus combines two sets of techniques – a methodology for framework development, and a multi-agent model. Several framework methodologies have been proposed, including GAIA (Mouratidis, M., et al., 2003), MaSE (Wood, M. & Deloach, S., 2000) and AoR (Erol, K. & Lang, J., 2000). We selected GAIA because it is designed for closed systems and FCVW is a closed system. Models of multi-agent systems include the RMIT agent model (Kendall, E., et al., 1997), CIR agent model (Shakshuki, E., et al., 2003) and RETSINA (Sycara, K., et al., 2003). Our framework (Shakshuki, E., et al., 2006) uses RMIT because it allows extension and changes of the environment, both essential in an application such as FCVW.

Our analysis of potential uses of agents in FCVW produced the description of various agent roles such as retriever agent, observer agent, garbage-collector agent, and document agent. The design phase resulted in the decision to implement the framework as an additional part of the existing FCVW architecture rather than a part of the object universe. We identified three basic framework components - communication, serialization, and agent-creator. The communication component allows an FCVW agent to communicate with the server using MCP commands. The serialization component serializes the agent to make it persistent. The agent-creator component allows the user to create a specific agent via the Agent Launcher tool described below. Following RMIT, we used the Layered Implementation model which defines the following seven agent layers, not all of them required by every type of agent:

1. A Sensory layer that allows the agent to gather information from the environment.
2. A Beliefs layer, which defines agent and environment models.
3. A Reasoning layer where agent goals, plans, models, and capabilities determine what the agent will do.
4. An Action layer that carries out intentions decided by the Reasoning layer.
5. A Collaboration defining services related to accepting or rejecting requests coming from other agents.
6. A Translation layer translating messages to other languages, for example, for communication with other agents.
7. A Mobility layer that helps to transport the agent from one place to another.

To interface with FCVW, our framework uses specialized repositories and components in the middleware architecture described earlier. Its main parts are the Agent pad that launches agents into the FCVW system and triggers their operation, and the Agent repository that stores agent information and provides access to it. Our Agent Framework is a black-box framework that uses inheritance from abstract classes to implement specific concrete agents. Each agent has an EventQueue class to support queuing of consecutive events, an AgentState class to maintain agent state, and an AgentTimer class to support time-related operations. Using the framework, users can either instantiate an existing agent type by selecting it in the Agent Virtual Environment Launcher (AVE-Launcher) (Figure 17), defining its properties, and launching it into FCVW, or create a new agent type by defining new classes that extend an existing agent class.

The Launcher connects to middleware repositories to retrieve or update information about users, maps, agents, and other objects in the universe, and instantiates the AgentManager class whose user interface (Figure 18) helps adding new agent classes to the framework.

To instantiate an existing agent type, the user selects an agent type and sets the properties that

Figure 17. AVE agent launcher user interface

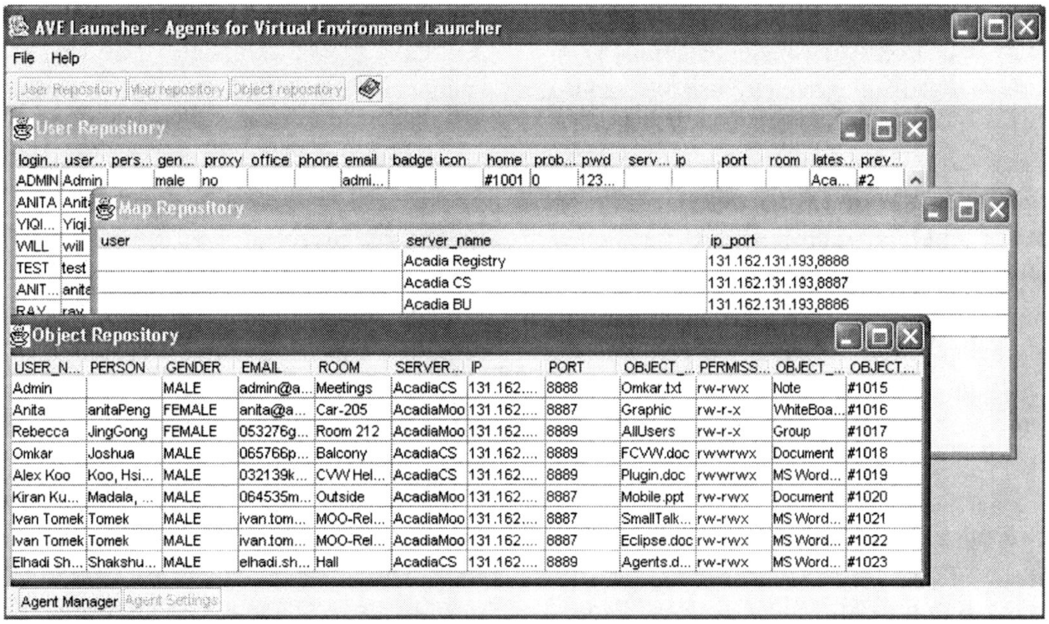

Figure 18. Agent manager interface

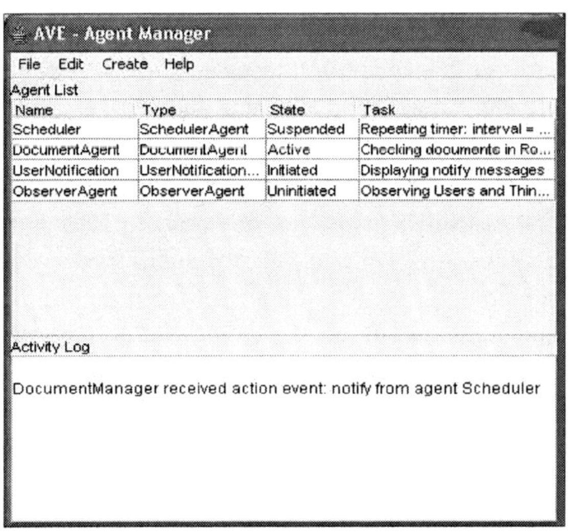

Figure 19. Class diagram of FCVW agent framework

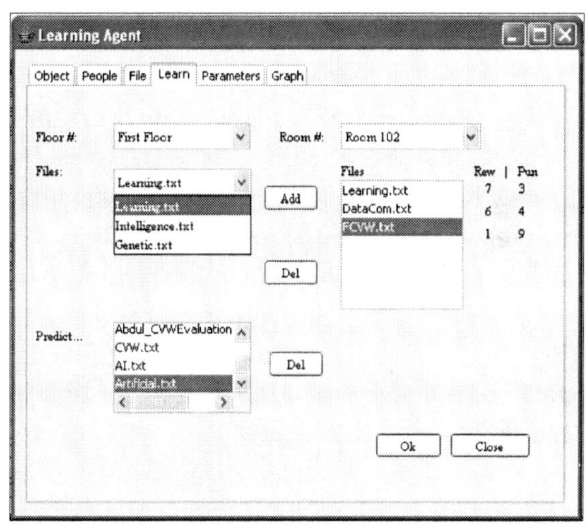

define the type of the desired agent such as reactive, knowledge-based, or broker agent. To add a new agent type the user declares new class definitions extending abstract classes in the framework (Figure 19) and adds their names to a property file. The Agent Manager then creates the desired agent via the Customizer class using Abstract Factory pattern, building agent layers using the AgentBuilder pattern (Fayad, M., et al, 1999).

As was already noted, the agent framework exists outside the FCVW virtual universe rather than inside it as other FCVW objects do. An alternative approach is to use the built-in LambdaMOO language to create agents in the form of agent hierarchies and this has been done in the past. Although this alternative is attractive and worth pursuing, its existing implementations have one serious disadvantage – they use the LambdaMOO programming language whose obscurity discourages users from implementing not only new agents but any new FCVW objects. To make the idea of building arbitrary objects (including agents) within the virtual universe attractive requires, in our opinion, an environment that can be programmed in one or more widely known programming languages. The development of such an environment is a part of our current research.

An Example: Learning Agent

To test the usability of the framework, we implemented several agents including an assignments agent, an object retriever agent, a scheduler agent, a garbage collector agent, and a learning agent. This section describes the learning agent to provide an example of the type of agents that can be created with the framework.

The purpose of the learning agent is to monitor Notes used by the FCVW user who owns the agent, and to try to predict Notes in which the user might be interested in the future. To create such an agent, the user sets various parameters and attributes using the learning agent interface (Figure 20) and the agent then uses these parameters and re-enforcement learning to perform its task (Shakshuki, E. & Matin, R., 2008).

Details of the procedure are as follows: After invoking the interface, the user first selects those Note files in FCVW rooms across the universe that are of either positive or negative interest and assigns a numerical reward or punishment value to each of them. The selection is stored in the agent's database value and the agent is launched. When the agent discovers a new Note, it analyzes its context and ranks it on the basis of its match with the files in its database. If the match value exceeds the existing threshold value, it presents the Note to the user who then decides whether the note is of interest. Depending on the result, the agent then updates its reward and punishment values and uses them to evaluate future Notes.

CONCLUSION

Collaboration and communication over the Internet is becoming an essential part of the computing milieu. Because the needs of different work teams and virtual communities are widely different, it is not possible to devise a single environment that will satisfy all needs. Moreover, needs continuously evolve and so even an environment designed specifically for one type of use will not remain satisfactory over a long period of time. Leaving the evolution and maintenance of such a product to a single group of developers is unsatisfactory because a closed group of developers will never be able to satisfy the diverse demands in a timely fashion. Because of all these considerations, we see the future of groupware and socialware in the development of a powerful framework that contains a sufficient foundation of widely required features and provides hot spots and APIs for new features. An essential aspect of such a framework should be a widely known built-in programming language that allows the wider user community to modify and extend the available features at any time with minimal effort. FCVW, the environment described in this paper, is an exploration of most of the features of such an environment. FCVW development has now been finished and certain conclusions drawn from this experience. The chief among these are as follows:

- Although CVW was very useful as a foundation for a variety of limited experiments, it is not suitable as the basis for an environment satisfying the criteria presented above. The main reason for this is the server, whose total lack of documentation, ob-

Figure 20. Defining properties via learning agent interface

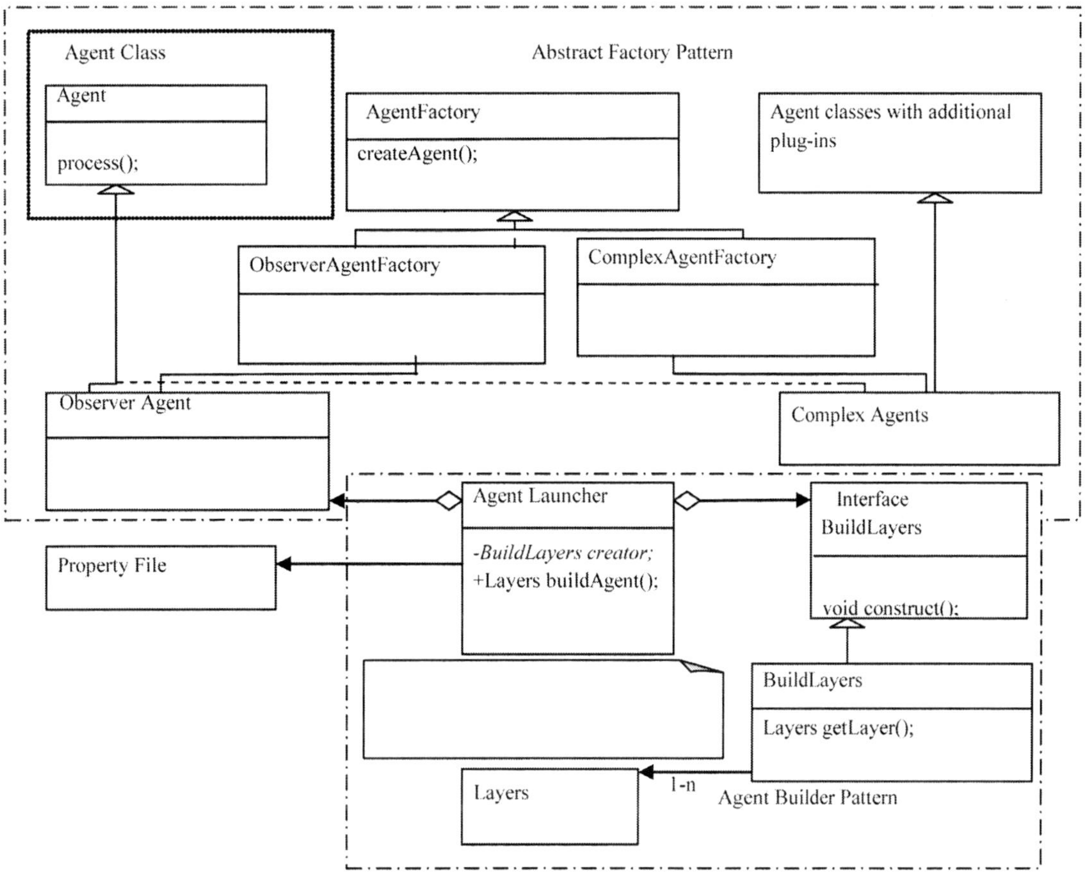

scure design, and technical obsolescence make substantial further development impossible.

- LambdaMOO is unsuitable as the built-in development language. It is old-fashioned and generally unknown and this makes extendibility of the environment only a theoretical but not practical possibility.

In view of these fundamental problems, we started developing a new environment called FVE (Federated Virtual Environment) (Tomek, I., et al., 2008). FVE has numerous new features but its major differences with respect to FCVW are as follows:

- The conceptual, design, and implementation approaches of both the server and the clients are new. Both of are completely re-written in Java and thoroughly documented.
- LambdaMOO and its interpreter have been replaced with the use of the 'factory' software pattern and this allows the use of any language running on the Java Virtual Machine (JVM) to script FVE objects. As a consequence, any programmer familiar with widely used languages such as Java, Ruby, and Python can now program FVE objects without having to learn the obscure LambdaMOO.

Although FVE is now a fully working prototype, a complete set of groupware features is not yet available and a full report on its functionality, design, and implementation will appear in the future.

ACKNOWLEDGMENT

This project is supported in part by an IBM Eclipse Innovation Award, a Teaching Innovation Fund from Acadia University, and NSERC (National Science and Engineering Research Council of Canada) grants. Development of FCVW would not have been possible without our graduate students Rebecca Gong (IDE and related middleware), Kiran Madala (thin client), Hsiang-Hwa Koo (user interfaces), Helen Lie (rich client), Anita Peng (most of middleware), and Omkar Prabhu (agents and related middleware). Li Di and Hai Zhang helped with the survey of existing groupware products and implemented the essential features of FVE, the successor of FCVW.

REFERENCES

Architecture OverviewC. V. W.http://cvw.sourceforge.net/cvw/info/docs32/ArchOverview.ph3

BlackBerry device simulators (2007). http://www.blackberry.com/btsc/search.do?cmd=displayKC&docType=kc&externalId=KB12436&sliceId=SAL_Public&dialogID=40874440&stateId=0%200%2040870665

BSCW. (2007). http://bscw.fit.fraunhofer.de/

Cheng, L.-T. Hupfer, S., Ross, S., Patterson, J., Clark, B., de Souza, C. (2003). Jazz: A Collaborative Application Development Environment, Proceedings of the 18th Annual ACM SIGPLAN Conference on Object-Oriented Programming Systems Languages and Applications (OOPSLA), demo.

CURE. (2007). Fern Universitaet Hagen, http://www.pi6.fernuni-hagen.de/en/CURE/

Curtis, P., & Nichols, D. A. (1993). *MUDs Grow Up: Social Virtual Reality in the Real World, Xerox Palo Alto Res* (pp. 193–200). CA: Center.

De Miguel, T. (1994). *ISABEL Experimental Distributed Cooperative Work Application over Broadband, Lecture Notes in Computer Science, Springer* (pp. 353–362). Volume.

Envy (2007). http://www.ibm.com/developerworks/blogs/page/stdt?tag=envy

Erol, K., & Lang, J. (2000). Levy, Designing Agents from Reusable Components, Proceedings of Fourth International Conference on Autonomous Agents, pp. 76-77.

Fayad, M., Schmidt, D., & Johnson, R. (1999). *Building Application Frameworks: Object-Oriented Foundations of Framework Design*. Wiley Computer Publishing.

Fitzpatrick, G. (2003). *The Locales Framework, Understanding and Designing for Wicked Problems* (1st ed.). Springer.

Foldera (2007). http://www.foldera.com/index.htm

Haake, J., Schümmer, T., Haake, A., Bourimi, M., & Landgraf, B. (2004). Supporting Flexible Collaborative Distance Learning in the CURE Platform, Proceedings HICSS.

Habanero (2007). http://www.isrl.uiuc.edu/isaac/Habanero/3.0/index.html

Java Server Pages Technology. (2007). Sun Developer Network. http://java.sun.com/products/jsp/

Java Servlet Technology. (2007). Sun Developer Network. http://java.sun.com/products/servlet/

Jazz Demo Videos. (2007). http://jazz.net/pub/learn/videos/videos.jsp

Kendall, E., Pathak, C., Krishna, M., & Suresh, C. (1997). The Layered Agent Pattern Language, Pattern Languages of Programming Conference (Plop), Monticello, Illinois.

Kindberg, T. (1996). Mushroom: a framework for collaboration and interaction across the Internet. Proc. CSCW & the Web, 5th ERCIM workshop, Busbach, U., Kerr, D., and Sikkel, K. (eds), GMD, pp. 43-53.

LambdaM. O. O.http://magrathea.cosmic.com/pub/moo/ProgrammersManual.html

Leuf, B., & Cunningham, W. (2001). *The Wiki Way: Collaboration and Sharing on the Internet*. Addison-Wesley.

Madala, K., Tomek, I., Shakshuki, E., & Giles, R. (2006). A Thin Mobile Client for a Groupware Application", COOP 2006 the 7th International Conference on the Design of Cooperative Systems, Provence, France, May 9-12, pp. 211-223.

Mansfield, T. & et al. (1997). Evolving Orbit: A Process Report on Building Locales, GROUP.

Marratech (2007). http://www.marratech.com/

Maybury, M. (2004). Collaborative Virtual Environments for Analysis and Decision Support.

Microsoft Office LiveMeeting. (2007). http://www.microsoft.com/windows/netmeeting/

Mouratidis, M., Giorgini, P., & Schumacher, M. (2003). Security Patterns for Agent Systems, Proceedings of Eighth European Conference on Pattern Languages of Programs, Germany, pp. 25- 29.

NotesL. (2007). http://www-306.ibm.com/software/lotus/

OverviewC. V. W. (2007). http://cvw.sourceforge.net/cvw/info/CVWOverview.php3

OverviewC. V. W. (2008). http://sourceforge.net/support/getsupport.php?group_id=12120

Roseman, M., & Greenberg, S. (1996). TeamRooms: Network Places for Collaboration, Proceedings of the 1996 ACM conference on Computer supported cooperative work pp. 325-333.

Schuckmann, C., et al. (1996). Designing object-oriented synchronous groupware with coast, Proceedings of the 1996 ACM conference on Computer supported cooperative work, pp. 30-38.

Schummer, T., & Schummer, J. (1999). *TUKAN: A Team Environment for Software Implementation*. OOPSLA.

Shakshuki, E., Ghenniwa, H., & Kamel, M. (2003). Agent-based System Architecture for Dynamic and Open Environments [IJITDM]. *International Journal of Information Technology and Decision Making*, *2*(1), 105–133. doi:10.1142/S0219622003000537

Shakshuki, E., Lei, M., & Tomek, I. (2006). A Rich Client for CSCW. IADIS International Conference on Applied Computing, February 25-28, San Sebastian, Spain, pp. 618-621.

Shakshuki, E., & Matin, R. (2008). Intelligent Learning Agent for Collaborative Virtual Workspace. International Journal of Ubiquitous Computing and Intelligence (JUCI) [ASP]. *American Scientific Publishers*, *2*, 1–18.

Shakshuki, E., Prabhu, O., & Tomek, I. (2006). FCVW Agent Framework. International Journal of Information and Software Technology, Elsevier Science, special issue, vol. 48, no. 6, pp. 385-392.

Sycara, K., Giampapa, J. A., Langley, B. K., & Paolucci, M. (2003). In Garcia, A., Lucena, C., Zambonelli, F., Omici, A., & Castro, J. (Eds.), *The RETSINA MAS, A Case Study. Research Issues and Practical Applications* (pp. 232–250). Springer-Verlag, Software Engineering for Large-Scale Multi-Agent Systems.

The Apache Software Foundation. (2007). Apache Tomcat. http://tomcat.apache.org/

Tomek, I., Giles, R., Zhang, H., & Di, L. (2008). FVE – a Virtual Environment for Collaboration. The 22nd International Conference on Advanced Information Networking and Applications (AINA), IEEE Computer Society, GinoWan, Okinawa, Japan, March 25-28, pp. 628-635.

Tomek, I., Shakshuki, E., Madala, K., & Giles, R. (2007). Rich and Thin Groupware Mobile Clients. The Thirteenth Australasian World Wide Web Conference, Charlesworth Bay, Australia, 30th June - 4th July, pp. 462-466.

Tomek, I., Shakshuki, E., Peng, A., Koo, H., Gong, R., & Giles, R. (2006). FCVW - A Federated Collaborative Virtual Workspace, Proceedings International Conference on Applied Computing, IADIS, February 25-28, San Sebastian, Spain, pp. 249-257.

Wood, M., & Deloach, S. (2000). An Overview of Multiagent Systems Engineering Methodology, The First International Workshop on Agent Oriented Software Engineering, pp. 207-222.

This work was previously published in International Journal of Mobile Computing and Multimedia Communications (IJMCMC) 1(1), edited by Ismail Khalil & Edgar Weippl, pp. 29-52, copyright 2009 by IGI Publishing (an imprint of IGI Global).

Chapter 19
A Model for Mobile Learning Service Quality in University Environment

Nabeel Farouq Al-Mushasha
Jerash Private University, Jordan

Shahizan Hassan
Universiti Utara Malaysia, Malaysia

ABSTRACT

It is generally known that accessibility and mobility are the main barriers for effective implementation of electronic learning. However, the advent of mobile technology could be a potential solution to remove the barriers. Nevertheless, there is a lack of research that addresses the issue of mobile learning service quality in a university environment. This study aims to propose a service quality model for m-learning in a university environment. A questionnaire survey was conducted which measured ten dependent variables and three independent variables. The dependent variables were meant to measure service quality, information quality, and system quality. The dependent variables were meant to measure the causal relationship between overall learners' perceived service quality, learner satisfaction, and learner behavioral intention to use the service in future. The findings revealed that the factors that lead to service quality of m-learning in a university environment were interface design, reliability, trust, content usefulness, content adequacy, ease of use, accessibility, and interactivity. The findings also indicates that there are causal relationships between learner satisfaction with overall service quality, and learner satisfaction with learner behavioral intention.

DOI: 10.4018/978-1-60960-563-6.ch019

INTRODUCTION

This rapid diffusion of the Internet and its deployment in learning, as well as on-line courses delivery is represented by Electronic Learning (e-learning). However e-learning much depends on the availability of a wired network connection to the Internet, there is still a limitation to this service among students and corporate professionals who are constantly on the move (Dye, 2003). In order to overcome such a limitation, there is an inclination to make use of wireless networking technologies which can provide potential learning at anytime and anywhere. Su et al. (2005) explain that according to the mobile learning theory, mobile learning learners can employ wireless network and mobile learning device to get simplicity, suitability and immediacy of mobile learning in a proper time and accessing appropriate learning content. Similar to other wired technologies; mobile wireless technologies have first been used in industry sectors such as business. The movement of mobile wireless technologies in education is a trend, and it is now becoming the newest technology in higher education (Al-Mushasha & Shahizan, 2007; Levine, 2002; McKenzie, 2001; McGhee & Kozma, 2001).

According to Grohmann et al. (2005), studies in m-learning can be divided into two main categories, which are technology and services. Research in m-learning technologies cover issues such as: small keyboards and screens, limited battery life, unreliable network connections and cellular systems. Meanwhile studies in m-learning services mostly focus on providing students access to learning resources, student support services and critical institution services. With the advent of mobile computing, institutes of higher learning have the opportunity to revitalize the process of teaching and learning via mobile learning. And due to this advancement the students of all academic levels (primary, secondary, college, and university) have developed much more sophisticated expectations, demands and study patterns than ever before (Al-Mushasha & Shahizan, 2007). Since mobile learning contents are delivered via wired or wireless Internet, they are information-oriented products and services (Kim & Ong, 2005). Thus each service sector should have service quality criteria that specifically fit its features and characteristics (Dedeke, 2003).

It has been well known that service quality, customer satisfaction and customer behavioral intention are becoming the most important factors for successful business competition for either manufacturers or service providers (Bolton & Drew, 1991; Parasurman, 1994; Parasurman et al., 1988, 1991; Zeithaml et al., 1996). Researchers in the field of ICT service-quality have proposed five dimensions for measuring the quality of ICT service which are: (1) Tangibles such as facilities, equipment, and personnel appearance; (2) Reliability in terms of ability to perform the promised service dependently and accurately; (3) Responsiveness in terms of willingness to help end users and provide prompt service, (4) Assurance represented in knowledge and courtesy of employees and their ability to inspire trust and confidence; and (5) Empathy such as caring and the individualized attention the service provider gives to the end users (Berry et al., 1994; Jiang et al., 2000; Kettinger & Lee, 1995;1997; kim & Ong, 2005; Pitt et al., 1995). From the customers' perspective, understanding customer satisfaction with service providers is useful in helping organizations assess current and potential ICT service providers.

The previous research on ICT service quality could be applied to the understanding of m-learning service quality. M-learning users do not just want a mobile device, rather, they seek the system that satisfies their mobile learning services and above all they demand for service quality that leads to their satisfaction as well as, behavioral intention to use the service in future (Kim & Ong, 2005). Nevertheless, there is a lack of research in the area of mobile learning service quality therefore; there is a need for a study that examines the factors that lead to service quality of m-learning in university environment. In addition, there is also a need to explore the causal relationships between service quality, learner satisfaction, and learner behavior intentions to use m-learning services in future. Such study would benefit universities and researchers in educational technology in their attempts to measure students' satisfaction and to understand its underlying dimensions. It also enables service providers (university) to benchmark their performance and identify areas that require improvement to ensure continues usage of the provided m-learning services.

This study aims to propose a service quality model for m-learning in a university environment. In order to accomplish this, a study was carried out to identify the factors that lead to service quality of m-learning in a university environment. The researchers administered a 57-item-questionnaire on 259 students representing different colleges of UUM. The questionnaire measured ten dependent variables and three independent variables. The dependent variables were meant to measure service quality (interface design, reliability, responsiveness, trust, and personalization), information quality (content usefulness, content adequacy), and system quality (ease of use, accessibility, and interactivity).The dependent variables were meant to measure the causal relationship between overall learners' perceived service quality, learner satisfaction, and learner behavioral intention to use the service in future. In order to analyze the obtained data, the researcher used multiple regression and exploratory factor analysis (EFA). The findings revealed that the factors that lead to service quality of m-learning in a university environment were interface design, reliability; trust, content usefulness, content adequacy, ease of use, accessibility, and interactivity.

Understanding the determinants of m-learning service quality provides valuable guidance to both vendors (universities) and customers (learners). M-learning vendors can benefit from this research by focusing on the factors that affect user satisfaction and their behavioral intention to use the service in future. Customers can benefit from this research in the selection of the m-learning vendors who can provide as many m-learning service quality as possible.

THEORETICAL BACKGROUND

Conceptualization of Service Quality

Service Quality (SQ) as a critical measure of efficiency and effectiveness in organizational performance that has become an important topic to practitioners and academics in the service marketing areas (Jensen & Markland, 1996). Many service organizations seek profitable ways to differentiate themselves from other organizations and gain a lasting competitive advantage by improving relative service quality. Therefore, an organization providing higher quality of service can have higher market share and higher return on investment and, finally, can gain long term-profitability (Buzzel & Gale, 1987; Ghobadian, Speller, & Jones, 1994; Margolies, 1988).Accordingly, it is important for practitioners and academics to understand what service quality consists of, its definition, and how service quality can be accurately measured (Asubonteng et al., 1996). In the modern marketing literature, most definitions of quality have focused on the consumer's perception of product or service excellence and the satisfaction of consumer requirements (Ghobadian et al., 1994; Gronroos, 1982; Parasuraman et al., 1985; 1988). From this perspective, perceived service quality has been defined as "a global judgment, or attitude, relating to the superiority of the service" (Parasuraman et al., 1988), and as "the consumer's overall impression of the inferiority/superiority of the organization and its service" (Bitner & Hubbert, 1994). In general, perceived service quality has been viewed as the degree and direction resulting from comparing consumer expectations with actual perceptions of service performance

(Parasuraman, et al., 1988; Zeithaml, Parasuraman, & Berry, 1990). According to Prasuraman et al. (1988), consumer expectation from the service quality perspective has been defined as the "desires or wants of consumers, what they feel a service provider should offer rather than would offer". Most research associated with service quality has been based on product satisfaction, especially on the disconfirmation paradigm. This paradigm has been used to evaluate the level of consumer satisfaction about the chosen product or service, and operationalized in terms of the comparison between consumer's expectation and perceived performance of the product or service.

Based on this concept of the disconfirmation paradigm, Parasuraman et al. (1988) have developed a SERVQUAL instrument for measuring the consumer's perception of overall service quality. They identify five dimensions of service quality: 1) tangibles, 2) reliability, 3) responsiveness, 4) assurance, and 5) empathy. Without question, this SERVQUAL instrument has been widely utilized and applied as a universal scale to measure the perceived service quality for a variety of service industries, by both managers and academics (Babakus & Boiler, 1992; Crompton & MacKay, 1989; Cronin & Taylor, 1992; Johnson, Dotson, & Dunlap, 1988; Webster, 1989). Subsequent research in a variety of service settings, however, suggests that the SERVQUAL instrument including five dimensions is not appropriate in accurately measuring service quality across industries (Asubonteng et al., 1996; Babakus & Boiler, 1992; Bowers, Swan, & Koehler, 1994; Brown, Churchill, & Peter, 1993; Buttle, 1996; Carman, 1990; Cronin & Taylor, 1992; Finn & Lamb, 1991; Rowley, 1996).

Electronic Service Quality (E-SQ): Conceptual Foundations

From the literature, the researchers found out that the terms website service quality and online service quality have been used interchangeably by researchers (Aladwani & Palvia, 2002; Lee & Lin, 2005; Piccoli et al., 2004; Riel et al., 2001; Zeithaml, 2002). Being one of the pioneers who introduced the concept of electronic service quality (e-SQ) and who examined the service quality of websites as well as their role in service quality delivery to customers, Zeithaml defined e-SQ and website service quality as "the extent to which a website facilitates efficient and effective shopping, purchasing and delivery of products and services" (2002). Based on the above definition, it is evident that it suggests that the quality of a website is to provide sufficient service to customers to comfortably and confidently do shopping, expecting fast delivery and reliable service. In order to achieve that, companies should understand customers' perceptions about service quality and how customers evaluate it (Zeithaml et al., 2002).

E-SQ dimensions as identified by previous researchers including Gounaris & Dimitriadis (2003), Loiacono et al. (2002) and Novak et al. (2000) have their origin in the Technology Acceptance Model (TAM) developed by Davis (1989). Davis defined perceived technology ease of use as "the degree to which the prospective user expects the target system to be free of effort". He further perceived technology usefulness as "the degree to which a person believes that using a particular system would enhance his or her job performance". These dimensions can help companies to predict consumers' behavior when they decide to use a specific technology. It is proposed that the ease of use and usefulness of using a particular system affect the customers' adoption of this system (Davis, 1989; Davis et al., 1989; Legris et al., 2003). Ziethaml et al. (2002) describe eight dimensions used by customers when they evaluate e- SQ and the quality of websites. These criteria are: 1) Information availability: It refers to the availability and sufficiency of information that can help consumers search any relevant information to any products they are interested to fully enquire about. 2) Ease of use/usability: It refers to the easiness of using the web site.This easiness can include downloading speed, design and organization. 3) Privacy/security: Privacy refers to the degree of protecting customers' personal information by not sharing their personal information with other websites (as in selling lists), protecting anonymity, and providing informed consent. As far as security is concerned, it refers to protecting users from the risk of fraud and financial loss when they use their credit card or any other financial information. Security also refers to providing data confidentiality, security auditing, encryption, and anti-virus protection. 4) Graphic style: It refers to the attributes of a website in terms of choice of colors, layout, print size and type, photographs, graphics, animation, 3D-effects and multimedia. 5) Fulfillment/reliability: It refers to the actual performance of the company, rather than with the website performance; defined as the provider's ability to deliver the service or product as promised. 6) Access: It refers to the availability of the contact information on the company's website. 7) Responsiveness: It refers to the promptness with which the company's personnel give feedback to customers via e-mails. 8) Personalization: It refers to the website's ability to address customers' preferences by providing personalized and customized services.

Zeithaml et al. (2002), in another exploratory study, categorized dimensions into four core and three recovery dimensions that can be used to measure customers' perceptions of e-SQ. The core dimensions include: 1) Efficiency: It refers to the customer's ability to effectively access the website and check any relevant information with minimal effort. 2) Fulfillment: It refers to the company's actual performance in respect to the accuracy of service promises, the availability of products in stock, and delivery time. 3) Reliability: It refers to the technical functioning of a website, such as the extent to which it is available and functions properly. 4) Privacy: It refers to the company's will and ability to maintain the integrity of customer data. The three recovery dimensions are mainly concerned with situations in which a problem needs to be solved and a "personal service" is required. These recovery dimensions include: 5) Responsiveness: It refers to the company's ability to provide an appropriate problem-solving mechanism, i.e. online complaint ability, handling returns mechanism, online guarantees. 6) Compensation: It refers to a money-back guarantee, return of shipping and handling costs. 7) Contact points: It refers to the company's ability to offer a live contact and customer support in real-time via online or other communication means. In the same line, Parasuraman (2004) suggests that there are eleven criteria of e-SQ that influence customers' perceptions about website quality and e-SQ. These criteria include Access, Ease of navigation, Efficiency, Customization/personalization, Security/privacy, Responsiveness, Assurance/trust, Price knowledge, Site aesthetics, Reliability, and Flexibility.

Cronin and Taylor (1992) examine the causal relationships among service quality, customer satisfaction, and behavioral intention. Each variable was measured by one item. There were 660 usable questionnaires that were randomly collected to examine the causal relationship among the above three variables. Dabholkar et al. (2000) also found out that customer satisfaction strongly mediated the effect of service quality on behavioral intentions. The data used in their study were systematically randomly collected from 397 churches. A test of discriminant validity revealed that the construct of service quality was different from the construct of customer satisfaction. The result of regression analysis in structural equations modeling supported their proposition that customer satisfaction had a stronger effect on behavioral intentions than service quality did (Dabholkar et al., 2000). Service quality literature indicated that perceptions of high service quality and high service satisfaction resulted in a very high level of purchase intentions (Cronin & Taylor, 1992; Zeithaml et al., 1996). Cöner & Güngör (2002) claim that customer' loyalty was affected by product quality, service quality, and retailer image. They also suggest "quality of product and service to be directly related to customer satisfaction, and leads to the loyalty of the customer". Customer satisfaction literature shows that the relationship between customer satisfaction and customer behavioral intention depends on the type of satisfaction. The positive impact of manifest satisfaction on customer behavioral intention was stronger than that of latent satisfaction on customer behavioral intention (Bloemer & Kasper, 1995; Bloemer & Ruyter, 1998). Based on empirical findings, service quality is one of the antecedents of satisfaction in the sense that if there is service quality, there is a possibility for obtaining the customers satisfaction (Cronin & Taylor, 1992, 1994; Spreng & Mackoy, 1996). This satisfaction can result in customer's loyalty intention to the service provider (Cöner & Güngör, 2002; Cronin & Taylor, 1992; 1994; Dabholkar et al., 2000).

CONCEPTUAL RESEARCH MODEL AND HYPOTHESES

A broad review of relevant literature in marketing, service quality, information systems service quality, e-service quality provide us with a foundation for developing the theoretical framework. As determined by previous research findings, both identify the variables that might be important. This study found that m-learning in academic environment is a new area of research, majority of the coming dimensions are new in the sense of this area of study, as it is very difficult to find related studies, supported by evidence, that focus on service quality, learner satisfaction, and their influences on learner behavior intentions in the m-learning environment but similarities between mobile and e-services can be expected since both are Internet-based. In this research, much attention is paid to the measurement model of service quality in m-learning in university environment based on the well-known SERVQUAL model. Many previous studies related to service quality suggest that it should be necessary to add and modify items of the SERVQUAL scale, developed by Parasuraman et al. (1985, 1988), and to create a unique and comprehensive conceptual model of service quality, depending on the nature of the service sector under investigation (Carman, 1990; Cronin & Taylor, 1992; 1994; Finn & Lamb, 1991; Parasuraman & Grewal, 2000). Based upon this suggestion, this study proposed and tested a multi-dimensional model of service quality for m-learning and examined the relationships between service quality and other significant constructs.

To understand and organize the m-learning services quality in university environment, for theoretical purposes, it was organized by dimensions and sub dimensions taking into account that mobile services can be considered a subgroup of electronic services. The sub dimensions adopted by this study were interface design, content usefulness, content adequacy, easy of use, accessibility, reliability, interactivity, responsiveness, trust, and personalization. The adoption of these dimensions that seems to have influenced the student's overall perception about m-learning service quality in university environment and student's satisfaction which are effects on the formation of his behavioral intention. While the overall service quality was determined by three dimensions: service quality; information quality; and system quality which are described next.

- Service quality: Learners' attitude formed by a long-term, overall evaluation of an m-learning system performance. The dimension of service quality is measured using five sub-dimensions: interface design; reliability; responsiveness; trust and personalization.
- Information quality: Learners' perception of the quality of information presented on a mobile portal. The dimension of information quality is measured by two sub-dimensions: content usefulness, content adequacy.

- System quality: This refers to learners' perception of a mobile portals performance in information retrieval and delivery. The dimension of system quality is measured by three sub-dimensions, including; ease of use, accessibility and interactivity.

Service Quality Dimensions

The first dimension interface design refers to the appearance of mobile portal and is consistent with the tangibility dimension in the SERVQUAL model. While Parasuraman et al. (1988) define a tangible dimension as the physical appearance, such as facilities, equipment, and personnel, many researchers replace this definition with the user interface required for adapting to the e-service context (Aladwania & Palvia, 2002; Lee & Lin, 2005; Loiacono et al., 2002; Riel et al., 2001; Wolfinbarger & Gilly, 2003). The reliability dimension in the SERVQUAL model is composed of the consistency, dependability, and accuracy of promised service performance (Parasuraman et al., 1988). Studies of new service-delivery options available with computer technology found that consistency and dependability of performance is an important dimension in the measurement of service quality, because of the user's consideration of performance risks based on new technology service (Cox & Dale, 2001; Dabholkar, 1996; Davis, Bagozzi, & Warshaw, 1992). This is very relevant for m-learning services considering the fact that learners are on the move and often in time-critical situations. The third dimension is responsiveness which is similar to the responsiveness dimension in the SERVQUAL model. The SERVQUAL model (Parasuraman et al., 1988) defines responsiveness as the willingness of employees to provide prompt service and to deal with consumer complaints. According to Wang (2003), "responsiveness" measures the company's ability to support customers with the appropriate information when a problem occurs; it also refers to the mechanism for handling returns, and the ability to carry out arrangement for online guarantees. A quick response to customers' request is an indication that the company is customer-oriented.This, by its turn, is going to overcome the issue of uncertainty and increase the perceived convenience of customers (Gummerus et al., 2004; Wolfinbarger & Gilly, 2003). According to Lynch and Lundquist (1996) trust is at the center of e-service with much academic discourse surrounding the security, privacy, and confidence which is similar to the assurance dimension in the SERVQUAL model. Kimery and McCard (2002) argue that "trust is the user willingness to accept the vulnerability of an online transaction based on their positive expectations regarding future online provider behaviors". Reichheld & Schefter (2000) point out that trust is a significant antecedent of participation in online settings because of the increased ease with which online transaction can behave opportunistically. Personalization in relation to service quality has been defined as caring, individualized attention for the consumer and subject knowledge of employees (Parasuraman et al., 1988). Riel et al. (2001) additionally define Personalization, in the e-service context, as the degree of customization of communication and service provider awareness of consumer needs in the e-service context. Personalization a key feature of most e-commerce and m-commerce business models because it offers real values for a customer and creates a perception of high-quality service. The heart of the personalization is to satisfy the individual customer's unique needs (Huang & Lin, 2005; Riel et al., 2001). Thus we can have the hypotheses as follows.

H1: *There will be a significant positive relationship between interface design and overall learner perceived service quality.*

H2: *There will be a significant positive relationship between reliability and overall learner perceived service quality.*

H3: *There will be a significant positive relationship between responsiveness and overall learner perceived service quality.*

H4: *There will be a significant positive relationship between trust and overall learner perceived service quality.*

H5: *There will be a significant positive relationship between personalization and overall learner perceived service quality.*

Information Quality Dimensions

Scholars in the area of traditional computing settings have established well-known models to measure information quality. In a study on the determinants of information system success, Delone and McLean (1992) assumed that the information system success is measured by service quality and information quality. Delone and McLean (1992) highlighted the importance of relevance, timeliness, and accuracy of information. In the same vein, Doll and Torkzadeh (1988; 1994) also emphasized three determinants of user satisfaction: content, accuracy, and timeliness. The dimensions suggested in these two studies and others (Aladwania & Palvia, 2002; Belcher et al. 2000; Bitner et al. 2000; Koller, 2001; Loiacono et al., 2002) could be classified into content usefulness and content adequacy. Content usefulness: This refers to the value, reliability, currency, and accuracy of information. Specifically, information value is concerned with relevancy and clearness. Information reliability refers to its accuracy, dependability, and consistency. Information currency is concerned with information timeliness and continuous update. Information accuracy describes the degree to which the system information is free from error. Content adequacy: This refers to the extent of completeness of information. Mobile portals need to provide information to facilitate learner understanding of the materials and services offered. In addition, learners need supplemental services, such as university information, professional advice, research reports, hyperlinks to relevant sites and contact information. In order to achieve high levels of the learners' perceived mobile service quality, universities should particularly focus on the content usefulness and content adequacy dimensions (Kim & Ong, 2005; Yang et al., 2004). Thus the following hypotheses can be proposed.

H6: *There will be a significant positive relationship between content usefulness and overall learner perceived service quality.*

H7: *There will be a significant positive relationship between content adequacy and overall learner perceived service quality.*

System Quality Dimensions

This refers to learners' perception of a mobile portals performance in information retrieval and delivery. Research on IS has resulted in various instruments to measure system quality. Among them, ease of use has been regarded as the most frequently used factor in measuring IS success or user satisfaction. In the context of Web sites, researchers have identified system quality factors such as interactivity, navigation, access, hyperlinks, and entertainment (Loiacono et al., 2002; Madu & Madu, 2002; McKinney et al., 2002; Riel et al., 2001; Zhang & Von Dran, 2002). The researchers studied the above dimensions and developed a conceptual model that categorized these dimensions into three major attributes: ease of use, accessibility, and interactivity. Davis (1989) defines ease of use as "the degree to which a person believes that using a particular system would be free of effort". Thompson (2001) defines it as "the degree to which the user expects the use of the system to be user friendly". In the m-learning environment, the likelihood of adoption and use of a system is increased when that system is easy to use in finding what learners are looking for, receiving course content, supporting services, and student-tutor interaction (Costabile et al., 2005; Uther, 2002). In summary, "ease of use" plays a vital role for learners' perceived mobile service quality (Chiu at al., 2004; Costabile et al., 2005, Papanikolaou & Mavromoustakos, 2006). Accessibility: In traditional marketing literature, speed of delivery is defined as the time it takes to actively perform the service (Dabholkar, 1996). Maister (1985) argues that if consumers perceive that service is delivered quickly, they are likely to evaluate the service more highly. In the e-services environment, speed of access may also be an important factor to lure users to a particular website (Sohn, 2000). The last dimension for measuring system quality is *interactivity*. Interactivity, in the current model, refers to the interactive relationship between the university, students, and other students on the mobile learning portal through e-mail communication methods and discussion group activities. According to Aldridge and Rowley (1998), Papanikolaou and Mavromoustakos (2006), Wang (2003), and Yang et al. (2004), communication methods like e-mail and discussion group are important tool for students to use in communicating complaints and inquiries to both university and teachers. Moreover, online community features, such as chat-rooms, guest books, talk forums, and message boards on the portals are effective methods for maintaining interactive relationships with learners (Khalifa & Shen, 2005). Therefore, the following hypotheses can be formed.

H8: *There will be a significant positive relationship between ease of use and overall learner perceived service quality.*

H9: *There will be a significant positive relationship between accessibility and overall learner perceived service quality.*

H10: *There will be a significant positive relationship between interactivity and overall learner perceived service quality.*

The Relationships between Service Quality, Satisfaction, and Behavioral Intention

In the research literature, there are three different relationships between service quality (SQ), consumer satisfaction, and behavioral intention has been reported. The first approach relates that satisfaction is an antecedent of perceived service quality (Satisfaction → SQ) and service quality directly influences behavioral intentions (SQ → Behavioral Intention) (Bitner, 1990; Bitner & Hubbert, 1994; Bolton & Drew, 1991; Oliver, 1981; Zeithaml et al., 1996). In the second approach, service quality is modeled as an antecedent of satisfaction (SQ → Satisfaction) and satisfaction directly influences behavioral intentions (Satisfaction → Behavioral Intention) (Anderson & Sullivan, 1993; Cronin & Taylor, 1992; Gotlieb et al., 1994; Taylor & Baker, 1994). In the final approach, there is a non-recursive relationship between service quality and satisfaction (SQ ←→ Satisfaction); that is, neither service quality nor satisfaction is an antecedent of the other (McAlexander et al., 1994; Taylor & Cronin, 1994). Because of the absence of empirical support for the first and third approaches, and the relatively strong empirical support for the second approach, the second approach (SQ → Satisfaction → Intention) is applied to the proposed conceptual model for this study. The direct paths from both overall service quality to learner satisfaction, and the direct path from learner satisfaction to behavioral intention on the m-learning services, are specified in the proposed conceptual model. Thus the following hypothesis can be proposed.

H11: *The learner perception of overall service quality will directly influence the level of learner satisfaction.*

H12: *The level of learner satisfaction will directly influence the behavioral intention.*

Figure 1. Hypotheses Research Model

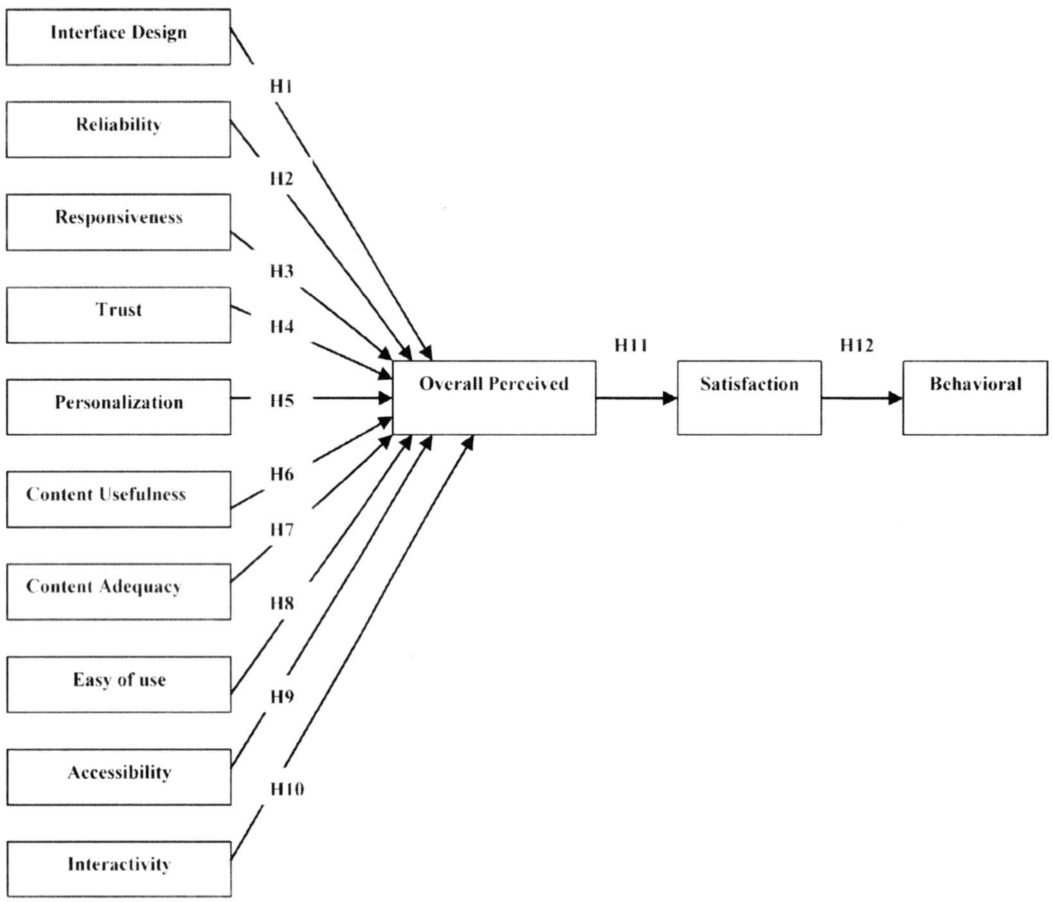

Based on the comprehensive literature review and the in-depth discussions mentioned above, provide us with a foundation for developing the theoretical framework for learner perceived service quality, learner satisfaction, and their impacts on behavioral intentions of learners is presented as Figure 1, in which the relevant hypotheses to be examined empirically are shown.

RESEARCH METHODOLOGY

Pre-Test

A pre-test of the questionnaire was conducted to assess the content validity of the measurement scales. Content validity is how well the instrument adequately measures the subject domain, (Sekaran, 2000). Adequate content validity is a matter of judgment and usually experts in the field are called upon to render a judgment. Content validity was supported several ways during the pre-test and pilot test. A copy of the questionnaire was circulated, and asked the reviewers and the participants to write down any comments needed to improve the quality of the questioner; some items were reworded, added or deleted based on their feedback. A copy of such feedback was kept for future reference. Academic domain experts with extensive experience and knowledge of service quality models, usability, and systems analysis were consulted several times at various points during the development of the instruments.

Pilot Test

The questionnaire was pilot testing to gain additional support for content validity and to obtain initial indications about construct validity and reliability, (Sekaran, 2000). The questionnaire was pilot tested with 30 students, 18 were graduate PhD and Master Information Technology students, and 12 graduate students were from various faculties. The sample was not only taken from IT students, as the concept of mobile learning service was thought to be universal and thus the data source not critical. The data were analyzed with reliability analysis using Cronbach's alpha, which is based on the average correlation of items within a test if the items are standardized. If the items are not standardized, it is based on the average covariance among the items. Some items were reworded, added or deleted based on the pilot study.

Measures

The final questionnaire consisted of six pages. Page one contains a brief statement about the research, objectives, instructions, definition of key terms, and a statement to the effect that completing and returning the survey constituted consent of the participant. The five inside pages were divided into 14 parts. Part 1 contains student background information. Part 2 contains 4 Likert 5-scale questions about the Interface Design of mobile learning portal. Part 3 contains 5 Likert 5-scale questions about the Reliability of mobile learning services. Part 4 contains 4 Likert 5-scale questions about the Responsiveness of mobile learning services.

Part 5 contains 6 Likert 5-scale questions about Trust in mobile learning services. Part 6 consists of 5 Likert 5-scale questions about Personalization of mobile learning services. Part 7 consists of 5 Likert 5-scale questions about Content Usefulness of mobile learning services. Part 8 contains 5 Likert 5-scale questions about Content Adequacy of mobile learning services. Part 9 contains 4 Likert 5-scale questions about Ease of Use mobile learning services. Part 10 consists of 4 Likert 5-scale questions about Accessibility of mobile learning services. Part 11 contains 4 Likert 5-scale questions about Interactivity of mobile learning services. Part 12 consists of 4 Likert 5-scale questions about learner perceived Service Quality of mobile learning. Part 13 consists of 4 Likert 5-scale questions about learner Satisfaction with mobile learning services and finally Part 14 contains 4 Likert 5-scale questions about learner Behavioral Intention to use in future.

Sample

The subjects for this study were confined to the mobile leaning users who have experienced mobile learning services. A screening question "Have you used mobile leaning before?" was asked before the self-administered surveys were given out. Out of the 341 students obtained, 82 were incomplete or contained unreliable answers. 259 students representing different colleges of University Utara Malaysia (UUM) were deemed usable. Of students responding, 50.8% were male and 49.2% female, 88.3% under the age of 26, while 11.3% between 26-40 years old. In terms of student's level, undergraduate made up the largest number with 76.4%, followed by master degree with 20.9%, and PhD students 2.7%. In terms of students' college, College of Art and Science (CAS) made up the largest group of respondents (58.5%), followed by College of Business (COB) with 30.2% and College of Law, Government and International Studies (COLGIS) 11.2%.

EMPIRICAL RESULTS

The validity and reliability of the instrument were evaluated. Construct validity was examined by factor analysis, using the principal components method with a varimax rotation. The final Exploratory Factor Analysis (EFA) solution resulted in an average of four items per dimension with one dimensions having only two items. Factor reliabilities, as represented with Cronbach's alpha in the last column in Table 1, were between 0.71 and 0.82 for each factor. The reliability coefficients above 0.60 are typically considered satisfactory (Pallant, 2001). Principal component factor analysis was performed and nine constructs were extracted one less than the number of constructs included in the model. As shown in table 1, there were no cross-loading items. Table 1 reports the factor loading, eigenvalue values, cumulative variance explained and Kaiser-Meyer-Olkin measure of sampling adequacy. Additionally, items intended to measure the same construct exhibited prominently and distinctly higher factor loadings on a single construct than on other constructs, suggesting adequate convergent and discriminant validity (Hair et al., 1998) jointly, the observed reliability and construct validity suggested adequacy of the measurements used in the study.

Despite face validity, the measures of responsiveness do not load on a single factor. Thus, this construct is not distinct aspects of m-learning service quality. Instead, it seems that the responsiveness construct is an attribute that load across the other dimensions. For instance, items within the construct "responsiveness" either loaded on wrong constructs or loaded across several other constructs such as reliability, and interface design, thereby making the construct problematic. Results were consistent with the findings of other studies related to traditional service quality, which have demonstrated the unstable nature of some SERVQUAL constructs (e.g., Llosa et al., 1998; Yang, 2001). The empirical evidence suggests that the other m-learning service quality dimensions require certain aspects of the above constructs. Therefore, this construct was eliminated from the m-learning service quality model.

Model Evaluation

Three models for each indicator of mobile learning services were estimated. Model-1 deals with the results of overall service quality of mobile learning, Model-2 deals with learner satisfaction with mobile learning service, and Model-3 deals with behavioral intention to use mobile learning services in future. Table 2 demonstrates that the correlations between "overall service quality" and nine mobile learning service quality factors were consistently high. The correlation coefficient indicates significant positive relationship between the independents and dependent variable at the 0.01 level. The regression analysis used overall service quality as the dependent variable and the nine mobile learning service quality factors as the independent variable. Pursuant to the initial regression run, the outliers were detected by examining the standardized residual. One outlier were found and eliminated. Missing values were handled by choosing the option "exclude cases pairwise", which means that only cases with complete data for the pair of constructs being correlated were used to compute the correlation coefficient on which the regression analysis is based. This procedure produced 258 effective samples. The mean of all scale items within a factor was used to represent that factor.

Table 3 shows the results of multiple regression analysis. Nine factors take account of 60 percentage of explained variance of overall service quality evaluation, which is significant as indicated by the F-value. F is the value calculated for the F statistic by SPSS. The significant value (Sig.) is the likelihood of committing a Type I error after rejecting the null hypothesis. In this case, Sig. =.000 ($p < .001$) and Sig. =.00 ($p < .01$) which is less than $P < .05$ (the criterion alpha level). Therefore, the regression equation as computed is statistically significant for all factors. Only the "personalization" factor do not significantly contribute to the overall service quality evaluation.

Table 1. Exploratory factor loadings

	Component									
	1	2	3	4	5	6	7	8	9	Alpha
Interface Design 1	.720									.788
Interface Design 4	.713									
Reliability 1		.760								.811
Reliability 2		.703								
Reliability 3		.736								
Reliability 4		.689								
Trust 1			.681							.783
Trust 2			.816							
Trust 3			.740							
Trust 4			.519							
Personalization 1				.581						.711
Personalization 2				.638						
Personalization 3				.791						
Content Usefulness 1					.635					.752
Content Usefulness 2					.722					
Content Usefulness 3					.756					
Content Adequacy 1						.634				.823
Content Adequacy 2						.637				
Content Adequacy 3						.590				
Content Adequacy 4						.594				
Content Adequacy 5						.606				
Easy to use 1							.772			.740
Easy to use 2							.581			
Easy to use 3							.779			
Easy to use 4							.651			
Accessibility 1								.593		.814
Accessibility 2								.688		
Accessibility 3								.654		
Accessibility 4								.542		
Interactivity 1									.721	.802
Interactivity 2									.722	
Interactivity 3									.727	
Interactivity 4									.515	

To identify the most important factors that determine the learner perception about m-learning service quality in university environment, a stepwise regression was used, where the number of independent variables entered and the order of entry are determined by statistical criteria gener-

Table 2. Correlations matrix between overall service quality and nine m-learning

	Service Quality	Interface Design	Reliability	Trust	Personalization	Content Usefulness	Content Adequacy	Ease to Use	Accessibility
Interface Design	.486								
Reliability	.591	.464							
Trust	.501	.305	.363						
Personalization	.374	.271	.359	.312					
Content Usefulness	.227	.209	.276	.267	.273				
Content Adequacy	.608	.381	.481	.395	.328	.309			
Ease of use	.505	.356	.459	.309	.386	.238	.334		
Accessibility	.429	.207	.365	.358	.366	.324	.288	.395	
Interactivity	.560	.258	.421	.406	.264	.340	.521	.435	.421

Table 3. Regression analysis results between overall service quality and m-learning service quality dimensions

Model		Unstandardized Coefficients		Standardized Coefficients		
		B	Std. Error	Beta	t	Sig.
1	Interface Design	.149	.045	.156	3.349	.001
	Reliability	.167	.046	.189	3.667	.000
	Trust	.167	.049	.158	3.402	.001
	Personalization	.026	.042	.028	.611	.542
	Content Usefulness	.112	.045	.109	2.485	.014
	Content Adequacy	.273	.054	.258	5.041	.000
	Ease of Use	.132	.050	.128	2.616	.009
	Accessibility	.092	.045	.098	2.062	.040
	Interactivity	.162	.048	.174	3.353	.001
Dependent Variable: Overall Service Quality (F = 43.804, P < =.000, R =.783, R^2 =.614, Adjusted R^2 =.600)						
Best Predictor: Content Adequacy (Beta.261, P < =.000, R2 =.369)						

ated by the stepwise procedure. This procedure found under the SPSS output heading Coefficients. This section of the output shows which of the variables statistically significant predictors of the dependent variable are.

To identify the most important factors that determine the learner perception about m-learning service quality in university environment, a stepwise regression was used, where the number of independent variables entered and the order of entry are determined by statistical criteria gener-ated by the stepwise procedure. This procedure found under the SPSS output heading Coefficients. This section of the output shows which of the

Table 4. Regression analysis results between learner satisfaction and overall service quality dimension

Model		Unstandardized Coefficients		Standardized Coefficients	t	Sig.
		B	Std. Error	Beta		
2	(Constant)	1.649	.183		9.007	.000
	Service Quality	.579	.047	.611	12.363	.000
Dependent Variable: Satisfaction (F = 152.480, P < =.000, R =.611, R^2 =.374, Adjusted R^2 =.371)						

Table 5. Regression Analysis Results between Behavioral Intention and Learner Satisfaction Dimension

Model		Unstandardized Coefficients		Standardized Coefficients	t	Sig.
		B	Std. Error	Beta		
3	(Constant)	.999	.237		4.220	.000
	satisfaction	.762	.060	.620	12.634	.000
Dependent Variable: Behavioral Intention (F = 159.628, P < =.000, R =.620, R^2 =.384, Adjusted R^2 =.382)						

variables statistically significant predictors of the dependent variable are. The critical components in this section are Beta and Sig. Beta; the standardized coefficient in is.261, which is the highest in beta values and significant at the.000, while this variable explains 36.9%.

To explore the relationship between overall service quality and learners' satisfaction as well as learners 'behavioral intention on m-learning services in the future several regression analyses were required. The regression analysis used overall service quality as the dependent variable and learner satisfaction as the independent variable. The mean of all scale items within a factor was used to represent that factor. Table 4 shows the results of multiple regression analysis. One factor takes into account the 37.4 percentage of explained variance of satisfaction evaluation, which is significant as indicated by the F-value. F is the value calculated for the F statistic by SPSS. The significant value (Sig.) is the likelihood of committing a Type I error after rejecting the null hypothesis. In this case, Sig. =.000 ($p < .001$) which is less than $P < .05$ (the criterion alpha level). Therefore, the regression equation as computed is statistically significant.

The regression analysis used the learner satisfaction as the dependent variable and behavioral intention as the independent variable. The mean of all scale items within a factor was used to represent that factor. Table 5 shows the results of multiple regression analysis. One factor takes into account the 38.4 percentage of explained variance of behavioral intention evaluation, which is significant as indicated by the F-value. F is the value calculated for the F statistic by SPSS. The significant value (Sig.) is the likelihood of committing a Type I error after rejecting the null hypothesis. In this case, Sig. =.000 ($p < .001$) which is less than $P < .05$ (the criterion alpha level). Therefore, the regression equation as computed is statistically significant.

DISCUSSION OF FINDINGS AND CONCLUSION

This section summarizes and discusses the results associated with the development of a suitable instrument for measuring student's perception about m-learning service quality, learner satisfaction with m-learning service, and learner behavioral intention to use m-learning services in future. This includes validity and reliability definition and analysis as well as content and face validity. The instrument was pilot tested, evaluated, and refined. Sample data was analyzed (typically with exploratory factor analysis) and rough assessments were made of validity and reliability. A major issue was adjusting the scales items: deleting, adding, and rewording. Also a full detailed of the instruments used for data analysis by exploratory factor analysis was provided, which was used to test the relationship between each set of manifest variables and the associated dimension. Those questions (manifest variables) found to have significant loadings, as they represent the common set of items that students agree constitute the m-learning service quality factors.

The results of this study suggest that interface design, reliability, trust, content usefulness, content adequacy, ease of use, accessibility, and interactivity are important factors of m-learning service quality in university environment. In addition, the "personalization" factor does not significantly contribute to the overall service quality evaluation. Previous research has also yielded the same result (Gefen, 2002; Carman, 1990). Personalization means the degree to which an m-learning system attends to, understands, and adapts to the needs of individual learners. Learners definitely expect to receive the information and services that they have requested on time or promptly from a knowledgeable m-learning system. However, it seems that they adopt a utilitarian orientation in online learning and do not care or expect much personalized service from an m-learning system. This might be due to the characteristic of online learning that human interaction between learners and m-learning systems is very minimal, which makes personalization a somewhat less experienced and less important aspect of service quality. Zeithaml et al. (2002) also argue that the personalization dimension is not critical in the transactional aspects of online service. For this reason, personalization dimension seems to have an insignificant role in m-learning service quality context.

The primary data obtained from university students supports the findings in recent studies. Content adequacy, reliability, and interactivity in particular, are the most relevant and significant motivators of m-learning service quality. In addition, the study provides support for the important role of the SERVQUAL's interface design reliability and trust constructs. Content usefulness, content adequacy, ease of use, accessibility, and interactivity were also found to exert strong influence in this study of academic environment. Moreover, the research results indicated that there are statistically positive causal relationship between learner satisfaction and overall service quality, these results imply that learner satisfaction increases when the perceptions of overall service quality are increased. In addition results indicated that the direct causal effect of learner satisfaction on learner behavioral intention was statistically significant and positive. This suggests that an increase in learner satisfaction improves the probability that the learner will use m-learning services in the future.

CONTRIBUTIONS AND FUTURE RESEARCH

This research investigated factors that lead to service quality of m-learning in university environment. It explored the causal relationship between service quality, learner satisfaction, and learner behavioral intention to use m-learning services in future, which potentially trigger a new stream of research. The following is a brief discussion of the most important contributions this research offered for theory and practice. These contributions add to the benefit of the students, academic organizations and hopefully to society and other non-academic organizations. This study emphasized that if a university has rewarded students with good academic services, it will contribute to the welfare of the student's achievements and therefore to the society. In addition the findings can be in much of use to m-government, m-commerce, and m-learning industry services implementations in general.

Although many researchers have conducted studies designed to understand the concept and develop the reliable and valid measurement of service quality in different areas, the conceptual gap of service quality still exists in the academic domain. Moreover, it is very difficult to find related studies, supported by evidence, that focus on service quality of m-learning in university environment. The major contribution of this research can be argued from the view that it is among the pioneers to investigate the factors that lead to service quality of m-learning, learner satisfaction and their influences on learner behavior intentions in university environment. Also as important contribution to the general evaluation of service quality research, this study has successfully developed the extended SERVQUAL model, in m-learning service quality context. It shows that the SERVQUAL is partially supported in m-learning services context. In addition to all achievements that were mentioned earlier, the results are in line with the overall findings across several studies in the information systems service quality area (e.g. Pitt et al., 1995; Kettinger & Lee, 1997; Jiang et al., 2000; Yang et al., 2004; Jiang et al., 2002). The results reinforced the role of interface design, reliability, and trust factors as the fundamental drivers for m-learning service quality. Another contribution goes beyond the traditional SERVQUAL and includes important additional variables such as content adequacy. Content adequacy factor has the most significant effect on m-learning service quality. Therefore, consideration of university with content adequacy is a prerequisite and not an afterthought for successful m-learning system implementation.

An additional investigation is needed for this stream of research. Numerous opportunities are available to study a wide range of topics. Some specific ideas are:

Investigation into the differences of perceptions about m-learning service quality for different organizations. Due to the scope of this study, which is, limited to the university environment, other organizations such as government and business enterprises should also be studied to find the influence of m-learning service in it is services enhancement.

Although this study attempts to cover all aspects of overall service quality for the m-learning, there may be some other aspects that may have been omitted from the proposed conceptual model or over looked. Any future study, therefore, needs to continuously refine the scale of overall service quality, proposed and supported in this study, and to incorporate new aspects of overall service quality into the proposed scale to verify the measure of overall service quality

The current study is limited to examining the causal relationships between overall service quality, learner satisfaction, and learner behavioral intention to use. In any future study, other significant variables, such as learner loyalty and value of m-learning services, may be included in the proposed causal structure model.

As the online learning has become mature, learners will shape clear expectations for e-service quality attributes. More and more industry-wide, service standards will be set forth and be accepted. Thus, future studies may utilize the expectation-disconfirmation paradigm to measure service quality and learner satisfaction.

A study on the effect of the globalization trends and the universal moiled interaction services based on geography and cultural differences of user preferences.

REFERENCES

Al-Mushasha, N. F., & Shahizan, H. (2007). *A study on the perception of students towards mobile learning services in university environment.* Paper presented at the Rural ICT Development Conference (RICTD '07) EDC, UUM, 20-21.

Aladwani, A. M., & Palvia, P. C. (2002). Developing and validating and instrument for measuring user-perceived Web quality. *nformation & Management, 39*(6), 467-476.

Aldridge, S., & Rowley, R. (1998). Measuring customer satisfaction in higher education. *Quality Assurance in Education, 6*(4), 197–204. doi:10.1108/09684889810242182

Anderson, E. W., & Sullivan, M. W. (1993). The antecedents and consequences of customer satisfaction for firms. *Marketing Science, 12*(2), 125–143. doi:10.1287/mksc.12.2.125

Asubonteng, P., McCleary, K. J., & Swan, J. E. (1996). SERVQUAL revisited: A critical review of service quality. *Journal of Services Marketing, 10*(6), 62–81. doi:10.1108/08876049610148602

Babakus, E., & Boiler, G. W. (1992). An empirical assessment of the SERVQUAL scale. *Journal of Business Research, 24*(2), 253–268. doi:10.1016/0148-2963(92)90022-4

Belcher, M., Place, E., & Conole, G. (2000). Quality assurance in subject gateways: creating high quality portals on the internet. *Quality Assurance in Education, 8*(1), 38–47. doi:10.1108/09684880010312686

Berry, L. L., Parasuraman, A., & Zeithaml, A. V. (1994). Improving Service Quality in America: Lessons Learned. *The Academy of Management Executive, 8*(2), 32–52.

Bitner, M. J. (1990). Evaluating service encounters: The effects of physical surroundings and employee responses. *JMR, Journal of Marketing Research, 45*(1), 69–82.

Bitner, M. J., Brown, S. W., & Meuter, M. L. (2000). Technology infusion in service encounters. *Journal of the Academy of Marketing Science, 8*(1), 138–149. doi:10.1177/0092070300281013

Bitner, M. J., & Hubbert, A. R. (1994). *Service Quality: New directions in theory and practice* (1 ed.): Thousand Oaks, CA: Sage Publications.

Bloemer, J., & Kasper, H. (1995). The complex relationship between consumer satisfaction and brand loyalty. *Journal of Economic Psychology, 16*(2), 311–329. doi:10.1016/0167-4870(95)00007-B

Bloemer, J., & Ruyter, K. (1998). Integrating Service Quality and Satisfaction: Pain in the Neck or Marketing Opportunity? *Journal of Consumer Satisfaction Dissatisfaction and Complaining Behavior, 8*(1), 44–52.

Bolton, R. N., & Drew, J. (1991). A Longitudinal Analysis of the Impact of Service Changes on Customer Attitudes. *JMR, Journal of Marketing Research, 55*(January), 1–9.

Bowers, M. R., Swan, J. E., & Koehler, W. F. (1994). What attributes determine quality and satisfaction with health care delivery. *Health Care Management Review, 19*(4), 49–55.

Brown, J., Churchill, A., & Peter, J. (1993). Improving the Measurement of Service Quality. *Journal of Retailing, 69*(1), 127–139. doi:10.1016/S0022-4359(05)80006-5

Buttle, F. (1996). SERVQUAL: Review, critique, research agenda. *European Journal of Marketing, 30*(1), 8–32. doi:10.1108/03090569610105762

Buzzel, R. D., & Gale, B. T. (1987). *The PIMSprinciples: Linking strategy to performance* (1 ed.): New York, NY: Free Press.

Carman, J. M. (1990). Consumer perceptions of service quality: An assessment of the SERVQUAL dimensions. *Journal of Retailing, 66*(1), 33–55.

Chiu, C., Hsu, M., Sun, S., Lin, T., & Sun, P. (2004). Usability, quality, value and e-learning continuance decisions. *Computers & Education*, *5*(2).

Cöner, A., & Güngör, M. (2002). Factors affecting customer loyalty in the competitive Turkish metropolitan retail markets. *Journal of American Academy of Business*, *2*(1), 189–195.

Costabile, M. F., De Marsico, M., Lanzilotti, R., Plantamura, V. L., & Roselli, T. (2005). *On the Usability Evaluation of E-Learning Applications*. Paper presented at the 38th Hawaii International Conference on System Sciences, Hawaii.

Cox, R., & Dale, B. G. (2001). Service quality and e-commerce: An exploratory analysis. *Managing Service Quality*, *77*(2), 121–131. doi:10.1108/09604520110387257

Crompton, J. L., & MacKay, K. J. (1989). Users' perceptions of the relative importance of service quality dimensions in selected public recreation programs. *Leisure Sciences*, *11*(3), 367–375. doi:10.1080/01490408909512233

Cronin, J. J., & Taylor, S. A. (1992). Measuring service quality: A reexamination and extension. *JMR, Journal of Marketing Research*, *56*(3), 55–68.

Cronin, J. J., & Taylor, S. A. (1994). Servperf Versus Servqual: Reconciling performance-Based And Perceptions-Minus-Expectations Measurement Of Service Quality. *Journal of Marketing*, *58*(3), 125–131. doi:10.2307/1252256

Dabholkar, A., Shepherd, C., & Thorpe, I. (2000). A Comprehensive Framework for Service Quality: An Investigation of Critical Conceptual and Measurement Issues Through a Longitudinal Study. *Journal of Retailing*, *76*(2), 139–173. doi:10.1016/S0022-4359(00)00029-4

Dabholkar, P. A. (1996). Consumer Evaluations of New Technology based Self-Service Options: An Investigation of Alternative Models of SO. *International Journal of Research in Marketing*, *13*(1), 29–51. doi:10.1016/0167-8116(95)00027-5

Davis, D., Bagozzi, R., & Warshaw, R. (1989). User Acceptance of Computer Technology: A Comparison of Two Theoretical Models. *Management Science*, *35*(8), 982–1003. doi:10.1287/mnsc.35.8.982

Davis, F. D. (1989). Perceived Usefulness, Perceived Ease of Use, and User Acceptance of Information Technology. *Management Information Systems Quarterly*, *13*(3), 319–339. doi:10.2307/249008

Davis, F. D., Bagozzi, R. P., & Warshaw, P. R. (1992). Extrinsic and intrinsic motivation to use computers in the workplace. *Journal of Applied Social Psychology*, *22*(14), 1109–1130. doi:10.1111/j.1559-1816.1992.tb00945.x

Dedeke, A. (2003). Service quality: a fulfillment-oriented and interaction-centred approach. *Managing Service Quality*, *13*(4), 276–289. doi:10.1108/09604520310484699

DeLone, W. H., & McLean, E. R. (1992). Information Systems Success: The Quest for the Dependent Variable. *Information Systems Research*, *3*(1), 60–95. doi:10.1287/isre.3.1.60

Doll, W. J., & Torkzadeh, G. (1988). The measurement of end-user computing satisfaction. *Management Information Systems Quarterly*, *12*(2), 259–274. doi:10.2307/248851

Doll, W. J., & Torkzadeh, G. (1994). The measurement of end-user computing satisfaction:Theoretical and methodological issues. *Management Information Systems Quarterly*, (March): 5–10.

Dye, A. (2003). Mobile Education – A Glance at the Future. Retrieved December 25,2005, from http://www.dye.no/articles/a_glance_at_the_future/index.html

Finn, D. W., & Lamb, C. W. (1991). An evaluation of the SERVQUAL scales in a retailing setting. *Advances in Consumer Research. Association for Consumer Research (U. S.)*, *18*(3), 483–490.

Gefen, D. (2002). Customer loyalty in e-commerce. *Journal of the Association for Information Systems*, *13*(1), 27–51.

Ghobadian, A., Speller, S., & Jones, M. (1994). Service quality concepts and models. *International Journal of Quality & Reliability Management*, *77*(9), 43–66. doi:10.1108/02656719410074297

Gotlieb, J. B., Grewal, D., & Brown, S. W. (1994). Consumer satisfaction and perceived quality: Complementary or divergent constructs? *The Journal of Applied Psychology*, *79*(6), 875–885. doi:10.1037/0021-9010.79.6.875

Gounaris, S., & Dimitriadis, S. (2003). Assessing Service Quality on the Web: Evidence from Business-to-Consumer Portals. *Journal of Services Marketing*, *17*(5), 529–548. doi:10.1108/08876040310486302

Grohmann, G., Hofer, A., & Martin, G. (2005). *ARIS MOBILE Helping to define the future of mobile learning*. Paper presented at the International Conference on Mobile Business (ICMB'05) 2005 IEEE.

Gronroos, C. (1982). An Applied Service Marketing Theory. *European Journal of Marketing*, *16*(7), 30–41. doi:10.1108/EUM0000000004859

Gummerus, J., Liljander, V., Pura, M., & Van Riel, A. (2004). Customer loyalty to content-based web sties: the case of an online health-care service. *Journal of Services Marketing*, *18*(3), 175–186. doi:10.1108/08876040410536486

Hair, J. F., Anderson, R. E., Tatham, R. L., & Black, W. C. (1998). *Multivariate data analysis*. Upper Saddle River, NJ: Prentice Hall.

Huang, E. Y., & Lin, C. Y. (2005). Customer-oriented financial service personalization. *Industrial Management & Data Systems*, *105*(1), 26–44. doi:10.1108/02635570510575171

Jensen, J. B., & Markland, R. E. (1996). Improving the application of quality conformance tools in service firms. *Journal of Services Marketing*, *10*(1), 35–55. doi:10.1108/08876049610147838

Jiang, J. J., Klein, G., & Carr, C. L. (2002). Measuring information system service quality: SERVQUAL from the other side. *Management Information Systems Quarterly*, *26*(2), 145–166. doi:10.2307/4132324

Jiang, J. J., Klein, G., & Krampton, S. (2000). A note on SERVQUAL reliability and validity in information system service quality measurement. *Decision Sciences*, *31*(1), 725–744. doi:10.1111/j.1540-5915.2000.tb00940.x

Johnson, L. L., Dotson, M. J., & Dunlap, B. J. (1988). Service quality determinants and effectiveness in the real estate brokerage industry. *Journal of Real Estate Research*, *3*, 21–36.

Kettinger, W. J., & Lee, C. C. (1997). Exploring a 'Gap' model of information services quality. *Information Resources Management Journal*, *8*(3), 5–16.

Kettinger, W. J., Lee, C. C., & Lee, S. (1995). Global measures of information service quality: A cross-national study. *Decision Sciences*, *26*(5), 569–588. doi:10.1111/j.1540-5915.1995.tb01441.x

Khalifa, M., & Shen, N. (2005). *Effects of Electronic Customer Relationship Management on Customer Satisfaction: A Temporal Model*. Paper presented at the Proceeding of the 38th Annual Hawaii International Conference on System Sciences (HICSS'05) Track 7.

Kim, G. M., & Ong, S. M. (2005). An exploratory study of factors influencing m-learning success. *ournal of Computer Information Systems, 46*(1), 92-97.

Kimery, K. M., & McCard, M. (2002). hird-party assurances: mapping the road to trust in e-retailing. *Journal of Information Technology Theory and Application, 4*(2), 63–82.

Koller, M. (2001). Tool offers personalization on the fly. *Internet Week, 857*(6), 19–22.

Lee, G. G., & Lin, H. F. (2005). Customer perceptions of e-service quality in online shopping. *International Journal of retail & Distribution Management, 33*(2), 161-176.

Legris, P., Ingham, J., & Collerette, P. (2003). Why Do People Use Information Technology? A Critical Review of the Technology Acceptance Model. *Information & Management, 40*(5), 191–204. doi:10.1016/S0378-7206(01)00143-4

Levine, L. M. (2002). Campus-wide mobile wireless: Mobility and convergence. *Syllabus, 5*(2), 18–24.

Llosa, S., Jean-Louis, C., & Chiara, O. (1998). An Empirical Study of SERVQUAL's Dimensionality. *The Service Industries Journal, 18*(2), 16–44. doi:10.1080/02642069800000017

Loiacono, E., Watson, R. T., & Goodhue, D. L. (2002). *WEBQUAL: a Measure of Website Quality.* Paper presented at the AMA Winter Conference, Austin, TX.

Lynch, D., & Lundquist, L. (1996). *Digital Money*. New York, NY: Wiley.

Madu, C. N., & Madu, A. A. (2002). Dimensions of e-quality. *International Journal of Quality & Reliability Management, 19*(3), 246–258. doi:10.1108/02656710210415668

Maister, D. H. (1985). *The service encounter: Managing employee/customer interaction in service business* (1 ed.): Lexington, MA: Lexington Books.

Margolies, J. M. (1988). *When good service isn't good enough* (2 ed.): Dallas, TX: Price Waterhouse.

McAlexander, J. H., Kaldenberg, D. O., & Koenig, H. F. (1994). Service quality measurement. *Journal of Health Care Marketing, 3*(2), 34–40.

McGhee, R., & Kozma, R. (2001). *New teacher and student roles in the technology-supported classroom.* Paper presented at the Annual meeting of the American Educational Research Association, Seattle.

McKenzie, J. (2001). The unwired classroom: Wireless computers come of age. *Educational Technology, 10*(4), 19–28.

McKinney, V., Yoon, K., & Zahedi, F. (2002). The measurement of Webcustomer satisfaction: an expectation and disconfirmation approach. *Information Systems Research, 13*(3), 279–296. doi:10.1287/isre.13.3.296.76

Novak, T., Donna, H., & Yung, Y. F. (2000). Measuring the Customer Experience in Online Environments: A Structural Modelling Approach. *Marketing Science, 19*(1), 22–42. doi:10.1287/mksc.19.1.22.15184

Oliver, R. L. (1981). Measurement and evaluation of satisfaction process in retail settings. *Journal of Retailing, 57*(6), 25–48.

Pallant, J. (2001). *SPSS Survival Maual: A step by step guide to data analysis using SPSS for windows (Version 10)*. Australia: Allen & Unwin.

Papanikolaou, K., & Mavromoustakos, S. (2006). *Critical Success Factors for the Development of Mobile Learning Applications.* Paper presented at the 24th IASTED International Multi-Conference Internet and Multimedia Systems and applications Innsbruck, Austria.

Parasuraman, A. (2004). Assessing and Improving Service Performance for Maximum Impact: Insights from a Two-Decade-Long research Journey. *Performance Measurement and Metrics*, *5*(2), 45–52. doi:10.1108/14678040410546064

Parasuraman, A., Berry, L. L., & Zeithaml, V. A. (1991). Refinement and Reassessment of SERVQUAL Scale. *Journal of Retailing*, *67*(4), 420–450.

Parasuraman, A., & Grewal, D. (2000). The impact of technology on the quality-value-loyalty chain: A research agenda. *Journal of the Academy of Marketing Science*, *28*(1), 168–174. doi:10.1177/0092070300281015

Parasuraman, A., Zeithaml, V., & Berry, L. (1994). Alternative Scales for Measuring Service Quality: A Comparative Assessment Based on Psychometric and Diagnostic Criteria. *Journal of Retailing*, *70*(3), 201–230. doi:10.1016/0022-4359(94)90033-7

Parasuraman, A., Zeithaml, V. A., & Berry, L. L. (1985). A Conceptual Model Of Service Quality And Its Implications For Future Research. *Journal of Marketing*, *49*(5), 41–50. doi:10.2307/1251430

Parasuraman, A., Zeithaml, V. A., & Berry, L. L. (1988). SERVQAL: A Multiple Item Scale For Measuring Consumer Perceptions Of Service Quality. *Journal of Retailing*, *64*(4), 35–48.

Piccoli, G., Broham, M., Watson, R., & Parasuraman, A. (2004). Net-Based Customer Service Systems: Evolution and Revolution in Web Site Functionalities. *Decision Sciences*, *35*(3), 423–455. doi:10.1111/j.0011-7315.2004.02620.x

Pitt, L. F., Watson, R. T., & Kavan, C. B. (1995). Service quality: a measure of information effectiveness. *Management Information Systems Quarterly*, *24*(3), 179–187.

Reichheld, F., & Schefter, P. (2000). E-loyalty: your secret weapon on the web. *Harvard Business Review*, *78*, 105–113.

Riel, A. C. R., Liljander, V., & Jurriens, P. (2001). Exploring consumer evaluations of e-services: A portal site. *International Journal of Service Industry Management*, *12*(4), 359–377. doi:10.1108/09564230110405280

Rowley, J. (1996). Retailing and shopping on the Internet. *Internet Research Electronic Networking Applications and Policy*, *5*(1), 81–91.

Sekaran, U. (2000). *Research methods for business: a skill-building approach* (4 ed.): NYC:John Wiley & Sons, Inc.

Sohn, C. A. U. (2000). *Customer evaluation of Internet-based service quality and intention to re-use Internet-based services.* Unpublished doctoral dissertation, Southern Illinois University, Carbondale, Illinois.

Spreng, R. A., & Mackoy, R. D. (1996). An empirical examination of a model of perceived service quality and satisfaction. *Journal of Retailing. Journal of Retailing*, *7*(2), 201–214. doi:10.1016/S0022-4359(96)90014-7

Su, Y., Chen, W., & Wang, T. (2005). *An Investigation of a Mobile Learning System in a Digital Filter Course.* Paper presented at the Fifth IEEE International Conference on Advanced Learning Technologies Washington, DC, USA.

Taylor, S. A., & Baker, T. L. (1994). An assessment of the relationship between service quality and customer satisfaction in the formation of consumers' purchase intentions. *Journal of Retailing*, *70*(2), 163–178. doi:10.1016/0022-4359(94)90013-2

Taylor, S. A., & Cronin, J. J. (1994). Modeling patient satisfaction and service quality. *Journal of Health Care Marketing*, *14*(1), 34–44.

Thompson, S. H. (2001). Demographic and motivation variables associated with Internet usage activities. *Internet Research: Electronic Networking Applications and Policy*, *11*(2), 125–137. doi:10.1108/10662240110695089

Uther, M. (2002). *Mobile Internet Usability: What can 'Mobile Learning' Learn From the Past?* Paper presented at the IEEE International Workshop on Wireless and Mobile Technologies in Education (WMTE'02).

Wang, M. (2003). Assessment of E-Service Quality via E-Satisfaction in E-Commerce Globalization. Retrieved December 25, 2006, from http://www.is.cityu.edu.hk/research/ejisdc/vol11/v11c10.pdf.

Webster, C. (1989). Can consumers be segmented on the basis of their service quality expectations? *Journal of Services Marketing*, *3*, 35–53. doi:10.1108/EUM0000000002485

Wolfinbarger, F., & Gilly, C. (2003). eTailQ: Dimensionalising, Measuring and Predicting eTail Quality. *Journal of Retailing*, *79*(1), 183–198. doi:10.1016/S0022-4359(03)00034-4

Yang, Z. (2001). *Measuring E-Service Quality And Its Linkage To Customer Loyalty*. New Mexico: Unpublished Doctor of Philosophy, New Mexico State University.

Yang, Z., Jun, M., & Tur, P. R. (2004). Measuring customer perceived online service quality. *International Journal of Operations & Production Management*, *24*(11), 1149–1174. doi:10.1108/01443570410563278

Zeithaml, V. (2002). Service Excellence in Electronic Channels. *Managing Service Quality*, *12*(13), 135–138. doi:10.1108/09604520210429187

Zeithaml, V. A., Berry, L. L., & Parasuraman, A. (1996). The behavioral consequences of service quality. *JMR, Journal of Marketing Research*, *60*(2), 31–46.

Zeithaml, V. A., Parasuraman, A., & Berry, L. (1990). Delivering quality service: Balancing customer perceptions and expectations.

Zeithaml, V. A., Parasuraman, A., & Malhotra, A. (2002). Service Quality Delivery Through Web Sites: A Critical Review of Extant Knowledge. *Academy of Marketing Science Journal*, *30*(4), 362–375. doi:10.1177/009207002236911

Zhang, P., & Von Dran, G. (2002). User Expectations and Rankings of Quality Factors in Different Web Sites Domains. *International Journal of Electronic Commerce*, *6*(2), 9–33.

This work was previously published in International Journal of Mobile Computing and Multimedia Communications (IJMCMC) 1(1), edited by Ismail Khalil & Edgar Weippl, pp. 70-91, copyright 2009 by IGI Publishing (an imprint of IGI Global).

Chapter 20
Evaluating E-Communities of Wireless Networks Worldwide

Theodoros I. Kavaliotis
University of Macedonia, Greece

Anastasios A. Economides
University of Macedonia, Greece

ABSTRACT

Many people communicate among themselves using wireless networks. They have developed e-communities in order to discuss issues about their network development, problems, opportunities, and wireless technology advances among others. The purpose of this article is to present an evaluation framework and analyze the current status of such Electronic Communities of Wireless Networks (ECWNs) in four continents: Africa, America, Europe and Oceania. The evaluation framework contains fifty criteria categorized into four categories: 1) Usability, 2) Technical Characteristics, 3) Community's Commitment, and 4) Members' Commitment. Then, fifty-seven ECWNs were evaluated using these criteria. The results show that there are large differences among ECWNs with respect to the forum structure, archives accessibility, interactivity, services, members' commitment, participation and relationships. In most ECWNs, two major drawbacks are the lack of online forums and a newsletter service. Finally, suggestions are made in order to improve current ECWNs.

INTRODUCTION

Currently, wireless networks have spread all over the world. Wireless networks could be used to provide various e-services such as mobile learning (Vasiliou & Economides, 2007) and mobile banking (Zarifopoulos & Economides, 2009). Using a wireless network protocol (e.g. IEEE 802.11x), antennas and web technology, people carrying wireless devices (Economides & Nikolaou, 2008) would communicate in a municipal or city-area (Gorp & Morris, 2008; Hampton & Gupta, 2008; Lawrence et al., 2007; Shankar, 2008; Szabo et al., 2007; Tapia et al., 2006). Such

DOI: 10.4018/978-1-60960-563-6.ch020

wireless networks first appeared in the late 1990s, when IEEE 802.11 devices became available to the public. Nowadays, they have spread in many cities all over the world.

People who use such wireless networks have created electronic communities on the Web where their members share ideas, experiences and suggestions about the improvement of their network, as well as other general issues. According to Leimeister et al. (2004) "A virtual community consists of people who interact together socially on a technical platform. The community is built on a common interest, a common problem or a common task of its members that is pursued on the basis of implicit and explicit codes of behavior. The technical platform enables and supports the community's interaction and helps to build trust and a common feeling among the members". The focus of this article is on Electronic Communities of Wireless Networks (ECWNs). Members of wireless network communities are interested in developing their wireless network, learning advances on wireless network technology, exchanging ideas and experiences related to wireless networks, etc. Members of many wireless networks have created ECWNs websites. Some of these communities are very well organized, but others are still at a very early level of development. The purpose of this article is to evaluate the ECWNs' status in four continents (Africa, America, Europe and Oceania) by investigating their websites. After identifying their inefficiencies, suggestions were made for overcoming these drawbacks.

After extensive Internet search during 2006, fifty seven ECWNs were selected through multiple lists and information sources (e.g., Free Global Wireless Community, FreeNetworks, Municipal wireless network, Wikipedia, Wireless Communities). The websites that appeared in the majority of the lists and sources were chosen. One restriction was the language. It was decided to examine communities that use English as the communication language. The number of selected communities from each continent was representative of the total number of communities in this continent. So, six ECWNs were selected from Africa, twenty five from America, eleven from Europe, and fifteen from Oceania.

In section 2, a framework for evaluating e-communities is developed. In section 3, ECWNs in the four continents are evaluated according to the evaluation framework. Finally, conclusions are made regarding the strengths and limitations of ECWNs and future research suggestions are made.

PREVIOUS RESEARCH

In order to evaluate the ECWNs, an evaluation framework is needed. Next, previous studies on evaluation frameworks are presented. For each previous study, the corresponding criterion of our evaluation framework (Table) is given in parenthesis.

Gregson and Ford (1998) recommended that both quantitative and qualitative approaches should be considered in evaluating community networks. Also, the community network goals should be considered. Unruh et al. (2002) provided an evaluation framework for digital community information systems. They emphasized that information should be objective and bias – free (Table: 3.1 Content). They also noted that the websites should support a wide range of access speeds and browsers (Table: 2.3 Openness). Koch et al. (2002a) showed how e – commerce can benefit from Internet communities. Communities' members provided all the needed personal data in order for suppliers to provide personalized offers to the members (Table1: 3.5 Services). Koch et al. (2002b) highlighted the importance of mobile access to the Internet. They underlined that the functionality of a community is based upon the participation of the largest part of its registered members (Table: 4.2 Participation). Borges and Baranauskas (2003) proposed ways to support the facilitator in e – learning communities. They emphasized on the interaction between all the members of a computer – based electronic environment. They supported that a facilitator (a user who initiates conversations) should motivate the other members to take part in the community's discussions (Table: 4.2 Participation). Leimeister et al. (2004) carried out a survey to identify the most important factors for an e-community's success, from the perspective of both the members and the communities' managers. According to members, handling their data sensitively was the most important factor (Table: 2.4 Security). Other important factors were the website's stability (Table: 2.1 Reliability & Maintainability), the frequent content updates (Table: 3.1 Content), the managers' appreciation of the members' contributions (Table: 3.4 Operators' Commitment) and the offerings to the members (Table: 3.5 Services). The most important factor according to the communities' operators was the wide participation (Table: 4 Members' Commitment). Mueller – Prothmann and Siedentopf (2003) explained the significance of usability for e–learning communities' success. Among other factors, they underlined the importance of the community website's easy navigation (Table: 1 Usability).

Preece et al. (2004) emphasized the significance of sociability and provided a usability/sociability heuristics table for online health support communities, from the members' perspective. Regarding the usability of an Internet community, they suggested usable websites (Table: 1 Usability) with accessible archives (Table: 1.4 Archives Accessibility) that are available 24/7 (Table: 2.1 Reliability & Maintainability). Regarding the community's sociability, they suggested that the members are mostly in charge of the community and that the operators nurture every new member (Table: 3.4 Operators' Commitment). An e-community is considered successful when there is a wide participation of the members (Table: 4.1 Users' Commitment, 4.2 Participation). Li (2004) reviewed thirty – six (36) academic papers on Internet communities and noticed the similarities between the papers regarding the ways in which a community can evolve. Regarding the members' participation in the community, he noted that anyone could register in a community and thereinafter participate as frequently as he wants to. The community's moderators should sometimes initiate discussions in which every member can participate, so that there is no dominance in the discussions by some members (Table: 4.2 Participation). Lurkers (users who do not post messages but are known to read postings regularly) should be able to read the archived discussions without registering (Table: 1.4 Archives Accessibility). Finally, he pointed out that the most important reason to register in an e-community is the development of online friendships (Table: 4.3 Relationships). Sorensen and O Murchu (2004) investigated e – learning communities, and suggested that even users that are not technology – experts should be able to navigate through an Internet community and participate broadly in the community's matters (Table: 1.2 Navigability). They also suggested that the members should feel comfortable in an e-community (Table: 4.3 Relationships) in order to participate actively (Table: 4.2 Participation). Lambropoulos (2005a; 2005b) presented a usability and sociability evaluation criteria catalogue for Internet communities and stresses the fact that both usability and sociability are important for a community's success. She argued about the importance of communities' operators help and support towards the members (Table: 3.2 Support), the privacy of personal data (Table: 2.4 Security), the necessity of the members' participation (Table: 4.1 Users' Commitment) and the significance of easily accessible communities' websites (Table: 1.1 User Interface, 1.3 Forum Structure). Schwabe and Prestipino (2005) conducted a research on travel communities' success factors and stated the critical ones. Regarding usability, they suggest easy – to – use and well – structured websites (Table: 1 Usability). Regarding commitment, they emphasized the significance of frequently updated and of high quality content

Table 1. Evaluation criteria for e-communities of wireless networks

1. Usability
1.1 User Interface
1.1.1 Simple & Easy to use
1.1.2 Aesthetic design
1.2 Navigability
1.2.1 Easy navigation
1.3 Forum Structure
1.3.1 Discussion board organized by topic
1.3.2 Discussion board organized by member
1.3.3 Discussion board organized by date
1.4 Archives Accessibility
1.4.1 Accessible archives
2. Technical Characteristics
2.1 Reliability & Maintainability
2.1.1 Continuous operation
2.1.2 Recoverability in case of error
2.1.3 Continuous improvement
2.2 Performance
2.2.1 Input / Output / Processing speed
2.2.2 Supports many concurrent users
2.3 Openness
2.3.1 Supports various types of users' connections
2.3.2 No need for users to have special software
2.4 Security
2.4.1 Certifications & Guarantees
2.4.2 User's privacy
2.4.3 No cookies
3. Community's Commitment
3.1 Content
3.1.1 Valid & Objective
3.1.2 Updated & of High quality
3.1.3 Quantity & Various Types of content
3.1.4 Tolerance (not censorship)
3.1.5 Bibliography, references & links to other sites
3.2 Support
3.2.1 Technical support and help, Frequently Asked Questions, How to use it
3.2.2 Statistical analysis of participation
3.3 Interactivity
3.3.1 Newsletter
3.3.2 Alerts for new content

continued on following page

Table 1. continued

3.4 Operators' commitment
3.4.1 Evolution of the community according to its members' ideas
3.4.2 Appreciation of the members' contributions by the operators
3.4.3 Members' integration in the administration of the community
3.5 Services
3.5.1 Quantity of services
3.5.2 Usefulness of services
4. Members' Commitment
4.1 Users' Commitment
4.1.1 Number of members registered in the community
4.1.2 Members committed to the community
4.1.3 Male / Female members
4.1.4 Variety of nationalities, cultures, religions etc. of members
4.1.5 Homogenous / Heterogeneous
4.1.6 Since when a member is registered
4.2 Participation
4.2.1 Number of messages per day
4.2.2 Number of messages per day per member
4.2.3 Number of answers per message
4.2.4 Length per message
4.2.5 Number of messages per topic
4.2.6 Number of messages per topic per day
4.2.7 Number of members per topic
4.2.8 Balance versus tolerance by some members
4.3 Relationships
4.3.1 Friendly versus hostile
4.3.2 Informal versus formal
4.3.3 Information exchange
4.3.4 Feelings & emotions exchange
4.3.5 Development of relationships in the real world

(Table: 3.1 Content). Finally, they discussed about the importance of personalization in Internet communities (Table: 3.5 Services).

So, most previous studies consider only some evaluation criteria. However, a comprehensive evaluation framework is needed in order to evaluate the e-communities in a holistic way. Next, such an evaluation framework is developed and is used to evaluate fifty seven (57) ECWNs. The results show large differences among these communities. Suggestions are made for improvements.

EVALUATION FRAMEWORK

In this section, a new evaluation framework is developed based on the authors' experience on websites' design, development and evaluation, on discussions with colleagues and on previously published papers regarding Internet communities. This evaluation framework combines criteria from previous research with new ones. Four evaluation categories are considered in the framework: 1) Usability, 2) Technical Characteristics, 3) Community's Commitment, and 4) Members' Commitment. The first two categories are related to the website of the community, while the next two categories are related to the community itself and its members. Each category contains several subcategories and each subcategory includes several evaluation criteria (Table). Next, these categories, subcategories and criteria are presented.

Usability

The usability of a website refers to how effective, efficient and satisfying is the website for its visitors (Mueller – Prothmann & Siedentopf, 2003). The Usability category contains four subcategories: User Interface, Navigation, Forum Structure, and Archives Accessibility.

- **User Interface:** The website of the community must be easy to use and its members should be able to perform basic tasks, like registering and reading or posting messages easily (Preece et al., 2004; Schwabe & Prestipino, 2005). However, its design also has to inspire those who are used to new technology (Sorensen & O Murchu, 2004).
- **Navigability:** The navigation through the website should be easy and fast (Preece et al., 2004).
- **Forum Structure:** A forum is the place where the community's members exchange their personal views on several topics. The presentation and the structure of the content in a community – in this case, the messages that the members have posted – should facilitate the users (Schwabe & Prestipino, 2005; Unruh et al., 2002). So, the discussion board should be organized by topic, by member and by date (Lambropoulos, 2005b).
- **Archives Accessibility:** The e-community should be open to all users (Li, 2004) and not only to its members. The archives of the posted messages and discussions should be accessible to anyone who is interested in them, so that he could make his own choice to join the community or not.

Technical Characteristics

The Technical Characteristics are related to the efficient operation of the community's website. Its subcategories are the following: Reliability & Maintainability, Performance, Openness, and Security:

- **Reliability & Maintainability:** The community must be accessible 24 hours a day, 7 days a week, so that its members can rely on its continuous availability (Preece et al., 2004). In case of a system error that may lead to a possible breakdown of the server, the reaction time of the operators should be rather quick (Leimeister et al., 2004). The managers of the ECWN should also make frequent improvements to the website, so that it remains up – to – date.
- **Performance:** The website should have fast input, output and processing speed and support many concurrent users.
- **Openness:** The website should be able to support various types of user connections. It would also be better if the members of the community do not need to use special software programs to access the community (Unruh et al., 2002).
- **Security:** The first and most important factor for the success of an e-community, from the perspective of its members, is the sensitive handling of their personal data (Leimester et al., 2004). The community's operators should certify and guarantee that

no personal data or profiles will be sent or sold to third parties. Only the member himself should be in control of his personal data. Finally, there should be no unauthorized monitoring (e.g. cookies) of the members' activities and navigation through the Internet.

Community's Commitment

The community should satisfy each one of its members, and their operators ought to take all appropriate steps to fulfill their expectations. It is essential that every new member is welcomed, supported and nurtured (Lambropoulos, 2005b; Preece et al., 2004). The subcategories in this category include the community's Content, the Support to the members, the Interactivity between community and members, the Operators' Commitment to the members, and the provided Services:

- **Content:** According to Powazek (2002), the most important subcategory of a community is its content. The posted information should be valid and accurate (Unruh et al., 2002). The content of the website should be of high quality and regularly updated (Leimeister et al., 2004). If a member expresses his personal opinion, which might be against the operators' opinion (within certain limits) he should not be censored (Li, 2004). The managers need to be tolerant to those kinds of messages and allow them to be posted on the discussion forum. Finally, it would be good if the operators provided a variety of links to other useful websites related to wireless technology.
- **Support:** A factor that should be taken into consideration (Lambropoulos, 2005a) – especially in ECWNs – is providing information about the community's subject (in this case, wireless technology). A lot of users are not very familiar with wireless technology, but are really interested in it. So, a section where they could learn how to use this kind of technology would be rather helpful. There would also be a Frequently Asked Questions section and a section for technical support. Finally, the operators could provide the statistical analysis of the members' participation, for those who want to learn about the community's members and their participation.
- **Interactivity:** The community's operators have to inform the members about every new project or event that takes place and concerns them. This could be achieved by alerting the members with frequent newsletters (Mueller – Prothmann & Siedentopf, 2003), RSS (Really Simple Syndication) feeds, IM (Instant Messaging) and pod casts. Similarly, automatic notifications would be sent to members according to their declared interests (Lambropoulos, 2005b). In this was, the operators show that they care about the community members and interact with them, keeping in mind their personal interests.
- **Operators' Commitment:** The members' opinion about the community and the various aspects having to do with it should be taken into account and the operators ought to seriously consider adopting ideas that could evolve the community (Leimeister et al., 2004). They should also recognize the members' contributions and reward them.
- **Services:** The range of services and privileges that the community's operators offer to its members make them contribute more to its evolution. The offered services and products ought to be many and useful. They would also be personalized, considering the fact that community managers have all the information about the members needed to select an appropriate service or product for them (Koch et al., 2002b; Schwabe & Prestipino, 2005).

Member's Commitment

According to Koch et al. (2002a), "a functioning community depends on the active participation of a significant percentage of its members … Only a broad participation in the community activities can sustain the functioning of the community". Subcategories in this category include the Users' Commitment, Participation [of the members], and Relationships [between the members]:

- **Users' Commitment:** A community is considered successful if the number of its registered members is high and growing (Leimeister et al., 2004; Preece et al., 2004). The members ought to be devoted and contribute to the community's evolution. Members could vary in terms of sex (male / female), nationality or culture. The community may be homogenous with members having similar points of views, or heterogeneous, with members having opposite points of views; both types are useful. Finally, the time that one has been a member of the community is also important. Older members are more familiar with the community's issues and are able to help new members acclimatize themselves in the community.
- **Participation:** A broad participation leads to better interaction between the members of the community. So, there should be a growing daily frequency of postings per member for each topic (Leimeister et al., 2004). The size of the conversations (the numbers of messages that each conversation contains) initiated about a topic could be rather long, too (Preece et al., 2004). An important issue that should be stressed is that there should be some kind of balance in messages posted by the members. That means that all the members of the community should post messages frequently, and not only a few of them. If some members do not participate actively, the operators could initiate a conversation with them and help them familiarize themselves with the community's issues (Borges & Baranauskas, 2003; Li, 2004).
- **Relationships:** According to Li (2004), a very important factor for users who participate in a virtual community is the development of online friendships. It would be better if the members discuss issues about not only wireless technology but also other topics. The online relationships would be also developed offline by arranging frequent real – world meetings.

So, the full evaluation framework consists of four major categories, sixteen subcategories, and fifty criteria (Table 1).

Members of a wireless networks community create a website in order to coordinate the development of their wireless network, exchange ideas and experience, and communicate in general. The proposed evaluation framework was used to evaluate ECWNs by looking at their websites.

The first author used this framework to evaluate fifty seven ECWNs. For every community, he assigned a score from 0 to 5 to each criterion (0 = non existence, 1 = very poor, 2 = poor, 3 = fair, 4 = good, 5 = very good). For example, if an ECWN lacked a discussion forum (which means that it was not mature enough or existed only to provide information about the wireless network and the offline meetings), the score for Forum Structure and Archives Accessibility was "0" ("non existence"). The scores were based on the author's personal view. It was trying to evaluate the attractiveness of these ECWNs to someone who is not a member in order to become one. Then, the score for each subcategory is calculated as the average of the scores given to the corresponding criteria.

Figure 1. User interface

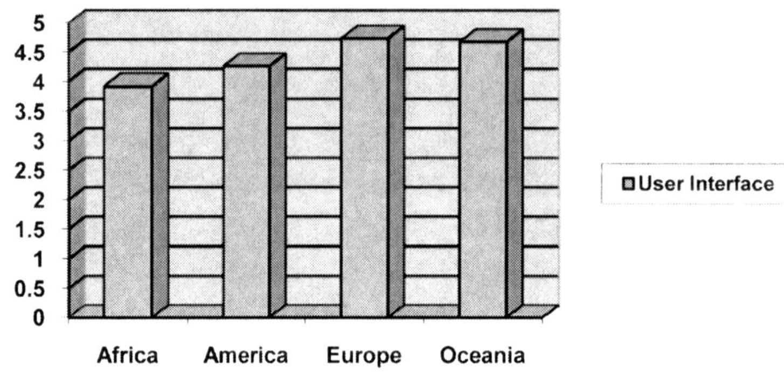

EVALUATION RESULTS

In this section, the status of the ECWNs is presented through sixteen figures, one for each subcategory. Each figure shows the average scores that correspond to the continents. The continent's score was calculated as the average score of the ECWNs in this continent for the particular subcategory.

Usability

User Interface

Most of the ECWNs' websites were easy to use, so the scores were high for User Interface (Figure 1). However, there were some websites, especially in Africa and America, whose design was not aesthetically pleasant and user friendly and a novice user would have difficulties in using it.

Navigability

The websites of all ECWNs were easy to navigate through, so they were all given a "very good" ("5") score for Navigability (Figure 2).

Forum Structure

A major drawback in some websites in America, Europe and Oceania was the lack of an online discussion forum (Figure 3). Some of these ECWNs were recently created so they may need more time to develop a well – organized discussion board and evolve to complete communities. However, there were some ECWNs that have existed for a long time, but whose managers were not committed enough to develop a forum. The website existed only to provide fixed information about the wireless network and wireless technology in general, or information about offline meetings of the networks' members. The score given to a website with this drawback for Forum Structure was "0" ("non existence"). The average scores vary from "2.09" to "3.91" in the three continents due to the number of websites in them that lacked a forum. Africa missed the "very good" ("5") score since in a few ECWNs the posted messages were not organized by member.

Archives Accessiblity

All of Africa's ECWNs had accessible archives, so they were given a "very good" ("5") score for Archives Accessibility (Figure 4). In the other three continents, the websites that lacked an online forum lacked archives as well, so they were scored with a "0" ("non existence") for the particular subcategory. In very few communities, especially in America and Oceania, if a user wanted to access the archives, he had to register first, a fact that is discouraging for most users.

Figure 2. Navigability

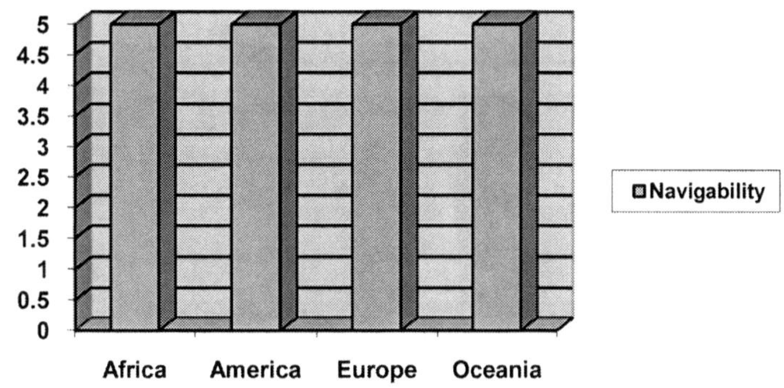

Figure 3. Forum Structure

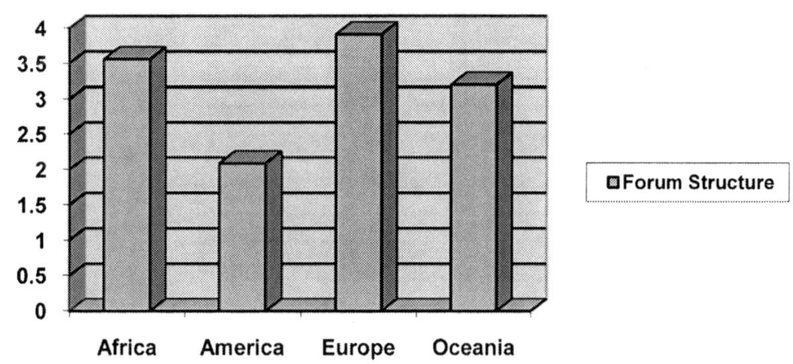

Figure 4. Archives accessibility

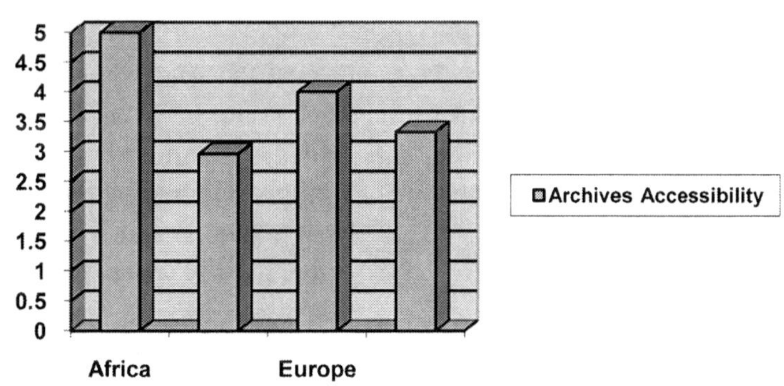

Technical Characteristics

Reliability & Maintainability

All the websites were reliable. Some ECWNs, though, were not recently updated, so they were given lower scores for this subcategory, which also includes maintainability (Figure 5). Furthermore, a few websites in Europe and Oceania, and more in America, had not been improved by their managers for a long time; that explains the fact that the average scores are lower in these continents than the score of Africa.

Figure 5. Reliability & Maintainability

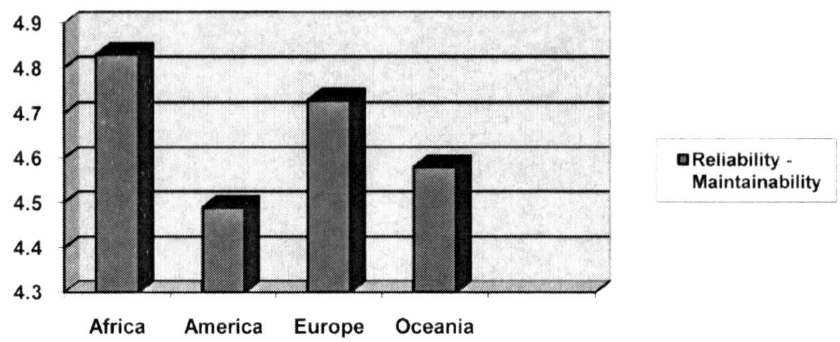

Figure 6. Performance

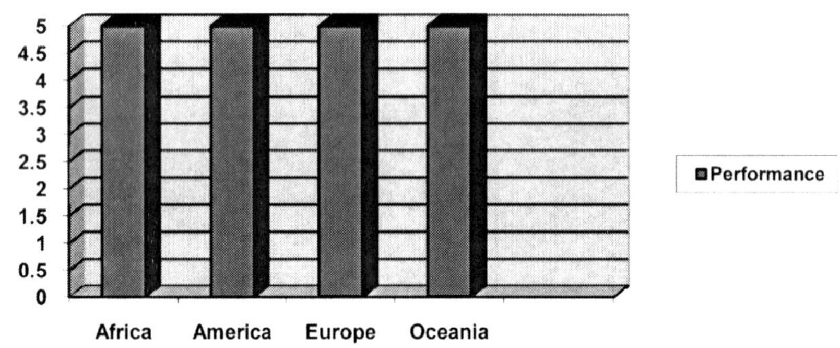

Performance

There were no performance problems in any of the evaluated ECWNs (Figure 6). The input, output and processing speed was fast and the websites could support many concurrent users. So, all the communities were scored with the highest mark, "5" ("very good").

Openness

No website presented an openness problem (Figure 7). All ECWNs were accessible without the need for special software, while they supported various types of Internet connection. All ECWNs scored "5" ("very good").

Security

In Africa, Europe and Oceania, the average score in Security was below "5" due to the fact that in some ECWNs, the operators did not explain thoroughly the guarantees of the personal data's security (Figure 8). In America, one ECWN informed the users that from the moment that they were going to register, cookies would monitor their navigation through the Internet, a fact that is quite discouraging for many people who do not want to be watched. The score given for Security in that ECWN was very low, so the average score for the specific subcategory in America dropped.

Community's Commitment

Content

The majority of the ECWNs contained valid, high quality and regularly updated content (Figure 9). Almost all of them had a "Links to other sites" section, where one could find many interesting links about wireless technology, like suppliers of hardware for wireless networks. In America,

Figure 7. Openness

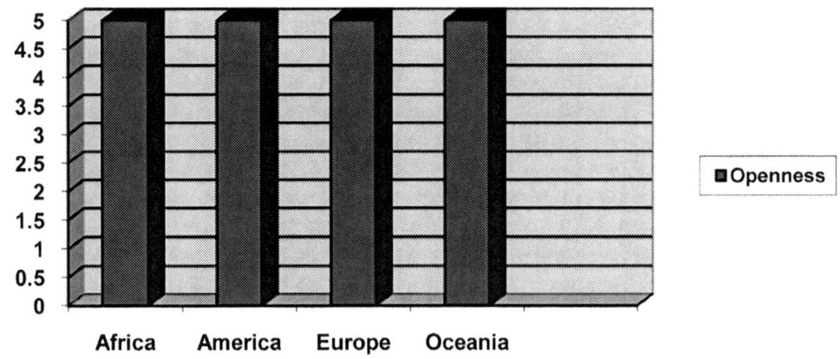

Figure 8. Security

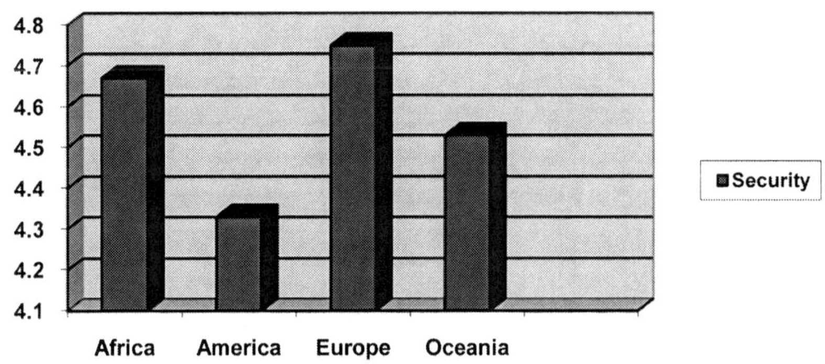

Europe and Oceania, the average score was lower than in Africa, because the managers of some ECWNs did not seem to care about their communities' content. The ECWNs were active for a long time, but their content had not been updated for a long time as well.

Support

Regarding the Support subcategory, most of the communities had a Frequently Asked Questions section, where all the basic questions about wireless technology and networks where thoroughly answered (Figure 10). The reason that the average score was almost "4" ("good") for Africa and Europe, "3.43" for Oceania and "2.36" for America is the lack of the statistical analysis of members' participation. Not many websites provided information about the participation of the registered members, which would help users to see the community's activity. Also, in very few communities in America and Oceania, there was hardly any help or support topics for the interested members or a Frequently Asked Questions section.

Interactivity

An important result of this research was the fact that a lot of communities ignore the Interactivity subcategory (Figure 11). In Africa, not one of the communities that were reviewed sent newsletters or alerts for new content to inform its members about a new interesting event or project. In Europe and Oceania, the average scores were "0.51" and "1.93" respectively, which means that very few communities sent newsletters. In America, more than half of the evaluated ECWNs achieved higher scores. However, this does not mean that there

Figure 9. Content

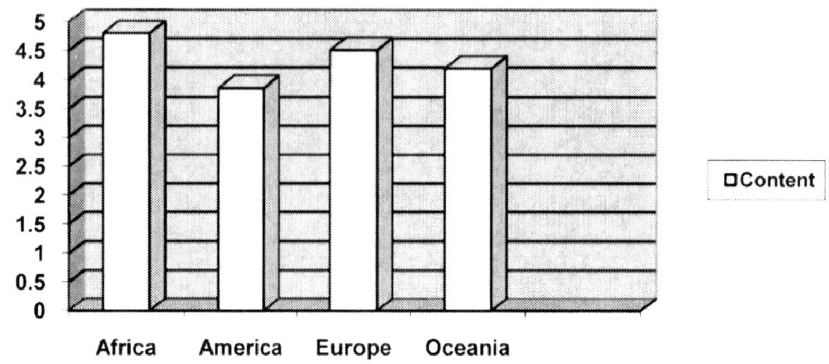

Figure 10. Support

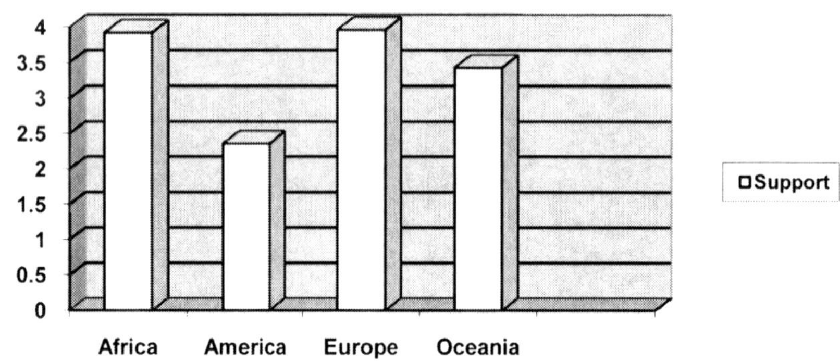

could not be any improvements to Interactivity in America; the status of the subcategory was better than in the other continents, but not good enough.

Operators' Commitment

In America, Europe and Oceania, the members had a fair part in the evolution of the majority of the communities, whereas their contributions were appreciated by their managers and were rewarded with a number of offers (Figure 12). However, there were some community managers that did not take much consideration of the members' ideas about its evolution and very few ECWNs whose managers seemed to be totally out of interest about them. These communities were given low marks, which resulted in the average score for the subcategory being less than "4", in all three continents. Finally, the operators' commitment towards the registered members was "fair" in Africa, where all the ECWNs were scored with the mark "3" for the subcategory.

Services

Europe leads in this factor, since the majority of its communities provided many and useful services to their members, such as links to websites regarding wireless technology as well as discounts in purchases of wireless equipment through certain sites (Figure 13). It was the only continent where almost all the ECWNs provided services. On the contrary, Africa's communities provided the fewest, still useful though, services. Oceania's drawback was the quantity of the services provided. They were useful for the members, but they were few, in comparison to European ECWNs. Finally, there were some communities

Figure 11. Interactivity

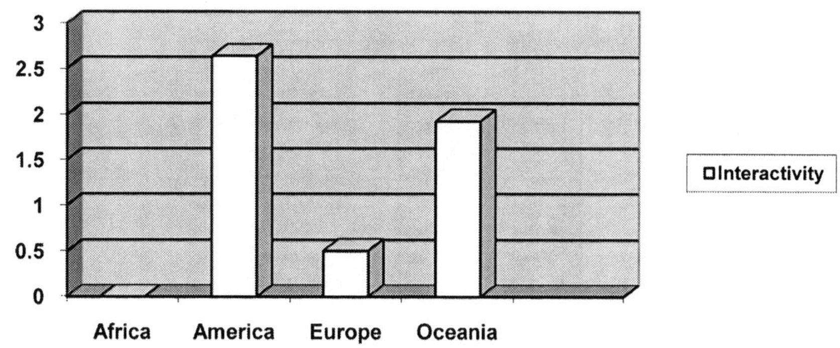

Figure 12. Operators' Commitment

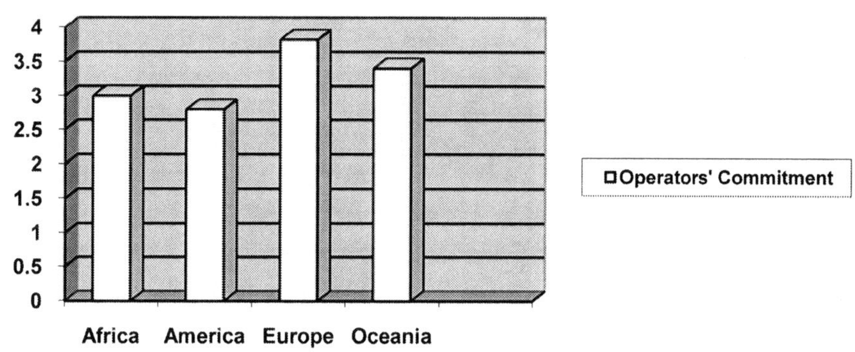

in America that provided no services at all and so its average score was lower than the other two continents' score.

Members' Commitment

Users' Commitment

The majority of the communities in Africa had a high number of loyal registered members, so Africa achieved the highest average score ("3.5") in Users' Commitment (Figure 14). The average score was not higher because in two communities the number of the registered members was not as high as in the other four. Europe and Oceania's ECWNs achieved the same average score, whilst America was the continent with the lowest score. As it was mentioned before, in all three continents, more in America and less in Europe and Oceania, various ECWNs had no online discussion board. The purpose of many websites' Internet existence was only to present information on how to get wireless access on wireless projects in the city that the website was referring to. However, the members of the wireless networks may have been interacting offline, but there should also be online participation as well.

Participation

The online participation of the members was "fair" in Africa and Europe (Figure 15). There were many messages posted for all the topics in the forums of the communities and the replies to the messages were numerous, too. The fact that lowered the average score was the dominance by some members in the conversations. In many communities, some of the first registered members

Figure 13. Services

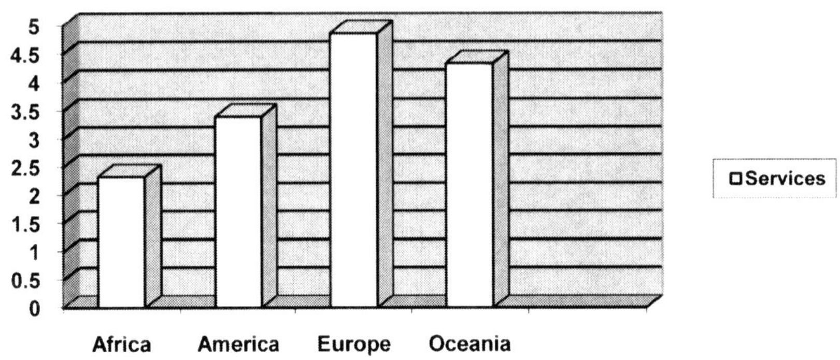

Figure 14. Users' Commitment

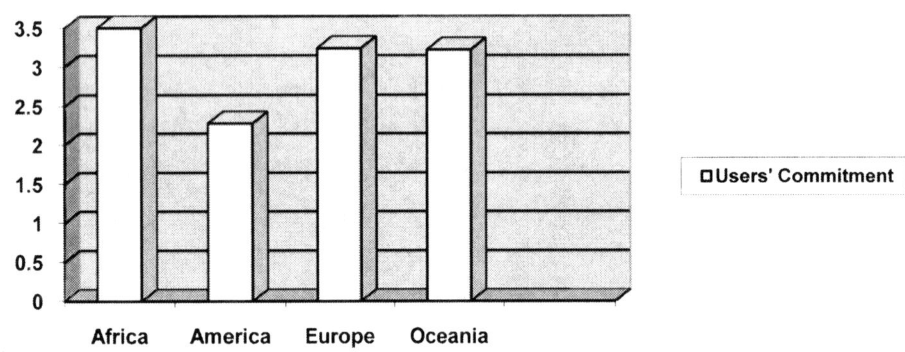

generated and carried on all of the discussions, while newer members did not participate actively.

Relationships

The relationships developed among the members were very friendly in all the continents (Figure 16). Besides the information about wireless technology and the network's development that they exchanged, they also exchanged feelings and emotions about general matters. Their online relationships were developed in the real world as well, due to the offline events that were regularly scheduled. There were meetings arranged even in websites that lacked an online forum, by posting announcements on the main page of the website. In that case, the ECWNs' members hoped to evolve the e-community by transferring their offline relationships to the Internet. America achieved a lower average mark than the rest because of very few ECWNs whose members not only did not interact online, but they also did not have any meetings planned for the near future.

CONCLUSION AND FURTHER RESEARCH

Wireless technology has many advantages. Many people believe that wireless networks will eventually replace wired ones. People should be informed about these new innovations. An efficient way to be informed, educated and exchange experiences is through Internet communities, where anyone can easily have access to. The ECWNs' managers should organize their websites properly, so that Internet users would be able to learn about

Figure 15. Participation

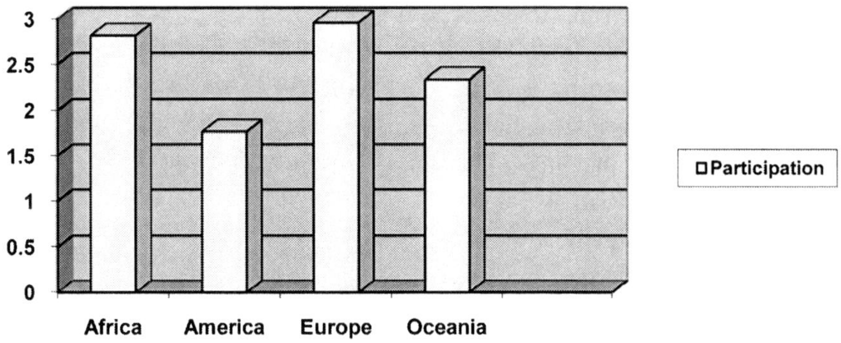

Figure 16. Relationships

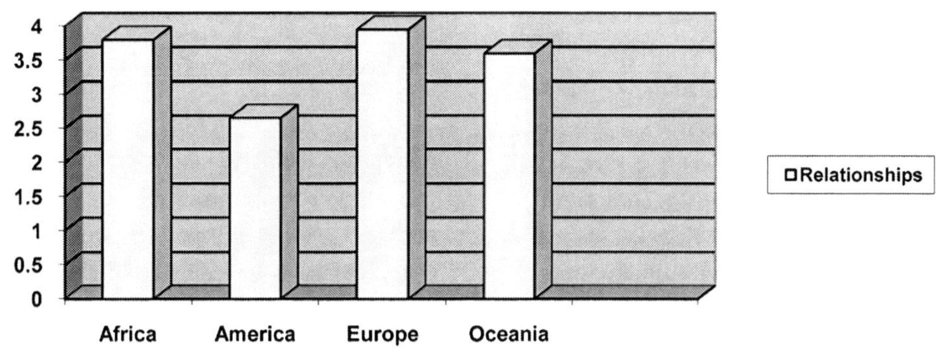

wireless networks advantages and make proper use of them.

In this article, an evaluation framework for ECWNs was developed. Using this framework, fifty seven ECWNs in four continents were evaluated. The results of the evaluation showed that there were differences among the continents and more work is needed to achieve a high level of efficiency.

More specifically, most of the European communities' websites were easy to use and did not have any technical problems. The communities' operators were committed to their members, by offering them various services. The members, on the other hand, had a satisfying online participation and had developed strong relationships among them. Although wireless networks have been developed only recently in Africa, its ECWNs achieved a high quality level. However, a very important drawback of its communities is the fact that none of them sent newsletters or alerts for new content to its members. Similarly, Oceania's ECWNs achieved a high quality level. However, some of its communities did not have an online forum, mostly because they were recently created. America was the first continent to build a wireless network and the number of existing ECWNs in America was very large. Many communities in America were highly sophisticated and could be considered as the state of the art. However, there were also several underdeveloped ECWNs that lacked some basic characteristics. This was due to the fact that either they were recently created, or their operators were not committed to their evolution.

The evaluation revealed two major drawbacks of the ECWNs. The first one is the lack of an online forum in many websites. There may be an excuse for the recently created communities since their operators need some time to manage them properly. However, there were other websites that have been operating for a long time and they only provided information about the wireless network and offline meetings. The managers of these ECWNs have to evolve them and make them complete e-communities, so that the offline relationships between the wireless network's members would be transferred online as well. Still, there were some communities whose operators were totally careless and also did not support any offline meetings.

The second important drawback was the lack of a newsletter service. The majority of the ECWNs did not send newsletters to inform their members about their community and network's progress. In Africa, no community provided a newsletter service. In Europe, the percentage of communities that sent newsletters was 20%, while in Oceania it was fairly higher (40%), but not high enough. In America, the percentage rose to 55%. More than half of the evaluated communities sent newsletters and alerts for new content to their members. The ECWNs that lack this service should start offering it.

The aim of this article was to present an evaluation framework and the average status of ECWNs. The selected e-communities were representative of ECWNs in the four continents. So, not only mature and best practice communities were evaluated, but also communities that were recently created. Further research could be carried out focusing only on mature communities, so that the status of best ECWNs could be identified. Thematic wireless communities would be also evaluated using this framework. For example, it would be interesting to investigate developments in tourism wireless communities. Finally, the ECWNs' members themselves could evaluate their and others' ECWNs.

REFERENCES

Borges, M. A. F., & Baranauskas, M. C. C. (2003). Supporting the facilitator in a collaborative learning environment. *International Journal of Continuing Engineering Education and Lifelong Learning*, *13*(1/2), 39–56.

Economides, A. A., & Nikolaou, N. (2008). Evaluation of handheld devices for mobile learning. [IJEE]. *International Journal of Engineering Education*, *24*(1), 3–13.

Free Global Wireless Community. http://global.freifunk.net/free_global_wireless_community

FreeNetworks. http://freenetworks.org/

Gorp, A., & Chris Morris, C. (2008). Serving under-serviced areas in South Africa: The potential for Wi-Fi community network deployment and the role of regulation. *Info*, *10*(1), 65–78. doi:10.1108/14636690810850166

Gregson, K., & Ford, C. (1998). Evaluation of community networks. In Wildemuth, B. M. (Ed.), *ASIS Midyear 1998 Proceedings, Collaboration across Boundaries: Theories*. Strategies, and Technology.

Hampton, K. N., & Gupta, N. (2008). Community and social interaction in the wireless city: Wi-fi use in public and semi-public spaces. *New Media & Society*, *10*(6), 831–850. doi:10.1177/1461444808096247

Koch, M., Groh, G., Hillebrand, C., & Fremuth, N. (2002). Mobile support for lifestyle communities. *Arbeitsberichte des Lehrstuhls fur Allgemeine und Industrielle Betriebswirtschaftslehre an der Technischen Universitat Munchen*, Nr. 34.

Koch, M., Moslein, K., & Schubert, P. (2002). Communities and personalization for individual products. *Proceedings of the Annual Conference of the British Academy of Management*, London, UK, September 2002.

Lambropoulos, N. (2005a). Sociability and usability for contribution based on situated informal learning and consensus knowledge building in online communities. *In the proceedings of the 1st Conference on Usability and Internationalization, in the 11th International Conference on Human – Computer Interaction 2005*, 22 – 27 July, Las Vegas, Nevada, USA.

Lambropoulos, N. (2005b). User – Designer HCI research model for online communities groupz – ware. *Proceedings of the 1st Conference on Usability and Internationalization, in the 11th International Conference on Human – Computer Interaction 2005*, 22 – 27 July, Las Vegas, Nevada, USA.

Lawrence, E., Bina, M., Culjak, G., & El-Kiki, T. (2007). Wireless community networks: Public assets for 21st century society. *Proceedings of the International Conference on Information Technology (ITNG'07),* pp. 166-174, IEEE.

Leimeister, J. M., Sidiras, P., & Krcmar, H. (2004). Success factors of virtual communities from the perspective of members and operators: An empirical study. *Proceedings of the 37th Hawaii International Conference on System Sciences (HICCS 37)*, IEEE, USA.

Li, H. (2004). Virtual community studies: a literature review, synthesis and research agenda. J. De Gross et al. (Eds.), *Proceedings of the Americas Conference on Information Systems*, 2708-2715, August 2004, New York, USA.

Mueller-Prothmann, T., & Siedentopf, C. (2003). Designing online knowledge communities: developing a usability evaluation criteria catalogue. *Proceedings 3rd European Knowledge Management Summer School*, 7 – 12 September, 2003, San Sebastian, Spain.

Powazek, D. (2002). *Design for community: the art of connecting real people in virtual places.* Indianapolis: New Riders.

Preece, J., Abras, C., & Maloney-Krichmar, D. (2004). Designing and evaluating online communities: research speaks to emerging practice. *International Journal of Web Based Communities*, *1*(1), 2–18.

Schwabe, G., & Prestipino, M. (2005). How tourism communities can change travel information quality. *Proceedings of the Thirteen European Conference on Information Systems (ECIS),* Regensburg, Germany.

Shankar, K. (2008). Wind, water, and Wi-Fi: New trends in community informatics and disaster management. *The Information Society*, *24*(2), 116–120.

Sorensen, E.K., & O Murchu D. (2004). Designing online learning communities of practice: A democratic perspective. *Journal of Educational Media*, *29*(3), 189–200.

Szabo, C., Horwath, Z., & Farkas, K. (2007). Wireless community networks: Motivations, design and business models. *Proceedings of the 3rd international conference on Wireless Internet, WICON 2007*, Austin, Texas, USA, ACM.

Tapia, A., Maitland, C., & Stone, M. (2006). Making IT work for municipalities: Building municipal wireless networks. *Government Information Quarterly*, *23*(3/4), 359–380.

Unruh, K. T., Pettigrew, K. E., & Durrance, J. C. (2002). Towards effective evaluation of digital community information systems. *Proceedings of the Annual Conference of the Association for Information Science & Technology (ASIST 2002)*, Philadelphia, USA.

Vasiliou, A., & Economides, A. A. (2007). Mobile collaborative learning using multicast MANETs. [IJMC]. *International Journal of Mobile Communications*, *5*(4), 423–444.

Wikipedia- List of wireless community networks by region, http://en.wikipedia.org/wiki/List_of_wireless_community_networks_by_region

Wireless Communities. http://wiki.personaltelco.net/WirelessCommunities

Zarifopoulos, M., & Economides, A. A. (2009). Evaluating mobile banking portals. [IJMC]. *International Journal of Mobile Communications*, 7(1), 66–90.

Chapter 21
Typology and Challenges in Developing Mobile Middleware Based Community Network Infrastructure

Vijayan Sugumaran
Oakland University, USA & Sogang University, South Korea

Shriram Raghunathan
B. S. Abdur Rahman University, India

ABSTRACT

The evolution of mobile devices has opened new opportunities for collaboration, communication and computation on the move. Increasing device capabilities, instant connectivity, portability, rapidly reducing costs, etc. are some of the drivers for this change. Though mobile devices have lower processing power, memory capabilities compared to the stationary computing devices and deal with varying network conditions with reduced power, the demand for anywhere anytime computing is a significant driver for change. There is great interest in application development for mobile devices which caters to different needs of the users. Applications can be easily developed as complete middleware systems which can provide a great deal of abstraction for users. There is great interest in designing novel and scalable mobile middleware especially utilizing the capabilities available for collaboration and communication. This work traces the factors contributing to the proliferation of mobile communities and places the middleware for mobile community networks in a current and future perspective. The architecture for middleware based mobile community network is proposed and the challenges in implementing such a network are also discussed.

DOI: 10.4018/978-1-60960-563-6.ch021

INTRODUCTION

In general, a middleware system resides between the operating system and the applications providing the user with a high degree of abstraction. In a mobile device based context, the role of the middleware is crucial due to the demands and constraints in the domain. While middleware itself is a software, the key difference from a software application and a middleware system is in the characteristic of allowing other applications to work on top of them. A middleware for mobile devices will allow other applications to use it whereas a software application may not. The middleware system can be installed on top of the operating environment, and in this, the term operating environment is defined loosely. In some cases, the operating environment is the hardware and in others it is the hardware, combined with the operating system.

Mobile middleware systems are widely used in recent years for a range of applications. This chapter focuses on the development of the middleware using community networks and especially in the domains of information retrieval and service composition. Currently, mobile community networks are receiving considerable attention due to the great strides made in the development of newer generation of hand held devices. Mobile community networks allow for devices with a reduced computational power to transfer the computation to a dedicated remote computational platform providing greater flexibility for users. The middleware systems allow this remote operation within the confines of the mobile phone-network environment. Thus the developments in mobile community networks can provide a convenient mode of communication and computation for mobile users. The main purpose of this work is to summarize and add to the literature on mobile middleware systems using mobile community networks in four specific areas: a) historical timeline tracing the evolution from communities to mobile community networks, b) typology for mobile community networks and c) the issues and challenges in developing mobile community networks.

In general, a community is composed of users with shared purpose, content and communication systems. Community services enable users who share common interests to join together in a closed user group (*community*) to have the means to communicate with each other via chat, whiteboards, or messaging services. The key terms are shared interests, communication ability and paradigms of interaction. Due to the steady progress in technologies and processes for information sharing, communication and collaboration, the work environment has gone in the direction of a more technology supported world. In future, the network, internet, mobile devices and the web of people – will become the workplace and the information gained from the community will become more relevant.

Community networks are different from communities in that community networks have a broad agenda of using technology to foster social objectives of community cohesion, enhanced education and strong democracy (Schuler, 1994). Community networks are defined as systems that use information and communications technologies (ICTs) to help revitalize, strengthen, and expand existing geographically-based human networks (Venkatesh, 2003). The phrase "community network" is often used in conjunction with the phrase "virtual community." We prefer the phrase community network over the virtual or online connotations, since the presence of internet and online portals are central to the goal of community relationships. The internet is a technology, not the goal. Ultimately, the traditional relationships are the most important.

Typically, community networks are also used to refer to physical location based entities. A search for the term community networks on the web yields results that show multiple case studies of communities which are in the same location, for e.g. the campus community network or the village community network. However, it need not be the case always as the term network communities are often used in tandem with community networks. Network communities are communities whose interactions are mediated mainly by the internet and the participants do not live close to each other (Carroll, & Rosson, 2003). Work-based communities (communities of practice) are characterized by similar work activities. The web abounds with forums and blogs where communities of practice proliferate. These communities of practice are characterized by ad-hoc task based interaction for solutions and purposeful access. Context based communities is an analogous term used. Context is a narrower version of the work based community with focus on a narrower domain of focus or reduced time frame of interaction.

Using mobile phones as the access mechanism to a virtual community gives an instant anytime-anywhere connectivity to the community, helps the users communicate in a variety of ways (using the audio, camera and video aspects of mobile phones) and emerge as a computing platform for developing countries (Adeya, 2005; Hammond, & Paul, 2006). From the sheer utility perspective, several features (battery operation, solid-state memory, wireless connectivity, affordable price) make it a better-suited device for rural developing world conditions than a conventional Personal Computer (Parikh, 2006).

A community which is a "network of interpersonal ties that provides sociability, support, information, a sense of belonging, and social identity," becomes a 'mobile community' (Wellman, 2005) when it always connects its members regardless of where they go. A mobile community brings together individuals who share a common interest through mobile technology. Eventually mobile community networks can supplement social arenas by offering universal channels of accessibility and providing enhanced opportunities for collaboration (Farooq, & Carrol, 2003).

The rest of the chapter is organized as follows. The next section traces the evolution of mobile community networks from a technological and personal network perspective and describes a timeline demarcating the various phases in the evolution. Then, the subsequent section discusses the typology for mobile community networks taking into account the mobile environments and future applications. Following that, the issues and challenges in mobile community networks along with architecture for mobile community networks are discussed. The last section summarizes and concludes the chapter.

TOWARDS MOBILE COMMUNITY NETWORKS

In this section we provide some background on mobile community networks and trace the time line of the evolution that led to mobile community networks.

Ubiquitous computing offers new possibilities for enlarging the reach of community support systems by opening new fields for community support, providing new functionalities and scenarios. The vision of mobile virtual communities is a combination of mobile technologies that facilitate information access anywhere and at any time for different user requirements. Porting community networks to different hardware platforms, like handheld devices, will afford universal channels of accessibility and enhanced opportunities for collaboration. Mobile community network applications can be used to develop and provide services dynamically using the mobile environment. The use of the word '*mobile*' in 'mobile community networks as opposed to 'community networks' is to emphasize the fact that community networks when 'mobile' bring an entirely different and new dimension to communities.

Studies in community networks such as Rheingold (2003), and Clodfelter, Buente, & Rosenbaum (2007) provide a historical timeline of key events in the community technology movement. These are taken further and placed in an interesting context

Figure 1. Evolution of Community Networks

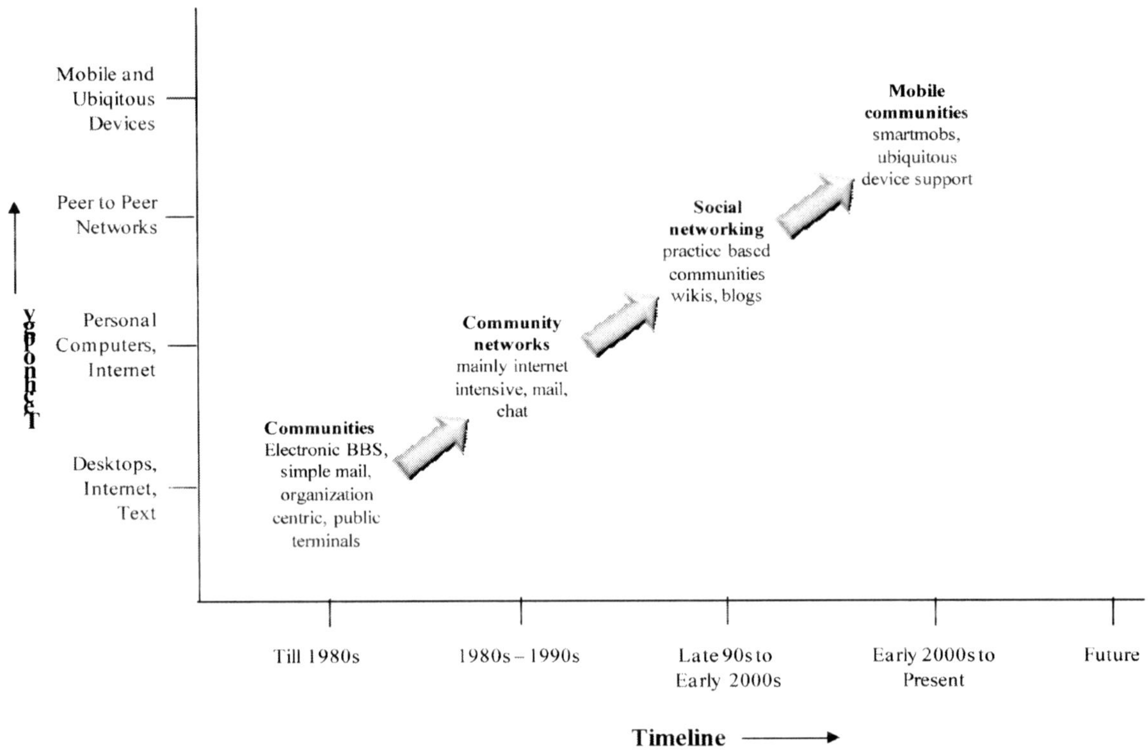

in this section. The pictorial representation of the time line is given in Figure 1.

In terms of computing technological level, as early as 1960s, the PLATO timesharing system had an early version of message boards, SF-LISTS (science fiction lists) were implemented as a version of mailing lists in the 1970s. Since the adoption of USENET news sharing program in 1973, online communities have evolved. Initially (till 1980s), community networking technology has been based on chats, electronic messaging, messenger based services, electronic bulletin boards, IRC channels, multi user dungeons (MUD) and newsgroups. The mode of interaction was mainly one to one. The technology interaction can be traced to the communications via telephones, desktops and two way radios etc., thus remaining 'neighborhood' communities. Each individual, thus, has a personal network and the combination of all the networks became a social network.

The period between the 1980s to 1990s saw the advent of internet allowing widespread use of web portals based assistance adding visual interfaces, the trend of online communities, and various forms of communication software like instant messengers, chat tools, etc. The internet dissolved the element of distance in personal communities and enabled communication between geographically distant people. This shifted the landscape in intensifying the personal communities and weakening the emphasis on 'Neighborhood' communities. However, the asynchronous nature of these interaction modes meant that internet was considered as a support system to mediate community interaction. While the communications were globalized, they still remained one to one, thus emphasizing 'networked' individualism.

The period in the late 1990s and 2000s has seen the rise of audio and video internet streaming providing a fillip to social networking, practice based communities and applications like blogs and wikis with the emphasis shifting to multiple interaction paradigms. The phenomenon also saw the intensification of ties posed by internet and perhaps removing the need for face to face interaction. In practice based communities, the interactions can be related to the single issue alone and may not necessitate any face to face contact. Examples of practice based communities are students studying online from different countries, but learning together without any face to face interaction. The gaming community has expanded the interaction as people across the world come together. The scope for interaction in personal communities also expanded as the opportunistic and spontaneous interaction became more and more possible. Hence the role of the 'individual' networks underwent a sea change as communities could be formed over specific issues, have a short life span and not entail any personal interaction at all.

The reason for the separation of the online community era and social networking era is that social networking and communities took off in a completely new reborn way after the advent of wiki's, blogs and practice based communities. While online or virtual communities have traditionally evolved as a way for people in a homogenous location to keep in touch, the social networking era allowed spontaneous interactions (often for very small periods of time) between completely different people to get in touch and form communities. Studies opine that by 2020, pervasive collaboration (anytime anywhere availability), context based proactive anticipative technologies, people centric and knowledge focused social computing paradigms which are scalable, will be the future of collaborative environments (Ballesteros, 2006).

Mobile communities differ in three main aspects from virtual communities as follows (Tasch, & Fremuth, 2002). The differences are in: a) access method (mobile phones, smart phones and PDAs), b) use of the additional service benefits (offered by mobile phone operators), and c) different usage patterns from virtual communities. Hence, the evaluation and research in mobile communities must be viewed through a completely different prism and, thus, our contention is that mobile communities be separated from their desktop based counterparts. The time line we envisage is that mobile community networks are already in vogue now and are very much a part of the present era. It is the future that holds very high potential.

The evolution of mobile community networks with ubiquitous device support in the coming years will fuel a big paradigm shift and necessitate a new hybrid (the combination of virtual communities and mobile communications) environment. This is because the traditional desktop/laptop based virtual communities will soon be supplemented by mobile phone based community interaction. The use of mobile devices for social interaction has been documented as early as 1997 (Maeda et al., 1997). Mobile communities may also give rise to 'Smartmobs' (Rheingold, 2002), a mobile technology mediated social organization that behaves intelligently or efficiently because of its exponentially increasing numbers and allows people to access information and other people. The integration of mobile phone and sensory environments will also provide interesting and hitherto unachievable avenues to communities.

To help understand the development of applications for mobile community networks, a typology is needed. 'Typology' is the systematic classification in multiple dimensions along multiple scales that have characteristics or traits in common (Ganning, Qin, Flint, & Gasteyer, 2007). The typology for mobile community networks is discussed in the next section.

TYPOLOGY OF MOBILE COMMUNITY NETWORKS

This work seeks to lay the foundation for modeling the community interaction between the current

Figure 2. Typology of Mobile Community Networks

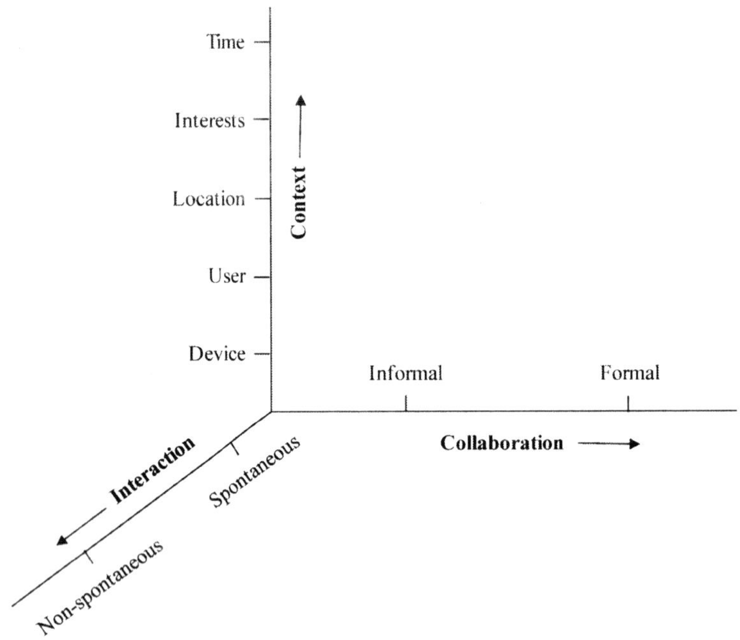

mobile and internet supported users. To do so, it develops a typology of the communication found in the mobile communities drawing on the current theoretical and empirical work. These mobile communities formed around a shared interest, support different types of communication and collaboration. The context of the users dictates the content that the users communicate or need. The existing literature in the relevant domain is first summarized in the following paragraph.

The typology for mobile virtual work consists of three dimensions (Vartiainnen, & Andriessen, 2005): a) physical location (onsite movers to nomads), b) communication mode (information access or voice communication), and c) type of work (individual work to collaborative teams). The typology proposed by Kawash, El Morr, & Itani (2007) consists of: a) degree of openness, b) degree of contextual information (contextual and non-contextual), and c) degree of interaction (synchronous and asynchronous). A three dimensional typology is also proposed in terms of networked working environments (Stanoevska-Slabeva, Högg, & Brodt, 2005) in terms of: a) level of human interaction (human centric vis-à-vis automated), b) collaboration (individuals, groups, intra organizational and inter organizational), and c) level of knowledge codified (implicit or explicit), as the major dimensions.

Our typology for mobile communities (Figure 2) has a similar three dimensional pattern but with different dimensions. We classify the mobile community networks based on:

1. Collaboration mode (informal, and formal)
2. Interaction (spontaneous and non-spontaneous)
3. Context (types of context).

While the above cited works are no doubt fundamentally sound in their categorization of mobile communities, the case for our typology is explained below.

The first dimension, i.e., the X axis, consists of the type of collaboration and ranges from informal communication, formal communication or hybrid communication. Mobile community networks need robust methods for collaboration and networking. The collaboration can be further classified into individual, group communication and formal communication. The communication method can be informal (SMS, mail, mobile chat) or formal (information storage & retrieval, collaborative mails, and transaction based communication) which by itself can be individual, collaborative, official to societal needs. Majority of communications in mobile community networks consists of informal and sporadic communication. In Porter (2004), the need for synchronous, asynchronous and hybrid interaction is discussed. In the mobile community network context, communication in turn can be a more apt dimension in terms of formal and informal dimensions with the synchronicity or lack of it being a process. The broader view is the way in which the communication is used.

In the Y axis, our proposed typology adapts the need for context from (Cao, 2007; Heitmann, Prykop, & Aschmoneit, 2004). The mobile community networks have been driven by the social activities such as blogs, exchanging pictures, SMS, chatting with each other. Mobile phones are an increasingly big source of context information. As mobile phones are personal devices, the scope for their use as instruments of community networks not just for communication, but for producing and consuming context based content is much higher. Hence, context is incorporated as a dimension with location, time, user details, device details, etc. as the parameters. Mobile communities have tremendous potential provided they facilitate context based personalization, location awareness, trust, personalization and immediacy of interaction. Hence, service providers and businesses seeking to leverage communities which provide these needs, have a great potential for success. Mobile context based marketing helps consumers by alleviating their need to memorize information in return for targeted and in some cases location based marketing strategies. Kuhn, & Wattenhofer (2006) show that leveraging the information available in mobile phones in their address books and personal profiles can help mobile applications be community aware. They also present a comprehensive list of community aware mobile systems which buttresses our conclusion about the applicability of mobile devices for community systems and the role of context and spontaneous networking.

The Z-axis consists of the type of interaction as spontaneous and non-spontaneous. This is supported by the typology for virtual communities where a multi-level formalism with type of initiation and participation is used as the primary classification and type of relationships as the secondary classification (Porter, 2004). Non-spontaneous interaction typically exists in organized communities (practice, process or people based) and works within the moral, philosophical and practical dimensions of the neighborhood of communities. The dimension of interaction extends with the overlap of spontaneous aspects and finally into the purely opportunistic. Broadly, the nature of interaction impacts the quality of service (QoS) we expect from the community network. QoS parameters such as real time metrics, security, reliability, scalability, etc. will all grow from this parameter. Hence a non-spontaneous interaction in an official, personal or group capacity would encompass initiation, participation and relationships. In a mobile community, the appropriate dimension will be the spontaneous or non-spontaneous interaction which includes whether the work is one-to-one, one-to-many or many-to-one or many-to-many. The interaction may be between people to people, people to devices and devices to devices.

Mobile communities will need facilities for spontaneous interaction between the people and devices to facilitate the dream of pervasive computing being gracefully integrated with the environment (Weiser, 1991). This will help the devices interact intelligently with the surround-

Figure 3. Client Server Architecture Model

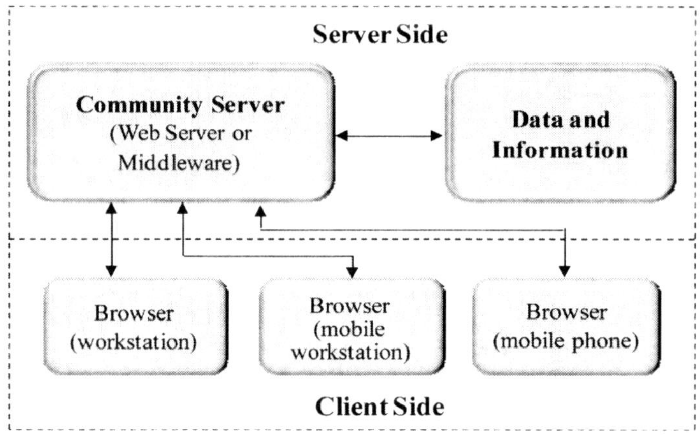

ing environment and support the ideal of distraction free computing. Spontaneous Interaction is defined as interaction between devices (and resources) that do not have prior knowledge of each other and do not have any administrative overhead [http://www.equator.ac.uk/var/uploads/ubi07-spontaneous-interaction.ppt]. Spontaneous interaction in ubiquitous computing has been studied for applications such as social interaction and game-playing in mobile user communities. 4. Architecture of Mobile Community Networks

The typology is of great importance in the design of a mobile community system as the knowledge acquisition process consists of coding the explicit knowledge of the community members using the implicit knowledge of the designers and thus enabling the reality of the community network system. Using the typology gives users flexibility and simplifies the process as the designers can run through the choices in order and design the community network system. The various design choices based on the typology are explained in this section. In general, the following aspects are important for users: a) technology, b) access, c) ease of use, and d) purpose. The users' description of needs should fit the profile of mobile communities. The choices in the architecture depend on the needs of the community and the environment the mobile community models. A client server model is shown in Figure 3.

This model allows data interoperability and access by a wide variety of devices. The community server (web server or middleware) provides the services requested by the mobile devices irrespective of their type and nature. The architecture provides for data to be represented in a wide variety of compatible formats. The community server is accessible from mobile devices with internet access and relies on the browsers in the devices to reformat the content according to the needs of the devices. Keeping this as the starting point, we examine other properties needed in the context of mobile communities and trace the evolution of the architecture.

The MILK project (Agostini, Albolino, De Michelis, De Paoli, & Dondi, 2003) proposes an architecture that uses various interaction managers at the middleware/web services level. Each requirement is modeled using a distinct interaction manager. The architecture is similar to the one proposed by Leimeister, Daum, & Krcmar (2004) and Hariharan (2008) in its three layer separation of access mechanism (mobile phone, mobile workstation, and stationary workstation), community server, and data. Our initial experiments on mobile community network based middleware

Figure 4. Architecture Model for Mobile Communities

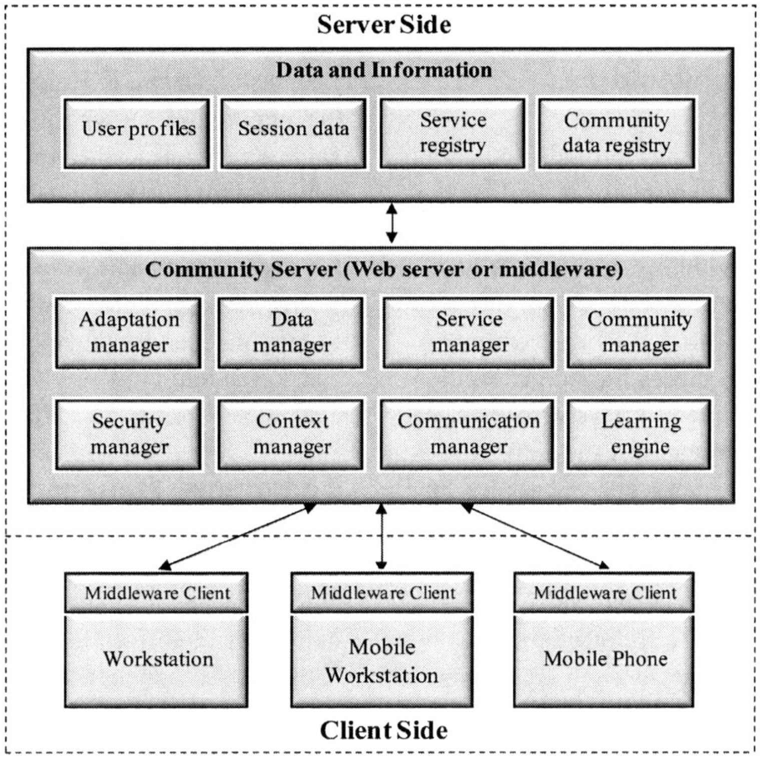

systems (Shriram, Sugumaran, & Vivekanandan, 2008; Sugumaran, Shriram, & Vivekanandan, 2008) have provided the basis for the vision on the components of the mobile community architecture. Based on the above architectures and the research challenges outlined below, an updated architecture is given in Figure 4.

The middleware clients shown in Figure 4 are an optional aspect (explained in the adaptation property section). Typically, when a middleware client is installed in the mobile device, it possesses routines for communication, security (encryption/decryption and access control), low-level data management to store (or modify) the personal profiles in the mobile phone and display management (Hariharan, 2008). The client request consists of device profile, network status, personal profile data and requested service/data. If the service/data is available, it is delivered to the client for display and further processing. Otherwise, the corresponding error message is sent. With this architecture, client applications can interact uniformly with community server uniformly, irrespective of the underlying interface. The profiles stored by the users help provide personalization and greater access to services in the immediate logical neighborhood of the users. Thus, mobile community systems are most effective when they help people reconnect to their physical surroundings by making them aware of events of relevance in their immediate neighborhood, taking advantage of mobile computing algorithms, wireless technologies, and real-time location systems. Mobile community network middleware can move computation away from the phones towards a trusted intermediary system that can perform a wide variety of tasks.

CHALLENGES IN DESIGNING MOBILE COMMUNITY NETWORKS

Light Weight Architectures

Light weight architectures provide the users flexibility for real time response in applications along with the ability to simultaneously interact within many communities. Light weight architectures built on top of mobile phones provides a common denominator for all the community users and a seamless means for communication in ubiquitous environments. They call for changes in user interaction mechanisms and architectures that will take into account properties such as low processing, memory, battery power and heterogeneity in displays and the additional add-ons such as camera, voice, graphics and SMS messages and ring tones etc. These architectures must be scalable from small groups to large organizations, provide self-organization, self-adaptation, and self deployment.

Collaboration Mode

The collaboration mode for mobile community networks varies from being spontaneous, issue based, unplanned interactions to structured and formal interactions. Formal communication requires secure interaction, constant monitoring by community moderators, metrics for trust and security. Examples for formal communications include access to information and practice based communities. Informal communications need facilities for spontaneous interaction and discovery methods for locating service providers and users. Three main collaboration applications (Bleecker, 2006) for mobile communities are: a) spatial annotations of objects, b) proximity interactions, and c) presence awareness. These pose high potential for collaborative applications and subsequent research.

Trust and Security

The importance of secure access is emphasized by Leimeister, Daum, & Krcmar (2004) who stress the importance of security in many levels: a) communication channel for access to community networks, b) trustworthy and transparent operators of community networks without any vested interests, c) mechanisms to foster trust in real life interactions, d) active community management to monitor member interaction and foster trust, and e) separation of service provider generated content and user generated content.

Community Management

Community management consists of policy planning when establishing the communities, formulating the framework for memberships, keeping track of and planning for the future based on current trends, actively fostering trust and security among the members, anticipating and planning for member requests and learning. The management aspects such as rules concerning access, communication modalities and content creation are equally important. Additionally, community managers must handle service discovery. It is not just the user and application data that is stored in databases, but also information about where the services are located. The decoupling of the location of services from services themselves provides a good abstraction for users. Each service location is stored along with the service description in the service registry. Thus, on receiving a user request, the community services locate the service(s) that match the requirements and provide the same to the users. This is in line with the traditional mechanism used in web services to provide content. Also, the location of services can use parameters such as relevance metrics, which consist of time, location, quality attributes or peer recommender information to help users make informed choices when more than one service or service provider is available.

Community management has to strive to increase the sense of community that members feel. The Sense of community is the key to the success of communities as it will dictate the progress of communities and continuity. The sense of community has the following distinct aspects (Blanchard, & Markus, 2004): a) membership, b) exchange of support, c) attachment, d) identity and identification of others, and e) relationship with specific members. The lessons learned from online communities are that people contribute when they think their uniqueness of contributions is valued in the group and when the perception of both unique contribution and benefits to the community are made salient. To foster the sense of community, the group has to offer additional loyalty based features to retain members. For example, a community website offers its members incentives in the form of gifts to participate in the community proceedings.

Context Awareness

The key issues for mobile community network systems to exploit context awareness are: a) acquisition and application of contextual data (static, dynamic, and external device dependent), b) dimensions of context applied (location, time and personal profile based), c) representation of contextual data, and d) learning modes based on community networks. The data acquisition and application can be static (based on calendars or planners), dynamic (based on capture of real time data and the processing) and external device dependent (cases where an add on card like attendance system provided to users alerts the community network system of the users presence in the specific location (Varshney, & Vitter, 2002). The dimensions of context modeled in the community network plays a crucial role in the performance.

Location based systems provide a wide variety of applications in tourism, leisure and collaborative work. Personal profile based systems help users in finding people with similar interests automatically increasing the opportunities for collaboration. Time based systems help users in planning their daily activities easily. The representation modes of contextual data play a crucial role in making use of the data, as good representations can shorten the time needed for data search, retrieval and association. As the mobile community networks evolve, so must the basic system used. The context aware system must have provisions for learning and adaptation. Such systems (for example, peer recommender systems) help the users exploit the power of the communities. Mobile virtual communities can offer services that alleviate the need for users to remember and track location based services and offer users marketing promos targeted at the context of users.

Adaptation

Boyera (2007) articulates the software needs of mobile phones as: a) standardized platform for service development, b) mini browser for rendering text and rendering the XHTML basic content, and c) local language rendering ability. The model used in Figure 3 relies on using browsers (in mobile devices) as the access mechanism. This helps in keeping the architecture simple and providing a seamless form of interaction, be it mobile phones or desktop workstations or mobile workstations. Other interaction mechanisms used may include installation of a mobile client at the access point. This allows:

1. personalization: User's personal preferences can be pre-stored as part of the profile of users. These data can then help in applications like automatic entry of forms, search for information, filtering product recommendations, and planning events (context),
2. user interface transformation: This helps in reformatting user content according to the device used. This can be very useful in image intensive applications like route map display and group visualization.

3. service execution: Depending on the devices used, the service can be executed in the application device or at the community server.

The mobile client will also help the use of avatars in mobile communities in a way similar to that of online communities improving the visual appeal and providing non-verbal communication. This architecture will also allow communities to consist of mobile devices as service providers. The access mechanism also includes the community server as an access media in that the community server can function as a proxy server for all the mobile clients filtering the access, reformatting the data according to the personal device and providing internet access in addition to the services in the community server itself.

Data and Information Management

A variety of data is encountered in mobile communities. They range from: a) user profiles consisting of user preferences and device specific data, b) session data consisting of user's participation in the mobile community (services requested and uncompleted so far, messages in queue, etc.) and information about the user's current context, c) community server service registry, d) community server data, e) user information describing number of users and their roles, and f) community control mechanisms. The key motivations in terms of social communities are: a) what to show to others, b) how to relate to others, and c) what to feel about others. 'What to show to others' is at the heart of data representation and abstraction. 'How to relate to others' and 'what to feel about others' will demand reasoning structures which categorize and react to the interaction with other devices and structures. While several representations (semantic and database oriented) have been mooted, it remains an active and fertile area of research.

Communication

Mobile devices need entirely newer forms of communication and information delivery, as ideally, not only should the users receive the right information at the right time and place, but also in a way that restricts neither their mobility nor their interaction with other people and the environment. The communication layer maintains each interaction session of communication and exchange of messages between the server and the client. The mobile community network must include facilities for the users to communicate in an orderly, secure and instant manner. The communication could range from exchanging instant messages and mails, to visual communication information in the form of photos, videos, information access, request for service or service providers, query responses, etc. The key parameters are timeliness of the communication and quality of responses (message delivery/failure, service messages etc.) as mobile phones have short switched on time, and often find it difficult to process remote requests due to network overload or disconnections. A robust communication manager should be tolerant towards such failures and degrade gracefully with increasing service/resource unavailability. Given the nature of mobile platform, good communication modes are essential for the success of the mobile community networks.

CONCLUSION

Research on mobile communities is essential to not only understand the mobile communities from a technological standpoint, but also the broad trends in the communities of users. In designing mobile community networks, it is very important to understand the nuances and practice of social networking that is reflected in the activities within a mobile community network. Mobile

Community networks allow users to access the community services anytime and anywhere. They can make positive contributions to social capital formation and informal learning. Hence, there is a great need to analyze how people respond to and confront social situations and the supporting technology viz. mobile community networks. As establishing what lies in the heart of community networking is key to understanding communities themselves, mobile community networks must be viewed as a key to understanding the connected era of pervasive and ubiquitous computing as a whole. This chapter has discussed the key aspects of mobile community networks starting from their evolution as a community to the present. A typology for mobile community has been discussed. The major contribution of the chapter has been to provide the architecture for mobile community networks and highlight the issues and challenges faced by developers. Further work on mobile community networks is needed to implement and experimentally validate the proposed architecture. Also, studies on mobile community networks as a social phenomenon is needed to help the sociologists and community informatics researchers understand and anticipate the challenges posed by the emerging technologies.

ACKNOWLEDGEMENT

Dr. Sugumaran's research has been partly supported by Sogang Business School's World Class University Program (R31-20002) funded by Korea Research Foundation.

The research between Dr. Sugumaran and Dr. Shriram is a part of the Research Collaboration between the Sogang Business School's World Class University Program and Faculty Training Academy – B.S. Abdur Rahman University.

REFERENCES

Adeya, C. N. (2005). Wireless Technologies and Development in Africa. *Wireless Communication and Development: A Global Perspective. Annenberg Research Network on International Communication Workshop.* Marina del Ray, California, USA, October.

Agostini, A., Albolino, S., De Michelis, G., De Paoli, F., & Dondi, R. (2003). Stimulating knowledge discovery and sharing. *Proceedings of the International ACM SIGGROUP conference on Supporting group work,* Sanibel Island, Florida, USA.

Ballesteros, I. L. (2006). New Collaborative Working Environments 2020. *Report on industry-led FP7 consultations and 3rd Report of the Experts Group on Collaboration@Work*. European Commission, February 28.

Blanchard, A. L., & Markus, M. L. (2004). The Experienced "Sense" of a Virtual Community: Characteristics and Processes. *SIGMIS Database*, *35*(1), 64–79. doi:10.1145/968464.968470

Bleecker, J. (2006). What's Your Social Doing In My Mobile? Design Patterns for Mobile Social Software. *In WWW2006 Workshop, MobEA IV—Empowering the Mobile Web*, May 23, Edinburgh, Scotland.

Botto, F. (2008). A socio-technical infrastructures framework for community networks. *Third International Conference on Communications and Networking, ChinaCom 2008*, Hangzhou, China, August 25-27.

Boyera, S. (2007). Can the Mobile Web Bridge the Digital Divide. *Interactions (New York, N.Y.)*, *14*(3), 12–14. doi:10.1145/1242421.1242433

Cao, Y. (2007). Mobile Social Software with Context Awareness and Data Uncertainty for Technology-Enhanced Learning. *Online-Proceedings of the 2nd Prolearn Doctoral Consortium in Technology Enhanced Learning (EC-TEL 2007)*, Crete, Greece, September.

Carpineto, C., Mizzaro, S., Romano, G., & Snidero, M. (2009). Mobile Information Retrieval with Search Results Clustering: Prototypes and Evaluations. *JASIST, 60*(5), 877–895. doi:10.1002/asi.21036

Carroll, J. M., & Rosson, M. B. (2003). A trajectory for community networks. *The Information Society, 19*(5), 381–393. doi:10.1080/714044685

Clodfelter, K., Buente, W., & Rosenbaum, H. (2007). CommNets 2.0: Using an Evolutionary Perspective to Examine a Statewide Community Networking Initiative. *Proceedings of the American Society for Information Science and Technology, 44*(1), 1–20. doi:10.1002/meet.1450440216

Farooq, U., & Carroll, J. M. (2003). Mobilizing Community Networks. *Conference on Home Oriented Informatics and Telematics (HOIT)*, April 6-8, Irvine, California.

Ganning, J., Qin, H., Flint, C. G., & Gasteyer, S. P. (2007). *Constructing Typologies in Community Research*. Paper presented at the annual meeting of the Rural Sociological Society, Marriott Santa Clara, Santa Clara, California.

Grasso, A., Koch, M., & Snowdon, D. (1999). Campiello - New user interface approaches for community networks. [August]. *ACM SIGGROUP Bulletin, 20*(2), 15–17.

Hammond, A., & Paul, J. (2006). *A New Model for Rural Connectivity*. Retrieved April 7, http://www.usaid.gov/our_work/ economic_growth_and_trade/ info_technology/tech_series/ Rural_Connectivity_508.pdf.

Hariharan, K. (2008). Best Practices: Extending enterprise applications to Mobile devices. *The Architecture Journal, 14*, 10–17.

Heitmann, M., Prykop, C., & Aschmoneit, P. (2004). Using Means-end Chains to Build Mobile Brand Communities. *Proceedings of the 37th Hawaii International Conference on System Sciences*, Big Island, HI, USA, January 4-8.

Hillebrand, C., Groh, G., & Koch, M. (2002). Mobile Communities – Extending Online Communities into the Real World. *Mobile and Collaborative Business*, 7–18, September.

Joanna, G., Hua, O., Courtney, F., & Stephen, G. (2007). Constructing Typologies in Community Research. *Annual meeting of the Rural Sociological Society*, Marriott Santa Clara, Santa Clara, California, August 02.

Kawash, J., El Morr, C., & Itani, M. (2007). A novel collaboration model for mobile virtual communities. *International Journal on Web Based Communities, 3*(4), 427–447. doi:10.1504/IJWBC.2007.015868

Kuhn, M., & Wattenhofer, R. (2006). Community-Aware Mobile Networking. *1st Workshop on Mobile Services and Personalized Environments (MSkuPE)*, Aachen, Germany, November.

Leimeister, J. M., Daum, M., & Krcmar, H. (2004). Towards mobile communities for cancer patients: the case of krebsgemeinschaft.de. *International Journal on Web Based Communities, 1*(1), 58–70.

Maeda, H., Kajihara, M., Adachi, H., Sawada, A., Takeda, H., & Nishida, T. (1997). Weak Information Structures for Community Information Sharing. *International Journal of Knowledge-Based Intelligent Engineering Systems, 1*(4), 225–234.

Parikh, T. S. (2006). Designing an Architecture for Delivering Mobile Information Services to the Rural Developing World. *Proceedings of 7th IEEE Workshop on Mobile Computing Systems and Applications*, Orcas, WA, 06-07 April.

Porter, C. E. (2004). A Typology of Virtual Communities: A Multi-Disciplinary Foundation for Future Research. *Journal of Computer Mediated Communication, 10*(1). Retrieved April 7, 2008 from http://jcmc.indiana.edu/vol10/issue1/porter.html.

Preece, J., Maloney-Krichmar, D., & Abras, C. (2003). History and Emergence of Online Communities. In Wellman, B. (Ed.), *Encyclopedia of Community*. New York: Berkshire Publishing Group.

Rheingold, H. (2002). *Smart Mobs: The Next Social Revolution*. New York: Perseus publishing.

Rheingold, H. (2003). *Mobile Virtual Communities*. Retrieved April 7, 2008 http://www.thefeaturearchives.com/ topic/Culture/ Mobile_Virtual_Communities.html.

Schuler, D. (1994). Community networks: Building a new participatory medium. *Communications of the ACM, 37*(1), 38–51. doi:10.1145/175222.175225

Shriram, R., Sugumaran, V., & Vivekanandan, K. (2008). A middleware for information processing in mobile computing platforms. *International Journal of Mobile Communications, 6*(5), 646–666. doi:10.1504/IJMC.2008.019327

Stanoevska-Slabeva, K., Högg, R., & Brodt, T. (2005). A Taxonomy for Networked Working Environments (NWEs). In *Innovation and the Knowledge Economy*. Amsterdam: IOS Press.

Sugumaran, V., Shriram, R., & Vivekanandan, K. (2008). Towards a Mobile Community Based Middleware for Spontaneous Interaction. *Decision Sciences Institute Annual Meeting*, Baltimore, November 22-25.

Tasch, A. E., & Fremuth, N. (2002). *Mobile Communities*. Retrieved April 7, 2008 from http://www.cosmos-community.org/ downloadFiles/icis_mobilecommunities_fremuthTasch.pdf.

Varshney, U., & Vetter, R. (2002). Mobile Commerce: Framework, Applications and Networking Support. *Mobile Networks and Applications, 7*(3), 185–198. doi:10.1023/A:1014570512129

Vartiainnen, M., & Andriessen, E. (2005). Mobile Virtual Work, What have we learned. In Vartiainen, M., & Andriessen, E. (Eds.), *Mobile Virtual Book: A new paradigm*. New York: Springer Publications.

Venkatesh, M. (2003). The Community Network Lifecycle: A framework for research and action. *The Information Society, 19*(5), 339–347. doi:10.1080/714044682

Weiser, M. (1991). The computer for the 21st Century. *Scientific American, 265*(3), 94–104. doi:10.1038/scientificamerican0991-94

Wellman, B. (2005). Community: from neighborhood to network. *Communications of the ACM, 48*(10), 53–55. doi:10.1145/1089107.1089137

Chapter 22
A Secure and Trustworthy Framework for Mobile Agent-Based E-Marketplace with Digital Forensics and Security Protocols

Qi Wei
University Kebangsaan Malaysia, Malaysia

Ahmed Patel
University Kebangsaan Malaysia, Malaysia

ABSTRACT

Mobile agents raise security issues such as the protection of platform/host that runs the mobile agent against attacks which can harm or use its resources without permission, and another is the need for protection to guard mobile agents and their supporting systems against the malicious attacks from a variety of intervening sources that might alter information it carries and processes when it visits the hosts in its transactions itineraries. In this article, the authors propose a framework which includes safe, secure, trusted and auditable services, as well as forensic mechanisms to provide audit trails for digital evidence of transactions and protection against illegal activities. The proposed framework and protocols provide a secure communication for mobile agents when they move to different security environments to deal with e-marketplace activities such as search information, negotiation and payments. This article is concluded by highlighting and discussing further research work to build viable systems.

INTRODUCTION

E-marketplace is a flexible and efficient approach to assist companies or corporations to extend their businesses to reach larger markets without regional boundaries via ad hoc networks. This must be performed in a safe, secure, trusted and auditable manner to ensure customer (buyer) and supplier (seller) confidence against all kinds of attacks and masquerades. It requires a new paradigm,

DOI: 10.4018/978-1-60960-563-6.ch022

improved integration architectures and services to bring this to fruition. Such e-marketplace is typically a cooperative distributed system composed of economically motivated software agents that interact cooperatively and/or competitively, find and process information, and disseminate it to humans and to other agents. In addition, it must provide supports for common economic services and transactions, such as dynamic pricing, negotiation, automated supply chains, as well as other e-marketplace service infrastructure to ensure secure, trusted and reliable transactions. Therefore, the mobile agent-based e-marketplaces are becoming more important in e-commerce applications. The mobility characteristic of mobile agent permits all the required operations to be performed locally in the e-marketplace without maintaining reliable connections with remote hosts and without bandwidth engagement. However, in e-marketplace environment, strong mobility will cause the high risk security threats; in contrary, low mobility causes low security threats. Although much have been dedicated to the design of agent-based marketplaces, the lack of standards for agent-based e-marketplace framework incorporating safe, secure, trusted and auditable services reflects that there are still many issues that need to be resolved before a standard could be defined for such an all encompassing unified framework.

Security of mobile agents, the most important technique in the e-marketplace applications is still in its infancy and research topic. There are different security approaches for mobile agents that have been proposed to protect the platform, host, agents and route. The security protocols such as Secure Socket Layer (SSL) (Freier, Karlton & Kocher, 1996) and Secure Electronic Transaction (SET) (Drew, 1999) are used for confidentiality and integrity to secure the communication between agents on different hosts. The SSL channel may be not secure for mobile agent since a mobile agent may move to an insecure platform or host to communicate with other agents. Also, it is dangerous for mobile agent to exchange sensitive information without the use of cryptography techniques because the information can be stolen or corrupted by malicious attacking agents. SET on the other hand offers better security than SSL since it uses Public Key Infrastructure (PKI) for privacy and X.509 digital certificates to authenticate participants in e-marketplace (Liu, 2003). More importantly, sensitive information is not seen by the merchant, and is not kept on the merchant's server to ensure buyer confidentiality, privacy and safety.

Furthermore, the e-marketplace should meet the requirements of mobile users whose expectation of e-marketplace services to be available in mobile computing devices and mobile networks that incur high communication cost and/or low bandwidth within bounded target security conditions to establish fully-fledged goal driven agent based e-marketplace (Jailani et al., 2008).

The goal of this article is, therefore, to define a framework together with all the safe, secure, trusted and auditable services, including forensic mechanisms for replying and proofing against illegal activities. The proposed protocols for the security issues are also defined and explained in this article.

The article is organized as follows. Section 2 represents the related works on secure framework for mobile agent-based e-marketplace. Section 3 illustrates the architecture of proposed framework mobile agent-based e-marketplace, followed by the description of major components and the working flow of the proposed framework. Section 4 gives the detailed explanation of the digital forensic for the proposed framework. The security protocols are defined, specified and proved in Section 5. Finally, we propose, discuss and conclude further research and development work to be done to in this subject area.

RELATED RESEARCH

One of the major weaknesses in today's mobile agent-based e-marketplace systems usage is the failure to accommodate the end-user or system processes to use mobile trading and its security, audit and forensic features in a seamless and transparent way as Jailani et al. (2008) have reported that there are many disparate frameworks but none has brought together all the central issues for a secure and auditable mobile agent-based e-marketplace into a single framework. For instance, Chan (2000) developed a shopping system based on mobile agent which called Shopping Information Agent System (SIAS). His research concentrated on the security and reliability issues of mobile agents and discussed about the possible security attacks by malicious hosts against mobile agent in the system. Specific solution to prevent these attacks was devised. Another mobile agent-based e-commerce framework proposed by Zhao et al. (2007) focused on the security, integrity and confidentiality of mobile agent to secure the trading transactions. They use the secure message digest and encryption to illustrate the mobile agent-based e-commerce model (MA-ECM). The implementation was done using Aglets mobile agent platform (IBM, 2004) and Java language.

The high risk security threats appear with strong mobility. To reduce the security threats of mobile agent, Zhang and Liu (2005) proposed a framework for e-commerce that uses both single hop and multi-hop to reduce the risks. The first phase uses multi-hop for information collection where mobile agent does not take any sensitive information. The second phase uses single hop to take the sensitive information to reduce the security risks. In another way, Wang Rui-yun et al. (2007) proposed a solution to protect the mobile agent in e-commerce. The article presented the secure authentication infrastructure using PKI and introduces an integrity protection and confidentiality protection solution's basing on Multiple Hop_Adaption (MH_A) integrity protection with regards to PKI limit.

It is important to ensure that the information carried by a mobile agent or stored on a platform is accessible only to authorized parties (Kannammal & Iyengar, 2007). To address such privacy and trust issues from both user and system perspectives, Yu Bin and Yi Xiang (2006) proposed a secure transmission mechanism for mobile agent-based system to overcome the weakness of traditional communications for the distribution of data encryption key without going through a third-party. While Wang et al. (2007) proposed a migration mechanism of security and fault tolerance for mobile agent that includes trust and travel services. The trust service can be duplicated as backup at several different trusted nodes in the e-marketplace environment. When the trust service expired, the system can select one backup from the trusted nodes as the trust service via the Election Algorithm (Krzyzanowski, 2000). However, this mechanism has some drawbacks such as the expiry of trust service and trusted nodes which may affect the backup of trust service, therefore causes the travel transmission problem of mobile agent.

The right of an Internet user acting anonymously conflicts with the rights of a server victim identifying the malicious user as exposed by Giannakis Antoniou et al. (2008). They proposed a security protocol for privacy and forensics investigation process which dramatically increases the level of the protocol's reliability that can address the reliability issue efficiently by defining what is considered as malicious and what is not at the beginning of a communication. The protocol to support digital forensic functions is divided into 3 phases: the initialization phase, main and forensic investigation phase. This model also has been partially used in Jailani et al. (2008) experimental framework and architecture.

The frameworks and systems described above include some aspects of security, trust, privacy and forensic issues. Jailani et al. proposed an extended secure and auditable agent-based framework for mobile users. They investigate the requirements and analyze the architecture needed to define a secure and auditable agent-based framework via the auditable and forensic investigations. They exemplified the framework and protocols with an auction e-marketplace as the commerce application and presented scenarios to make their point. They defined the essential requirements for the secure and auditable agent-based e-marketplace as well as requirements for mobile commerce environment which an agent-based e-marketplace framework should includes the features such as security, privacy, trust, audit and digital forensic. Thus, the proposed framework requires the safe, secure, trusted and auditable service, including forensic mechanisms for reply and proofing against illegal activities from outside or within the system.

PROPOSED FRAMEWORK

The mobile agent-based e-marketplace is taking place within the autonomous agents such as the buyer agent and supplier agent. Maes et al. (1999) surveyed the buying and selling agents and discussed several agent-mediated electronic commerce systems in context of a general model of the buying process. According to their research, the mobile agent-based e-marketplace trading normally can be categories into six steps:

1. Information request;
2. Search product;
3. Matching buyer and suppliers;
4. Negotiation between buyer and supplier;
5. Purchase and payment;
6. Delivery the services and evaluation.

However, during this working process, the mobile agent-based e-marketplace is facing four types of security problems:

1. Protect the data that carries by mobile buyer agent during the data transmission.
2. Protect the agent from malicious supplier agent host.
3. Protect the buyer agent host from the malicious agent.
4. Protect the agent from other mobile buyer agent harm or use its resources without permission.

To solve such security problems, we propose a secure, trustworthy mobile agent-based e-marketplace framework with the central issues such as security, privacy, trust, audit and digital forensic as part of a *total* solution. The security protocols are designed for the proposed framework such as the migration protocol, negotiation protocol, payment protocol and the protocol for the digital forensic. In addition, the cryptography technique and encryption algorithm are also provided. Figure 1 shows the architecture of Secure and Trustworthy Mobile Agent-based E-marketplace called STMAE.

The Major Components of Framework

Based on the architecture of proposed framework shown in Figure 1, the roles of the mobile agent, the activities and behaviours of the major components are designed to perform the transactions in the e-marketplace as components are as follows:

Buyer Agent (BA)

Buyer agent is a static application agent that stays at the local platform with its security parameters like private key, etc. The BA can create the mobile agent according to the different requirements of user. BA has ability to manage these mobile agents to execute missions such as search information, negotiation and trade payment in the e-marketplace environment.

Figure 1. Architecture of secure and trustworthy mobile agent-based e-marketplace

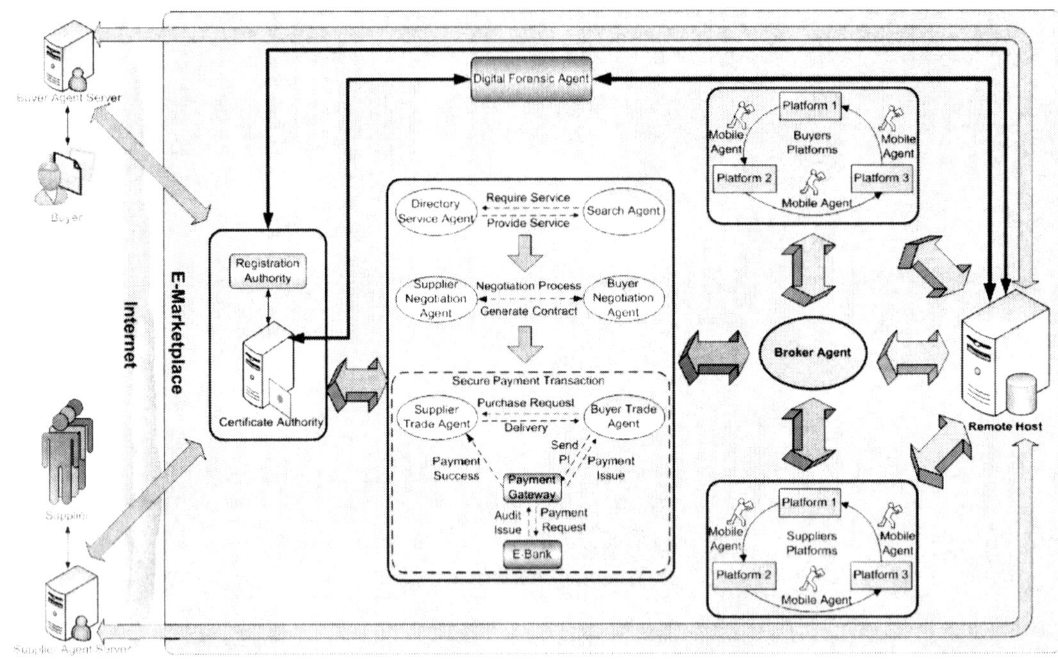

Supplier Agent (SA)

Supplier agent can be represented as the companies or organizations which decided to register to the e-marketplace. The working principle is similar as MA. It also has ability to create the mobile agent based on the different requirements of supplier. It dispatches the mobile agent into e-marketplace for products/services providing, negotiation, payment and delivery products/services in the e-marketplace.

Registration Authority (RA)

Registration authority handles the registration for buyers or suppliers to join into the e-marketplace. A master mobile agent (MA or SA) dispatches worker mobile agent who carries all the request information including its certificate and necessary related information to registration authority. The registration authority performs verification of mobile agents' certificates through the certificate authority.

Certificate Authority (CA)

Certificate authority acts as a TTP (trusted third party) that issues and manages public keys and thus guarantees the link between the user and the cryptographic key used. The CA assumed to create only valid and reliable certificates. CA performs the authentication between the trading parties throughout a transaction for a proof of identity.

Guard Agent (GA)

A Guard Agent is concerned with the process of migrating or transmitting a mobile agent from one node to another. In the marketplace, GA is situated in components which become the entry and exit points to the marketplace. Therefore, GA will be installed in the RA and BroAs.

Information Search Agent (ISA)

It is an agent which seeks out information on behalf of their users. It carries the request information from buyer to search from the products/services catalogues maintained by the broker. The matched products/services will be returned after the filtering.

Directory Services Agent (DSA)

Directory Services Agent carries the owner's information, products/services information and price and response for its products/services to be advertised and sold. It will return back when the trading partner has been found.

User Interface Agent (UIA)

The user interface agent is an important component which used to monitor the interaction of their users with information systems and regulates the interaction in ways that helps the users.

Negotiation Agent (NA)

Negotiation agent is an essential component in the e-marketplace. The negotiation requirements include the price (highest/lowest/discount prices), time limitation, negotiation strategy, etc. According to the requirements, BA dispatches the Buyer Negotiation Agent (BNA) into the e-marketplace to negotiate with the Supplier Negotiation Agent (SNA) who dispatches by SA until they achieve the desired price. The contract will be generated after the negotiation.

Payment Gateway (PG)

Payment gateway is a trusted entity that responsible for process payment authorization and payment capture requests from suppliers. The Secure Electronic Transaction is used for the payment model in this framework. The protocol for the payment process is based on the existing SET protocol to facilitate safe and reliable business among the cardholders and merchants. After a buyer and a supplier have been matched by the Broker Agent, both parties will be informed through their corresponding mobile agent. The proposed payment protocol will be discussed in section 5.

Broker Agent (BroA)

Broker agent is a key component of bargain in the e-marketplace. A broker can register with a certain auction type as the price negotiation mechanism to match suppliers with buyers. Both buyer and supplier can register with a number of brokers. Broker agent will record all the information in the auction database which include the information of buyers and suppliers with their associated buyers' request.

Delivery Agent (DA)

Delivery Agent is a TTP (Trusted Third Party) that used to deliver the products and services. DA carries the information of the location, type of transportation, delivery time, etc. SA will announce DA to give the delivery information from buyers such as the buyer's corresponding address, time, transportation and cost incurred. Buyer will receive the delivery detail from DA.

Information Collection Agent (ICA)

Information Collection Agent is used to record the information related to the transactions carried out by buyer and supplier mobile agents based on the mechanism it provided. The log files are used to record the information that would be used for forensic investigations. Events related to transactions also will be recorded.

Digital Forensic Agent (DFA)

Digital forensic service provides a mechanism of identifying users, communicating and performing forensic which incorporates Privacy Enhancing Technologies (PETs). A PET offers privacy to the Internet users within the communication ano-

Figure 2. Architecture of digital forensics

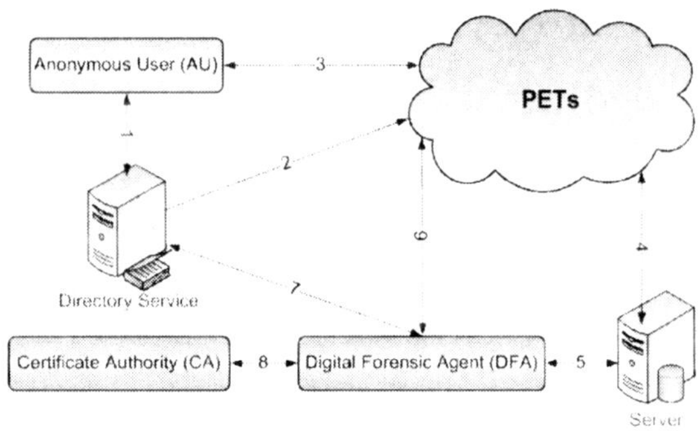

nymity. However, a PET may hide the identity of attackers masquerading as a *normal* user. Thus, protocols are required for digital forensic which can solve this problem of the perpetrators while protecting the anonymity of genuine, bona fide, users. Figure 2 shows the architecture of digital forensic.

Digital forensic agent is used to investigate a particular sequence of transactions received by the anonymous user through PET thus forms the initial evidence. If the Server has enough evidence to prove that somebody (AU) did try to attack the server, then it will further investigate and obtain the identity of the abuser from the Privacy Enhancing Technologies (PET) and the DFA can acquire more information about the abuser from the CA to produce solid evidence against the abuser's non-repudiated actions. We will give a detailed explanation of digital forensic in section 4.

Process Flow of Secure and Trustworthy Framework

According to the architecture of secure and trustworthy mobile agent-based e-marketplace that shown in Figure 1, the general process flow for the proposed framework is described as follows:

Buyer creates a BA as a master mobile agent at buyer local host and distributes the specific requirements of the products/services to be requested to BA such as the name, model and type of products/services. The requirements may also include the weight customization of each item. BA creates and addresses the tasks to ISA which will be dispatched to the marketplace to perform the tasks. Not only the information of requested products/services that ISA carries but also buyer's certificate.

It will inform a GA for authorization before ISA leaves from local host and carries the tasks and migrates to the marketplace. After that, ISA will ask for the certificate verification from the RA. RA will perform the verification through CA and issue the authority to ISA after ISA passed the identification. ISA performs a search from the products/services catalogues maintained by the broker. SA dispatches DSA to the marketplace which carries the information of products/services. Broker agent will record the information in auction database. ISA will return an evaluated and filtered list of products/services based on buyer's specified requirements and preferences.

ISA returns back to buyer local host and reports the results to BA. Buyer will decide which product he/she wants to buy. BA will dispatch a number of BNA in parallel which carries the information such as the price, time limitation and negotiation strategy. Once the BNA enter into the marketplace, it will negotiate with SNAs. The negotiation deals autonomously according to a set of user constraints and strategic guidelines. The negotiation will be done until both the buyer and supplier achieve the desired price. The contract will be generated after the negotiation and there should be an agreement between buyer and supplier.

Upon the successful results of negotiation, buyer will choose one of the best prices offered for product/service and decided to purchase it. BA dispatches the BTA which carries the purchase information to marketplace for the purchase request. STA will acknowledge the purchase confirmation. The payment is based on Secure Payment Transaction card payment model which through a TTP, PG is a trusted entity for the payment model. To reduce and control the risks involved in auditing, digital forensic should be used to make the audit service more secure.

Once the payment process has been done, SA dispatches the DA to delivery the purchased products which carries the information of corresponding address, time, mode of delivery and transportation. The delivery information will be recorded at both supplier and buyer database.

Since the characteristic of mobile agent is mobility, the mobile agent can fast and efficiently collect the information in the network. Therefore, we use the multi-hop (serial itinerary) for mobile agent to travel for searching from one node to another on the network within the specific requirements. During the information collection, mobile agent does not take any sensitive information such as price, negotiation strategy, etc. The databases that mobile agent travelled are read-only databases. For the negotiation and payment process, we are using the parallel dispatching of mobile agents in which all the mobile agents will take the sensitive information for the processing. We are using single-hop (parallel itinerary) to reduce the risks for negotiation and payment process and provide the secure transactions. Thus, the proposed framework solves all central issues and the process efficiency, low cost and high communication.

Cryptography for Proposed Framework

Above we discussed the general process flow of the proposed framework. Here, we discuss the proposed framework from the security point of view.

Nowadays, security of every distributed system relies on some cryptographic techniques. Mobile agent-based e-marketplaces are no exception, as security requirements of these systems overlap with main goals of cryptography are integrity, confidentiality and authentication. The proposed framework embeds the cryptography technique such as message digest and encryption algorithm for the protection of data integrity and confidentiality. Specifically, the algorithm used to calculate message digest is SHA-1(Secure Hash Algorithm) (FIPS PUB 180-1, 1995) and the encryption algorithm used in the process is RSA and AES (Advanced Encryption Standard) algorithms.

SHA-1 (Secure Hash Algorithm)

SHA-1 was published in year 1995 in FIPS PUB 180-1. It is designed to work with the Digital Signature Algorithm (DSA) (Barker et al., 2006). SHA-1 is a one-way hash function. One-way functions are characterized by two properties. The first is that they are one-way. This means that you can take a message and compute a hash value, but you cannot take a hash value and recreate the original message. It is also collision-free, thus, no two messages can hash to the same value. SHA-1 produces a 160-bit (20 byte) message digest. Although slower than MD5 (Message Digest 5) which was designed by Rivest in 1991 (Preneel, 2005), this larger digest size makes it stronger against brute force attacks. The reason that we do not select MD5 is because it has been deciphered by Wang Xiaoyun (Yang & Ding, 2004).

RSA Algorithm

RSA (Rivest, Shamir & Alderman, 1978) algorithm is a well-known public key cryptography. It is the first algorithm known to be suitable for signing as well as encryption, and one of the first great advances in public key cryptography (Tyoviriyakul & Kittitornkun, 2008). RSA is suitable for the encryption of the data, digital signatures and digital envelope. RSA is widely used in electronic commerce protocols, and is believed to be secure given sufficiently long keys and the use of up-to-date implementations. The RSA algorithm involves three stages: key generation, encryption and decryption. RSA involves a public key and a private key. The public key can be known to everyone and is used for encrypting messages.

Figure 3. Process of digital signature by using SHA-1 and RSA

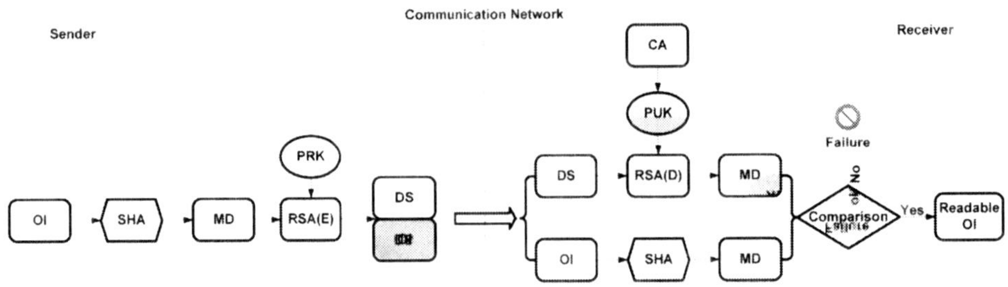

Messages encrypted with the public key can only be decrypted using the private key.

AES Algorithm

AES (Advanced Encryption Standard) is a symmetric key cryptography which adopted by U.S. government and announced by National Institute of Standards and Technology in year 2002.

AES was developed based on Rijndael (Courtois & Pieprzyk, 2002). It has more precise mathematical algorithms and the data only encrypted once. AES is designed for high-speed, strong security features and also able to support a variety of small equipment. Due to the limitation of DES, AES has been introduced to replace it (NIST, 2000). The performance comparison between AES and DES are described as follows:

- The major weakness of DES is the key size. It encrypts data in 64-bit clear text blocks, with a key size of 56 bits. Therefore, an exhaustive search through 56-bit key space can be done in weeks or moths, even with the common equipment. The key size of AES is 1021 times safe than key size with DES since it has a fixed block size of 128 bits and a key size of 128, 192, or 256 bits.
- AES algorithm has simple design, fast key installation, less memory space, has ability to run on different platforms and support the parallel processing. It is also has ability to resist all known attacks.

Secure Data Transmission of Proposed Framework

The main goal of using digital signatures in mobile agent-based e-marketplace is to ensure the authenticity of inter-platform messages and migrated agents.

The digital signature scheme comprises the creation of the digital signature and its verification (e.g. *Sign(OI)* = *E* and *Verify(E)* = *{0, 1}*). Sender can use the SHA-1 algorithm to generate a hash-value of the original message. After calculating the hash-value, sender can use the RSA algorithm to encrypt the hash-value with its private key. Verification of the digital signature calculated from the hash-value is two-stage process. The original hash-value is decrypted from the digital signature using the receiver's public key. In addition, the new hash-value is calculated from the signed message by using the same hashing algorithm (SHA-1) as the original one. Then, these two hash value will be compared to each other. If they are equal, the message is authenticated and its integrity is proved. Otherwise, the message has been changed during the transmission and neither its sender's identity nor message content can be trusted.

Figure 3 illustrate the process of digital signature by using SHA-1 and RSA.

From the cryptography algorithms mentioned, we propose combining cryptography algorithms (SHA-1, RSA and AES) for data encryption, decryption, digital signatures and verification to make a secure data transmission in our framework.

Figure 4. Data transmission by using the combined cryptography algorithms

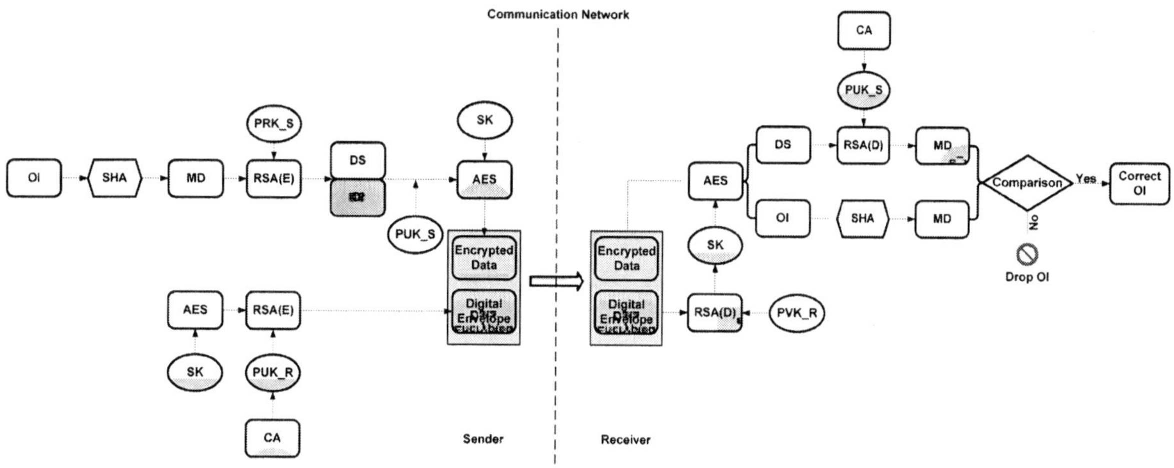

Figure 4 shows a simple data transmission from a sender to a receiver by using these algorithms. The steps involved are:

1. Sender(S) generates a message digest (h(MD)) by using SHA-1.
2. MD is encrypted with sender's private key by using the RSA algorithm which attached with the original information. Thus, the sender's digital signature has been generated.
3. Sender (S) creates a random symmetric key (SK) through the symmetric algorithm-AES which is used to encrypt the message, digital signature and public key of S. Thus, the encrypted data has been generated.
4. Before S sends the encrypted data, it must receive the public key from R and encrypted by SK to make a digital envelope. Then, S sends both the encrypted data and the digital envelope to R.
5. Once R receives the secret message, digital envelope and the S's symmetric key (SK) inside the DE are decrypted with the R's private RSA key.
6. The encrypted data will be decrypted and recovered by SK. Thus, R gets the message, digital signature and the public key of S.
7. The digital signature will be decrypted by using S's public key and get the message digest (h(MD)); at the same time, a new MD has been generated from the original message that encrypted by SHA-1 algorithm.
8. Make a comparison between the decrypted MD and new generated MD, if they are equal means the message is authenticated and its integrity is proved. Otherwise, the message has been changed and R won't accept the message.

This cryptography services is used in STMAE to make a secure data transmission.

DIGITAL FORENSICS FOR PROPOSED FRAMEWORK

Digital forensics is an interdisciplinary of law, intrusion detection and computer science. Hence, it is an important role in e-marketplace. In e-marketplace environment, digital forensic is used to investigate the cyber crimes, establish the identity of abusers and gather digital evidence of malicious activities for presentation to law.

To make the proposed framework all encompassing and comprehensive in its coverage to

Figure 5. Activities sequence of initialization phase and main phase

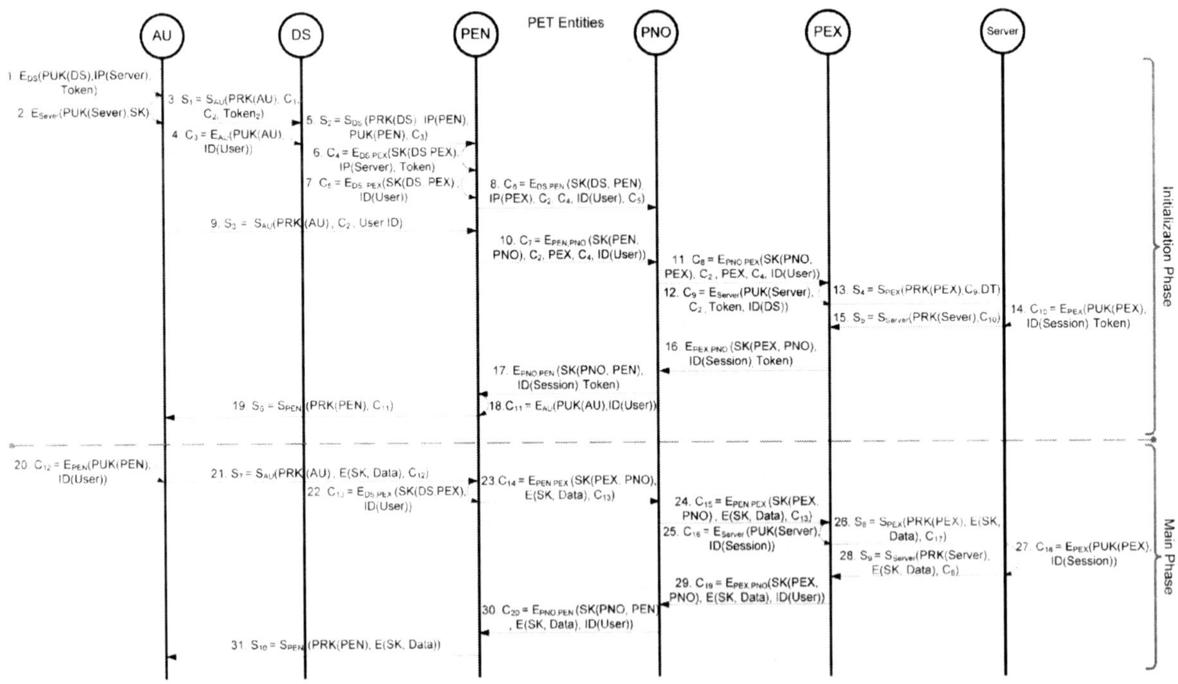

ensure auditability and traceability, we define a protocol for the digital forensics activities based on the terminology of such protocol designed by Giannakis Antoniou et al. (2008).

There are three phases of the protocol supporting the digital forensic function: the initialization phase, main phase and the investigation phase. Different participating entities support different phase. During the initialization and main phase, the participating entities are the Anonymous User (AU), Privacy Enhancing Technology entities (PET) and a Server. During the investigation phase, the participating entities are the DS, PET, Server, DFA and CA. The protocol is named Digital Forensic Protocol (DFP). Figure 5 and Figure 6 illustrate the activities sequence of three phases of DFP.

Initialization Phase

During the initialization phase, both AU and Server need to agree with the security policy to determine which malicious host is and which is not. Initialization phase is required for the AU and Server to exchange a secret key in order to provide end-to-end encryption. The secret key is valid only during that session; otherwise, the key is not valid and a new pair of key need to be generated for future session, even if the participating entities are the same. The Directory Service (DS) nominates the entry point PEN and the exit point PEX, and each node PNO chooses the next node PNO. DS is used to verify the validity of $Token_2$ based on the given Token. DS stores the messages in order to prove that specific AU communicates with Server by using the secret key which has the specific $Token_2$. DS should not accept the $Token_2$ of the AU if it has been used by another AU. The DS must explicitly request the AU to perform a new request with a different $Token_2$. Figure 5 shows the activities sequence of initialization phase.

Figure 6. Activities sequence of investigation phase

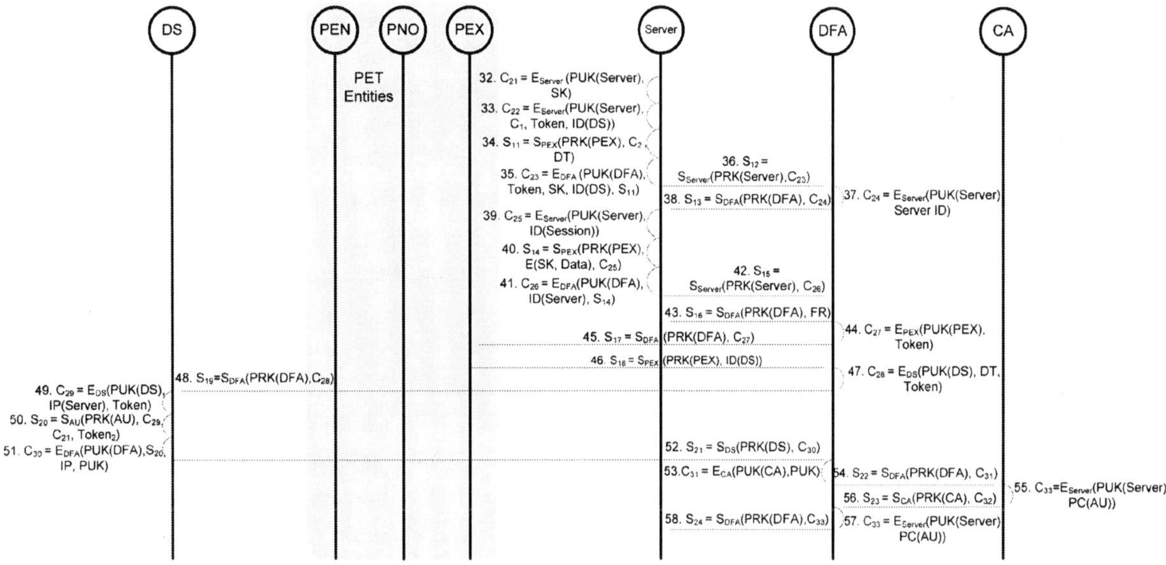

Main Phase

The actual communication is taking place under the main phase. The AU and Server use the session public key as a secret key during their communication in order to offer message in a confidential way. Thus, AU can safely enjoy the confidentiality or integrity of the exchanged transaction messages and the anonymity offered by PET, while Server is ensured that the identity of the AU will be revealed if the user is an abuser.

Investigation Phase

For the investigation phase, it begins with the server detects the malicious messages sent by an abuser. The server informs the DFA who will investigate the incident and identify the abuser forensically. The digital evidence of abuser's action is required by the DFA in order to continue the investigation. This digital evidence must be provided by the Server. The communication transaction messages between the PET Entity (PET) and Server should be saved by the Server in order to prove that the transaction messages came from a specific PET. The Server basically accuses the PET entity until the PET can proof otherwise, that is, it must provide evidence that an AU with the following unique identity generated these transaction messages. To make this scheme work, the investigation phase must take place while the Server is still functioning without having to move to off-line mode for a thorough investigation by the DFA to gather the evidence. The feature of this scheme is that none of the entities need to reveal their private key in order to prove their claims.

The Forensic Receipt (FR) is used to guarantee that DFA received the data which sent from the Server. DFA does not have right to deny investigation of the transaction messages where the Server had asked. The DFA sends the DT (Digest of Token) and the Token to DS to announce that the investigation is taking place at the Server request. The DFA cannot trick the DS, request and the identification of an AU without the cooperation of the Server, because the Token and the DT are only known by the Server, the DS and the AU. Figure 6 illustrate the activities sequence of Investigation Phase.

We explain the protocol in detail through a scenario that AU (BA) tries to attack the Server via the PET in e-marketplace by masquerading as a supplier in order to change the suppliers' privacy policy therefore to gain the lowest selling price.

Firstly, BA generates a pair of keys and contacts DS to get the ticket. DS issues the ticket by verifying the $Token_2$ is hash value of Token. After the initialization phase, when the Server realizes and identifies the abuser's action, it exits from the Main Phase and enter the digital forensics Investigation Phase. The DFA checks the transactions when the Server requests the DFA to provide the identification of the abuser. DFA will request the PET for the identification of the abuser if the checked messages contain evidence pertaining to a malicious action. Then, the evidence is found to indicate the PET has received the suspect transaction messages from specific AU since PET proves "AU→DS: S_{AU}(PRK(AU), C_1, C_2, $Token_2$)" and the PET reveals the identity of the abuser. Therefore, DFA calculates the hash value of $Token_2$ from the parameters of given Token received by Server. If the calculated hash value is same as the value signed by AU, that is, the malicious transaction message is not from that user. If the hash value is different, it implies that the PET tries to blame an innocent AU instead of the real target abuser. Therefore, DFA requests more information about the abuser from the CA to declare its final verdict.

The scenario discussed above, however, is just the theoretical explanation. We will implement, test and evaluate this protocol in order to verify and validate it.

SECURITY PROTOCOLS FOR PROPOSED FRAMEWORK

Based on the proposed framework, not only the digital forensic protocol is required but also several security protocols are required such as migration protocol, negotiation protocol and secure payment protocol.

Migration Protocol

In the mobile agent-based e-marketplace, the migration of an agent from one node to another is called a migration step. The proposed migration protocol is a secure protocol for mobile agent transmission based on the terminology of a protocol for transmission model was adapted from Neuenhofen and Thomson (1998).

In our proposed framework, we are using GA as an interacter between the nodes and helps mobile agents to migrate and transmit from one node to another. It is situated in components which become the entry and exit points to the marketplace. Hence, GA will be installed in the RA and BroAs.

When BA dispatches the worker mobile agent to the marketplace, it must inform a GA which administers the site it is currently situated in. The local GA will send a request to the remote GA who will issue authorization based on the credentials supplied by the mobile agent. The mobile agent is put into the inactive state when the authorization is granted; the initiating GA wraps up the mobile agent's code, data information and the state that needed for its regeneration on the remote side. A frozen copy of mobile agent will be generated by initiating GA while sent to the responding GA who will inspect and reinstantiate it. The GA then is terminated.

Few preconditions need to be set initially such as mobile agent (MA) trusts initiator A, responder (B) and MA trusts CA, while all other parties do not trust each other. The operations sequence of Migration Protocol is illustrated in the following in Figure 7.

Negotiation Protocol

Negotiation is a critical activity in e-commerce which two or more parties with different criteria, constraints and preferences, jointly reach an agreement on the terms of a transaction in e-commerce. Mobile agent-based negotiation is more efficient and objective than human-based negotiation.

Figure 7. Steps of migration protocol

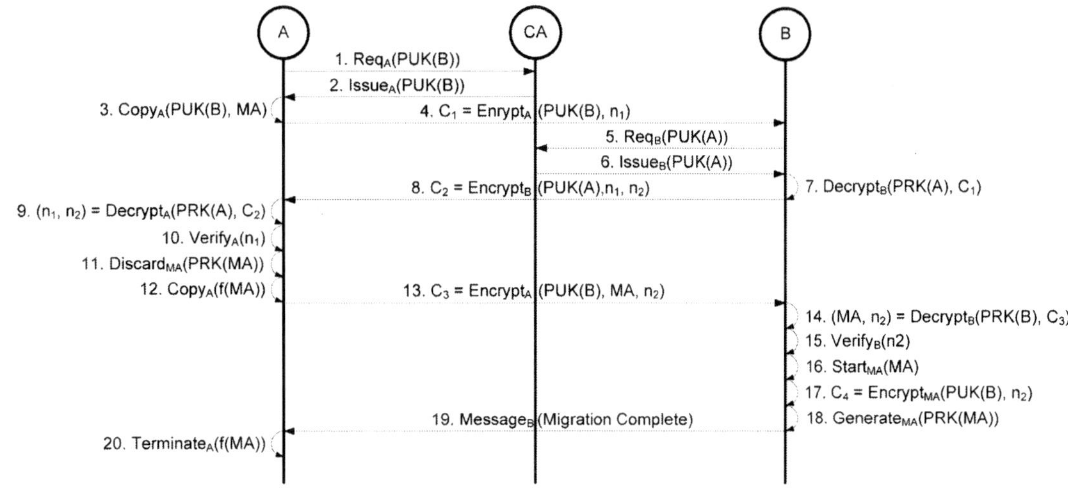

Negotiation can be one-to-one, one-to-many or many-to-many, as a result, many different protocols have been designed for negotiation.

Common issues involved in the negotiation mechanism when a mobile agent decides to enter into a negotiation process are the negotiation space, negotiation protocol and negotiation strategy. A negotiation protocol is a set of rules that determines the process of negotiation, such as how and when a proposal can be generated, when a deal has been struck, when the negotiation should be terminated.

Auction is an efficient method of allocating goods/services in dynamic situations to the entities that value them highly (Amin & Ballard, 2000). An auction is a one-to-many negotiation, between buyer and many suppliers. One-to-many protocols include the Dutch auction, the English auction, and the Contract Net. The auction process in our proposed framework is the type of bilateral auction, where buyer agent is the auctioneer and supplier agents are bidders.

In the proposed framework, the buyer negotiation agent is responsible for negotiating with multiple suppliers' agents simultaneously and proposes counter offers based on the response from suppliers. Suppliers will complete with each other in the negotiation process to bid for the best price offer in order to make the deal. Figure 8 shows the process flow of the negotiation protocol. The steps of the negotiation protocol are as the follows:

1. The worker mobile agent SNA proposes the initialized offer based on the results of the offer evaluation. BNA calls for a proposal from SNA.
2. BNA sends back the offer details of the product, such as the price, weight of products and warranty services to master mobile agent BA.
3. Buyer uses his/her decision function to decide whether accept supplier's proposal or not. Similarly, the supplier also can use his/her decision function to make decision whether accept buyer's counter proposal or not.
4. If the proposal is accepted, the negotiation process stops and a contract will be generated; or the negotiation will be end when the termination conditions are satisfied such as time exceeds the deadline or a contract is settled or failed.
5. If the proposal is not accepted, a few negotiation rounds may take place, where

Figure 8. Process flow of negotiation protocol

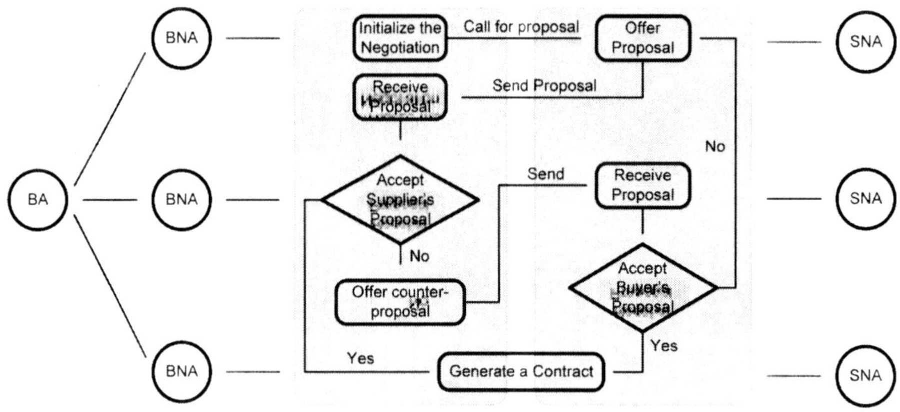

several proposals and counter-proposal are exchanged.

6. Collect the negotiation results and process offer evaluation.

The evaluation algorithm and offer generation algorithm will be discussed in another article.

Secure Payment Protocol

Electronic Payment System (EPS) as a service is a very important component of electronic marketplace. The EPS model is based on the existing card payment model which is named as Secure Electronic Transaction (SET) and has been adapted from Lee et al. (2001). SET provides confidentiality of order information (OI) and payment information (PI). To ensure the integrity of the transmitted data and also provides the authentication of both buyer and supplier. However, the proposed payment model does not use the original SET. The payment protocol is designed to be as compatible with the original SET protocol as possible with some modification on the transaction role. It is because in the original SET protocol, the verification of certificate needs 13 steps (times) and the modified protocol needs just 7 steps (times), which translates in efficiency improvement of approximately 2.1 times (Sun, Xu & Zhu, 2007). Based on this selection, the protocol which we identify as Secure Payment Protocol-0 (SET-0) is shown in Figure 9 to illustrate the certificate issuance scenario.

BA dispatches the BTA which carries the request, public key PUK(BA), and the account information to CA. CA receives the request from BTA and returns the appropriate application form where buyer need to filled up. Then, BTA carries fill-in form and PUK(BA) with account information and moves to CA. CA verifies form and generates the corresponding certificate and returns back to server with this certificate.

The card payment instrument incurs the highest transaction cost whereas such cost associated with the cash payment instrument is the lowest (Lee et al., 2001). This makes the cash payment instrument the best candidate for payment with amount on the cent base, otherwise the transaction cost would have accounted for a significant percentage over the total amount of payment.

To make a purchase request, the BA dispatches BTA to the supplier's site bringing along the request, certificate of buyer and the necessary information for generating the order information (OI) and payment information (PI). We identify a protocol which called Secure Payment Protocol-1 (SET-1) for the payment. Figure 10 illustrates the steps of purchase request phase.

Figure 9. Certificate issuance phase of SET-0

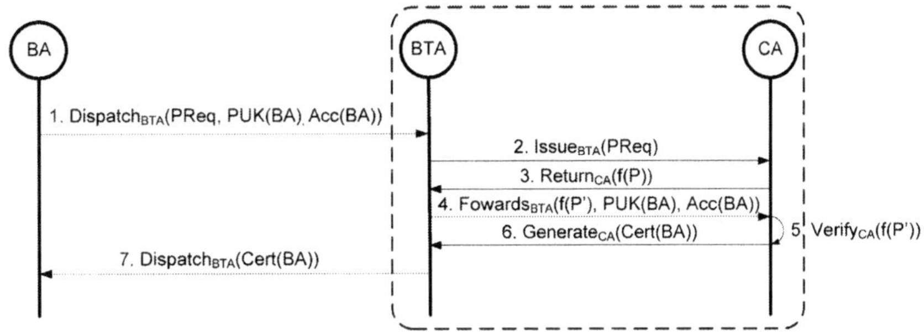

Figure 10. Steps of purchase request phase

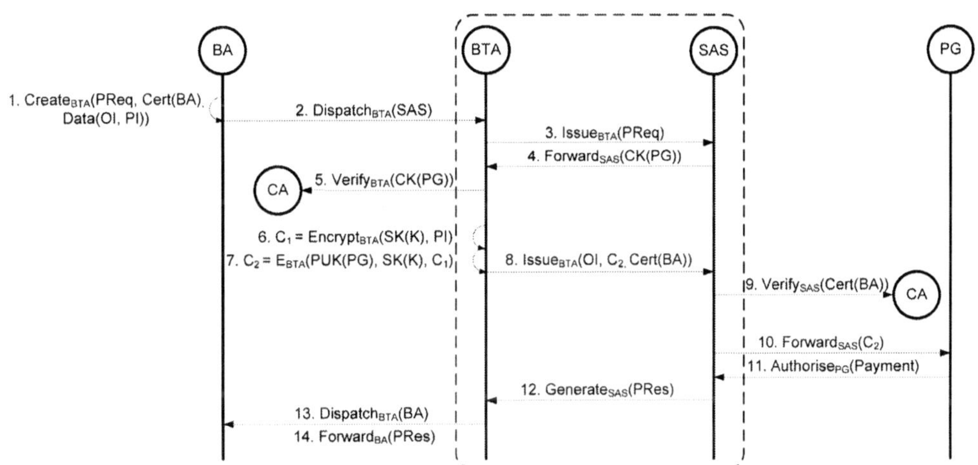

BA creates a worker mobile agent which called BTA and dispatches it which carries the request, certificate Cert(BA) of buyer and the information of OI and PI to the supplier server (SAS). The SAS receives the request and forwards the key exchange certificate CK(PG) of payment gateway to BTA. The agent verifies the CK(PG) through CA and generates the ciphertext of OI and PI. Particularly, the PI is encrypted with a randomly generated symmetric key SK. A digital envelope is generated by encrypting both SK and the encrypted PI with public key of PG. Afterward, it is sent to the SAS together with the OI and the buyer's signature certificate. Then, the SAS forwards the digital envelope to PG for authorization. After that, SAS receives authorization of payment from PG. The SAS then generates and returns a purchase response to BTA and BTA returns to BA with the purchase response and the UIA will take over further actions on behalf of the buyer.

DISCUSSION

From what have been mentioned above, the current mobile agent-based frameworks and systems have their weaknesses especially from security point of view. Most researches focus on these issues specifically in the area of privacy, trustworthy and safety, and less attention has been paid on the requirements of digital forensics of transactions that happens in e-marketplaces.

In this article, the proposed framework offers the digital forensic for auditability and traceability, trustworthiness, safety and confidence for secure and trustworthy mobile agent-based e-marketplace. We are firstly applying the mobility of mobile agents in order to search information efficiently from one node to another on the network within the specific requirements and then dispatching the number of mobile agents in parallel within the sensitive information for the negotiation and payment processes. The framework offers the cryptography algorithms like the SHA-1, RSA and AES for data integrity, authentication and confidentiality. These algorithms are selected according to the comparison between other cryptography algorithms and the possibility for the mobile agent-based e-marketplace. The cryptography offers the features such as the quick encryption, simple secret key management and secure data transmission. It is only used to prove the concept of STMAE. Basically, we do not concentrate on cryptography.

The security protocols include the migration protocol, negotiation protocol, secure payment protocol and digital forensic protocol. By using these kinds of protocols, the mobile agents can efficiently migrate to the e-marketplace; buyer agents can get the best offer through the auction process; payment is more secure through the TTP that offers integrity of the transmitted data and the authentication of both the buyers and suppliers; the framework can audit and trace the malicious actions through the digital forensic investigation. The verification and validation for protocols correctness are outside the scope of this article but important and it will be presented in another article.

For the further research of this subject area, it would be to design a functional architecture for the implementation based on the proposed framework. Use UML method (UML Org, 2009) to illustrate the functional architecture. The class diagram and use case diagram are also required. The prototype development is included to demonstrate the system to show the proposed framework is feasible. For implementation, Java-based technologies are applied in mobile agent system i.e. Aglets as the system platform, J2EE for web-based system have been identified to realise prototype e-marketplace system with XML and security components such as the proposed security protocols, the cryptography algorithms. The experiment will be done through the comparison between the proposed system and MA-ECM proposed by Zhao et al. (2007) that focused on the security, integrity and confidentiality of mobile agent therefore to secure the trading transactions. MA-ECM combines the MD5, RSA and Rijindael as the cryptography algorithms. As what have been mentioned above, the cryptography algorithms in our model are more secure and efficient than MA-ECM's to prove that our model is better.

CONCLUSION AND FURTHER RESEARCH

In this article, we have defined a secure, trustworthy mobile agent-based e-marketplace framework with the central issues such as security, privacy, trust, audit and digital forensic. The security protocols are defined for the proposed framework such as the migration protocol, negotiation protocol, payment protocol and the protocol for the digital forensic. In addition, the cryptography technique and encryption algorithms are also provided provide the secure transmission. The proposed framework also has the features of the process efficiency, lower cost and high communication.

We are currently implementing a prototype in order to analyze, evaluate, verify and validate the framework. We also intend to expand the scope of the research topic: firstly, we can enhance the user interface with intelligent features that can be used in the mobile devices such as PDA. Secondly, make the implementation automatically reusable for further development. Thirdly, try out different cryptography algorithms in order to measure the performance and benefits. Fourthly, verify and validate the protocols to address its reliability. We hope further research can improve the worthy of the secure and trustworthy mobile agent-based e-marketplace.

REFERENCES

Amin, M., & Ballard, D. (2000). Defining new markets for intelligent agents. *IT Professional*, *2*(4), 29–35. doi:10.1109/6294.869380

Antoniou, G., Sterling, L., Gritzalis, S., & Udaya, P. (2008). Privacy and forensics investigation process: The ERPINA Protocol. *Journal of Computer Science & Interface*, *30*(4), 229–236. doi:10.1016/j.csi.2007.10.008

Barker, E., Barker, W., Burr, W., Polk, W., & Smid, M. (2006). Recommendation for Key Management. *Special Publication 800-57*.

Bin, Y., & Xiang, Y. (2006). *Mobile Agent-based Encryption Transmission System*. Unpublished Thesis. The GuiZhou University, China.

Chan Hing–wing. (2000). *Mobile A gent Security and Reliability Issues in Electronic Commerce.* Unpublished Master Thesis. The Chinese University of Hong Kong, China.

Chen, Y., & Zhen, D. (2004). International Cryptography Conference (Crypto' 2004) upload to good news: Chinese scholars crack MD5. *Journal of Netinfo Security, 10*.

Courtois, N., & Pieprzyk, J. (2002). *Cryptanalysis of Block Ciphers with Overdefined Systems of Equations* (pp. 267–287). ASIACRYPT.

Da-lu, Z., & Chen, L. (2005). Security Model of Mobile Agent in E-commerce. *China Academic Journal Electronic Publishing House*, *25*(6), 1271–1273.

Dongsheng, L. (2003). Research of the Two Electronic Commerce Payment Protocols: SSL and SET. *Security & Safety Magzine*, *4*, 61–63.

Drew, G. N. (1999). *Using SET for Secure Electronic Commerce*. New Jersey: Prentice-Hall, Inc.

Freier, A. O., Karlton, P., & Kocher, P. C. (1996). *The SSL Protocol.* Retrieved August 27, 2008, from http://home.netscape.com/eng/ssl3/ssl- toc. hml.

IBMAgletWorkbench [EB/OL]. (2004). Retrieved August 12, 2008, from http://www.trl.ibm.co.jp/aglets/.

Jailani, N., Mohd Yatim, N. F., Yahya, Y., Patel, A., & Othman, M. (2008). Secure and auditable agent-based e-marketplace framework for mobile users. *Journal of Computer Science & Interface*, *30*(4), 237–252. doi:10.1016/j.csi.2007.10.009

Kannammal, A., & Iyengar, N. Ch. S. N. (2007). A Model for Mobile Agent Security in E-Business Applications. *International Journal of Business and Information*, *2*(2), 185–198.

Krzyzanowski, P. (2000). Process Synchronization and Election Algorithms. *Lectures note on distributed systems.* Retrieved October 16, 2008, from http://www.cs.rutgers.edu/~pxk/rutgers/notes/content/06- mutex.pdf.

Lee, T. O., Yip, Y. L., Tsang, C. M., & Ng, K. W. (2001). An Agent-based Micropayment System for E-Commerce. In Liu, J., & Ye, Y. (Eds.), *E-Commerce Agents. Lecture Notes in Artificial Intelligence* (*Vol. 2033*, pp. 247–263). Springer-Verlage Berlin Heidelberg.

Ling-fang, S., Jinbo, X., & Yun, Z. (2007). A Secure Technology of E-commerce Based on SET. *Commercial Research, 2*.

Maes, P., Guttman, R. H., & Moukas, A. G. (1999). Agent that buy and sell [J]. *Communications of the ACM*, *42*(3), 81–91. doi:10.1145/295685.295716

National Institute of Standards and Technology. (2000). *Advanced Encryption Standard (AES) Questions and Answers.* Retrieved December 13, 2008, from http://www.nist.gov/public_affairs/releases/aesq&a.htm.

Neuenhofen, K., & Thompson, M. (1998, May 10-13). A Secure Marketplace for Mobile Java Agents. *Proceedings of the 2nd International Conference on Autonomous agents,* Minneapolis, MN USA (pp. 212-218).

Org, U. M. L. (2009). *UML® Resource Page.* Last Updated on Thursday, January 08, 2009. Retrieved January 10, 2009, from http://www.uml.org/.

Preneel, B. (2005). A Hash of the Crypto Year. *Information Security Solutions Europe Conference.* Retrieved 22 March, 2009, from http://www.certification.tn/fileadmin/Ecolecrypto/Preneel/preneel_isse05v1.pdf.

Rivest, R., Shamir, A., & Adleman, L. (1978). A Method for Obtaining Digital Signatures and Public-Key Cryptosystems. *Communications of the ACM, 21*(2), 120–126. doi:10.1145/359340.359342

Rui-yun, W., Li, L., & Wei-feng, L. (2004). Research of Mobile Agent-based Security in the E-commerce. *Journal of Application Research of Computers, 11*, 137-138, & 141.

Shu-hai, Z., Feng-qin, X., & Jin-zhu, M. (2007). *Research on Secure Mobile agent-based Electronic Commerce* (pp. 10–11). China Academic Journal Electronic Publishing House.

Tyoviriyakul, P., & Kittitornkun, S. (2008). Optimizing RSA encryption for ARM microprocessor. *Electrical Engineering/Electronics, Computer, Telecommunications and Information Technology, 2008. ECTI-CON 2008. 5th International Conference* (Vol 1) (pp. 117-120).

Yong, W., Zhong-qun, W., & Liang-fen, W. (2007). A Migration Mechanism of Mobile Agent System Supporting Security and Fault-Tolerance. *Journal of Computer Technology And Development, 17*(3), 169–175.

This work was previously published in International Journal of Mobile Computing and Multimedia Communications (IJMCMC) 1(3), edited by Ismail Khalil & Edgar Weippl, pp. 8-26, copyright 2009 by IGI Publishing (an imprint of IGI Global).

Compilation of References

Aboba, B., Adrangi, F., Jones, M., Lior, A., & Tschofenig, H. (2008, November). *IETF Geopriv: Carrying Location Objects in RADIUS and Diameter draft-ietf-geopriv-radius-lo-20.txt.*

Adeya, C. N. (2005). Wireless Technologies and Development in Africa. *Wireless Communication and Development: A Global Perspective. Annenberg Research Network on International Communication Workshop.* Marina del Ray, California, USA, October.

Agboma, F., and Liotta, A. (2007). Addressing user expectations in mobile content delivery. *Mobile Information Systems, 3*(3/4), 153:164.

Agostini, A., Albolino, S., De Michelis, G., De Paoli, F., & Dondi, R. (2003). Stimulating knowledge discovery and sharing. *Proceedings of the International ACM SIGGROUP conference on Supporting group work,* Sanibel Island, Florida, USA.

Alheraish, A. (2004). Autoregressive video conference models. *International Journal of Network Management, 14*, 329–337. doi:10.1002/nem.528

Alia, M., Eide, V. S. W., Paspallis, N., Eliassen, F., Hallsteinsen, S. O., & Papadopoulos, G. A. (2007). A Utility-Based Adaptivity Model for Mobile Applications. *International Conference on Advanced Information Networking and Applications Workshops, 2,* (pp. 556-563).

Al-Karaki, J., & Kamal, A. E. (2008, February). Efficient Virtual-Backbone Routing in Mobile Ad Hoc Networks. *Elsevier Computer Networks, 52*(2), 327–350. doi:10.1016/j.comnet.2007.09.007

Al-Majeed, S.S., & Fleury, M. (2010). Options for WiMAX uplink media streaming. *International Journal of Mobile Computing and Multimedia Communications, 2*(2), 49:66

Al-Suhail A.G., (2008). An efficient error-robust wireless video transmission using link-layer FEC and low-delay ARQ schemes. *Journal of Mobile Multimedia, 3 & 4*(3), 275:292.

Amin, M., & Ballard, D. (2000). Defining new markets for intelligent agents. *IT Professional, 2*(4), 29–35. doi:10.1109/6294.869380

Amis, A. D., Prakash, R., Vuong, T., & Huynh, D. (2000). Max-min-d-cluster formation in wireless ad hoc networks. In *IEEE INFOCOM* (pp. 1293–1302). Washington, DC: IEEE.

Andrews, J. G., Ghosh, A., & Muhamed, R. (2007). *Fundamentals of WiMAX.* Upper Saddle River, NJ: Prentice Hall.Bajie I. V., (2006). Efficient error control for wireless video multicast. In *8th IEEE workshop on Multimedia Signal Processing,* (pp. 306-309).

Antoniou, G., Sterling, L., Gritzalis, S., & Udaya, P. (2008). Privacy and forensics investigation process: The ERPINA Protocol. *Journal of Computer Science & Interface, 30*(4), 229–236. doi:10.1016/j.csi.2007.10.008

Arkko, J., & Haverinen, H. (2006). *IETF Network Working Group: RFC 4187 Extensible Authentication Protocol Method for 3rd Generation Authentication and Key Agreement.* EAP-AKA.

Athans, M., & Falb, P. L. (1966). *Optimal Control.* McGraw-Hill, Inc.

Azarian, K., Gamal, H. E., & Schniter, P. (2005). On the achievable diversity-multiplexing tradeoff in half-duplex cooperative channels. *IEEE Transactions on Information Theory, 51*(12), 4152–4172. doi:10.1109/TIT.2005.858920

Bake, S. J., Bae, M. J., & Sung, K. M. (1999). A fast vector quantization encoding algorithm using multiple projection axes. *Signal Processing*, *75*, 89–92. doi:10.1016/S0165-1684(99)00035-3

Bake, S. J., Jeon, B. K., & Sung, K. M. (1997). A fast encoding algorithm for vector quantization. *IEEE Signal Processing Letters*, *4*(2), 325–327. doi:10.1109/97.650035

Baker, D. J., & Ephremides, A. (1981). The architectural organization of a mobile radio network via a distributed algorithm. *IEEE Transactions on Communications*, *29*(11), 1694–1701. doi:10.1109/TCOM.1981.1094909

Balakrishnan, H., Padmanabhan, V., Sehan, S., & Katz, R. (1997). A comparison of mechanisms for improving TCP performance over wireless links. *IEEE/ACM Transactions on Networking*, *5*(6), 756–769. doi:10.1109/90.650137

Balakrishnan, H., & Katz, R. (1998). Explicit loss notification and wireless Web performance. In *IEEE GLOBCOM Internet Mini-Conference*.

Balan, R. K., Lee, B. P., Kumar, K. R., & Jacob, L. (2001). TCP header checksum option to improve performance over lossy links. In *IEEE INFOCOM* (pp. 309–318). TCP HACK.

Ballesteros, I. L. (2006). New Collaborative Working Environments 2020. *Report on industry-led FP7 consultations and 3rd Report of the Experts Group on Collaboration@Work*. European Commission, February 28.

Barker, E., Barker, W., Burr, W., Polk, W., & Smid, M. (2006). Recommendation for Key Management. *Special Publication 800-57*.

Barman, D., Matta, I., Altman, E., & El Azouzi, R. (2004). TCP optimization through FEC, ARQ, transmission power tradeoff. In. *Proceedings of Wired-Wireless Internet Communication, LNC, 2957*, 87–98. doi:10.1007/978-3-540-24643-5_8

Basagni, S., Mastrogiovanni, M., Panconesi, A., & Petrioli, C. (2006). Localized Protocols for Ad Hoc Clustering and Backbone Formation: A Performance Comparison. *IEEE Transactions on Parallel and Distributed Systems*, *17*(4), 292–306. doi:10.1109/TPDS.2006.52

Basagni, S., Turgut, D., & Das, S. K. (2001). Mobility adaptive protocols for managing large ad hoc networks. *In IEEE International Conference on Communications, 5*, 1539–1543.

Basu, P., Khan, N., & Little, T. D. C. (2001). *A mobility based metric for clustering in mobile ad hoc networks*. In Proceedings IEEE ICDCW 01, (pp. 413–18).

Ben, L., & Zygmunt, J. H. (2006). Hybrid Routing in Ad Hoc Networks with a Dynamic Virtual Backbone. *IEEE Transactions on Wireless Communications*, *5*(6), 454–463.

Benchmarks, S. P. E. C. (Standard Performance Evaluation Corporation). (2006). Retrieved from http://spec.org/ benchmarks.html.

Bender, M. A., Farach-Colton, M., Kuszmaul, B. C., & Leiserson, C. E. (2004, April). Adversarial Analyses of Window Backoff Strategie. *Proceedings of the 18th International Parallel and Distributed Processing Symposium, 203*, 26-30.

Ben-Kiki, O., Evans, C., & Dot, I. (2008). *YAML 1.2 Specification*, (Available at http://yaml.org/spec/1.2/).

Bennani, Y., & Gallinari, P. (1995). Neural Networks for discrimination and modelization of speakers. *Journal of Speech Communication*, *17*(1-2), 159–175.. doi:10.1016/0167-6393(95)00014-F

Bennani, Y. (1992). *Approches connexionnistes pour la reconnaissance du locuteur: modélisation et identification*. Phd thesis, Paris XI University, France.

Bergman, A., & Sidi, M. (2006). Energy efficiency of collision resolution protocols. *Computer Communications*, *29*(17), 3397–3415. doi:10.1016/j.comcom.2006.01.016

Bettstetter, C., Resta, G., & Santi, P. (2001). The Node Distribution of the Random Waypoint Mobility Model for Wireless Ad hoc Networks. *IEEE Transactions on Mobile Computing*, *2*(3), 257–269. doi:10.1109/TMC.2003.1233531

Bihannic, N., Boutaba, R., Javaid, U., Meddour, D-E., & Rasheed, T. (Manuscript submitted for publication). Completing the Convergence Puzzle-from the Survey to a tentative Roadmap. *IEEE Wireless Communications Magazine*.

Bimbot, F., Magrin-Chagnolleau, I., & Mathan, L. (1995). Second-Order Statistical Measures for text-independent Broadcaster Identification. *Journal of Speech Communication, 17*(1-2), 177–192..doi:10.1016/0167-6393(95)00013-E

Bin, Y., & Xiang, Y. (2006). *Mobile Agent-based Encryption Transmission System*. Unpublished Thesis. The GuiZhou University, China.

BlackBerry device simulators (2007). http://www.blackberry.com/btsc/search.do?cmd=displayKC&docType=kc&externalId=KB12436&sliceId=SAL_Public&dialogID=40874440&stateId=0%200%2040870665

Blanchard, A. L., & Markus, M. L. (2004). The Experienced "Sense" of a Virtual Community: Characteristics and Processes. *SIGMIS Database, 35*(1), 64–79. doi:10.1145/968464.968470

Bleecker, J. (2006). What's Your Social Doing In My Mobile? Design Patterns for Mobile Social Software. *In WWW2006 Workshop, MobEA IV — Empowering the Mobile Web*, May 23, Edinburgh, Scotland.

Bonastre, J. F. (2005). *L'authentification biométrique vocal*. Bordeaux, France: Tutorial PARISTIC Bordeaux.

Bonastre, F., & Besacier, L. (1997). *Traitement Indépendant de Sous-bandes Fréquentielles par des méthodes Statistiques du Second Ordre pour la Reconnaissance du Locuteur*. In Proceedings of the 4th « Congrès Français d'Acoustique », Marseille April 14-18, 357-360.

Borges, M. A. F., & Baranauskas, M. C. C. (2003). Supporting the facilitator in a collaborative learning environment. *International Journal of Continuing Engineering Education and Lifelong Learning, 13*(1/2), 39–56.

Boslaugh, S., & Watters, P. (2008). *Statistics in a Nutshell: A Desktop Quick Reference*. (pp. 118-119). Sebastopol, CA: O'Reilly Media, Inc.

Botto, F. (2008). A socio-technical infrastructures framework for community networks. *Third International Conference on Communications and Networking, ChinaCom 2008*, Hangzhou, China, August 25-27.

Boukerche, A., & Jing, F. (2006, November 14-16). Buffer Management for 3D Image-based Rendering over Wireless Network with QoS Adaptation. *Proceedings of 31st IEEE Conference on Local Computer Networks* (pp. 55-62). Tampa, Florida, USA: IEEE Press.

Boyera, S. (2007). Can the Mobile Web Bridge the Digital Divide. *Interactions (New York, N.Y.), 14*(3), 12–14. doi:10.1145/1242421.1242433

Bright, C. (2005). *Improving IEEE 802.11 Performance with Power Control and Distance-Based Contention Window Selection*. Master's Thesis, University of Illinois, Urbana-Champaign

BSCW. (2007). http://bscw.fit.fraunhofer.de/

Burney, K. (2009). *New Year's Bustle? Vertical Market Expectations for 2009 ICT Spending in the US- From Crisis to Slow, Long-term Recovery*. Compass Intelligence.

Burt, P. J., & Adelson, E. (1983). The Laplacian pyramid as a compact image code. *IEEE Transactions on Communications, COM-31*, 532–540. doi:10.1109/TCOM.1983.1095851

Cai, Ghandeharizadeh, Girardot, Schmidt, & Song. (2002, September). *A Comparison of Alternative Encoding Mechanisms for Web Services*. In 13th International Conference on Database and Expert Systems Applications, Aix en Provence, France.

Camarillo, G., & Garcia-Martin, M. A. (2008). *The 3G IP Multimedia Subsystem (IMS): Merging the Internet and the Cellular Worlds* (3rd ed.). Chippenham, England: John Wiley & Sons, Ltd.doi:10.1002/9780470695135

Canonical, X. M. L. (2001, March). Retrieved from http://www.w3.org/TR/xml-c14n

Cao, H. Q., & Li, W. (2000). A fast search algorithm for vector quantization using a directed graph. *IEEE Transactions on Circuits and Systems for Video Technology, 10*, 585–593. doi:10.1109/76.845003

Cao, Y. (2007). Mobile Social Software with Context Awareness and Data Uncertainty for Technology-Enhanced Learning. *Online-Proceedings of the 2nd Prolearn Doctoral Consortium in Technology Enhanced Learning (EC-TEL 2007)*, Crete, Greece, September.

Cao, Y., & Li, V. O. K. (2002). Utility-oriented adaptive QoS and bandwidth allocation in wireless networks. *IEEE International Conference on Communications* (pp. 3071-3075). Anchorage, AK.

Cardinal, J. (1999, September). Fast search for entropy-constrained VQ. In *Proceeding of (ICIAP) the 10th International Conference on Image Analysis and Processing* (pp. 1038-1042).

Carpineto, C., Mizzaro, S., Romano, G., & Snidero, M. (2009). Mobile Information Retrieval with Search Results Clustering: Prototypes and Evaluations. *JASIST*, *60*(5), 877–895. doi:10.1002/asi.21036

Carroll, J. M., & Rosson, M. B. (2003). A trajectory for community networks. *The Information Society*, *19*(5), 381–393. doi:10.1080/714044685

Cen, S., Cosman, P., & Voelker, G. (2003). *End-to-end differentiation of congestion and wireless losses, 11*(5), 703-717.

Certicom-Research. (2000). *Standards for Efficient Cryptography, sec 1: Elliptic Curve Cryptography (Tech. Rep.)*. Foster City, CA: Certicom Research.

Chakeres, I. D., & Royer, E. M. B. (2003, July). Utilizing Resource rich Nodes in Ad hoc Networks. *ACM SIGMOBILE Mobile Computing and Communications Review*, *7*(3), 17–18. doi:10.1145/961268.961271

Chalmers, D., & Sloman, M. (1999). A Survey of Quality of Service in Mobile Computing Environments. *IEEE Communications Surveys, 2* (2).

Chan, S.-H., Zheng, X., Zhang, Q., Zhu, W., & Zhang, Y. (2006). Video loss recovery with FEC and stream replication. *IEEE Transactions on Multimedia*, *8*(2), 370–381. doi:10.1109/TMM.2005.864340

Chan Hing–wing. (2000). *Mobile A gent Security and Reliability Issues in Electronic Commerce.* Unpublished Master Thesis. The Chinese University of Hong Kong, China.

Chatterjee, M., Das, S. K., & Turgut, D. (2002). WCA: a weighted clustering algorithm for mobile ad hoc networks. *Cluster Computing*, *5*(1), 193–204. doi:10.1023/A:1013941929408

Cheikh, S., Cluade, C., Guillaume, C., & Guerin, L. I. (2008). Bandwidth estimation for IEEE 802.11-based ad hoc networks. *IEEE Transactions on Mobile Computing*, *7*(10), 1228–1241..doi:10.1109/TMC.2008.41

Chen, L., & Heinzelman, W. B. (2005, March). QoS-Aware Routing Based on Bandwidth Estimation for Mobile Ad Hoc Networks. *IEEE Journal on Selected Areas in Communications*, *23*(3), 561–572. doi:10.1109/JSAC.2004.842560

Chen, T. S., Chang, C. C., & Hwang, M. S. (1998). A virtual image cryptosystem based on vector quantization. *IEEE Transactions on Image Processing*, *7*(10), 1485–1488. doi:10.1109/83.718488

Chen, M., & Zakhor, A. (2005). Rate control for streaming video over wireless. *IEEE Wireless Communications*, *12*(4), 32–44. doi:10.1109/MWC.2005.1497856

Chen, M., & Zakhor, A. (2006). Multiple TFRC connection based rate control for wireless networks. *IEEE Transactions on Multimedia*, *8*(5), 1045–1062. doi:10.1109/TMM.2006.879837

Chen, W. C., Hsieh, C. T., & Lai, E. (2004). Multiband Approach to Robust Text-Independent Speaker Identification. *Journal of Computational Linguistics and Chinese Language Processing*, *9*(2), 63–76.

Chen, G., Nocetti, F., Gonzales, J., & Stojmenovic, I. (2002). *Connectivity based k-hop clustering in wireless networks.* In HICSS 02 Proceedings of the 35th Annual Hawaii International conference on System Sciences. IEEE Computer Society, (pp. 2450-2459).

Chen, G., Vijaykrishnan, N., & Irwin, M. J. (2005). PennBench: A benchmark suite for embedded Java. In *The IEEE 5th Annual Workshop on Workload Characterization.*

Chen, Y., & Lemin, L. (2005, September 21-23). A random early expiration detection based buffer management algorithm for real-time traffic over wireless networks. *Proceedings of the IEEE International Conference on Computer and Information Technology (CIT 2005) (pp. 507–511). Shanghai, China: IEEE Press.*

Chen, Y., & Zhen, D. (2004). International Cryptography Conference (Crypto' 2004) upload to good news: Chinese scholars crack MD5. *Journal of Netinfo Security, 10.*

Cheng, L.-T. Hupfer, S., Ross, S., Patterson, J., Clark, B., de Souza, C. (2003). Jazz: A Collaborative Application Development Environment, Proceedings of the 18th Annual ACM SIGPLAN Conference on Object-Oriented Programming Systems Languages and Applications (OOPSLA), demo.

Chiang, Ching-Chuan & Gerla, M. (1997). *Routing in clustered multi-hop, mobile wireless networks with fading channel.* In Proc. IEEE SICON 97 (pp. 197-211).

Chiasserini, C.-F., & Meo, M. (2002). A reconfigurable protocol setting to improve TCP over wireless. *IEEE Transactions on Vehicular Technology Multimedia*, *51*(6), 1608–1620. doi:10.1109/TVT.2002.804863

Chin, H. S., Goodge, J. W., Griffiths, R., & Parish, D. J. (1989). Statistics of video signals for viewphone-type pictures. *IEEE Journal on Selected Areas in Communications*, *7*(5), 826–832. doi:10.1109/49.32346

Choi, W., & Woo, M. (2006). *A distributed weighted clustering algorithm for mobile ad hoc networks*. In Proc. IEEE ICIW. (pp.73-78).

Chou, W. (2003). *Elliptic Curve Cryptography and Its Applications to Mobile Devices (Tech. Rep.).* College Park, MD: University of Maryland.

Chowj, P. S., Cioffi, M., & Binghama, J. A. C. (1995). Practical discrete multitone transceiver loading algorithm for data transmission over spectrally shaped channels. *IEEE Transactions on Communications*, *43*, 773–775. doi:10.1109/26.380108

Chu, S. C., Lu, Z. M., & Pan, J. S. (2007). Hadamard transform based fast codeword search algorithm for high-dimensional VQ encoding. *Elsevier Science Inc.*, *177*(3), 734-746

Clark, K. G. (2002, July). *Look Ma, No Tags*. Retrieved from http://www.xml.com/pub /a/2002/07/24/yaml.html.

Clausen, T. (2004). Comparative study of routing protocols mobile Ad hoc Networks. *INRIA Report Research* RR-5135.

Clodfelter, K., Buente, W., & Rosenbaum, H. (2007). CommNets 2.0: Using an Evolutionary Perspective to Examine a Statewide Community Networking Initiative. *Proceedings of the American Society for Information Science and Technology*, *44*(1), 1–20. doi:10.1002/meet.1450440216

Cohen, D. M., & Heyman, D. P. (1993). Performance modeling of video teleconferencing in ATM networks. *IEEE Transactions on Circuits and Systems for Video Technology*, *3*(6), 408–422. doi:10.1109/76.260197

Colbourn, C. J., Cui, M., Syrotiuk, V. R., & Lloyd, E. L. (2007). A Carrier Sense Multiple Access Protocol with Power Backoff (CSMA/PB). *Ad Hoc Networks*, *5*(8), 1233–1250. doi:10.1016/j.adhoc.2007.02.017

Courant, R., & Hilbert, D. (1962). *Method of Mathematical Physics, 11*, 167-180. New York: Wiley.

Courtois, N., & Pieprzyk, J. (2002). *Cryptanalysis of Block Ciphers with Overdefined Systems of Equations* (pp. 267–287). ASIACRYPT.

Cover, T., & Gamal, A. E. (1979). Capacity theorems for the relay channel. *IEEE Transactions on Information Theory*, *25*(5), 572–584. doi:10.1109/TIT.1979.1056084

Cover, T., & Leung, C. (1981). An achievable rate region for the multiple-access channel with feedback. *IEEE Transactions on Information Theory*, *27*(3), 292–298. doi:10.1109/TIT.1981.1056357

Cox, D. R., & Miller, H. D. (1965). *The Theory of Stochastic Processes*. London, UK: Chapman & Hall.

Crockford, D. (2006, July). *The application/json Media Type for JavaScript Object Notation (JSON).* Retrieved from http://tools.ietf.org/ html/rfc4627

Cui, L., Koh, S. J., Cui, X., & Kim, Y. J. (2007). Adaptive increase and decrease algorithm for wireless TCP. In *International Conference on Natural Computation*, (pp. 392-398).

CURE. (2007). Fern Universitaet Hagen, http://www.pi6.fernuni-hagen.de/en/CURE/

Curtis, P., & Nichols, D. A. (1993). *MUDs Grow Up: Social Virtual Reality in the Real World, Xerox Palo Alto Res* (pp. 193–200). CA: Center.

Czylwik, C. (1996). Comparison of the channel capacity of wideband radio channels with achievable data rates using adaptive OFDM. *Proceedings of the 5th European Conference on Fixed Radio Systems and Networks ECRR '96, Bologna,* (pp. 238-243).

Czylwik, C. (1997). Adaptive (OFDM for wideband radio channels. *Proceedings of the GLOBECOM '96, London,* (pp. 713-718).

Da-lu, Z., & Chen, L. (2005). Security Model of Mobile Agent in E-commerce. *China Academic Journal Electronic Publishing House*, *25*(6), 1271–1273.

Das, B., & Bhagghavan, V. (1997). Routing in Ad Hoc Networks Using Minimum Connected Dominating Sets. *International Conference on Communications*, Montréal (pp. 376–380).

Dasarathy, B. V. (1994). *Decision Fusion. In proceedings of IEEE Computer Society Press*. CA: Los Alamitos.

Datta, A., Mukherjee, S., & Viguier, I. R. (1999). Buffer management in real-time active database systems. *IEEE Transactions on, Systems, Man and Cybernetics. Part A*, *29*(2), 216–224.

Daugman, J. (2007). New Methods in Iris Recognition. *Journal of the IEEE Transactions on Systems. Man. Cybernetics*, *3*(5), 1167–1175. doi:10.1109/TSMCB.2007.903540

de Groot, G. J., Karrenberg, D., Lear, E., Moskowitz, B., & Rekhter, Y. (1996). *IETF Network Working Group: RFC 1918 Address Allocation for Private Internets.*

De Miguel, T. (1994). *ISABEL Experimental Distributed Cooperative Work Application over Broadband, Lecture Notes in Computer Science, Springer* (pp. 353–362). Volume.

Degrande, N., Laevens, K., & De Vleeschauwer, D. (2008). Increasing the user perceived quality for IPTV services. *IEEE Communications Magazine*, *46*(2), 94–100. doi:10.1109/MCOM.2008.4473090

Devaram, & Andresen. (2003). *SOAP Optimization via Parameterized Client-Side Caching*. Retrieved from http://java.sun.com/ developer/ technicalArticles/ xml/ fastinfoset/

Dianhui, W., Wong, A. K. Y., & Dillon, T. S. (2006, December 2-5). Heuristic rule based neuro-fuzzy approach for adaptive buffer management for Internet-based computing. *Proceedings of 31st IEEE Conference on Local Computer Networks, 1*, 296 – 299. Melbourne, Victoria, Australia: IEEE Press.

DiBurro, L., Huttunen, A., Swander, B., Stenberg, M., & Volpe, V. (2005). *IETF Network Working Group: RFC 3948 UDP Encapsulation of IPsec ESP Packets.*

Dimitrakakis, C. (2006). Nearly optimal exploration-exploitation decision thresholds. *IDIAP Research Report* (IDIAP-RR-06-12). *Artificial Intelligence*, 1–10.

Dongsheng, L. (2003). Research of the Two Electronic Commerce Payment Protocols: SSL and SET. *Security & Safety Magzine*, *4*, 61–63.

Doolan, D. C., & Tabirca, S. (2007). The Need for Speed: coding styles for Mobile Devices. In *Elmar* (pp. 193-196).

Doulamis, N. D., Doulamis, A. D., Konstantoulakis, G. E., & Stassinopoulos, G. I. (2000). Efficient Modeling of VBR MPEG-1 Coded Video Sources. *IEEE Transactions on Circuits and Systems for Video Technology*, *10*(1), 93–112. doi:10.1109/76.825864

Drew, G. N. (1999). *Using SET for Secure Electronic Commerce*. New Jersey: Prentice-Hall, Inc.

Du, X., & Pomalaza-Ráez, C. (2004). Delay Sensitive QoS Routing For Mobile Ad Hoc Networks. *Proc. of the MILCOM 2004 Conf. Monterrey.*

Economides, A. A., & Nikolaou, N. (2008). Evaluation of handheld devices for mobile learning. [IJEE]. *International Journal of Engineering Education*, *24*(1), 3–13.

Edoh, K. D. (2004). *Elliptic Curve Cryptography: Java Implementation*. In INFOSECCD '04: Proceedings of the 1st Annual Conference On Information Security Curriculum Development (pp. 88–93). New York, NY: ACM.

Edrich, M., & Schmalenberger, R. (2002). Combined DSSS/FHSS approach to interference rejection and navigation support in UAV communications and control. *IEEE Seventh International Symposium on Spread Spectrum Techniques and Applications, 3*, 687-691. Prague, Czech Republic.

Elhilali, M., & Shamma, S. (2008). Information-bearing components of speech intelligibility under babble-noise and bandlimiting distortions. In *Proceedings of Acoustics, Speech and Signal Processing (ICASSP).* March 31 -April 4, pp 4205 - 4208.

Envy (2007). http://www.ibm.com/developerworks/blogs/page/stdt?tag=envy

Ephremides, A., Wieseltheir, J. E., & Baker, D. J. (1987). A design concept for reliable mobile radio networks with frequency hopping signaling. *Proceedings of the IEEE, 75*(1), 56–73. doi:10.1109/PROC.1987.13705

Er, I. I., & Seah, W. K. G. (2006). Performance analysis of mobility-based d-hop (mobdhop) clustering algorithm for mobile ad hoc networks. *Computer Networks, 50*(17). doi:10.1016/j.comnet.2005.12.013

Erol, K., & Lang, J. (2000). Levy, Designing Agents from Reusable Components, Proceedings of Fourth International Conference on Autonomous Agents, pp. 76-77.

Esa, J., & Virtamo, J. (2007). Random waypoint mobility model in cellular networks. *Wireless Networks, 13*(2), 177–188..doi:10.1007/s11276-006-4600-3

Exclusive, X. M. L. Canonicalization. (2002, July). Retrieved from http://www.w3.org/ TR/xml-exc-c14n/).

Falavigna, D. (1995). *Comparison of Different HMM Based Methods For Speaker Verification.* In proceedings of Eurospeech'95- Fourth European Conference on Speech Communication and Technology, Madrid, *Spain,* September 18-21.

Farooq, U., & Carroll, J. M. (2003). Mobilizing Community Networks. *Conference on Home Oriented Informatics and Telematics (HOIT),* April 6-8, Irvine, California.

Fayad, M., Schmidt, D., & Johnson, R. (1999). *Building Application Frameworks: Object-Oriented Foundations of Framework Design.* Wiley Computer Publishing.

Feng, L. (2004). *Speaker Recognition.* IMM thesis, Technical University of Denmark, Denmark.

Fisher, W., Zue, V., Bernstein, J. & Pallet, D. (1986). An acoustic-phonetic database. *Journal JASA, suppl. A, 81(S92).*

Fitzpatrick, G. (2003). *The Locales Framework, Understanding and Designing for Wicked Problems* (1st ed.). Springer.

Flinn, J., & Satyanarayanan, M. (1999). PowerScope: A Tool for Profiling the Energy Usage of Mobile Applications. *In WMCSA '99: Proceedings of the Second IEEE Workshop on Mobile Computer Systems and Applications* (p. 2).

Floyd, S., & Fall, A. (1999). Promoting the use of end-to-end congestion control in the Internet. *IEEE/ACM Transactions on Networking, 7*(4), 43–56. doi:10.1109/90.793002

Forde, L., & Doyle, L. (2006). Ad hoc Innovation: Distributed Decision Making in Ad Hoc Networks. *IEEE Communications Magazine, 44*(4), 131–137. doi:10.1109/MCOM.2006.1632660

Fornber, B., & Whitham, G. B. (1978). A numerical and theoretical study of certain non-linear wave phenomena. *Phil. Trans. R. Soc London, SeriesA, 289*, 373–404.

Freier, A. O., Karlton, P., & Kocher, P. C. (1996). *The SSL Protocol.* Retrieved August 27, 2008, from http://home.netscape.com/eng/ssl3/ssl- toc. hml.

Friis, H. T. (1946). A note on a simple transmission formula. *Proceedings of IRE, 34*(5), 254–256. doi:10.1109/JRPROC.1946.234568

Frodigh, M., Johansson, P., & Larsson, P. (2000). Wireless *ad hoc* networking—The art of networking without a network. *Ericsson Review, 4*, 248–263.

Fu, Y., Hu, R., Tian, G., & Wang, T. (2006). TCP-Friendly Rate Control for streaming service over 3G network. In *International Conference on Wireless Communications., Networking and Mobile Computing*, 4 pages.

Fuemmeler, J. A., Vaidya, N. H., & Veeravalli, V. V. (2004). *Selecting Transmit Powers and Carrier Sense Thresholds for CSMA Protocols. Technical report.* University of Illinois at Urbana Champaign.

Furui, S. (1981). -a. Cepstral analysis technique for automatic speaker verification. *Journal of IEEE Transactions on Acoustics, Speech, and Signal Processing, 29*(2), 254–272..doi:10.1109/TASSP.1981.1163530

Furui, S. (1981). -b. Comparison of speaker recognition methods using statistical features and dynamic features. *Journal of IEEE Transactions on Acoustics, Speech, and Signal Processing, 29*(3), 342–350..doi:10.1109/TASSP.1981.1163605

Furui, S. (1991). Vector-quantization-based speech recognition and speaker recognition techniques. In *Proceedings of Conference Record of the Twenty-Fifth Asilomar Conference on Signals, Systems and Computers,* November 4-6, 954-958.

Ganchev, T., Tsopanoglou, A., Fakotakis, N., & Kokkinakis, G. (2002). Probabilistic Neural Networks Combined with GMMs for Speaker Recognition Over Telephone Channels. *14th International Conference on Digital Signal Processing (DSP2002), 2*, Santorini, Greece, July 1- 3, 1081-1084.

Ganning, J., Qin, H., Flint, C. G., & Gasteyer, S. P. (2007). *Constructing Typologies in Community Research*. Paper presented at the annual meeting of the Rural Sociological Society, Marriott Santa Clara, Santa Clara, California.

Garcia, C., & Tziritas, G. (1999). Face detection using quantized skin color regions merging and wavelet packet analysis. *IEEE Transactions on Multimedia, 1*(3), 264–277. doi:10.1109/6046.784465

Gersho, A., & Gray, R. M. (1991). *Vector quantization and signal compression*. Boston: Kluwer.

Ghosh, R., & Basagni, S. (2005). *Limiting the impact of Mobility on Ad Hoc Clustering*. Proceedings of Second ACM International Workshop on Performance Evaluation of Wireless Ad Hoc, Sensor and Ubiquitous Network, 197-204.

Girardot, & Sundaresan. (2000). *Millau: An Encoding Format For Efficient Representation And Exchange Of XML Over The Web*. In Proceedings of the 9th International World Wide Web Conference On Computer Networks (p. 747-765). Amsterdam: North-Holland Publishing Co. Retrieved from citeseer.ist.psu.edu/article /widener01open.html

Girod, B., Steinbach, E., & Faerber, N. (1997). Performance of the H.263 Video Compression Standard. *Journal of VLSI Signal Processing Systems for Signal, Image and Video Technnology. Special issue on recent development in video: Algorithms* [Springer Editions.]. *Implementation and Applications, 17*, 101–111.

Gish, H. (1990). Robust discrimination in automatic speaker identification. In *Proceedings of IEEE International Conference on Acoustics Speech and Signal Processing, New Mexico*, 289-292.

Gokran, P. (2007, November). *Making your Low-Cost Handsets Plans Succeed*. In Telecom Era (pp. 23–32).

Goldsmith, A. J., & Chua, S. G. (1997). Variable-rate variable-power MQAM for fading channel. *IEEE Transactions on Communications, 45*, 1218–1230. doi:10.1109/26.634685

Görkemli, B., Sunay, M. O., & Tekalp, A. M. (2008). Video streaming over wireless DCCP. In *IEEE International Conference. on Image Processing*, (pp. 2028-2031).

Gorp, A., & Chris Morris, C. (2008). Serving underserviced areas in South Africa: The potential for Wi-Fi community network deployment and the role of regulation. *Info, 10*(1), 65–78. doi:10.1108/14636690810850166

GPP TS 23.207 V8.0.0. (2008). *3rd Generation Partnership Project; Technical Specification Group Services and System Aspects; End-to-end Quality of Service (QoS) concept and architecture* (Release 8).

GPP TS 26.071 V7.0.1. (2007). *3rd Generation Partnership Project; Technical Specification Group Services and System Aspects; Mandatory speech CODEC speech processing functions; AMR speech CODEC; General description* (Release 7).

GPP TS 43.318 V7.2.0. (2007). *3rd Generation Partnership Project; Technical Specification Group GSM/EDGE Radio Access Network; Generic access to the A/Gb interface; Stage 2* (Release 7).

Grasso, A., Koch, M., & Snowdon, D. (1999). Campiello - New user interface approaches for community networks. [August]. *ACM SIGGROUP Bulletin, 20*(2), 15–17.

Gray, R. M. (1984). Vector quantization. *IEEE Acoustics, Speech, and Signal Processing, 1*(2), 4–29.

Gregson, K., & Ford, C. (1998). Evaluation of community networks. In Wildemuth, B. M. (Ed.), *ASIS Midyear 1998 Proceedings, Collaboration across Boundaries: Theories*. Strategies, and Technology.

Grindrod, P. (1991). *Patterns and Waves: The Theory and Applications of Reaction-Diffusion Equations* (p. 177). Oxford: Clarendon Press.

Groot, M. H. (1970). *Optimal Statistical Decisions*. New York: McGraw-Hill.

Grossman, D. (2002). *IETF Network Working Group: RFC 3260 New Terminology and Clarifications for Diffserv.*

Grossman, D., & Heinanen, J. (1999). *IETF Network Working Group: RFC 2684 Multiprotocol Encapsulation over ATM Adaptation Layer 5.*

Grüneberg, K., Hellge, C., Mirta, S., Schierl, T., & Wiegand, T. (2007). Using H.264/AVC-based Scalable Video Coding (SVC) for Real Time Streaming in Wireless IP Networks. *IEEE International Symposium on Circuits and Systems*, ISCAS 2007.(pp. 3455-3458).

Grunenfelder, R., Cosmas, J. P., Manthorpe, S., & Odinma-Okafor, A. (1991). Characterization of Video Codecs as Autoregressive Moving Average processes and Related Queuing System Performance. *IEEE Journal on Selected Areas in Communications*, *9*(3), 284–293. doi:10.1109/49.76626

Guan, L., & Kamel, M. (1992). Equal-average hyperplane partitioning method for vector quantization of image data. *Pattern Recognition Letters*, *13*(10), 693–699. doi:10.1016/0167-8655(92)90098-K

Guha, S., & Khuller, S. (1998). Approximation Algorithms for Connected Dominating Sets. *Algorithmica*, *20*, 374–387. doi:10.1007/PL00009201

Haag, M. (2002). *Stationary and Nonstationary Random Processes, Version 2.1*. Online available at http://cnx.rice.edu/content/m10684/latest/.

Haake, J., Schümmer, T., Haake, A., Bourimi, M., & Landgraf, B. (2004). Supporting Flexible Collaborative Distance Learning in the CURE Platform, Proceedings HICSS.

Habanero (2007). http://www.isrl.uiuc.edu/isaac/Habanero/3.0/index.html

Haitao, L., & Rajiv, G. (2004, October 24-27). Selective Backbone Construction for Topology Control in Ad Hoc Networks. *Proceedings of MASS*, Florida, USA, (pp. 41-50).

Haleem, P. P. A., & Sebastian, M. P. (2008). An Alternative Approach For Slicing Down The Message Size And Enhancing The Security In Wireless Mobile Network. *The Mediterranean Journal of Computers and Networks*, *5*, 148–149.

Hammond, A., & Paul, J. (2006). *A New Model for Rural Connectivity*. Retrieved April 7, http://www.usaid.gov/our_work/ economic_growth_and_trade/ info_technology/tech_series/ Rural_Connectivity_508.pdf.

Hampton, K. N., & Gupta, N. (2008). Community and social interaction in the wireless city: Wi-fi use in public and semi-public spaces. *New Media & Society*, *10*(6), 831–850. doi:10.1177/1461444808096247

Han, Y., Yang, X., Wei, P., Wang, Y., & Hu, Y. (2006). ECGSC. *Elliptic Curve Based Generalized Signcryption, LNCS4159*, 956–965.

Han, J.-H., Kim, Y.-J., Jun, S.-I., Chung, K.-I., & Seo, C.-H. (2002). *Implementation Of ECC/ECDSA Cryptography Algorithms Based On Java Card.* In ICDCSW '02: Proceedings of the 22nd International Conference On Distributed Computing Systems (pp. 272–278). Washington, DC: IEEE Computer Society.

Handley, M., Pahdye, J., Floyd, S., & Widmer, J. (2003). *TCP-Friendly Rate Control (TFRC): protocol specification.* RFC 3448.

Haque, S., Togneri, R., & Zaknich, A. (2006). *Zero-Crossings with Adaptation for Automatic Speech Recognition.* Paper presented at the Eleventh Australasian International Conference on Speech Science and Technology 2006, University of Auckland, Auckland, New Zealand.

Harai, H., & Murata, M. (2006). High-speed buffer management for 40 GB/s-based photonic packet switches. *IEEE/ACM Transactions on Networking*, *14*(1), 191–204. doi:10.1109/TNET.2005.863450

Harbin Institute of Technology (2006). *Almost Periodic Type and Remotely Almost Periodic Type Solutions to Parabolic Inverse Problems* [Doctor]. Yang, F. L.

Hariharan, K. (2008). Best Practices: Extending enterprise applications to Mobile devices. *The Architecture Journal*, *14*, 10–17.

Hasib, A., & Fapojuwo, A. O. (2008). A mobility model for heterogeneous wireless networks. *2008 IEEE Radio and Wireless Symposium,* pp. 815-818. doi:10.1109/RWS.2008. 4463617

Haskell, B. G. (1972). Buffer and channel sharing by several interframe picturephone coders. *The Bell System Technical Journal*, *51*(1), 261–289.

Hassanein, H., & Zhou, A. (2001). Routing with Load Balancing in Wireless Ad hoc Networks. *International Workshop on Modeling Analysis and Simulation of Wireless and Mobile Systems* (pp. 89-96), Rome, Italy.

Haßlinger, G., & Hohlfeld, O. (2008). The Gilbert-Elliott model for packet loss in real time services on the Internet. In *14th GI/ITG Conference on Measurement, Modelling, and Evaluation of Computer and Communications Systems.*, (pp. 269–283).

Haverinen, H., & Salowey, J. (2006). *IETF Network Working Group: RFC 4186 Extensible Authentication Protocol Method for Global System for Mobile Communications (GSM) Subscriber Identity Modules*. EAP-SIM.

He, Y. N. (2005). Taylor Expansion Method for Nonlinear Evolution Equations. *Applied Mathematics and Mechanics*, *26*(4), 522–529. doi:10.1007/BF02465392

Hean, K. O., Hean, L. O., Tee, E. R., & Sureswaran, R. (2006). Scalability study of ad-hoc wireless mobile network routing protocol in sparse and dense networks. *Proc. of DFMA Int. Conf. on Distributed Frameworks for Multimedia Applications* (pp. 29-36). doi:10.1109/DFMA.2006.296891

Heitmann, M., Prykop, C., & Aschmoneit, P. (2004). Using Means-end Chains to Build Mobile Brand Communities. *Proceedings of the 37th Hawaii International Conference on System Sciences*, Big Island, HI, USA, January 4-8.

Heyman, D. P., Tabatabai, A., & Lakshman, T. V. (1992). Statistical analysis and simulation study of video teleconference traffic in ATM networks. *IEEE Transactions on Circuits and Systems for Video Technology*, *2*(1), 49–59. doi:10.1109/76.134371

Hillebrand, C., Groh, G., & Koch, M. (2002). Mobile Communities – Extending Online Communities into the Real World. *Mobile and Collaborative Business*, 7–18, September.

Hirosakai, B. Yoshida, Tanakas, Hasegawak, Inoue, & Watanabea, K. (1985). 191.2 kbps voice band data modem based on orthogonally multiplexed QAM techniques. *IEEE International Conference on Communications* (pp. 661-665).

Hole, D. P., & Tobagi, F. A. (2004). Capacity of an IEEE 802.11b Wireless LAN supporting VoIP. *2004 IEEE International Conference on Communications, 1,* 196-201.

Home Gateway Technical Requirements: Residential Profile Version 1.0. (2008, April). *Home Gateway Initiative.*

Hou, Y. T., Dapeng, W., Yao, J., & Takafumi, C. (2000, November 8-10). A core-stateless buffer management mechanism for differentiated services Internet. *Proceedings of 25th Annual IEEE Conference on Local Computer Networks (LCN 2000),*(pp. 168 – 176). Tampa, Florida, USA: IEEE Press.

Hu, Y., & Loizou, P. C. (2007). A Comparative Intelligibility Study of Speech Enhancement Algorithms. In *proceedings of IEEE International Conference on Acoustics, Speech and Signal Processing: Vol. 4,* April 15-20.

Huang, D. S., & Zhao, W. B. (2002). *A Comprehensive Understanding for Radial Basis Probabilistic Neural Networks* (pp. 1237–1241).

Huang, C. J., Chuang, Y. T., Lai, W. K., Sun, Y. H., & Guan, C. T. (2007). Adaptive resource reservation schemes for proportional DiffServ enabled fourth-generation mobile communications system. *Computer Communications Journal*, *30*(7), 1613–1623. doi:10.1016/j.comcom.2007.01.015

Huang, C. M., Bi, Q., Stiles, G. S., & Harris, R. W. (1992). Fast full search equivalent encoding algorithms for image compression using vector quantization. *IEEE Transactions on Image Processing*, *1*(3), 413–416. PubMeddoi:10.1109/83.148613

Hughes-Hartogs, D. (1987). *Ensemble modem structure for imperfect transmission media.* U. S. Patent 4,679,227

Huttunen, A., Kivinen, T., Swander, B., & Volpe, V. (2005). *IETF Network Working Group: RFC 3947 Negotiation of NAT-Traversal in the IKE.*

Hwang, W. J., Jeng, S. S., & Chen, B. Y. (1997). Fast codeword search algorithm using wavelet transform and partial distortion search techniques. *Electronics Letters*, *33*(5), 365–366. doi:10.1049/el:19970249

IBM AgletWorkbench [EB/OL]. (2004). Retrieved August 12, 2008, from http://www.trl.ibm.co.jp/aglets/.

IEEE, 802.16e-2005. (2005). IEEE Standard for Local and Metropolitan Area Networks. Part 16: Air Interface for Fixed and Mobile Broadband Wireless Access Systems.

IEEE P802.21/D9.0. (2008). Draft Standard for Local and Metropolitan Area Networks: Media Independent Handover Services.

IEEE Standard for Information technology. (2007). *Telecommunications and information exchange between systems- Local and metropolitan area networks- Specific requirements- Part 11: Wireless LAN Medium Access Control (MAC) and Physical Layer (PHY) Specifications.*

IEEE Standard for Information technology. (2008). Telecommunications and information exchange between systems-Local and metropolitan area networks-Specific requirements- Part 11: Wireless LAN Medium Access Control (MAC) and Physical Layer (PHY) Specifications Amendment 1: Radio Resource Measurement of Wireless LANs IEEE 802.11k.

Imamura, K., Swilem, A., & Hashimoto, H. (2003, May). Fast codeword search algorithm for ECVQ using hyperplane decision rule. Proceeding of (ISCAS) International Symposium on Circuits and Systems, vol. 2, (pp. 476-479).

Imamura, K., Swilem, A., & Hashimoto, H. (2004, October). Fast VQ encoding algorithms using angular constraint. Proceeding of (ICIP) International Conference on Image Processing, (pp. 3161-3164).

Ippoliti, D., Xiaobo, Z., & Liqiang, Z. (2007, August 13-16). Packet Scheduling with Buffer Management for Fair Bandwidth Sharing and Delay Differentiation. *Proceedings of 16th International Conference on Computer Communications and Networks (ICCCN 2007), (pp. 569 - 574). Honolulu, Hawaii, USA: IEEE Press.*

Issa, O., Li, W., & Liu, H. (2009). Performance evaluation of TV over broadband wireless access networks. *IEEE Transactions on Broadcasting*, *56*(2), 201–210. doi:10.1109/TBC.2010.2046979

Jabloun, F., & Enis Cetin, A. (1999). *The Teager energy based feature parameters for robust speech recognition in car noise.* In proceedings of IEEE International Conference on Acoustics, Speech and Signal Processing, 15-19 Mar 1999, (Vol. 1, pp 273 – 276).

Jailani, N., Mohd Yatim, N. F., Yahya, Y., Patel, A., & Othman, M. (2008). Secure and auditable agent-based e-marketplace framework for mobile users. *Journal of Computer Science & Interface*, *30*(4), 237–252. doi:10.1016/j.csi.2007.10.009

Jain, R. (1991). *The Art of Computer Systems Performance Analysis: Techniques for Experimental Design, Measurement, Simulation, and Modeling* (p. 36). New York: John Wiley & Sons.

Jain, A. K., Ross, A., & Prabhakar, S. (2004). An Introduction to Biometric Recognition. *Journal of IEEE Transactions on Circuits and Systems for Video Technology*, *14*(1), 4–20..doi:10.1109/TCSVT.2003.818349

James, S., Hou, F., & Ho, P. H. (2007, May, 21-23). An Application-Driven MAC-layer Buffer Management with Active Dropping for Real-time Video Streaming in 802.16 Networks. Proceedings of the IEEE International Conference on Advanced Information Networking and Applications (AINA '07), *(pp 451–458). Niagara Falls, Ontario, Canada: IEEE Press.*

Jammeh, E., Fleury, M., & Ghanbari, M. (2004). Smoothing transcoded MPEG-1 video streams for Internet transmission. *IEE Proceedings. Vision Image and Signal Processing*, *151*(4), 298–306. doi:10.1049/ip-vis:20040747

Jankowski, C., Kalyanswamy, A., Basson, S., & Spitz, J. (1990). NTIMIT: a phonetically balanced, continuous speech, telephone bandwidth speech database. In proceedigs of IEEE International Conference on Acoustics, Speech, and Signal Processing,3-6 Apr. (pp 109 – 112).

Jardosh, S., Zunnun, N., Ranjan, P., & Srivastava, S. (2008, December 15-18). Effect of network coding on buffer management in wireless sensor network. *Proceedings of International Conference on Intelligent Sensors, Sensor Networks and Information Processing (ISSNIP 2008)*, (pp.157 - 162). Sydney, NSW, Australia: IEEE Press.

Java Server Pages Technology. (2007). Sun Developer Network. http://java.sun.com/products/jsp/

Java Servlet Technology. (2007). Sun Developer Network. http://java.sun.com/products/servlet/

Jazz Demo Videos. (2007). http://jazz.net/pub/learn/videos/videos.jsp

Jbenchmark, New Estimator Software Measures True Mobile Phone Performance. (2004). Retrieved from http:// www.jbenchmark.com/ jbacepr.jsp.

Jespersen, M.O., Nielsen, K.D., & Frølund, J. (2003, May 13). *Optimising performance in AOMDV with pre-emptive routing*. University of Aarhus, Computer Science Ny Munkegade DK-8000 Århus C.

Joanna, G., Hua, O., Courtney, F., & Stephen, G. (2007). Constructing Typologies in Community Research. *Annual meeting of the Rural Sociological Society,* Marriott Santa Clara, Santa Clara, California, August 02.

Johansson, P., Larson, T., Hedman, N., Mielczarek, B., & Degermark, M. (1999). Scenario-Based Performance Analysis of Routing Protocols for Mobile Ad hoc Networks. *The ACM/IEEE MOBICOM'99*, Seattle, WA (pp. 195-206).

Johnson, & Menezes. (1999, August). *The Elliptic Curve Digital Signature Algorithm (ECDSA) (Tech. Rep.)*. Technical report CORR 99-34, Department of C&O, University of Waterloo, 1999.

Johnson, D., & Maltz, D. (1996). Dynamic source routing ad hoc wireless networks. *Mobile Computing, 1*(1):153–181. Boston, MA: Kluwer Academic Publishers

Johnson, M. H., Ladner, R., & Riskin, E. A. (1996, September). Fast nearest neighbor search for ECVQ and other modified distortion measures. In *Proceeding of (ICIP) International Conference on Image Processing, 3*, 423-426.

Johnson, M. H., Ladner, R., & Riskin, E. A. (2000). Fast nearest neighbor search of entropy-constrained vector quantization. *IEEE Transactions on Image Processing, 9*, 1435–1437. PubMeddoi:10.1109/83.855438

Junhua, T., Gang, F., Chee-Kheong, S., & Liren, Z. (2008). Providing Differentiated Services Over Shared Wireless Downlink Through Buffer Management. *IEEE Transactions on Vehicular Technology, 57*(1), 548–555. doi:10.1109/TVT.2007.905246

Kadel, G. (1996). Diversity and equalization in frequency domain - a robust and flexible receiver technology for broadband mobile communication systems. In *Proceedings of the IEEE Vehicular Technology Conference '97, Phoenix*

Kalil, M. A., Al-Mahdi, H., & Mitschele-Thiel, A. (2008, July 27-August 1). Performance Evaluation of Hop-Aware Buffer Management Scheme in Multihop Ad Hoc Networks. *Proceedings of the Fourth International Conference on Wireless and Mobile Communications (ICWMC '08)*, (pp. 31–36). Athens, Greece: IEEE Press.

Kangasharju, J. (2005). *Mobile XML Messaging (Tech. Rep.). Department of Computer Science*. Finland: University of Helsinki.

Kangasharju, J., Lindholm, T., & Tarkoma, S. (2008). XML Security with Binary XML for Mobile Web Services. *International Journal of Web Services Research, 5*, 1–19. doi:10.4018/jwsr.2008070101

Kannammal, A., & Iyengar, N. Ch. S. N. (2007). A Model for Mobile Agent Security in E-Business Applications. *International Journal of Business and Information, 2*(2), 185–198.

Katsianis, D., Rokkas, T., Sphicopoulos, T., & Varoutas, D. (2007). Fixed Mobile Convergence For an Integrated Operator: A Techno-Economic Study. *18th Annual IEEE International Symposium on Personal, Indoor and Mobile Radio Communications* (pp. 1-5).

Kaufman, C. (2005). *IETF Network Working Group: RFC 4306 Internet Key Exchange (IKEv2)*. Protocol.

Kawash, J., El Morr, C., & Itani, M. (2007). A novel collaboration model for mobile virtual communities. *International Journal on Web Based Communities, 3*(4), 427–447. doi:10.1504/IJWBC.2007.015868

Keijzers, J., den Ouden, E., & Lu, Y. (2008). Usability benchmark study of commercially available smart phones: cell phone type platform, PDA type platform and PC type platform. In *MobileHCI '08: Proceedings of the 10th international conference on Human computer interaction with mobile devices and services* (pp. 265-272).

Kendall, E., Pathak, C., Krishna, M., & Suresh, C. (1997). The Layered Agent Pattern Language, Pattern Languages of Programming Conference (Plop), Monticello, Illinois.

Kim, W., & Stern, R. M. (2006). *Band-Independent Mask Estimation for Missing-Feature Reconstruction in the Presence of Unknown Background Noise*. In proceedigs of IEEE International Conference on Acoustics, Speech, and Signal Processing, Toulouse, France.

Kindberg, T. (1996). Mushroom: a framework for collaboration and interaction across the Internet. Proc. CSCW & the Web, 5th ERCIM workshop, Busbach, U., Kerr, D., and Sikkel, K. (eds), GMD, pp. 43-53.

Kinnunen, T. (2003). *Spectral Features for Automatic Text-Independent Speaker Recognition.* Licentiate's thesis, University of Joensuu, Department of Computer Science, Joensuu, Finland.

Kittler, J. (2005). *Multiple classifier systems in decision-level fusion of multimodal biometric experts.* 1st BioSecure residential workshop, Paris, France, August 1- 26.

Klaue, J., Rathke, B., & Wolisz, A. (2003). EvalVid - A framework for video transmission and quality evaluation. In *International Conference on Modeling Techniques and Tools for Computer Performance*, (pp. 255–272).

Koch, M., Groh, G., Hillebrand, C., & Fremuth, N. (2002). Mobile support for lifestyle communities. *Arbeitsberichte des Lehrstuhls fur Allgemeine und Industrielle Betriebswirtschaftslehre an der Technischen Universitat Munchen*, Nr. 34.

Koch, M., Moslein, K., & Schubert, P. (2002). Communities and personalization for individual products. *Proceedings of the Annual Conference of the British Academy of Management*, London, UK, September 2002.

Kohler, E., Handley, M. & Floyd, S. (2006). Datagram Congestion Control Protocol. *Internet Engineering Task Force*, RFC 4340.

Koreman, J., Wu, D., & Morris, A. C. (2007). Enhancing Speaker Discrimination at the Feature Level. *Journal of Lecture Notes In Artificial Intelligence: Springer Berlin / Heidelberg Volume 4343,* 260-277.

Krachodnok, P., & Benjapolakul, W. (2001). Buffer management for TCP over GFR service in an ATM network. In *Proceedings of Ninth IEEE International Conference on Networks,* (pp. 302 - 307). Washington, DC, USA: IEEE Press.

Kramer, G., Gastpar, M., & Gupta, P. (2005). Cooperative strategies and capacity theorems for relay networks. *IEEE Transactions on Information Theory*, *51*(9), 3037–3063. doi:10.1109/TIT.2005.853304

Kravets, R., & Krishnan, P. (1998). Power Management Techniques for Mobile Communication. In *MobiCom '98: Proceedings of the 4th annual ACM/IEEE international conference on Mobile computing and networking* (pp. 157-168).

Krifa, A., Baraka, C., & Spyropoulos, T. (2008). Optimal Buffer Management Policies for Delay Tolerant Networks. *Proceedings of 5th Annual IEEE Communications Society Conference on Sensor, Mesh and Ad Hoc Communications and Networks (SECON '08)*, (pp. 260-268). San Francisco, CA, USA: IEEE Press.

Krzyzanowski, P. (2000). Process Synchronization and Election Algorithms. *Lectures note on distributed systems.* Retrieved October 16, 2008, from http://www.cs.rutgers.edu/~pxk/rutgers/notes/content/06- mutex.pdf.

Kuhn, M., & Wattenhofer, R. (2006). Community-Aware Mobile Networking. *1st Workshop on Mobile Services and Personalized Environments (MSkuPE)*, Aachen, Germany, November.

Kwan, Julian, S. H., & Sun, H. W. (2006). An improved landslide mobility model. *Canadian Geotechnical Journal*, *43*(5), 531–539..doi:10.1139/T06-010

Lai, Y.-X., Chen, Y.-L., & Chen, H. (2008, November). PEJA: Progressive Energy-Efficient Join Processing For Sensor Networks. *Journal of Computer Science and Technology*, *23*(6), 957–972. doi:10.1007/s11390-008-9191-2

Lai, J. Z. C., & Lue, C. C. (1996). Fast search algorithms for VQ codebook generation. *Journal of Visual Communication and Image Representation*, *7*(2), 163–168. doi:10.1006/jvci.1996.0016

Lai, Y. L., Tseng, Y. Y., Lin, C. W., Zhou, Z., & Sun, M. T. (2005). H.264 Encoder Speed-UP via Joint Algorithm/ Code-Level Optimization. *Proceedings of VCIP 2005*, Beijing, China.

LambdaM. O. O.http://magrathea.cosmic.com/pub/moo/ProgrammersManual.html

LAN MAN Standards Committee of the IEEE Computer Society. (1999). *IEEE Standard 802.11-1999, Part 11: Wireless LAN Medium Access Control (MAC) and Physical Layer (PHY) Specifications*. IEEE.

Laneman, J. N., Tse, D. N. C., & Wornell, G. W. (2004). Cooperative diversity in wireless networks: Efficient protocols and outage behavior. *IEEE Transactions on Information Theory*, *50*(12), 3062–3080. doi:10.1109/TIT.2004.838089

Lawrence, E., Bina, M., Culjak, G., & El-Kiki, T. (2007). Wireless community networks: Public assets for 21st century society. *Proceedings of the International Conference on Information Technology (ITNG '07)*, pp. 166-174, IEEE.

Lee, C. H., & Chen, L. H. (1994). Fast closest codeword search algorithm for vector quantization. *IEE Proceedings. Vision Image and Signal Processing*, *141*(3), 143–148. doi:10.1049/ip-vis:19941140

Lee, C.-H., & Chen, L.-H. (1995). A fast search algorithm for vector quantization using mean pyramids of codewords. *IEEE Transactions on Communications*, *43*, 1697–1702. doi:10.1109/26.380218

Lee, T., Chan, S., Zhang, Q., Zhu, W., & Zhang, Y. (2002). Allocation of layer bandwidths and FECs for video multicast over wired and wireless networks. *IEEE Transactions on Circuits and Systems for Video Technology*, *12*(2), 1059–1070. doi:10.1109/TCSVT.2002.806816

Lee, H. S., & Tsoi, A. C. (1995). Application of multilayer perceptron in estimating speech / noise characteristics for speech recognition in noisy environment. *Journal of Speech Communication*, *17*(1-2), 59–76.. doi:10.1016/0167-6393(95)00018-J

Lee, T. O., Yip, Y. L., Tsang, C. M., & Ng, K. W. (2001). An Agent-based Micropayment System for E-Commerce. In Liu, J., & Ye, Y. (Eds.), *E-Commerce Agents. Lecture Notes in Artificial Intelligence* (*Vol. 2033*, pp. 247–263). Springer-Verlage Berlin Heidelberg.

Lee, B., Yoon, S., & Kang, H. (2006). *On the Use of Voting Methods for Speaker Identification Based on Various Resoulution Filterbanks.* In Proceedings of ICASSP'06, Toulouse, France.

Lee, C. H., & Chen, L. H. (1994). Fast closest codeword search algorithm for vector quantization. *IEEE Proceedings Vision, Image &. Signal Processing*, *141*(3), 143–148.

Lee, C.-H., & Chen, L.-H. (1995). A fast search algorithm for vector quantization using mean pyramids of codewords. *IEEE Transactions on Communications*, *43*, 1697–1702. doi:10.1109/26.380218

Lee, S. J., Youn, J. S., & Jung, S. H. Lee, J., & Kang, C. H. (2005). Selective route-request scheme for on-demand routing protocols in ad hoc networks. *Lecture Notes in Computer Science, v 3738 LNCS, Ad-Hoc, Mobile, and Wireless Networks-4th International Conference, ADHOC-NOW* (pp. 153-163).

Leimeister, J. M., Daum, M., & Krcmar, H. (2004). Towards mobile communities for cancer patients: the case of krebsgemeinschaft.de. *International Journal on Web Based Communities*, *1*(1), 58–70.

Leimeister, J. M., Sidiras, P., & Krcmar, H. (2004). Success factors of virtual communities from the perspective of members and operators: An empirical study. *Proceedings of the 37th Hawaii International Conference on System Sciences (HICCS 37)*, IEEE, USA.

Leuf, B., & Cunningham, W. (2001). *The Wiki Way: Collaboration and Sharing on the Internet*. Addison-Wesley.

Li, C. M. (2001). Random exploration and optimal exploration. *Military Operations Research and Systems Engineering*, *2*(4), 10–20.

Li, L. L., & Zhou, D. H. (2004). Robust Fault Diagnosis of Nonlinear System Based on Analytical Models: a Surevey. *Information and Control*, *33*(4), 451–456.

Li, H. (2004). Virtual community studies: a literature review, synthesis and research agenda. J. De Gross et al. (Eds.), *Proceedings of the Americas Conference on Information Systems*, 2708-2715, August 2004, New York, USA.

Li, M. (1993).Determination of optimal buffer sizes in static buffer management. *Proceedings of IEEE Region 10 Conference on Computer, Communication, Control and Power Engineering (TENCON '93)*, (pp. 479–486). Beijing, China: IEEE Press.

Liao, H. Y. M., Chen, D. Y., Su, C. W., & Tyan, H. R. (2006). Real-time event detection and its applications to surveillance systems. Proceedings IEEE International Symposium on Circuits and Systems, (pp. 509-512).

Lifecycles, C., Fundamentals, A., & Guide, A. D. (2009). Retrieved from http://dev.android.com/ guide/topics/fundamentals.html.

Lihua, S., Xiaotong, Z., Qin, W., & Yanfei, G. (2007). MYBUF: A High Performance Buffer Management Mechanism. *Proceedings of International Conference on Wireless Communications, Networking and Mobile Computing (WiCom 2007),* (pp. 3007 - 3010). Shanghai, China: IEEE Press.

Lim, H. H., & Qiu, B. (2001). Predictive fuzzy logic buffer management for TCP/IP over ATM-UBR and ATM-ABR. *Proceedings of the IEEE International Conference on Global Telecommunications (GLOBECOM '01),* (Vol. 4) (pp. 2326–2330). San Antonio, Texas, USA: IEEE Press.

Lin, Y.-B., & Pang, A.-C. (2005). *Wireless and Mobile: All-IP networks*. Chichester, UK: Wiley and Sons.

Lin, C., & Gerla, M. (1997). Adaptive clustering for mobile wireless networks. In *IEEE JSAC, 15*, 1265–75.

Lin, S. J., Chung, K. L., & Chang, L. C. (2001). An improved search algorithm for vector quantization using mean pyramid structure. *Pattern Recognition Letters*, *22*(3/4), 373–379. doi:10.1016/S0167-8655(00)00136-7

Linde, Y., Buzo, A., & Gray, R. M. (1980). An algorithm for vector quantizer design. *IEEE Transactions on Communications*, *COM-28*(1), 84–95. doi:10.1109/TCOM.1980.1094577

Liu, H., Zhang, W., & Yang, X. (2006). Cross-layer conditional retransmission for layered video streaming over cellular networks. *Computer Communications*, *29*(11), 2066–2073. doi:10.1016/j.comcom.2006.01.002

Liu, C., & Fu, Q. (2007). *Effects of Bandwidth Extension on Telephone Speech Recognition by Cochlear Implant Users.* Paper presented at the conference on Implantable Auditory Prostheses, Granlibakken, Conference Center Lake Tahoe, California.

Liu, Q., Zhou, S., & Giannaki, G. B. (2004). TCP performance in wireless access with adaptive modulation and coding. *IEEE International Conference on Communications*, pp. 3989-3993, 2004.

Liu, Z., Wu, Z., Liu, H., & Stein, A. (2007). A layered hybrid-ARQ scheme for scalable video multicast over wireless networks. In *Asilomar Conference on Signals, Systems and Computers*, (pp. 914-919).

Lo, A., Heijenk, G., & Niemegeers, I. (2005). Performance evaluation of MPEG-4 video streaming over UMTS network using an integrated tool environment. *International Symposium on Performance Analysis of Wireless Networks and Communications Systems.*

Lu, S. Q., Zhu, X. Q., Zhao, W. Q., & Bao, C. G. (1981). Eikonal-Born Expansion And Its Application In Large-Angle Scattering. *Physica Energiae Fortis Et Physica Nuclearis*, *5*(1), 121–126.

Lu, Z. M., & Sun, S. H. (2003). Equal-average equal-variance equal-norm nearest neighbor search algorithm for vector quantization. *IEICE Transactions on Information and systems. E (Norwalk, Conn.)*, *86D*(3), 660–663.

Lu, Z. M., Pan, J. S., & Sun, S. H. (2000). Efficient codeword search algorithm based on hadamard transform. *Electronics Letters*, *36*(16), 1364–1365. doi:10.1049/el:20000972

Lucantoni, D.M., Neuts, M.F., & Reibman, A.R. (1994). Methods for performance evaluation of VBR video traffic models. *IEEE/ACM Transactions on Networking*, *2*(2), 176–180.

Madala, K., Tomek, I., Shakshuki, E., & Giles, R. (2006). A Thin Mobile Client for a Groupware Application", COOP 2006 the 7th International Conference on the Design of Cooperative Systems, Provence, France, May 9-12, pp. 211-223.

Maeda, H., Kajihara, M., Adachi, H., Sawada, A., Takeda, H., & Nishida, T. (1997). Weak Information Structures for Community Information Sharing. *International Journal of Knowledge-Based Intelligent Engineering Systems*, *1*(4), 225–234.

Maes, P., Guttman, R. H., & Moukas, A. G. (1999). Agent that buy and sell [J]. *Communications of the ACM*, *42*(3), 81–91. doi:10.1145/295685.295716

Maglaris, B., Anastassiou, D., Sen, P., Karlsson, G., & Robbins, J. D. (1988). Performance models of statistical multiplexing in packet video communications. *IEEE Transactions on Communications*, *36*(7), 834–843. doi:10.1109/26.2812

Magrin-Chagnolleau, I., Bonastre, J. F., & Bimbot, F. (1995). *Effect of Utterance Duration and phonetic Content on Speaker identification Using Second-Order Statistical Methods*. In Proceedings of the ESCA EUROSPEECH'95, Madrid, 1, 337-340.

Magrin-Chagnolleau, I., Wilke, J., & Bimbot, F. (1996). *A Further Investigation on AR-Vector Models for Text-Independent Speaker Identification*. In Proceedings of the ICASSP'96 conference, Atlanta, GA.

Mansfield, T. & et al. (1997). Evolving Orbit: A Process Report on Building Locales, GROUP.

Marco, P. D., Rinaldi, C., Santucci, F. M., Johansson, K. H., & Moller, N. (2006). Performance analysis and optimization of TCP over adaptive wireless links. In *17th Annual IEEE International Symposium on Personal, Indoor and Mobile Radio Communications*.

Marpe, D., Schwarz, H., & Wiegand, T. (2007). Overview of the Scalable Video Coding Extension of the H.264 / AVC Standard. *IEEE Transactions on Circuits and Systems for Video Technology*, *17*(9), 1103–1120. doi:10.1109/TCSVT.2007.905532

Marratech (2007). http://www.marratech.com/

Master, J. (1992). *Pratical Neural Network Recipes in C*. New York: Academic press.

Matsui, T., & Furui, S. (1993). *Concatenated Phoneme Models for Text-Variable Speaker Recognition*. In Proceedings of ICASSP, Vol. 2, Minneapolis, 391-394.

Maxell, P., Seamone, B., & Verstraete, J. (2007). Independent dominating sets and Hamiltonian cycles. *Journal of Graph Theory*, *54*(3), 233–244. doi:10.1002/jgt.20205

Maybury, M. (2004). Collaborative Virtual Environments for Analysis and Decision Support.

Mc Donald, A. A., & Znati, T. F. (1999). Mobility Based framework for Adaptive Clustering in Wireless Ad Hoc Networks. *IEEE Journal on Selected Areas in Communications*, *17*(8), 1466–1486. doi:10.1109/49.780353

McKenzie, E. (1986). Autoregressive moving average processes with negative binomial and geometri marigi-nal distributions. *Advances in Applied Probability*, *18*, 679–705. doi:10.2307/1427183

Meignier, S. (2002). *Indexation en locuteurs de documents sonores: Segmentation d'un document et Appariement d'une collection*. PhD thesis, LIA Avignon, France.

Michael, M. P. (2005). *Research Seminar On Energy Awareness, University Of Helsinki*. Energy Awareness for Mobile Devices.

Microsoft Office LiveMeeting. (2007). http://www.microsoft.com/windows/netmeeting/

Ming, Y., Timoney, J., Doyle, L., & O'Mahony, D. (2001, November 27). Evaluation of Channel Fairness Models for Ad-Hoc Networks. *Proceedings of the First Joint IEI/IEE Symposium on Telecommunications Systems Research* (p. 4), Dublin.

Mishra, P., & Saksena, M. (1998). Designing buffer management policies at an IP/ATM gateway. In *Proceedings of IEEE ATM Workshop*, (pp.144–153). Fairfax, Virginia, USA: IEEE Press.

Miyajima, C., Hattori, Y., Tokuda, K., Masuko, T., Kobayashi, T., & Kitamura, T. (2001). Text-independent speaker identification using Gaussian mixture models based on multi-space probability distribution. *Journal of IEICE Transactions on Information and Systems. E (Norwalk, Conn.)*, *84-D*(7), 847–855.

Mnif, K., & Kadoch, M. (2006). Construction and Maintenance of Backbone for Routing Protocols Enhancement in Mobile Ad hoc Networks. *10th IEEE International Conference on Communication Systems 2006, ICCS'06*, Singapore (pp. 239-243).

Morris, A., Wu, D., & Koreman, J. (2005). GMM based clustering and speaker separability in the Timit speech database. *Journal of IEICE Transactions Fundamentals/ Commun./ Electron./ Inf. & Syst., Vol. E85- A/ B/ C/ D, No. 1*.

Mouratidis, M., Giorgini, P., & Schumacher, M. (2003). Security Patterns for Agent Systems, Proceedings of Eighth European Conference on Pattern Languages of Programs, Germany, pp. 25- 29.

Mueller-Prothmann, T., & Siedentopf, C. (2003). Designing online knowledge communities: developing a usability evaluation criteria catalogue. *Proceedings 3rd European Knowledge Management Summer School*, 7 – 12 September, 2003, San Sebastian, Spain.

Mühlera, R., Ziesea, M. & Rostalskib, D. (2009). Development of a Speaker Discrimination Test for Cochlear Implant Users Based on the Oldenburg Logatome Corpus. *Journal of oto-rhino-laryngology (ORL), 71*(1), pp 14-20.

Muqattash, A., & Krunz, M. (2003). Power Controlled Dual Channel (PCDC) Medium Access Protocol for Wireless Ad Hoc Networks. Twenty-Second Annual Joint Conference of the IEEE Computer and Communications Societies. *IEEE INFOCOM 2003, 1,* 470-480.

Nanda, S., Balachandran, K., & Kumar, S. (2000). Adaptation techniques in wireless packet data services. *IEEE Communications Magazine, 38,* 54–64. doi:10.1109/35.815453

Nasipuri, A., Zhuang, J., & Das, S. R. (1999). A multichannel CSMA MAC protocol for multihop wireless networks. *Proc. IEEE Wireless Comm. and Networking Conf. (WCNC).*

National Institute of Standards and Technology Information Technology Laboratory. (2002, March). *The Keyed-Hash Message Authentication Code (HMAC).* Federal Information Processing Standards Publication.

National Institute of Standards and Technology. (2000). *Advanced Encryption Standard (AES) Questions and Answers.* Retrieved December 13, 2008, from http://www.nist.gov/public_affairs/releases/aesq&a.htm.

Nedjah, N., & Mourelle, L. D. M. (2006). *Swarm Intelligent Systems.* Berlin, Heidelberg: Springer-Verlag. doi:10.1007/978-3-540-33869-7

Neuenhofen, K., & Thompson, M. (1998, May 10-13). A Secure Marketplace for Mobile Java Agents. *Proceedings of the 2nd International Conference on Autonomous agents,* Minneapolis, MN USA (pp. 212-218).

Ng, P. C., & Liew, S. C. (2007, April). Throughput Analysis of IEEE802.11 Multi-hop Ad hoc Networks. [TON]. *IEEE/ACM Transactions on Networking, 15*(2), 309–322. doi:10.1109/TNET.2007.892848

Noh, K. J., & Bae, C. S. (2007). A QoS RED Buffer Management Scheme of Intelligent Gateway Gadget based on a Wireless LAN. *Proceedings of the IEEE International Conference on Advanced Communication Technology (ICACT '07),* (Vol.1) (pp. 286-291). Gangwon-Do, South Korea: IEEE Press.

Nomura, M., Fujii, T., & Ohta, N. (1989). Basic characteristics of variable rate video coding in ATM environment. *IEEE Journal on Selected Areas in Communications, 7*(5), 752–760. doi:10.1109/49.32338

NotesL. (2007). http://www-306.ibm.com/software/lotus/

Ohsumi, A. (1986). Optimal Searching for a Markovian Target and Relation to Optimal Stochastic Control. *Theory and Applications of Nonlinear Control Systems,* CByrnes, A Lindquist North Holland, Amsterdam (pp. 569-583).

Online Internet Assigned Numbers Authority. (2009). *Radius Types* [WWW page]. URL http://www.iana.org/assignments/radius-types

Open Node Syntax. (2005, June). Retrieved from http://www.seairth.com/ web/onx/onx.html

Orchard, M. D. (1991). A fast nearest-neighbor search algorithm. In *Proceeding of (ICASSP) International Conference on Acoustics, Speech, and Signal Processing, 4,* 2297-2300.

Org, U. M. L. (2009). *UML® Resource Page.* Last Updated on Thursday, January 08, 2009. Retrieved January 10, 2009, from http://www.uml.org/.

Ouamour, S., Sayoud, H., & Guerti, M. (2009, April-June). Optimal Spectral Resolution in Speaker Authentication, Application in noisy environment and Telephony. [IJMCMC]. *International Journal of Mobile Computing and Multimedia Communications, 1*(2), 36–47. doi:10.4018/jmcmc.2009040103

OverviewC. V. W. (2007). http://cvw.sourceforge.net/cvw/info/CVWOverview.php3

OverviewC. V. W. (2008). http://sourceforge.net/support/getsupport.php?group_id=12120

Padhye, J., Firoiu, V., Towsley, D. F., & Kurose, J. F. (2000). Modeling TCP Reno performance: a simple model and its empirical validation. *IEEE/ACM Transactions on Networking, 8*(2), 133–145. doi:10.1109/90.842137

Pan, J. S., Lu, Z. M., & Sun, S. H. (2000). Fast codeword search algorithm for image coding based on mean-variance pyramids of codewords. *Electronics Letters, 36*(3), 210–211. doi:10.1049/el:20000237

Pan, J. S., Lu, Z. M., & Sun, S. H. (2003). An efficient encoding algorithm for vector quantization based on sub-vector technique. *IEEE Transactions on Image Processing*, *12*(3), 265–270. PubMeddoi:10.1109/TIP.2003.810587

Pan, Z., Kotani, K., & Ohmi, T. (2004). An improved fast encoding algorithm for vector quantization using 2-pixel-merging sum pyramid data structure. *Pattern Recognition Letters*, *25*, 459–468. doi:10.1016/j.patrec.2003.12.009

Pan, Z., Kotani, K., & Ohmi, T. (2006). Subvector-based fast encoding method for vector quantization without using two partial variances. *Optical Review*, *13*(6), 410–416. doi:10.1007/s10043-006-0410-1

Paninski, L. (2006). *The Spike-Triggered Average of the Integrate-and-Fire Cell Driven by Gaussian White Noise.* Neural computation, 2006 - MIT Press, pp 2592-2616.

Parekh, A. (1994). *Selecting routers in ad hoc wireless networks*. In Proceedings of IEEE international Telecommunications symposium (Pp. 420-424).

Parikh, T. S. (2006). Designing an Architecture for Delivering Mobile Information Services to the Rural Developing World. *Proceedings of 7th IEEE Workshop on Mobile Computing Systems and Applications*, Orcas, WA, 06-07 April.

Park, N., Kim, H., Chung, K., Sohn, S., & Won, D. (2006). *Xml-Signcryption Based LBS Security Protocol Acceleration Methods In Mobile Distributed Computing*. In ICCSA 2006: International Conference On Computational Science And Its Applications (LNCS 3984, pp. 251–259). Glasgow, UK: Springer-VerlagPeng, C., & Li, X. (2005, September). *Threshold Signcryption Scheme Based On Elliptic Curve Cryptosystem And Verifiable Secret Sharing.* In 2005 International Conference On Wireless Communications, Networking And Mobile Computing (Vol. 2, pp. 1182–1185).

Paternò, F., Russino, A., & Santoro, C. (2007). *Remote Evaluation of Mobile Applications* (pp. 155–169). Task Models and Diagrams for User Interface Design.

Pei Y. & Modestino J.W. (2004). Interactive video coding and transmission over heterogeneous wired-to-wireless IP networks using an edge proxy. *EURASIP Journal on Applied Signals*, 2004 volume [online], 253-264.

Perkins, C., & Bhagwat, P. (1994). Highly dynamic destination-sequenced distance vector routing (DSDV) for mobile computers. *Computer Communication Review*, *1*(1), 234–244. doi:10.1145/190809.190336

Perkins, C. E., Royer, E. B., & Das, S. (2003). *Ad hoc On-Demand Distance Vector (AODV) Network Working Group, Routing. Draft-ietf-manet-aodv-13.txt*. The Internet Society.

Perkins, C. (2002). *IETF Network Working Group: RFC 3344 IP Mobility Support for IPv4.*

Perkins, C., & Royer, E. M. (1999). *Ad hoc on demand distance vector routing*. In Proceedings 2nd IEEE Workshop on Mobile Computing Systems and Applications(pp. 90-100).

Perkins, C., Belding-Royer, E., & Das, S. (2003, July). Ad hoc On-Demand Distance Vector (AODV) Routing. *Internet Experimental RFC, 3561*.

Porter, C. E. (2004). A Typology of Virtual Communities: A Multi-Disciplinary Foundation for Future Research. *Journal of Computer Mediated Communication, 10*(1). Retrieved April 7, 2008 from http://jcmc.indiana.edu/vol10/issue1/porter.html.

Potlapally, N., Ravi, S., Raghunathan, A., & Jha, N. (2003, August). *Analyzing The Energy Consumption Of Security Protocols.* In Proceedings Of The 2003 International Symposium Of Low Power Electronics And Design, Seoul, Korea.

Powazek, D. (2002). *Design for community: the art of connecting real people in virtual places*. Indianapolis: New Riders.

Preece, J., Abras, C., & Maloney-Krichmar, D. (2004). Designing and evaluating online communities: research speaks to emerging practice. *International Journal of Web Based Communities*, *1*(1), 2–18.

Preece, J., Maloney-Krichmar, D., & Abras, C. (2003). History and Emergence of Online Communities. In Wellman, B. (Ed.), *Encyclopedia of Community*. New York: Berkshire Publishing Group.

Preneel, B. (2005). A Hash of the Crypto Year. *Information Security Solutions Europe Conference.* Retrieved 22 March, 2009, from http://www.certification.tn/fileadmin/Ecolecrypto/Preneel/preneel_isse05v1.pdf.

Proakis, J. G. (1995). *Digital Communications*. New York: McGraw-Hill.

Qiu, X., & Chawla, K. (1999). On the performance of adaptive modulation in cellular systems. *IEEE Transactions on Communications*, *47*, 884–895. doi:10.1109/26.771345

Radzievskij, V. G., & Ufaev, V. A. (2003). Primary signal processing in digit6al panoramic detectors-direction finders. *Radiotekhnika*, *2*, 26–31.

Ramamurthy, G., & Sengupta, B. (1990). Modeling and analysis of a variable bit rate video multiplexer. *Proceedings of the 7th Internat. Teletraffic Congress Seminar*, Morristown, NJ.

Razavi, R., Fleury, M., & Ghanbari, M. (2008). *Unequal protection of video streaming through adaptive modulation with a tri-zone buffer over Bluetooth enhanced data rates*. EURASIP Journal on Wireless Communications and Networking. [online]

Razik, J., Fohr, D., Mella, O., & Parlangeau-Vallès, N. (2004). *Segmentation Parole/Musique pour la Transcription Automatique*. Paper presented at the JEP'04, Journées d'Etude sur la Parole, Fez, Morocco.

Razouqi, Q., Seong-Soon, J., & Ghosh, S. (2000). Performance analysis of fuzzy thresholding-based buffer management for a large-scale cell-switching network. *IEEE Transactions on Fuzzy Systems*, *8*(4), 425–441. doi:10.1109/91.868949

Reynolds, D. A. (1995). Speaker identification and verification using Gaussian mixture speaker models. *Journal of Speech Communication*, *17*(1-2), 91–108.. doi:10.1016/0167-6393(95)00009-D

Reynolds, D. A., & Rose, R. C. (1995). Robust test-independent speaker identification using Gaussian mixture speaker models. *Journal of IEEE Transactions on Speech and Audio Processing*, *3*(1), 72–83.. doi:10.1109/89.365379

Reynolds, D. (2008). *Speaker and language recognition: a guided safari odyssey workshop*. Paper presented at the Odyssey Speaker and Language Characterization Interest Group, Stellenbosch.

Rheingold, H. (2002). *Smart Mobs: The Next Social Revolution*. New York: Perseus publishing.

Rheingold, H. (2003). *Mobile Virtual Communities*. Retrieved April 7, 2008 http://www.thefeaturearchives.com/ topic/Culture/ Mobile_Virtual_Communities.html.

Rickenbacn, P. V., Schmid, S., Wattenhofer, R., & Zollinger, A. (2005). A robust interference model for wireless ad-hoc networks. *Proceedings of the 19th IEEE International Parallel and Distributed Processing Symposium* (pp. 1-8). doi:10.1109/ IPDPS.2005.65

Rivest, R., Shamir, A., & Adleman, L. (1978). A Method for Obtaining Digital Signatures and Public-Key Cryptosystems. *Communications of the ACM*, *21*(2), 120–126. doi:10.1145/359340.359342

Rodriguez-Dagnino, R. M., Khansari, M. R. K., & Leon-Garcia, A. (1991). Prediction of bit rate sequences of encoded video signals. *IEEE Journal on Selected Areas in Communications*, *9*(3), 305–314. doi:10.1109/49.76628

Roseman, M., & Greenberg, S. (1996). TeamRooms: Network Places for Collaboration, Proceedings of the 1996 ACM conference on Computer supported cooperative work pp. 325-333.

Rubenstein, D., Kurose, J., & Towsley, D. (1998). *Real-time reliable multicast using proactive forward error correction. Technical Report 98-19*. Amherst, MA: Department of Computer Science, University of Massachusetts.

Rui-yun, W., Li, L., & Wei-feng, L. (2004). Research of Mobile Agent-based Security in the E-commerce. *Journal of Application Research of Computers*, *11*, 137-138, & 141.

Ryu, J., Song, S., & Cho, D.-H. (2001). *New clustering schemes for energy conservation in two tiered mobile ad-hoc networks*. In Proc. IEEE ICC 01, pages 862–66, Vol.3. IEEE.

Ryu, Y. S., & Koh, K. (1996). A dynamic buffer management technique for minimizing the necessary buffer space in a continuous media server. In *Proceedings of the third IEEE International Conference on Multimedia Computing and Systems (MMCS 1996)*, (pp. 181-185). Hiroshima, Japan: IEEE Press.

Sanchez, M. (2006). Mobility Models. http://www.dis-ca.upv.es/misan/mobmaodel.htm

Sandoz, P., & Pericas-Geertsen, S. (2005). *Fast Infoset @ Java.net*. In Xtech 2005: XML, The Web And BeyondRetrieved from http://idealliance.org/ proceedings/ xtech05/ papers/04-01-01/

Sari, H., Karama, G., & Jeanclaudea, I. (1994). An analysis of orthogonal frequency-division multiplexing for mobile radio applications. *Proceedings of the VTC '94 in Stockholm*, (pp. 1635-1639).

Sari, H., Karama, G., & Jeanclaudea, I. (1994). Frequency-domain equalization of mobile radio and terrestrial broadcast channels. *Proceedings of the Globecom '94 in San Francisco* (pp. 1-5).

Sarr, C., Chaudet, C., Chelius, G., & Lassous, I. G. (2005). A node-based available bandwidth evaluation in IEEE 802.11 ad hoc networks. *Proceedings of the 11th International Conference on Parallel and Distributed Systems - Workshops, 2*, 68-72.

Satyanarayanan, M. (2001). Pervasive computing: Vision and challenges. *Personal Communications, IEEE, 8*(4), 10–70. doi:10.1109/98.943998

Satyanarayanan, M. (1996). Fundamental challenges in mobile computing. In *PODC '96: Proceedings of the fifteenth annual ACM symposium on principles of distributed computing* (pp. 1-7).

Sayoud, H., Ouamour, S., & Boudraa, M. (2003). 'ASTRA' An Automatic Speaker Tracking System based on SOSM measures and an Interlaced Indexation. *Journal of Acta Acustica, 89*(4), 702–710.

Sayoud, H. (2003). *Automatic speaker recognition using neural approaches*. PhD thesis, USTHB University, Algiers, Algeria.

Sayoud, H., & Ouamour, S. (2000). *Reconnaissance Automatique du Locuteur en Milieu Bruité*. In Proceedings of JEP'00, 345-348, Aussois, France.

Sayoud, H., & Selmane, M. K. (1998). *Méthodes comparatives en reconnaisance du locuteur*. Paper presented at the JTEA'98 conference, Nabeul, Tunis, Tunisia.

Sayoud, H., Ouamour, S., & Guerti, M. (2007). *Authentification Discriminative du Locuteur Basée sur une Fusion Statistique – Connexionniste*. Paper presented at the 4rth International Conference on Sciences of Electronic, Technologies of Information and Telecommunications, Tunisia.

Schmelzer, R. (2004, November). *Will Binary XML Solve XML Performance Woes?* Retrieved from http:// searchwebservices.techtarget.com /tip1,289483, sid,26 gci1027726,00.html

Schuckmann, C., et al. (1996). Designing object-oriented synchronous groupware with coast, Proceedings of the 1996 ACM conference on Computer supported cooperative work, pp. 30-38.

Schuler, D. (1994). Community networks: Building a new participatory medium. *Communications of the ACM, 37*(1), 38–51. doi:10.1145/175222.175225

Schulzrinne, H. (2002). *IETF Network Working Group: RFC 3361 Dynamic Host Configuration Protocol (DHCP-for-IPv4) Option for Session Initiation Protocol (SIP)*. Servers.

Schummer, T., & Schummer, J. (1999). *TUKAN: A Team Environment for Software Implementation*. OOPSLA.

Schwabe, G., & Prestipino, M. (2005). How tourism communities can change travel information quality. *Proceedings of the Thirteen European Conference on Information Systems (ECIS)*. Regensburg, Germany.

Schwarz, H., Sullivan, G., Wiegand, T., & Wien, M. (2007). *Text of ISO/IEC 14496-10:200X/FDIS Advanced Video Coding* (4th ed). International Organisation For Standardisation ISO/IEC JTC1/SC29/WG11 Coding of Moving Pictures and Audio.

Sen, P., Maglaris, B., Rikli, N. E., & Anastasiou, D. (1989). Models for packet switching of variable bit rate video in ATM networks. *IEEE Journal on Selected Areas in Communications, 7*(5), 865–869. doi:10.1109/49.32350

Sendonaris, A., Erkip, E., & Aazhang, B. (2003). User cooperation diversity-Part II, Implementation aspects and performance analysis. *IEEE Transactions on Communications, 51*(11), 1939–1948. doi:10.1109/ TCOMM.2003.819238

Seo, J. H., Im, C. H., Heo, C. G., Kim, J. K., Jung, H. K., & Lee, C. G. (2006). Multimodal function optimization based on particle swarm optimization. *IEEE Transactions on Magnetics, 42*(4), 1095–1098. doi:10.1109/ TMAG.2006.871568

Shakshuki, E., Ghenniwa, H., & Kamel, M. (2003). Agent-based System Architecture for Dynamic and Open Environments [IJITDM]. *International Journal of Information Technology and Decision Making, 2*(1), 105–133. doi:10.1142/S0219622003000537

Shakshuki, E., & Matin, R. (2008). Intelligent Learning Agent for Collaborative Virtual Workspace. International Journal of Ubiquitous Computing and Intelligence (JUCI) [ASP]. *American Scientific Publishers, 2*, 1–18.

Shakshuki, E., Lei, M., & Tomek, I. (2006). A Rich Client for CSCW. IADIS International Conference on Applied Computing, February 25-28, San Sebastian, Spain, pp. 618-621.

Shakshuki, E., Prabhu, O., & Tomek, I. (2006). FCVW Agent Framework. International Journal of Information and Software Technology, Elsevier Science, special issue, vol. 48, no. 6, pp. 385-392.

Shan, X. C., Fang, W. L., Wei, L. Z., & Tian, Q. Z. (2008). Fast search algorithm for vector quantization based on subvector technique. *IEICE Transactions on Information & Systems. E (Norwalk, Conn.), 91-D*(7), 2035–2040.

Shankar, K. (2008). Wind, water, and Wi-Fi: New trends in community informatics and disaster management. *The Information Society, 24*(2), 116–120.

Shapiro, S., Wilk, M., & Chen, H. (1968). A Comparative Study of Various Tests for Normality. *American Statistical Association, 63*(324), 134–137. doi:10.2307/2285889

Shi, H. L., Deng, W., & Qi, L. (2007). Properties of maximal mathematical expectation. *Journal of Shandong University. Natural Science, 42*(2), 92–94.

Shim, C., Ryoo, I., Lee, J., & Lee, S. (1994). Modeling and call admission control algorithm of variable bit rate video in ATM Networks. *IEEE Journal on Selected Areas in Communications, 12*(2), 332–344. doi:10.1109/49.272884

Shinfeng, D. L., Shih, C. S., & Chi, Y. H. (2003). A Fast Codebook Search Algorithm Based on Sum-Norm Pyramid of Codewords. In *Proceeding of (CVGIP) 16th IPPR Conference on Computer Vision, Graphics and Image Processing.*

Shokrollahi, A. (2006). Raptor Codes. *IEEE Transactions on Information Theory, 52*(6), 2551–2567. doi:10.1109/TIT.2006.874390

Shriram, R., Sugumaran, V., & Vivekanandan, K. (2008). A middleware for information processing in mobile computing platforms. *International Journal of Mobile Communications, 6*(5), 646–666. doi:10.1504/IJMC.2008.019327

Shu-hai, Z., Feng-qin, X., & Jin-zhu, M. (2007). *Research on Secure Mobile agent-based Electronic Commerce* (pp. 10–11). China Academic Journal Electronic Publishing House.

Shuhui, Y., Jie, W., & Fei, D. (2008). Efficient Directional Network Backbone Construction in Mobile Ad Hoc Networks. *IEEE Transactions on Parallel and Distributed Systems, 19*(12), 1601–1613. doi:10.1109/TPDS.2008.43

Simple Outline, X. M. L. SOX. (2002). Retrieved from http://www.langdale.com.au/ SOX/

Skianis, C., Kontovasilis, K., Drigas, A., & Moatsos, M. (2003). Measurement and Statistical Analysis of Asymmetric Multipoint Videoconference Traffic in IP Networks. *Kluwer. Telecommunication Systems, 23*, 95–122. doi:10.1023/A:1023620903222

Smith, J. E. (1988). Characterizing computer performance with a single number. *Communications of the ACM, 31*(10), 1202–1206. doi:10.1145/63039.63043

Soap-tutorial. (2007). Retrieved from http://www.w3schools.com/ soap/default.asp

Song, B. C., & Ra, J. B. (2002). A fast search algorithm for vector quantization using L2-norm pyramid of codewords. *IEEE Transactions on Image Processing, 11*(1), 10–15. PubMeddoi:10.1109/83.977878

Song, G., & Li, Y. (2005). Cross-layer optimization for OFDM wireless networks-part I: theoretical framework. *IEEE Transactions on Communications, 4*, 614–624. doi:10.1109/TWC.2004.843065

Sorensen, E.K., & O Murchu D. (2004). Designing online learning communities of practice: A democratic perspective. *Journal of Educational Media, 29*(3), 189–200.

Spohn, M. A., & Garcia-Luna-Aceves, J. J. (2007). Bounded-distance multi-clusterhead formation in wireless ad hoc networks. *Ad Hoc Networks, 5*, 504–530. doi:10.1016/j.adhoc.2006.01.005

Stanoevska-Slabeva, K., Högg, R., & Brodt, T. (2005). A Taxonomy for Networked Working Environments (NWEs). In *Innovation and the Knowledge Economy*. Amsterdam: IOS Press.

Steinbach, E., Faerber, N., & Girod, B. (1997). Standard Compatible Extension of H.263 for robust video transmission in Mobile environments. *IEEE Transactions on Circuits and Systems for Video Technology*, *7*(6), 872–881. doi:10.1109/76.644067

Stewart, R. (Ed.). (Sept. 2007). Stream Control Transmission Protocol. *Internet Engineering Task Force*, RFC 4960.

Strathmeyer, J. (2005). *Automatic Text Categorization. Internal Project Report, Pittsburgh FLAC_Projects Computer Science Department*. USA: Carnegie Mellon University.

Stylianou, Y., Pantazis, Y., Calderero, F., Larroy, P., Severin, F., Schimke, S., Bonal, R., Matta, F., & Valsamakis, A. (2005). *GMM- Based Multimodal Biometric Verification*. Final Project Report 1, Enterface'05, Mons, Belgium, JULY 18th- August 12th.

Su, Y., & Sun, M. T. (2005). Encoder Optimization for H.264/AVC Fidelity Range Extensions. *Proceedings of VCIP 2005*, Beijing, China.

Sugumaran, V., Shriram, R., & Vivekanandan, K. (2008). Towards a Mobile Community Based Middleware for Spontaneous Interaction. *Decision Sciences Institute Annual Meeting*, Baltimore, November 22-25.

Suhoen, J., et al. (2002). Video transfer control protocol for a wireless video demonstrator. In *International Conference on Information Technology: Coding and Computing*, (pp. 462–467).

Sullivan, G. (2005). The H.264/MPEG-4 AVC vdeo coding standard and its deployment status. *Proceedings of VCIP 2005*, Beijing, China.

Sullivan, G. J., Topiwala, P., & Luthra, A. (2004). The H.264/AVC Advanced Video Coding Standard: Overview and Introduction to the Fidelity Range Extensions. *SPIE Annual Conference on Applications of Digital Image Processing XXVII, Special Session on Advances in the New Emerging Standard H.264/AVC* (pp. 454-474).

Swilem, A., Imamura, K., & Hashimoto, H. (2002). A fast search algorithm for vector quantization using hyperplane decision rule. *Journal of the Institute of Image Information and Television Engineers*, *56*(9), 1513–1517.

Swilem, A., Imamura, K., & Hashimoto, H. (2004). A fast codebook design algorithm for ECVQ based on angular constraint and hyperplane decision rule. IEICE Transactions on Fundamentals of Electronics, Communications and Computer Sciences. *E (Norwalk, Conn.)*, *87-A*(3), 732–739.

Swilem, A., Imamura, K., & Hashimoto, H. (2005). A high-speed closest codeword search algorithm using the pyramid structure of codewords. (IIEEJ). *Journal of the Institute of Image Electronics Engineers of Japan*, *34*(5), 653–662.

Sycara, K., Giampapa, J. A., Langley, B. K., & Paolucci, M. (2003). In Garcia, A., Lucena, C., Zambonelli, F., Omici, A., & Castro, J. (Eds.), *The RETSINA MAS, A Case Study. Research Issues and Practical Applications* (pp. 232–250). Springer-Verlag, Software Engineering for Large-Scale Multi-Agent Systems.

Sysmark2007 preview. (2007). Retrieved from http://www.bapco.com/ techdocs/SYSmark2007 Preview_WhitePaper.pdf.

Szabo, C., Horwath, Z., & Farkas, K. (2007). Wireless community networks: Motivations, design and business models. *Proceedings of the 3rd international conference on Wireless Internet, WICON 2007*, Austin, Texas, USA, ACM.

Tamine-Lechani, L., Boughanem, M., & Daoud, M. (2009, July). *Evaluation Of Contextual Information Retrieval Effectiveness: Overview Of Issues And Research*. Knowledge and Information Systems.

Tapia, A., Maitland, C., & Stone, M. (2006). Making IT work for municipalities: Building municipal wireless networks. *Government Information Quarterly*, *23*(3/4), 359–380.

Tappayuthpijam, K., Liebl, G., Stockhammer, T., & Steinbach, T. (2009). Adaptive video streaming over a mobile network with TCP-Friendly Rate Control. In. *International Conference on Wireless Communications and Mobile Computing*.

Tasch, A. E., & Fremuth, N. (2002). *Mobile Communities*. Retrieved April 7, 2008 from http://www.cosmos-community.org/ downloadFiles/ icis_mobilecommunities_fremuthTasch.pdf.

The Apache Software Foundation. (2007). Apache Tomcat. http://tomcat.apache.org/

The Network Simulator-ns-2. (*n. d.*) Retrieved June 20, 2006, from http://www.isi.edu/nsnam/ns/

Thompson, D. V., Hamilton, R. W., & Rust, R. T. (2005). Feature Fatigue: When Product Capabilities Become Too Much of a Good Thing. *JMR, Journal of Marketing Research*, *42*, 431–442. doi:10.1509/jmkr.2005.42.4.431

Tickoo, O., Subramanian, V., Kalyanaraman, S., & Ramakrishnan, K. K. (2001). LT-TCP: End-to-end framework to improve TCP performance over networks with lossy channels. In *13th International Workshop on Quality of Service*.

Tishby, N. Z. (1991). On the application of mixture AR hidden Markov models to text independent speaker recognition. *Journal of IEEE Transactions on Signal Processing*, *39*(3), 563–570..doi:10.1109/78.80876

Tomek, I., Giles, R., Zhang, H., & Di, L. (2008). FVE – a Virtual Environment for Collaboration. The 22nd International Conference on Advanced Information Networking and Applications (AINA), IEEE Computer Society, GinoWan, Okinawa, Japan, March 25-28, pp. 628-635.

Tomek, I., Shakshuki, E., Madala, K., & Giles, R. (2007). Rich and Thin Groupware Mobile Clients. The Thirteenth Australasian World Wide Web Conference, Charlesworth Bay, Australia, 30th June - 4th July, pp. 462-466.

Tomek, I., Shakshuki, E., Peng, A., Koo, H., Gong, R., & Giles, R. (2006). FCVW - A Federated Collaborative Virtual Workspace, Proceedings International Conference on Applied Computing, IADIS, February 25-28, San Sebastian, Spain, pp. 249-257.

Tsai, F. C.-D., et al. (2006). The design and implementation of WiMAX module for ns-2 simulator. In *Workshop on NS2: The IP Network Simulator*, article no. 5.

Tseng, Y.-C., NI, S.-Y., CHEN Y.-S., & SHEU J.-P. (2002). The Broadcast Storm Problem in Mobile Ad Hoc Networks. *Wireless Networks*, *8*, 153–167. doi:10.1023/A:1013763825347

Tyoviriyakul, P., & Kittitornkun, S. (2008). Optimizing RSA encryption for ARM microprocessor. *Electrical Engineering/Electronics, Computer, Telecommunications and Information Technology, 2008. ECTI-CON 2008. 5th International Conference* (Vol 1) (pp. 117-120).

UI/Application Exerciser Monkey, Android Developer Guide. (2009). Retrieved from http://dev.android.com/ guide/developing /tools/monkey.html.

Unruh, K. T., Pettigrew, K. E., & Durrance, J. C. (2002). Towards effective evaluation of digital community information systems. *Proceedings of the Annual Conference of the Association for Information Science & Technology (ASIST 2002)*, Philadelphia, USA.

Varshney, U., & Vetter, R. (2002). Mobile Commerce: Framework, Applications and Networking Support. *Mobile Networks and Applications*, *7*(3), 185–198. doi:10.1023/A:1014570512129

Vartiainnen, M., & Andriessen, E. (2005). Mobile Virtual Work, What have we learned. In Vartiainen, M., & Andriessen, E. (Eds.), *Mobile Virtual Book: A new paradigm*. New York: Springer Publications.

Vary, P., & Martin, R. (2007). Digital Speech Transmission: Enhancement, Coding, and Error Concealment. *The Journal of the Acoustical Society of America*, *121*(1), 10–11. doi:10.1121/1.2400671

Vasiliou, A., & Economides, A. A. (2007). Mobile collaborative learning using multicast MANETs. [IJMC]. *International Journal of Mobile Communications*, *5*(4), 423–444.

Venkatesh, M. (2003). The Community Network Lifecycle: A framework for research and action. *The Information Society*, *19*(5), 339–347. doi:10.1080/714044682

Verbiest, W., Pinnoo, L., & Voeten, B. (1988). The impact of the ATM concept on video coding. *IEEE Journal on Selected Areas in Communications*, *6*(9), 1623–1632. doi:10.1109/49.12890

Verlinde, P. (1999). *Contribution à la vérification multimodale d'identité en utilisant la fusion de décisions*. PhD thesis, Ecole Nationale Supérieure des Télécommunications, Paris, France, September 17th.

Vivaracho, C. E. Ortega-Garcia, J., Alonso, L., & Moro, Q. I. (2001). A Comparative Study of MLP- based Artificial Neural Networks in Text- Independent Speaker Verification against GMM- based Systems. In *Proceedings of Eurospeech 2001- Scandinavia, volume 3*, 1753-1756, Aalborg-Denmark, September.

Walsh, W. E., Tesauro, G., Kephart, J. O., & Das, R. (2004). Utility Functions in Autonomic Systems. In *International Conference on Autonomic Computing*. (pp. 70-77).

Wan, P.-J., Alzoubi, K. M., & Frieder, O. (2004, April). *Distributed Construction of Connected Domination Set in Wireless Ad hoc and communications, 9*(2), 141-149.

Wang, Z. (2001). *Internet QoS Architectures and Mechanisms for Quality of Service*. San Francisco: Morgan Kaufmann Publishers.

Wang, C.-Y., Wu, C.-J., Chen, G. N., & Hwang, R.-H. (2005, July4-7). p-MANET Efficient Power Saving Protocol for Multi-Hop Mobile Ad Hoc. *Third International Conference on Information Technology and Applications, ICITA 2005, 2*, 271-276.

Wang, H., & Moayeri, N. (1995). Finite-state Markov channel-a useful model for radio communication channels. *IEEE Transactions on Vehicular Technology, 44*, 163–171. doi:10.1109/25.350282

Wang, W., & Zeng, G. S. (2007). Trusted dynamic level scheduling based on Bayes trust model. *Science in China (Ser F-: Inf Sci), 50*(3), 456-469

Wang, Y., Wang, W., & Li, X.-Y. (2005) *Distributed Low-Cost Backbone formation for Wireless Ad hoc Networks "6th ACM symposium(Mobihoc '05)* USA, (pp 2-13)

WAP Binary XML Content Format. (1999, June). Retrieved from http://www.w3.org/ TR/wbxml/

Ware, C., Judge, J., Chicharo, J., & Dutkiewicz, E. (2000). Unfairness and capture behaviour in 802.11 Ad hoc networks. *2000 IEEE International Conference on Communications, 1*, 159-160.

Weiser, M. (1991). The computer for the 21st Century. *Scientific American, 265*(3), 94–104. doi:10.1038/scientificamerican0991-94

Wellman, B. (2005). Community: from neighborhood to network. *Communications of the ACM, 48*(10), 53–55. doi:10.1145/1089107.1089137

Wen-Zhan, S., Wang, Y., Xiang, Y., & Frieder, L. O. (2004, December). Localized Algorithms for Energy Efficient Topology in Wireless Ad hoc networks. *Mobile Networks and Applications, 10*(6), 911–923.

Widmer, J., Denda, R., & Mauve, M. (2001). A survey on TCP-friendly congestion control. *IEEE Network, 15*(3), 28–37. doi:10.1109/65.923938

Wiegand, T., Sullivan, G., Bjontegaard, G., & Luthra, A. (2003). Overview of the H.264/AVC Video Coding Standard. *IEEE Transactions on Circuits and Systems for Video Technology*, Special Issue in H.264.

Wisitpongphan, N., & Tonquz, O. K. (2006). Scalable broadcast strategies for ad hoc routing protocols. *2006 1st International Symposium on Wireless Pervasive Computing*, pp. 1-6.

Wood, M., & Deloach, S. (2000). An Overview of Multiagent Systems Engineering Methodology, The First International Workshop on Agent Oriented Software Engineering, pp. 207-222.

Woodland, P. C., Gales, M. J. F., Pye, D., & Young, S. J. (1997). *The Development of the 1996 HTK broadcast news transcription system*. In: DARPA Speech Recognition Workshop, 97-99.

Wu, K. S., & Lin, J. C. (2000). Fast VQ encoding by an efficient kick-out condition. *IEEE Transactions on Circuits and Systems for Video Technology, 10*(1), 59–62. doi:10.1109/76.825859

Wu, H., Claypool, M., & Kinicki, R. (2005). Adjusting Forward Error Correction with quality scaling for streaming MPEG. In *International Workshop on Network and Operating Support for Digital Audio and Video*, (pp. 111-116).

Wu, J., & Li, H. (n.d.). *On calculating connected dominating set for efficient routing in ad hoc wireless networks.* In Proc. of Proceedings of the 3rd ACM International Workshop on Discrete Algorithms and Methods for Mobile Computing and Communications, pages 7–14.

Wu, K. S., & Lin, J. C. (2000). Fast VQ encoding by an efficient kick-out condition. *IEEE Transactions on Circuits and Systems for Video Technology*, *10*(1), 59–62. doi:10.1109/76.825859

Xebu Serialization Format. (2007, September). Retrieved from https://hoslab.cs.helsinki.fi/ homepages/xebu/

Xi'an University of Electronic Science and Technology. (2006). *Schema Theory and Convergence Theory for Genetic Algorithms* [Doctor]. Xi'an: Wang, Y. P.

Xiaojiang, D., & Dapeng, W. (2007, January). Joint Design of Routing and Medium Access Control for Hybrid mobile Ad Hoc Networks. *Mobile Networks and Applications*, *12*(1), 57–68. doi:10.1007/s11036-006-0006-9

Xiaojiang, D., Dapeng, W., Wei, L., & Yuguang, F. (2006, January). Multiclass Routing and Medium Access Control for Heterogeneous Mobile Ad Hoc networks. *IEEE Transactions on Vehicular Technology*, *55*(1), 270–277. doi:10.1109/TVT.2005.861183

XML Encryption Syntax and Processing. (2002, December). Retrieved from http://www.w3.org/TR/xmlenc-core/

XML-Signature Syntax and Processing. (2008, June). Retrieved from http://www.w3.org/ TR/xmldsig-core/

Xu, S., & Hung, Z. (1996). A gamma autoregressive video model on ATM networks. *IEEE Transactions on Circuits and Systems for Video Technology*, *8*(4), 138–142.

Ya, X., Heidemann, J., & Estrin, D. (2001). Geography informed Energy Conservation for Ad Hoc Routing. *International Conference on Mobile Computing and Networking* (pp. 70-84), Rome, Italy.

YAML. (2005)Retrieved from http://yaml.org/

Yaml. (2007, November). Retrieved from http://en.wikipedia.org/ wiki/YAML)

Ye, S., Blum, R. S., & Cimini, L. J. (2002, Spring). Adaptive modulation for variable-rate OFDM systems with imperfect channel information. *IEEE Veh. Technol. Conf.* [Birmingham, AL.]. *VTC*, *2*, 767–771.

Yegenoglu, F., Jabbari, B., & Zhang, Y.-Q. (1993). Motion-classified autoregressive modeling of variable bit rate video. *IEEE Transactions on Circuits and Systems for Video Technology*, *3*(1), 42–53. doi:10.1109/76.180689

Yerima, S. Y., & Al-Begain, K. (2007). Buffer Management for Multimedia QoS Control over HSDPA Downlink. *Proceedings of 21st International Conference on Advanced Information Networking and Applications Workshops (AINAW '07)*, (Vol.1) (pp. 912 – 917). Niagara Falls, Ontario, Canada: IEEE Press.

Yogesh, Y., Bose, S., & Kannan, A. (2006). Effective TCP Friendly Rate Control scheme for video streaming over wireless networks. *Asian Journal of Information Technology*, *5*(4), 442–447.

Yong, W., Zhong-qun, W., & Liang-fen, W. (2007). A Migration Mechanism of Mobile Agent System Supporting Security and Fault-Tolerance. *Journal of Computer Technology And Development*, *17*(3), 169–175.

Yoo, T., Lavery, R. J., Goldsmith, A., & Goodman, D. J. (2004). *Throughput optimization using adaptive techniques. Technical Report*. US: Stanford University.

Yousefi'zadeh, H., & Jonckheere, E. A. (2005). Dynamic neural-based buffer management for queuing systems with self-similar characteristics. *IEEE Transactions on Neural Networks*, *16*(5), 1163–1173. doi:10.1109/TNN.2005.853417

Youssif, A. A. A., Chowdhury, M. U., Ray, S., & Nafaa, H. Y. (2007). Fingerprint Recognition System Using Hybrid Matching Techniques. In *Proceedings of International Computer and Information Science (ICIS 2007). 6th IEEE/ACIS International Conference*, July 11-13, 234-240.

Yu, M., Malvankar, A., Su, W., & Foo, S. Y. (2007). A link availability-based QoS-aware routing protocol for mobile ad hoc sensor networks. *Computer Communications*, *30*, 3823–3831. doi:10.1016/j.comcom.2007.09.009

Yu, J. Y., & Chong, P. (2003). *3hbac(3-hops between adjacent clusterheads): a novel non-overlapping clustering algorithm for mobile ad hoc networks*. In Proc. IEEE Pacrim 03, (Vol. 1, pp. 318–21).

Yuan, W., & Nahrstedt, K. (2003). Energy-Efficient Soft Real-Time CPU Scheduling for Mobile Multimedia Systems. In *SOSP '03: Proceedings of the Nineteenth ACM Symposium on Operating Systems Principles* (pp. 149-163).

Zarifopoulos, M., & Economides, A. A. (2009). Evaluating mobile banking portals. [IJMC]. *International Journal of Mobile Communications*, *7*(1), 66–90.

Zhang, Y., & Gulliver, T. A. (2005, August 22-24). Quality of Service for Ad hoc On-demand Distance Vector Routing. *IEEE International Conference on Wireless And Mobile Computing, Networking. Communications*, *3*, 192–196.

Zhang, Z., CI, Y. W. & Yang, X. Z. (2005). Design of wearable computer software platform applying to multimedia information collection based on ad hoc network. *Computer Engineering*, *31*(21), 173–175.

Zhang, X., Tang, J., Chan, H., Ci, S., & Guizani, M. (2006). Cross-layer-based modeling for quality of service guarantees in mobile wireless networks. *IEEE Communications Magazine*, *44*(1), 100–106. doi:10.1109/MCOM.2006.1580939

Zhang, F., Mu, Y., & Susilo, W. (2005). *Reducing Security Overhead For Mobile Networks*. In AINA '05: Proceedings Of The 19th International Conference On Advanced Information Networking And Applications (pp. 398–403). Washington, DC: IEEE Computer Society.

Zhang, Q. T. (1999). A general AR-based technique for the generation of arbitrary gamma VBR video traffic in ATM networks. *IEEE Transactions on Circuits and Systems for Video Technology*, *9*(10), 1130–1137. doi:10.1109/76.795063

Zheng, J., & Hu, M. (2006). An anomaly intrusion detection system based on vector quantization. IEICE Transactions on Information and Systems. *E (Norwalk, Conn.)*, *89-D*(1), 201–210.

Zhu, X. G., Hong, B. R., & Wang, D. M. (1999). Implementation of Time\Scale Transformation Based on Continuous Wavelet Theory. *Systems Engineering and Electronics*, *21*(12), 11–13.

Zhu, Q. X. (2005). Optimal Search Theory and Some Applications. *World Sci-tech R & D*, *27*(4), 39-49.

About the Contributors

Ismail Khalil is one of the key people in the field of mobile multimedia. He started the Advances in Mobile Computing and Multimedia (MoMM) series of conferences, He is the editor of the handbook of research on Mobile Multimedia series, the book Mobile Multimedia: Communication Engineering Perspective and the book Multimedia Transcoding in Mobile and Wireless Networks. He holds a PhD in computer engineering and received his habilitation degree in applied computer science on his work on agents' interactions in ubiquitous environments in May 2008. He currently teaches, consults, and conducts research in Mobile Multimedia, Agent Technologies, and the Semantic Web and is also interested in the broader business, social, and policy implications associated with the emerging information technologies. Before joining Johannes Kepler University of Linz in 2002, he was a research fellow at the Intelligent Systems Group at Utrecht University, Netherlands from 2001-2002 and the project manager of AgenCom project at the Software Competence Center Hagenberg - Austria from 2000-2001. Dr. Khalil has authored around 100 scientific publications, books, and book chapters. He serves as the editor-in-chief of the International Journal on Web Information Systems (IJWIS), Journal of Mobile Multimedia (JMM), International Journal of Mobile Computing and Multimedia Communication (IJMcMc), Advances in Next Generation Mobile Multimedia book series, and Atlantis Ambient and Pervasive Intelligence book series. He is on the editorial board of several international journals. He is a distinguished speaker at conferences, workshops, tutorials and seminars, author of several publications.

Edgar R. Weippl is Research Director of Secure Business Austria and Priv.-Doz. at the Vienna University of Technology. His research focuses on applied concepts of IT-security and e-learning. Edgar is member of the steering committee of the ED-MEDIA conference; he organizes the ARES conference (as PC chair 2007, 08; panel and workshop chair 2009). After graduating with a Ph.D. from the Vienna University of Technology, Edgar worked for two years in a research startup. He then spent one year teaching as an assistant professor at Beloit College, WI. From 2002 to 2004, while with the software vendor ISIS Papyrus, he worked as a consultant for an HMO (Empire BlueCross BlueShield) in New York, NY and Albany, NY, and for Deutsche Bank (PWM) in Frankfurt, Germany. In 2004 he joined the Vienna University of Technology and founded together with A Min Tjoa and Markus Klemen the research center Secure Business Austria.

* * *

Ibrahim Abualhaol received the B.S. and M.S. degrees in electrical engineering from Jordan University of Science and Technology, Irbid, Jordan, in 2000 and 2004, respectively, and the Ph.D. degree in electrical engineering from the University of Mississippi, Oxford, MS, in August 2008. He was a Research Assistant with the Center for Wireless Communication at the University of Mississippi, MS. From May 2008 to Sep 2008, he was a Wireless System Engineer Intern with Qualcomm Incorporation, San Diego, CA. He is currently with Broadcom Corporation, San Diego, CA. His research interests include OFDM, space–time coding, cooperative networks, and MIMO wireless communications. Dr. Abualhaol is a member of IEEE, Phi Kappa Phi, and Sigma Xi.

Amin Anjomshoaa is a university assistant at the "Institute of Software Technology and Interactive Systems" (ISIS) of Vienna University of Technology (TU Wien). He has received his Bachelor in Mathematics applied in Computer Science from Kerman University, Iran. Afterward he has worked as Software Manager, Network Manage and Chief Technology Officer till 2003 when he joined the ISIS as project assistant and has been involved in some international and national research projects. He has received one master degree in Software engineering and Internet computing and a second one in Information management. In 2009 he received his Ph.D. from Vienna University of technology and now he is perusing a second Ph.D. degree in Building Science. His research areas include Semantic Web, Knowledge management, Web 2.0, Internet Computing and Building Information Models.

Nabeel F. Al-Mushasha is an assistant professor at the computer sciences department and computer center manager at Jerash Private University at Jerash, Jordan. He received his B.S. in computer engineering from National Technical University of Ukraine in 1996, his M.S. computer and intellectual systems and networks from National Technical University of Ukraine in 1996, and his Ph.D. from University Utara Malaysia (UUM) in 2008. Dr. Al- Mushasha's research focuses on three areas in the field of information systems: mobile learning, service quality and human-computer interaction. His current research centers on the role of service quality on learner satisfaction and his behavioral intentions to use mobile learning services in the future.

Ghaida A. Al-Suhail is an Assistant Professor in the Department of Computer Engineering at the University of Basrah, in Basrah, Iraq. She received her B.E. and M.E. degrees in electrical engineering from the University of Basrah, Iraq, in 1984 and 1989, respectively. Her Ph.D. degree was also in electrical engineering, multimedia communications from the University of Basrah in 2007. Dr. AL-Suhail was a Visiting Academic of Endeavour Award 2008 for 6 months in the Research School of Information Sciences and Engineering (RSISE) at the Australian National University, Canberra, Australia. She has been serving on a number of review committees of workshops and conferences like MoMM2009, ICSPCS2010, AusCTW2010, and EPC-IQ01. Her recent research interests include mobile multimedia, Internet protocols and IP networks, wireless networks, computer communication networks, wireless sensor networks, cross-layer design & quality of service, optical communications, Telemedicine and m-heath systems and image processing.

Mohammad Anbar received his B.Tech. in Electronics Engineering from Tishreen University, Lattakia, Syria, in the year 2003, and M.Tech. in Computer Science from Jawaharlal Nehru University, New Delhi, India in the year 2007. Currently, Mohammad Anbar, is Ph.D. student at the School of Computer & Systems Sciences, Jawaharlal Nehru University, New Delhi. His research interest includes Wireless Communication, Mobile computing, Particle Swarm Optimization.

V. S. Anitha has received her Master's degree in Computer Science and Engineering from Indian Institute of Technology, Madras in 1997. She is a Professor in the Department of Computer Science and Engineering, Govt. Engineering College, Wayanad, Kerala and currently pursuing her Ph.D at National Institute of Technology Calicut in the field of self configuring ad hoc networks. Her research interests are in the areas of ad hoc and sensor networks, network security, design and implementation of mobile ad hoc routing protocols and QoS management.

Qin Danyang was born in 1983. She received the B.S. and M.S. degrees in Information and Communication Engineering from Harbin Institute of Technology, P.R.China, in 2006 and 2008 respectively. She is currently a candidate for the Ph.D. degree in School of Electronic and Information Engineering School from Harbin Institute of Technology, P.R.China. Her research interests include mobile computing, cooperative communication techniques, ad hoc wireless networks model and routing algorithms design.

Anastasios A. Economides received the Dipl.Eng. degree in electrical engineering from the Aristotle University of Thessaloniki. Holding a Fulbright and a Greek State Fellowship, he received the M.Sc. and the Ph.D. degrees in computer engineering from the University of Southern California, Los Angeles. At graduation, he received the Outstanding Academic Achievement Award from the University of Southern California. Currently, he is an Associate Professor and Chairman of the Information Systems Department at the University of Macedonia, Thessaloniki, Greece. He is the director of CONTA (COmputer Networks and Telematics Applications) Laboratory. His research interests include networking technologies and e-services. He has published over one hundred fifty peer-reviewed papers. He has been the plenary speaker in two International Conferences.

Martin Fleury holds a degree in Modern History (Oxford University, UK) and a Maths/Physics-based degree from the Open University, Milton Keynes, UK. He was awarded an MSc in Astrophysics from QMW College, University of London, UK in 1990 and an MSc from the University of South-West England, Bristol in Parallel Computing Systems in 1991. He holds a PhD in Parallel Image Processing Systems from the University of Essex, Colchester, UK. He is currently employed as a Senior Lecturer at the University of Essex. Martin has authored over one hundred and eighty articles, book chapters and a book on high-performance computing for low-level image- and signal-processing algorithms. His current research interests center upon video communication over wired and wireless networks.

Péter Fülöp was born in Mosonmagyarovar in 1980. He received his M.Sc. degree in Technical Informatics from the Budapest University of Technology and Economics in 2005. Currently he is a PhD student on the faculty of Informatics on the Department of Telecommunication and fellow reserarcher in the Mobile Innovation Centre of BUTE. He is working as a system test engineer at Ericsson Telecommunications Hungary at Research and Development Department. His research interest is IP mobility, interworking of heterogeneous mobile networks and movement modeling in cellular, mobile networks. The PhD thesis is going to be written on Complex Mobility Management Systems and Applications.

Mamdouh M. Gomaa received his B.Sc. Degree in Computer Science from Faculty of Science, Minia University, Egypt in 2003. His research interests include Vector quantization, data compression and image processing.

Olivier Le Grand has been involved in the telecommunication industry for nearly twenty years. He joined France Telecom Orange Labs, France, in 1990. Involved in standardization since 1992, he has been very active in ITU-T SG11 and ETSI in the area of ATM related signalling standardization and now in NGN and IPTV related standardization activities of ITU-T SG13. At WTSA 2008 he has been reappointed as SG13 vice-chairman for the 2009-2012 study period. He is currently acting within SG13 as chair of Working Party 3 (Frameworks and functional architectures) and also as co-editor of edition 2 of Recommendation ITU-T Y.2012 (Functional requirements and architecture of the NGN). In 1987, he received an Engineer degree in Information Technology from the Institut National des Télécommunications Evry, France.

Mhania Guerti is a full professor at the National Polytechnic College of Algiers: ENSP (www.enp.edu.dz), who got her PhD from the INPG institute of Grenoble, France in 1993. She is specialized in Speech and language processing. Pr Guerti has supervised a lot of students in Master and PhD degree. Her research works concern all the aspects of the speech: synthesis, recognition, analysis, etc. She is a motivated professor at the department of electronics and also a research director of the speech processing team at the national polytechnic college of Algiers.

P. P. Abdul Haleem is a research scholar in the Department of Computer Science & Engineering, National Institute of Technology Calicut, India. His research interests include wireless mobile computing, XML technologies, and Semantic Web.

Shahizan Hassan, a PhD degree holder from the University of Newcastle upon Tyne, United Kingdom, is an Associate Professor in Information Systems at the College of Business, University Utara Malaysia. He is the editor for Malaysian Management Journal (MMJ) and has published three books and ten articles in national and international journals. He is very active in research projects in the area of Information Systems Evaluation, Web design and evaluation, Knowledge Management, electronic government, and wireless applications. To date, he had completed a total of ten research projects funded by various agencies. He has supervised more than 100 undergraduate and post-graduate students' projects in various IT disciplines. Currently, he is the Director for University-Industry Link (UIL), Universiti Utara Malaysia and can be reached at *shahizan@uum.edu.my*.

Sándor Imre was born in Budapest in 1969. He received the M.Sc. degree in Electronic Engineering from the Budapest University of Technology (BUTE) in 1993. Next he started his Ph. D. studies at BUTE and obtained dr. univ. degree in 1996, Ph.D. degree in 1999 and DSc degree in 2007. Currently he is carrying his teaching activities as Head of the Dept. of Telecommunications of BUTE. He was invited to join the Mobile Innovation Centre of BUTE as R&D director in 2005. His research interests include mobile and wireless systems. Especially he has contributions on different wireless access technologies, mobility protocols and reconfigutable systems.

Michel Kadoch (S'67, M'77, SM'04) received the B. Eng from Sir George Williams University (Canada) in 1971, the M. Eng from Carleton (Canada) in 1974, MBA from McGill (Canada) in 1983 and the Ph.D from Concordia (Canada) in 1991. He is a full professor at Ecole de technologie supérieure ETS (Canada) and the director of the Master Program in engineering. He is active in research mostly in performance analysis and network management and control in wired as well as wireless networks. He is the director of the research laboratory LAGRIT at ETS. He is also an adjunct professor at Concordia University (Canada). He is presently working on Cross layer, and on Reliable multicast in wireless Ad hoc and WiMax networks. Professor Kadoch has published many articles and is the author of a book « Protocoles et réseaux locaux » (Edition ETS). He is serving as a reviewer for journals and conferences and for grants for NSERC as well as track TPC for ICCCAS 08, WiMob '06, '07, '08. He has been involved for many years at ITU-T as a special reporter and with the industry namely Teleglobe Canada, CAE, and Communication Canada. He has been a consultant with Harris, Bell South, BC Tell, Concert and British Telecom UK, as well as the CTO (Commonwealth Telecommunication Organization) on issues related to installation and optimization of SONET, telephone and high speed data networks and services.

Inderjeet Kaur has done her B.Tech (CSE) from Kurushetra University, Kurushetra (India) in 2002 in which she topped the list of the successful candidates. She successfully completed her M.Tech (I.T.) from Indraprastha University, New Delhi (India) in 2007 with excellent grade of marks. Her M.Tech dissertation entitled "Implementation of OFDM Transceivers" received high appraisal from all the examiners of her M.Tech dissertation. Her area of research interest is wireless communication in which she has published large number of papers which appeared in the proceedings of various International conferences. At present Inderjeet Kaur is working as Assistant Professor at the Department of CSE of the Ajay Kumar Garg Engineering College, Ghaziabad, India and alongside pursuing her Ph.D. in the area "Performance evaluation of OFDM systems in wireless applications" from Delhi College of Engineering, University of Delhi, Delhi, India. Inderjeet Kaur intends to build a career with leading corporate of committed people and willing to work as a key player in challenging environment by exploiting her potentials to full extents, and to develop congenial environment.

Theodoros I. Kavaliotis received the Bachelor of Science degree in Economics from the University of Macedonia, Thessaloniki, Greece, in 2007. He received a Master of Arts degree in Management and Leadership from Webster University, London, England, in 2008. He has done undergraduate and postgraduate research focused on electronic commerce and on the interaction in electronic communities.

Harilaos Koumaras was born in Athens, Greece. He received his BSc degree in Physics in 2002 from the University of Athens, Physics Department, his MSc in Electronic Automation and Information Systems in 2004, being scholar of the non-profit organization Alexander S Onassis, from the University of Athens, Computer Science Department and his PhD in 2007 at Computer Science from the University of Athens, Computer Science Department, having granted the four-year scholarship of National Centre of Scientific Research "Demokritos". He has received twice the Greek State Foundations (IKY) scholarship during the academic years 2000-01 and 2003-04. He has also granted with honors the classical piano and harmony degrees from the classical music department of Attiko Conservatory. Since 2004 he is a principal lecturer at the Business College of Athens (BCA) teaching modules related to Information Technology and Mathematics, Data Networks and Local Area Networks. From 2009, he has been elected as the Head of the Computer Science Department of BCA and Course leader of the respective franchised course of London Metropolitan University. At the same time, since 2003 is a research associate of the Digital Communications Lab at the National Centre of Scientific Research "Demokritos" by participating in numerous EC-funded and national funded projects. His research interests include objective/subjective evaluation of the perceived quality of multimedia services, video quality and picture quality evaluation, video traffic modeling, digital terrestrial television and video compression techniques. Currently, he is the author or co-author of more than 30 scientific papers in international journals, technical books and book chapters, numbering 52 non-self citations. He is an active editorial board member of Telecommunications Systems Journal and he has served many conferences as TPC member and has acted as reviewer for various journals, such as IEEE Transactions on Broadcasting and IEEE Transactions on Image Processing. Dr. Koumaras is a member of IEEE, SPIE and National Geographic Society.

Anastasios Kourtis received his BS degree in Physics in 1978 and his PhD degree in Telecommunications in 1984, from the University of Athens. From 1986 he is a member of the research stuff in the Institute of Informatics and Telecommunications of the Greek National Centre for Scientific Research "Demorkitos" and currently he is a Research Director. His technical and research activities have been in the area of digital modulation techniques, spread spectrum systems, multimedia applications, Quality of Services (Perceived and Network), broadband wireless networks, Wireless Local Area Networks (WLAN), digital interactive TV, satellite communications and networks convergence. He has participated in a number of EU-funded research and development projects: ESPRIT (FCPN, OFSES), ACTS (CRABS, WATT), IST/FP5 (WIN, MAMBO, SOQUET, REPOSIT), IST/FP6 (ATHENA, IMOSAN, ENTHRONE-I, ENTHRONE-II, ADAMANTIUM). He has been Project Manager in three of them and he has also co-ordinated a number of National funded projects. He is the author of several papers on telecommunications, broadband wireless networks and digital interactive TV.

Tayeb Lemlouma is an associate professor at the University of Rennes, France, since 2006. His work is centered around multimedia and networks, multimedia and mobile computing, content adaptations in heterogeneous environments. From 2001 to 2005, he works at the INRIA (Institut National de Recherche Informatique et Automatique). His work focused on content adaptations, Web and XML technologies, mobile users in heterogeneous environments. From 2003 to 2005, he was a member of the Device Independence working group of the World Wide Web Consortium. From 2001 to 2002, he was a member of the Opera team at the INRIA laboratory. In 2004, he received a Ph.D. degree in Computer Science from the Institut National Polytechnique of Grenoble, France.

Ma Lin was born in 1980. He is currently a docent in School of Electronic and Information Engineering School of Harbin Institute of Technology, P.R.China. Dr. Ma received his B.S., M.S. and Ph.D. degrees from HIT in 2003, 2005 and 2009 respectively. His areas of interest include Broadband wireless transmission, special communication network and wireless location technology. He has remained active in his areas of interest by obtaining grants and authoring several papers in refereed conferences and journals.

Mustafa M. Matalgah received his Ph.D. in Electrical and Computer Engineering in 1996 from The University of Missouri, Columbia. From 1996 to 2002, he was with Sprint, Kansas City, MO, USA, where he led various projects dealing with optical transmission systems and the evaluation and assessment of 3G wireless communication emerging technologies. In 2000, he was a Visiting Assistant Professor at The University of Missouri, Kansas City, MO, USA. Since August 2002, he has been with The University of Mississippi in Oxford where he is now an Associate Professor in the Electrical Engineering Department. In Summer 2008, he was a Visiting Professor at Chonbuk National University in South Korea. His current technical and research experience is in the fields of emerging wireless communications systems and communication networks. He also previously published in the fields of signal processing and optical binary matched filters. He has published more than 70 refereed journal and conference proceeding papers, five book chapters, and more than 20 technical industrial applied research reports in these areas. Dr. Matalgah received several certificates of recognition for his work accomplishments in the industry and academia. He is the recipient of the Best Paper Award of the IEEE ISCC 2005, La Manga del Mar Menor, Spain. He is the recipient of the 2006 School of Engineering Junior Faculty Research Award at The University of Mississippi. He served on several international conferences technical program and organization committees.

Kais Mnif received the B.Eng. Electrical Engineering from the École Nationale d'Inégineurs de Sfax (ENIS), University of Sfax, Tunisia in 1994, M.Sc form the Institut National de la Recherche Scientifiquel,University of Québec, Canada in 1998 and Ph.D from the Ecole de technologie supérieure, University of Quebec, Canada in 2006. From 2000 to 2003, he worked for Harris-MCD in Montreal. He was with the System integration group for the Broadband Wireless Access (BWA) system. He is currently assistant professor at the Institute of Electronics and Telecommunications in Sfax (ISCES), Tunisia. His current research interests include wireless Ad hoc, sensor Networks, vehicular wireless Networks, QoS and routing protocol, unicast and multicast routing protocols, management and security system for wireless and mobile networks,. He serves as a reviewer for several conferences such as SSD'09, Globecom'07, WCNC'07, ICC'06, Wimob'06, ISCC'05.

Siham Ouamour graduated from the USTHB University where she got her MSc in 2002, she got her PhD at the National Polytechnic College of Algiers (ENSP) in 2009. She has been lecturer at the USTHB University since 2002 and has several works on speaker characterization and discrimination. Her research interests include research on artificial neural networks, applied to speaker diarization and discrimination. Mrs Ouamour has published many articles and papers in international journals and conference proceedings, which are related to the field of speaker characterisation. Now she is an Assistant Professor at the USTHB University of Algiers (www.usthb.dz).

Ahmed Patel received his M.Sc. and Ph.D. degrees in Computer Science from Trinity College Dublin (TCD), specializing in the design, implementation and performance analysis of packet switched networks. He is a Lecturer and Consultant in ICT and Computer Science. He is a Visiting Professor at Kingston University in the UK and currently lecturing at Universiti Kebangsaan Malaysia. His research interests span topics concerning high-speed computer networking and application standards, network security, forensic computing, autonomic computing, heterogeneous distributed computer systems and distributed search engines and systems for the Web. He has published well over hundred and seventy technical and scientific papers and co-authored two books on computer network security and one book on group communications, co-edited a book distributed search systems for the Internet. He is a member of the Editorial Advisory Board of the following International Journals: (i) Computer Communications, (ii) Computer Standards & Interface, (iii) Digital Investigations, (iv) Cyber Criminology, and (v) Forensic Computer Science.

Joel Penhoat has been involved in the telecommunication industry for ten years. He joined France Telecom Orange Labs, France, in 2000. His work is centered around security and QoS in radio networks. Since 2006, his current research interests are in the area of Fixed Mobile Convergence, content-aware networks, network-aware contents, IP Multimedia Subsystem and the Future Internet. In 1985, he received a Master degree in Electronics from the University of Brest, France. In 2008, he received a Master degree in Information Technology from the Ecole Nationale Supérieure des Télécommunications Paris, France. He is currently contributing to the networking design of the Future Internet to obtain a Ph.D. degree from the University of Rennes, France.

Manu J. Pillai received his B.E. in Computer Science and engineering from Manonmaniam Sundarnar University,Tirunelveli,Tamil nadu, India in 2002 and M.E in Computer science and engineering from Anna University, Chennai, Tamil nadu India In 2004. He is doing Ph.D. in Computer Science and Engineering in National Institute of Technology, Calicut, India as part time. His research interest includes Routing and Medium access control protocols for heterogeneous mobile ad hoc networks, trust manangement in ad hoc networks and Cryptography. He has been working as Assistant Professor and Head of the Department in Travancore Engineering College, Oyoor,Kollam, Kerala, India from 2004 onwards.

Ahmed A. A. Radwan is the Head of Computer Science Department, Faculty of Science, Minia University, Egypt, and the Head of Quality Assurance and Accreditation Project unit – Faculty of Computer & Information Science-Minia University He received his B.Sc.1975 and M.Sc. 1981 Degrees from Assiut University, Faculty of science, Egypt in mathematics and Ph.D. degree from Liverpool University, England in Computer Science. His main areas of interests are parallel processing, multicast communication, Algorithm design and analysis, Optimal Routing Algorithms for Computer Networks, Parallel Systems, Graph Drawing and Mobile Networks.

Shriram Raghunathan is Professor of Computer Science and Engineering, and Director of the Faculty Training Academy at B.S. Abdur Rahman University, Chennai, Tamil Nadu, India. He received his Ph.D in Computer Science and Engineering from Bharathiar University, Coimbatore, Tamil Nadu, India. His research interests are in the areas of Mobile Computing, Pervasive Computing and Web Information Retrieval. He has published over 10 articles in peer reviewed Journals and Conferences.

Reza Rawassizadeh is a Ph.D. student at University of Vienna. He received his MS.c. from the Technical University of Vienna and his Bs.c. from Azad Tehran University- Central Branch. Formerly he is working for Siemens PSE in telecommunication branches and AIT (Austrian Institute of Technology) in digital preservation projects. His research interests include Mobile and Ubiquitous Computing, life logging, Personal information, Computer Mediated Communication and Social Networks.

Mikael Salaun has been involved in the telecommunication industry for ten years. He is a senior expert in telecommunications at France Telecom, and an assistant professor at the University of Rennes, France. He joined France Telecom Orange Labs, France, in 1999. He was responsible for security projects against DoS and DDos attacks with active networks. Since 2002, his current research interests are in the area of autonomic networking. He is responsible for several projects such as Autonomic Information Networking, Home Autonomic Networking and Autonomic IP Multimedia Subsystem. In 2002, he received a Ph.D. degree in Computer Science from the University of Brest, France.

Salah Saleh was awarded a BSc from Baghdad University, Iraq in 1992. He went on to take an MSc in Mobile and Satellite Communication at Westminster University, London, UK. His research interests are cross-layer design of control congestion, quality-of-service issues, scheduling algorithms, bandwidth allocation mechanisms and multi-connection techniques, especially in WiMAX broadband wireless. He is editor-in-chief of the Computer Science and Engineering: International Journal.

Halim Sayoud is an associate professor at the FEI-USTHB (Faculty of Electronics and Informatics). He received his MSc in 1994, and his PhD in 2003 from the USTHB University in collaboration with the LIA laboratory of Avignon (in France). He was a research visitor at several universities in France and Greece: IRIT of Toulouse, LIA of Avignon, ENST of Paris and ILSP of Athens. His research works focus on the area of speaker identification and text mining. He has been investigating the effects of different types of reduced features in speaker characterization. He is also interested in developing novel research and methods for text mining and author verification. Personal homepage: http://www.sayoud.dr.ag/

M. P. Sebastian received his masters and Ph.D degrees from the Indian Institute of Science, Bangalore. Prior to joining the Indian Institute of Management Kozhikode as Professor of Information Technology & Systems, he served at National Institute of Technology Calicut as Professor & Head of the Department of Computer Science & Engineering. He served at Space Applications Centre (ISRO), Ahmedabad and Bharat Electronics Ltd., Bangalore also for short durations. His areas of teaching/ research interest include Mobile Computing, Cloud Computing, Enterprise Resource Computing, Information Security, and Software Project Management.

Elhadi Shakshuki is an associate professor at the Jodrey School of Computer Science at Acadia University, Canada. He is the founder and the head of the Cooperative Intelligent Distributed Systems Group at the Computer Science Department, Acadia University. He received his M.A.Sc. and Ph.D. degrees in systems design engineering from the University of Waterloo, Canada. Dr. Shakshuki is an Adjunct Professor at Dalhousie University, Canada. Dr. Shakshuki has authored and co-authored more than 120 research papers in international journals, conferences and workshops. He is the co-founder and co-chair of the annual International IEEE Symposium on Web and Mobile Information Systems (WAMIS), the annual International IEEE Workshop on Intelligent Systems Techniques for Ad hoc and Wireless Sensor Networks (IST-AWSN), the annual IEEE International Workshop on Cooperative Distributed Systems (CDS), and the annual IEEE Workshop on Applications of Ad hoc and Sensor Networks. He manages several research projects in his research expertise in the area of distributed artificial intelligence and its applications. Dr. Shakshuki is a member of APENS, IEEE, ACM and AAAI.

Charalabos Skianis is currently Assistant Professor in the Department of Information and Communication Systems at the University of the Aegean in Samos, Greece. He holds a PhD degree in Computer Science, University of Bradford, United Kingdom and a BSc in Physics, Department of Physics, University of Patras, Greece. His work is published in journals, conference proceedings and as book chapters and has also been presented in numerous conferences and workshops. He acts within Technical

Program and Organizing Committees for numerous conferences and workshops (e.g., IFIP Networking 2006, IEEE Globecom 2006, IEEE ICC 2006) and as a Guest Editor for scientific journals (e.g., IEEE Networks magazine). He is at the editorial board of journals (e.g., IEEE Wireless Communications), a member of pronounced professional societies (senior member of IEEE) and an active reviewer for several scientific journals. He is an active member of several Technical Committees within the IEEE ComSoc (e.g., vice chair for the CSIM technical committee), member of the education board of IEEE ComSoc, and member of IEEE BTS; IEEE TVT and IEEE CS. He is currently a Technical Manager for ICT FP7 VITAL++ project and representative for the University on ICT FP7 HURRICANE project.

Ahmed Swilem Received B. S., M. S. degree in computer science from Minia University (Egypt), in 1992 and 1998, respectively and Ph.D. degree in information and system engineering in 2005 from Kanazawa University (Japan). He is now an assistant Prof. in Computer Science Department Minia University. His research interests include vector quantization, data compression and image processing.

Vijayan Sugumaran is Professor of Management Information Systems in the department of Decision and Information Sciences at Oakland University, Rochester, Michigan, USA. He is also WCU Professor of Service Systems Management and Engineering at Sogang University, Seoul, South Korea. He received his Ph.D in Information Technology from George Mason University, Fairfax, VA. His research interests are in the areas of Service Science, Ontologies and Semantic Web, Intelligent Agent and Multi-Agent Systems, Component Based Software Development, and Knowledge-Based Systems. He has published over 140 peer-reviewed articles in Journals, Conferences, and Books. He has edited eight books and two journal special issues. He is the editor-in-chief of the *International Journal of Intelligent Information Technologies* and also serves on the editorial board of seven other journals. He was the program co-chair for the *13th International Conference on Applications of Natural Language to Information Systems* (NLDB 2008). In addition, he has served as the chair of the *Intelligent Agent and Multi-Agent Systems* mini-track for Americas Conference on Information Systems (AMCIS 1999 - 2011) and *Intelligent Information Systems* track for the Information Resources Management Association International Conference (IRMA 2001, 2002, 2005 - 2007). He served as Chair of the E-Commerce track for Decision Science Institute's Annual Conference, 2004. He was the Information Technology Coordinator for the Decision Sciences Institute (2007-2009). He also regularly serves as a program committee member for numerous national and international conferences.

Sándor Szabó was born in 1977. He received the M.Sc. degree in Electronic Engineering from the Budapest University of Technology (BUTE) in 2000. Next he started his Ph. D. studies at BUTE. Currently he is carrying his teaching activities as an assistant professor at the Dept. of Telecommunications of BUTE. He leads research projects, and also a project leader at the Mobile Innovation Centre of BUTE. His research interests include mobile and wireless systems, mobility modeling, mobility management and the integration of heterogeneous wired and wireless networks.

Tamás Szálka was born in Budapest in 1982. He received the M.Sc. degree in Computer Science from Budapest University of Technology and Economics (BUTE) in 2006. Next he started his Ph.D. studies at BUTE, on the Department of Telecommunications. Beside his Ph.D. studies, he is doing engeneering research as Ph.D. student and fellow researcher in the Mobile Innovation Centre of BUTE. As planned, he will obtain his Ph.D. degree in 2011. His work is relatead to the area of mobililty modeling and location based services of cellular networks.

A. Min Tjoa is full professor at the Vienna University of Technology and director of the institute for software technology and interactive systems (www.isis.tuwien.ac.at). His main research interests are in databases, data warehouses, semantic web and software engineering. He was member of the IF-IPTC2 (Software, Theory and Practice) and president of the Austrian Computer Society; the Programme Committee Chairperson of the International conference on the Entity Relationship Approach in 1992, General Chairperson of the IEEE conference on Data Engineering in 1993 and Programme Committee Chairperson of the 8th Int. Conf. on Database and Expert Systems Applications in 1997. He is member of the Russian Academy of Technical Sciences and member of the Austrian UNESCO Commission. He was guest professor at the University of Zurich, the Technical University in Lausanne and the University of Kyushu. He was member of the EU-Commission's ISTAG (Information Society Technologies Advisory Group) Task Group for Ambient Intelligence of the 5th Framework Programme. He was General Chairman of the Joint 8th European Software Engineering Conference (ESEC) and 8th ACM SIGSOFT International Symposium on the Foundations of Software Engineering (FSE-7) in 2001 and of the IEEE Int. Conf. on Distributed Systems in 2002. Since 2000, A Min has been Vice-Chairman of the DEXA-Association (Database and Expert Systems Applications) and recently chairperson for the international conference series "Availability, Reliability, and Security" (ARES, 2006 to 2009) and "Complex, Intelligent and Software Intensive Systems"□ (CISIS, 2008 to 2009). He has published 25 books and more than 200 peer-reviewed articles in the field of databases, information systems, semantic technologies, and software engineering.

Ivan Tomek is an Adjunct Professor, retired from full time employment after 30 years at the School of Computer Science at Acadia University. His most recent interests are in socialware and groupware but in the past, he published articles in the areas of pattern recognition, computer architecture, learning and teaching computer science, hardware simulation, and others. He also wrote book chapters and authored and co-authored several books on computer organization and architecture, the Pascal programming language, and a book on a programming language that he designed for teaching Computer Science. In the past, Dr. Tomek chaired several international conferences, served on multiple program committees, and is a member of an Editorial Board of an international Computer Science journal.

Deo Prakash Vidyarthi, received his Masters Degree in Computer Application from MMM Engineering College Gorakhpur and PhD in Computer Science from Jabalpur University (work done at Banaras Hindu University, Varanasi). Taught UG and PG students in the Department of Computer Science of Banaras Hindu University, Varanasi for more than 12 years. Joined JNU in 2004 and currently working as Associate Professor in the School of Computer & System Sciences, Jawaharlal Nehru University, New Delhi. Dr. Vidyarthi has published around 35 research papers in various international journals and transactions (including IEEE, Elsevier, Springer, World Scientific etc.) and around 25 papers in proceedings of various peer-reviewed conferences. He has contributed chapters in many edited books. He is in the editorial board of two international journals and in the reviewer's panel of many international journals. Also, he has co-authored a book (research monograph) entitled "Scheduling in Distributed Computing Systems: Design, Analysis and Models" published by Springer, USA released in December,

2008. Dr. Vidyarthi is member of IEEE, International Society of Research in Science and Technology (ISRST), USA and senior member of the International Association of Computer Science and Information Technology (IACSIT), Singapore. Research interest includes Parallel and Distributed System, Grid Computing, Mobile Computing.

K. Vivekanandan is the Reader, School of Management and Entrepreneur Development, Bharathiar University, Coimbatore, Tamil Nadu, India. His research interests are in the areas of management information sciences, data mining and warehousing, computer simulation and computational intelligence. He has published over ten peer reviewed articles in national/international journals and presented papers in 30 national/international conferences. He has produced five PhDs and is currently guiding eight candidates for doctoral studies.

Qi Wei received her Bachelor of Applied Science in Computer Science from Royal Melbourne Institute of Technology, Australia in year 2008. She did a small research project on operating system security during her final year of the bachelor. She is currently pursuing her master degree at Universiti Kebangsaan Malaysia (UKM), Department of Computer Science. She's interested in distributed computing, security of mobile agent-based systems, forensic investigation of transactions in agent marketplace and network security.

Xu Yubin was born in 1954. He received his B.S. degree in radio measurement from Harbin Institute of Technology in1986, M.S. and Ph.D. degrees in Electronic and Communication System from Harbin Institute of Technology, P.R.China, in 1993 and 2005 respectively. Currently he is a Professor and doctoral supervisor in Information and Communication Engineering of HIT. Prof. Xu is senior member of Institute of Electronics, senior member of china institute of communications, executive director of Heilongjiang Province Institute of Communications, and member of Professionals I Canada and China Association. Prof. Xu's research interests include communication networks, special mobile communication system, navigation and positioning technique and multimedia communication technology.

Index

D

E

S

T

U

V

W

Y

Z

CPSIA information can be obtained at www.ICGtesting.com
260174BV00008BA/19-26/P

9 781609 605636